...onary of

...**g**...**aphy**

Susan Mayhew is a geography teacher and the author of *Masterstudies Geography*. Her current research interests are in feminist and historical geography and she is dedicated to the demystification of her favourite subject.

(⊕) SEE WEB LINKS

For recommended web links for this title, visit www.
oxfordreference.com/page/geog when you see this sign.

The most authoritative and up-to-date reference books for both students and the general reader.

Accounting
Agriculture and Land Management
Animal Behaviour
Archaeology
Architecture
Art and Artists
Art Terms
Arthurian Literature and Legend
Astronomy
Bible
Biology
Biomedicine
British History
British Place-Names
Business and Management
Chemical Engineering
Chemistry
Christian Art and Architecture
Christian Church
Classical Literature
Computer Science
Construction, Surveying, and Civil Engineering
Cosmology
Countries of the World
Critical Theory
Dance
Dentistry
Ecology
Economics
Education
Electronics and Electrical Engineering
English Etymology
English Grammar
English Idioms
English Language
English Literature
English Surnames
Environment and Conservation
Everyday Grammar
Film Studies
Finance and Banking
Foreign Words and Phrases
Forensic Science
Fowler's Concise Modern English Usage
Geography
Geology and Earth Sciences

Hinduism
Human Geography
Humorous Quotations
Irish History
Islam
Journalism
Kings and Queens of Britain
Law
Law Enforcement
Linguistics
Literary Terms
London Place-Names
Marketing
Mathematics
Mechanical Engineering
Media and Communication
Medical
Modern Poetry
Modern Slang
Music
Musical Terms
Nursing
Philosophy
Physics
Plant Sciences
Plays
Political Quotations
Politics and International Relations
Popes
Proverbs
Psychology
Quotations
Quotations by Subject
Rhyming
Rhyming Slang
Saints
Science
Scottish History
Shakespeare
Social Work and Social Care
Sociology
Statistics
Synonyms and Antonyms
Weather
Weights, Measures, and Units
Word Origins
World Mythology
Zoology

Many of these titles are also available online at www.oxfordreference.com

A Dictionary of

Geography

SIXTH EDITION

SUSAN MAYHEW

OXFORD
UNIVERSITY PRESS

Great Clarendon Street, Oxford, OX2 6DP,
United Kingdom

Oxford University Press is a department of the University of Oxford.
It furthers the University's objective of excellence in research, scholarship,
and education by publishing worldwide. Oxford is a registered trade mark of
Oxford University Press in the UK and in certain other countries

First published 1992 as *The Concise Oxford Dictionary of Geography*
First edition (with Anne Penny) 1992
Second edition 1997
Third edition 2004
Fourth edition 2009
Fifth edition 2015
Sixth edition 2023

Published in the United States of America by Oxford University Press
198 Madison Avenue, New York, NY 10016, United States of America

British Library Cataloguing in Publication Data
Data available

Library of Congress Control Number: 2022923762

ISBN 978-0-19-289639-1

Printed and bound in the UK by
Clays Ltd, Elcograf S.p.A.

How to use this dictionary

This dictionary provides coverage in one volume of the terms used in both human and physical geography. There are over 3,000 definitions across the following fields: cartography, surveying, remote sensing, statistics, meteorology, climatology, biogeography, ecology, simple geology, soils, geomorphology, population, migration, settlement, agriculture, industry, transport, development, and diffusion.

Headwords are printed in bold type and appear in alphabetical order. However, some entries contain further definitions. These have been included under the headword to avoid unnecessary repetition and to indicate some of the wider applications of the headword. Many terms have been marked with an asterisk. This points the reader to cross-references. In a very few cases the cross-reference as indicated does not have the exact wording as in the entry, but is close enough to make further reference possible.

References to geographical authors include surname, initials, and date. The use of these details in a card or electronic library catalogue will yield the book referred to.

Most journals are easily identifiable by their abbreviations, less simple abbreviations include:

AAAG   *Annals of the Association of American Geographers*
ESPL   *Earth Surface Processes and Landforms*
TIBG   *Transactions of the Institute of British Geographers*

To find recommended web links for this title, visit www.oxfordreference.com/page/geog when you see this sign.

 SEE WEB LINKS

Acknowledgement

I would like to thank my editor Jamie Crowther for his patience and understanding and Abigail Humphries Robertson, my production editor, for her moral support and humour during difficult times.

aa *See* BLOCK LAVA.

abiotic Not living, non-biological, usually describing factors in an ecosystem: atmospheric gases, humidity, salinity, soil mineral particles, water, and so on.

ablation Loss of snow and ice from a glacier by *sublimation, melting, and evaporation; and from the *calving of icebergs, and avalanches. In temperate and subpolar regions melting is the major form of ablation; in the Antarctic, calving. The rate of loss varies with air temperature, relative humidity, wind speed, insolation, aspect, and the nature of the surface. In snowfields, ablation includes snow removed by the wind, and is affected by aspect, depth of snow, and the nature of the underlying surface.

The **ablation sub-system**, where annual ablation exceeds annual accumulation, lies between the *firn line and the glacier snout. Where ablation occurs at the edges of glaciers, debris accumulates to form **ablation moraine/ablation till**. An **ablation valley** is a subsidiary valley formed beneath the crest of a lateral moraine and the valley side. True ablation valleys can result from differences in *insolation, and may act in the development of lateroglacial moraine valleys. See Iturrizaga (2001) *Geogr. J.* 54.

aborigine A member of an indigenous people existing in a land before invasion or colonization from outside. For Canadian aboriginal peoples, see *Atlas of Urban Original Peoples*; for Australian, the Aboriginal Environments Research Centre, U. Adelaide. There are 38 recognized peoples in Bolivia, the majority in the Andes. *See also* INDIGENOUS.

(⊕) SEE WEB LINKS

- IWGIA website – The International Work Group for Indigenous Affairs

abrasion Also known as corrasion, this is the grinding away of bedrock by fragments of rock which may be incorporated in ice (*glacial abrasion), water (marine abrasion, *fluvial abrasion), or wind (*aeolian abrasion). In fluvial environments, the main agent of abrasion is the *bed load. The mass of solid material removed varies with the size, density, and velocity of the particles, and the density of the vector carrying these particles. Try G. Domokos and G. W. Gibbons 2013 on geometrical and physical models of abrasion if you are of a mathematical bent.

absolute humidity The density of the water vapour present in a mixture of air and water vapour. Cold air cannot contain as much water vapour as warm air, so has a lower absolute humidity than warm air. Low absolute humidity results in more evaporation than from air with high absolute humidity. *See also* RELATIVE HUMIDITY.

absolute plate motion The movement of a crustal plate in relation to a fixed point, such as latitude and longitude.

abundance The total number of individuals of a certain species present in an area, See Yin and He *Methods Ecol. & Evol.* (2014) 5, 4.

An **index of abundance** is a relative measure of the size of a population or sub-unit of the population, by comparison with a chosen benchmark, such as a year. See Orensanz et al. (2016) *Developments in Aquaculture and Fisheries Science* on capture/recapture methods used to assess the abundance of fish stocks.

abyssal, abysso- Of or belonging to the ocean depths, especially below 2000 m. The **abyssopelagic zone** is that part of deep lakes, oceans, or seas characterized by specific forms of plankton and nekton which inhabit open water, and the **abyssobenthic zone** is the bottom of a deep lake, ocean, or sea. An **abyssal plain** is the deep sea-floor, with a gradient of less than 1 in 10000. The Canada Abyssal Plain, lying between Canada and the Alpha ridge, is the largest of the Arctic sub-basins, with an average depth of 3658 m. **Abyssal hills**, which interrupt an abyssal plain, are 50–250 m high.

accelerator A factor which increases the momentum of a boom or slump in an economy, so that a small change in demand, for example, may lead to a greater industrial growth (or decline). Bone, Allan, and Haley (2017) BEIS research paper number 7 is strongly recommended.

access Connection or entry to a service. A. Sen (1999) measures **access to advantage** by one's access to basic needs: satisfying goods—like food; freedoms—as in a labour market; and capabilities. S. W. Allard (2004), in an examination of access to social services in three American cities, finds that poor populations in urban centres generally have greater spatial access to social services than poor populations in the suburbs.

accessibility The ease of approach to one location from other locations: in terms of the distance travelled, the cost of travel, or the time taken. Improved infrastructure improves accessibility, and that increases the competitiveness of the economy and favours specialization benefits and scale economies. Regions with good access to their input materials and markets tend to be more productive and competitive than peripheral and remote regions; see Gutiérrez in R. Kitchin and N. Thrift, eds (2009). Accessibility also refers to the ability to connect with information, knowledge, and people.

Topological accessibility is measured in relation to a system of nodes and paths (transportation network), assuming that accessibility is significant only to specific elements of a transport system, such as airports or ports. **Physical accessibility** is the spatial separation of people from the supply of goods and services; see, for example, Orcao and Diez-Cornago (2007) *Area* 39, 3 on physical access to health services in Spain, and Gage and Calixte (2006) *Pop. Studs* 60, 3 on maternal health services in Haiti. **Social accessibility** is the ability of an individual to reach a resource or location, as affected by class structures, income, age, educational background, gender, or race. See Foresight (Government Office for Science) Inequalities in Mobility and Access in the UK Transport System (2018). Condeço-Melhorado, Reggiani, and Gutiérrez (2018) New Data and Methods in Accessibility Analysis, *Netw. Spat. Econ.* 18 is also a must.

accordant Complying with; thus, **accordant drainage** has evolved in conformity with the underlying geological structure: domes show a radial pattern, for example.

accreting margin *See* CONSTRUCTIVE MARGIN.

accretion 1. The growth of land by the offshore deposition of sediment, forming *spits and *tombolos. Accretion is most active in *estuaries, particularly within the tropics. Vertical accretion in estuarine marshes depends on rates of sediment deposition and is a complex function of different interacting variables. See Butzeck et al. (2015) *Estuaries and Coasts* 38, 434–50.

2. The increase in size of a continent by the addition of **accretion terranes**.

3. The growth of a landform by the addition of deposits; *seif dunes grow by accretion. See H. S. Edgell (2006).

4. The increase in size of particles by additions to the exterior, as in the formation of *hailstones. See Zheng and List (1995) *Procs Conf. Cloud Phys.*

acculturation The adaptation to, and adoption of, a new culture. This may occur simultaneously as two cultures meet, but occurs more often as an immigrant group takes to the behaviour patterns and standards of the receiving group. A major example is acculturation among African Americans; see H. Landrine and E. Klonoff (1996), and Sclemper (2007) *J. Hist. Geog.* 33, 2. Acculturation can be a two-way street encompassing change within the 'incomers' and the receiving society; see F. Pedrazza (2014) *Political Research Quarterly* 67, 4 (pp. 889–904).

accumulated temperature One of the key factors for crop growth, this is the length of time, from a specific date, for which mean daily temperatures have been above, or below, a stated temperature; the total time for which temperatures varied from that standard.

accumulation 1. The input of ice to a glacier. It is at its greatest in shaded uplands. The **accumulation zone** (**accumulation sub-system**) is between the source and the *firn line, where the input of snow, firn, and ice exceeds *ablation. For snowfall in an accumulation zone to emerge as ice at the glacier snout takes about 100 years (Tuffen et al. (2002) *Sed. Geol.* 149).

2. The reinvestment of surplus value, in the form of capital, in order to increase that capital. Accumulation is a key feature of *capitalism because, in order to remain in business, capitalists have not only to preserve the value of their capital, but to add to it; accumulation of a firm's net worth determines the growth rate of capital and the growth rate of the economy. D. Harvey (1982) argues that the obligatory accumulation of capitalism has fostered *uneven development.

acidification A soil-forming process whereby organic acids (from humus) increase hydrogen ion concentration, as in the transformation of *brown-earth soils into acid brown earth. Acidification has a strong negative impact on soil, possibly suppressing soil carbon emission. **Ocean acidification** results from interactions between sea water and increasing atmospheric CO_2 concentrations. For a clear examination of the chemistry of ocean acidification, and its environmental impacts, go to Barker and Ridgwell (2012) 'Ocean Acidification', *Nature Education Knowledge* 3(10), 21.

acid rain When *fossil fuels are burned, dioxides of sulphur and nitrogen are released into the air; these dissolve in atmospheric water to form acid rain. In addition, nitrogen oxides combine with volatile organic compounds to form ground-level smog. These pollutants lead to the acidification of lakes and streams (making some of them incapable of supporting aquatic life), impair visibility, weaken forests, and degrade buildings.

SEE WEB LINKS
• The US Environmental Protection Agency provides a clear, but basic, site on this topic.

acid soil Soil with a pH under 7, such as *podzols and *brown earths. Acidity in a soil may be due to the *leaching out of cations when *precipitation exceeds *evapotranspiration. Other factors include the nature of the vegetation (and thus the *humus) and of the parent rock. Adams and Evans (1989) *Eur. J. Soil Sci.* 40, 3 propose a model of soil acidification.

actant Usually in *actor–network theory, something that acts or to which activity is granted by others. The term is not restricted to humans in general; it can truly be anything that is the source of an action.

action research *See* PARTICIPATORY ACTION RESEARCH.

action space (activity space) The area in which an individual moves and makes decisions about her or his life. Wong and Shaw (2011) *J. Geog. Syst.* 13, 2, 127 use this concept to assess segregation in south-east Florida, and include statistical methods of determining activity space. A useful and powerful paper.

active layer A highly mobile layer, periodically thawing, located above the *permafrost in *tundra regions, and ranging in depth from a few centimetres to 3 m. Its thickness depends on slope angle and aspect, drainage, rock and/or soil type, depth of snow cover, and ground-moisture conditions; clearing vegetation will increase the depth and mobility of the active layer. The mobility

of the active layer is due to the tundra vegetation, which scarcely binds it together, so that it moves on slopes as gentle as 2°. Thawing may occur daily or only in summer; on refreezing, the active layer may expand, especially if *silt-sized particles predominate.

*Periglacial processes in the active layer include *frost heaving, *frost thrusting, *ice wedging, *gelifluction, and the formation of *patterned ground. *See* THERMOKARST.

active margin Also called Pacific margin, this is a type of destructive *plate margin characterized by *ocean trenches, *earthquakes, andesitic *volcanic chains, and young *fold mountains.

active volcano Usually agreed to be a volcano known to have erupted in the last 10 000 years. Look for the US Geological Survey, which assesses and monitors hazards at volcanoes within the United States and its territories. Good sources for information about volcanoes outside the United States include Oregon State University's Volcano World and the Smithsonian Institution's Global Volcanism Program.

activist research A research approach whereby geographers (and other scholars) work with local communities in producing knowledge *together*. In addition this approach stresses the responsibility of the academic institutions to channel resources in support of social, economic, environmental, and human rights. Torres (2018) *Professional Geographer* 71, 1 describes two community-engaged and activist research projects: (1) establishing a Spanish–English dual-language programme in partnership with a low-income public school in rural North Carolina, and (2) collaborating on an action research project with a community-based worker's centre to document construction workers' conditions in Austin, Texas, and to lobby for protections. *See also* ACTION RESEARCH.

actor–network theory (ANT) This theory explores the interconnectedness of all things, arguing that the basis of every action and decision depends on the actor's subjective interpretations, which are made and mutually adjusted in interaction with others.
Actor–network theory emphasizes and considers all surrounding factors, whether human or non-human. It claims that all objects and things exhibit consciousness, and the closer in space they are to one another, the more essential they are to each other. As an example, De Laet and Mol (2000, *Social Studies of Science* 30) focus on the spread and functioning of a hand water pump in rural Zimbabwe and how it '*does* all kinds of things … it *acts* as an actor'. An admirable summary of this work comes from Müller (2015) *Prog. Hum. Geog.* 39: 'Rather than being a passive object, the pump is bound up in a socio-material network: it reshapes village life and, through its presence throughout the country and its key role in tying the government together with communities, also helps build a nation. De Laet and Mol show that the pump's spread is due to its being a fluid object: flexible and responsive to changing circumstances and thus able to circulate and adapt easily. The pump thus has a life of its own: it is not the docile subject of its creator, but it evolves and mutates as it circulates.'

adaptive management A management strategy that emphasizes the use of feedback in planning strategies: a systematic approach for improving resource management by learning from management outcomes.

Adaptive management has proved difficult to achieve in practice; however, the US Fish and Wildlife Service has used adaptive management over the past 20 years for the annual sport hunting regulations of mallard ducks: see Nichols et al. (2015) *Journal of Applied Ecology* 52, 539–43.

adaptive radiation The diversification of species to fill a wide range of ecological *niches. Perhaps the most famous example is that of Darwin's finches in the Galapagos Islands: see K. T. Grant and G. B. Estes (2009).

adiabat A line plotted on a thermodynamic diagram, usually on a *tephigram, showing as a continuous sequence the temperature and pressure states of a parcel of air with changing height. An **adiabatic change** is a change in temperature, pressure, or volume, involving no transfer of energy to or from another material or system. In an adiabatic process, compression is accompanied by warming, and expansion by cooling. An **adiabatic temperature change** thus results from a pressure change. The speed at which the temperature of rising air falls with altitude is the **adiabatic lapse rate**. Dry, rising air expands with height. The energy needed for this expansion comes from the air itself in the form of heat.

The resulting change in temperature is expressed in the equation:

$$\frac{dT}{dz} = \frac{-g}{Cp}$$

where dT is the temperature change, g is the acceleration due to gravity, dz is the height change, and Cp is the specific heat of the air parcel. This change is the **dry adiabatic lapse rate** (DALR): 9.84 °C/1000 m. The temperature change sustained by any parcel of dry air is calculated using Poisson's equation.

If the rising air becomes saturated to *dew point, condensation of vapour will begin. This condensation is accompanied by the release of *latent heat, which partly offsets the cooling with height, so that the rate of cooling of moist air—the **saturated adiabatic lapse rate** (SALR)—is lower than the DALR. In the lower *troposphere, the vapour content of air is high so the latent heat of condensation is high; SALRs may be as low as 5 °C/1000 m. In the cold, dry, high troposphere, though, there is little vapour ready for condensation, so the SALR may be close to the DALR. Quantitative expressions of the SALR are therefore quite complex. See B. Haurwitz (2007).

adsorption In *soil science, the addition of ions or molecules to the electrically charged surface of a particle of clay or humus. In this way, minerals become bonded to soil particles (M. Ashman and G. Puri 2002). An **adsorption complex** is any substance in soil that can absorb ions or molecules.

advanced capitalism From the 1970s onwards, a shift away from industrial production towards services, notably financial services, in the more developed countries. This shift caused de-industrialization prompting liberalization, privatization, deregulation, intensified global competition, and increased unemployment. A key feature of advanced capitalism is the possession of capital by fewer and fewer owners. Torbern Iverson and David Soskice (2019) are terrific on this.

advanced economy (economically advanced economy) 'A term used by the International Monetary Fund (IMF) for the top group in its hierarchy: advanced economies, countries in transition, and developing countries.

It includes: Australia, Austria, Belgium, Canada, Denmark, Finland, France, Germany, Greece, Hong Kong, Iceland, Ireland, Israel, Italy, Japan, South Korea, Luxembourg, Netherlands, NZ, Norway, Portugal, Singapore, Spain, Sweden, Switzerland, Taiwan, UK, and USA' (CIA (2007) *World Factbook, appendix B*). The IMF also includes Cyprus and Slovenia (IMF (2007) *WEO Groups & Aggregates Info.*).

advection The 'horizontal transport of energy and mass by parcels of air in convective flow systems' (Sivertsen (2007) *EMS7/ECAM8 Abstracts* 4). In meteorology, the term usually means the horizontal transfer of heat. **Advective transfers** occur in air streams, ocean currents, and surface *run-off, partly redressing the *insolation imbalance between the tropics and the poles. For **advection fog**, *see* FOG.

aeolian Of the wind: aeolian processes include erosion, transport, and deposition, and work best with sparse or absent vegetation. **Aeolian landforms** include desert pavements *reg, *ergs, *nebkhas, *dunes, *zeugen, and *yardangs. See Maghsoudhi (2020) *Aeolian Landforms*, available online.

aeration zone The zone between the soil-moisture zone and the capillary zone, just above the water-table, which stores and gradually releases water, but is not wholly saturated. The depth of the aeration zone varies with the seasons, as a result of precipitation and (upward) capillary action. The Groundwater Foundation provides a very useful glossary.

aerobic Describing any living organism which depends on atmospheric oxygen to release energy from foodstuffs during respiration. *Compare with* ANAEROBIC.

aerological diagram A chart plotting the factors which determine the movement of air. Variations of temperature, pressure, dry and saturated *adiabatic lapse rates, and *saturated mixing ratio lines are plotted against height in a *tephigram. An adiabatic chart may be used to predict the *convective condensation level.

aerosol A suspension of droplets or particles in a gas; more precisely, of particles with a maximum diameter of 1 μm (*fog and *mist are thus aerosols). In meteorology, the term is often used to describe the particles suspended within the air, such as minute fragments of sea-salt, dust, organic matter, and smoke. These enter the atmosphere by natural processes such as vulcanicity, and by human agency such as burning *fossil fuels. Aerosols have a role in the radiative forcing of climate; they can absorb and scatter both infrared and solar radiation in the atmosphere; and they can affect and change the processes that control cloud and precipitation formation through their role as condensation nuclei. See Myhre, Samset, and Storelvmo (2013), Aerosols and their Relation to Global Climate and Climate Sensitivity. *Nature Education Knowledge* 4(5).

SEE WEB LINKS
- A daily global aerosol map from NASA.

affect This can be a process of change within a person; a person *is* affected and *can* affect. Alternatively, affect is a feeling, disposition, or mood that exists before understanding or rational thought. In other words, affect is our unconscious response which precedes our conscious feelings and decisions. What has this to do with geography? Put simply, decisions may often be made on the basis of individuals' unconscious affect, and these 'primitive' decisions

may have profound impacts on places and spaces. See Lin and Minca (2020) *TIBG* 45, 4.

afforestation Planting trees. Tree-planting can stabilize soils by increasing *interception and reducing *run-off; reduce flooding through the reduction of silting; improve soil fertility; and provide timber, charcoal, and firewood, and habitats. But there are drawbacks: afforested land can no longer be used for residential developments and agriculture, and afforestation can reduce *biodiversity. See Hert, Muys, and Hansen, eds (2007) on the environmental effects of afforestation in NW Europe.

aftershock A smaller tremor or series of tremors after an earthquake. Aftershocks result from static and dynamic stresses than weaken faults. The literature on aftershocks in academic journals is quite technical, and may be daunting to the beginner. Try Lui, Hang, Tuyen, Giang, and Hai (2019) *Ekológia* (Bratislava) 38, 189–200.

A smaller tremor or series of tremors occurring after an *earthquake. Aftershocks result from static stress changes and 'dynamic' stresses (transient, oscillatory stress changes), that can weaken faults. Such dynamically weakened faults may fail after the seismic waves have passed, and might even cause earthquakes that would not otherwise have occurred (Kilb et al. (2003) *Nature* 408). See also Johnson et al. (2008) *Nature* 451. See Abe and Suzuki (2012) *Acta Geophysica* 60, 3, 547 on aftershocks in modern perspectives.

age dependency The dependency of people too young or too old to be employed full-time on the contributions of those in full employment. The UK government's *Examining the* *relationship between population ageing, economic dependency and international migration in the UK* is comprehensive, clear, accurate, and regularly updated.

ageing, geographies of Spatial variations in demographic ageing, and the distribution of the elderly at local, regional, and national scales, are key issues in geographies of ageing: see the United Nations Population Division *World Population and Ageing: 1950–2050* (2001) and Ferry et al. (2006) *EPRC* on variations within the EU.

Papers such as that of Skinner, Cloutier, and Andrews (2014) *Prog. Hum. Geog.* will tell you of the changing ways some geographers have thought about geographies of ageing, but give not one example of a geography of ageing in real life.

agency The capacity of the individual to act independently: a capacity that is meaningful, and has both intent and purpose. Agency is used in pursuing a particular goal, whether as individuals or as groups, and agency is important to geography because it creates spaces. Clearly, the degree of agency each individual can exert varies with social class and geographical location.

The **agency–structure debate** refers to the relative merits of agency and *structure—the relative importance of social structure or human agency—in producing particular outcomes. For the human geographer, these outcomes include spaces and places.

Social structures and human agency suffuse and create each other; structure is the context within which geographical agency takes place. Agency is related to geopolitics, for example when a nationalist group tries to create a nation-state, while actions are located within spaces.

agent An entity which can take a role—a function, a service, or an identity. *Neoliberalism envisages people as autonomous, individualized, self-directing, decision-making agents whose interactions are constrained by social and spatial structure, but, over time, structure and agency suffuse and create each other. Fundamentally, an **agent-based model (ABM)** constitutes a network of agents and the relationships between them. Centrally located agents are ones connected to many other agents, and remote agents, far from the network's centre, are unconnected or just connected to a very few other agents. Centrality here means connectivity, and not location based on latitude and longitude.

An agent-based model is thus made up of a number of real-world agents—people, households, institutions, and so on—which follow an algorithm (set of rules) controlling their interactions with other adjacent agents. The benefits of ABM over other modelling techniques are that an ABM captures emergent phenomena, provides a natural description of a system, and is flexible. Go to Macal and North (2010) *J. Simulation* 4, 151–62 for a tutorial on agent-based modelling and simulation.

age–sex pyramid A set of two histograms set back to back on a vertical axis, depicting the numbers of the two sexes by age group. In many less developed countries the pyramid will have a very wide base; these are termed **progressive pyramids** because they suggest future population growth. **Regressive pyramids** characteristic of advanced economies, where there are fewer children and more old people, are more cylindrical; population is likely to fall.

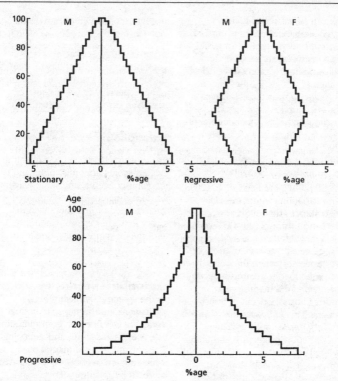

Age–sex pyramid

age structure The composition of a nation by age groups. The age structure of the UK population has become older, the median age rising from 34.1 years in 1971 to 38.4 in 2003 and projected to rise to 43.3 in 2031 (*UK National Statistics*). Europe, with more than 15% of the population over 60, suffers from *age dependency, while South Asia, with 45–55% of its population under 16, is concerned to limit population growth. *See* AGE–SEX PYRAMID.

agglomerate A coarse-grained volcanic rock, made of sharp/sub-angular lava fragments, set in a fine matrix. Some

are *pyroclastic; others are deposits from *mud flows or *lahars.

agglomeration 1. Within a geographical area, a concentration of economic activity brought about by, among others, *external economies (such as a pool of skilled labour), economies of scale, *cumulative causation, local authority planning, and chance. **Agglomeration economies** are the benefits to people or firms that come from locating near one another in cities and industrial clusters, such as California's Silicon Valley. For an industry, these are the benefits of locating in a densely peopled and highly industrialized situation, where the market is large, but spatially concentrated. Transport costs are thus low, and specialized industries can evolve, since local demand is high. Further benefits arise because of *functional linkages and *external economies. 'The agglomeration of economic activities is a phenomenon as old as cities themselves' Clark et al., eds (2000).

2. In meteorology, the process by which cloud droplets grow by assimilating other droplets.

aggradation The build-up of a surface or soil through the addition of organic and inorganic materials such as flood sediments; *aeolian, marine, or *fluvial processes, when the quantity or calibre of the load is greater than the competence of the medium that transports it; or as a result of *mass movement. This increases the elevation of the land. Certain soils develop *beneath* an episodically or continually aggrading land surface, in a process that is known as 'upbuilding' and 'cumulization'. This is **aggradational pedogenesis**: see A. Mason and P. M. Jacob (2013).

aggregate A group of soil particles held together by electrostatic forces, *polysaccharide gums, and *cementation by carbonates and iron oxides.

aggregation A way of combining a collection of very large datasets into classes, both to make subsequent analysis simpler and to provide a more generalized account of a phenomenon. For any analyses based on areal units, the problem is one of *scale: that plotting geographical information on the wrong scale can conceal information that only becomes apparent at other scales and

resolutions. See Openshaw (1977) *TIBG* 2, 4, 459.

aggressivity In *karst geomorphology, the ability of water to dissolve calcium carbonate. This term refers especially to water containing dissolved carbon dioxide (carbonic acid).

agoraphobia A disorder whereby the sufferer avoids places or situations where he or she previously had a panic attack; an agoraphobic typically avoids public places. Between 70 and 90% of people diagnosed with agoraphobia are female. Some geographers have termed agoraphobia a 'space fear' which ultimately results in the inhibition of movement. See J. Davidson (2003) *Phobic Geographies*.

agricultural geography The study of the spatial relationship between agriculture and humans, including agricultural influences on populations and the environment, and the cultural, political and environmental processes which govern the way we use land for agricultural purposes. This includes topics such as habitat management, crop production, and agricultural policy. Related subjects and specialisms in this field include agroforestry, rural sustainability, land management, and sustainable agriculture. Among the models used in agricultural geography are: the input–output, decision-making, diffusion (Hagerstrand), von Thunen, and Jonasson models. For a superb overview of recent topics in agricultural geography, see G. Robinson (2018) 'New frontiers in agricultural geography: Transformations, food security, land grabs and climate change', *Boletín de la Asociación de Geógrafos Españoles*.

agriculture, organic An approach to farming and ranching that avoids the use of herbicides, pesticides, growth hormones, and other similar synthetic inputs. An excellent summary by the Soil Association UK concludes that 'collectively, the evidence supports the hypothesis that organically grown crops are significantly different in terms of food safety, nutritional content, and nutritional value from those produced by non-organic farming' and that 'feeding trials have shown significant improvements in the growth, reproductive health, and recovery from illness of animals fed organically produced feed'. Organic intensive areas tend to have more diverse farm operations, with more women operators and direct sales to people and the community; see Kuo and Peters (2017) The socioeconomic geography of organic agriculture in the United States, *Agroecology and Sustainable Food Systems* 41. See also Žiga Malek et al. (2019) *Agricultural Systems*, 176 for an explanation of the global spatial distribution of organic crop producers.

⊕ SEE WEB LINKS

- A pdf of *Organic farming, food quality and human health: A review of the evidence*, by the Soil Association.

agriculture, sustainable

Agriculture strategies that are profitable, environmentally sound, and supportive to communities. *Sustainable development 'meets the needs of the present without compromising the ability of future generations to meet their own needs' (1987 UN Convention on Environment and Development). The go-to source here is Lichtfouse et al. (2009) *Sustainable Agriculture.*

agrobiodiversity

A subset of *biodiversity that includes all forms of life directly relevant to agriculture. See Pacicco et al. (2018) PLoS ONE on a methodological approach to identify agro-biodiversity.

agroclimatology

The study of the interactions between atmospheric variables and biological systems in agriculture, and the application of this knowledge to increase food production and improve food quality. Agroclimatological data are used to understand how climate can affect agricultural activities, crop productivity, and sustainability, thus informing agricultural decisions. More recently, agroclimatologists have been working on the impact of agriculture on climate—agriculture as a component of the climate system. See Dev Niyogi (2017).

agroecology

A scientific discipline that uses ecological theory to study, design, manage, and evaluate agricultural systems that are productive but also resource conserving. It is concerned with the maintenance of a productive agriculture that sustains yields and optimizes the use of local resources while minimizing the negative environmental and socio-economic impacts of modern technologies.

agro-food system

A system that delivers food from the farm to our tables. This can include credit for the farmer, fertilizer, seeds, and animals, and the complex system from producer to supermarket: 'the interdependent sets of enterprises, institutions, activities, and relationships which collectively develop and deliver material inputs to the farming sector, produce primary commodities, and subsequently handle, process, transport, market, and distribute food and other agro-based products to consumers' (World Bank, Africa Reg. Working Paper Series 44, 2003).

agroforestry

See AFFORESTATION.

aid

Transferring resources from developed to less developed countries. Every economically advanced country gives foreign aid, and each foreign aid scheme is aimed at improving the development of the inhabitants of the poorest countries. In 1984, Peter Bauer commented that 'the one common characteristic of the Third World is not poverty, stagnation, exploitation, brotherhood or skin colour. It is the receipt of foreign aid'. Incoming aid is often a major economic input to the receiving states. For example, Bangladesh has a *GDP of US$ 5.1 billion, and received a total foreign aid of US$ 2.6 billion (2012 data). Many economically less developed countries would face serious political, economic, and social upheaval if aid flows were to dry up; for example the World Bank (2018) estimates that Afghanistan is dependent on international aid to maintain state functions. There are two major approaches to the donation of foreign aid; the 'humanitarian' lobby, and the feeling that states donate foreign aid out of self-interest—that aid gives the donor political and commercial advantages. **Bilateral aid** is from one donor to a recipient country, while **multilateral aid** comes from a group of countries. **Emergency aid** is short-term, generally given in response to disasters, while **structural aid** is given for long-term *development. A donor of **tied aid** can transfer resources more cheaply, as its aid is tied to its own exports.

The WTO-sponsored **Aid for Trade** programme assists developing countries in expanding their trade to help growth and reduce poverty, since many developing countries face internal constraints such as a lack of productive capacity, poor infrastructure, inefficient customs management, excessive red tape, and difficulties to meet technical standards in high-value export markets. The Aid for Trade initiative can include support for building new transport, energy, or telecommunications infrastructure, investments in agriculture, fisheries, and services, as well as assistance in managing any balance of payments shortfalls due to changes in the world trading environment.

( ) SEE WEB LINKS
- Explanation of the European Union's Aid for Trade strategy.

aiguille French for a *pyramidal peak, sharply pointed due to *nivation and *frost wedging.

air mass An area of the *atmosphere extending for hundreds of kilometres that, horizontally, has generally uniform properties, especially of temperature and *humidity, and with similar vertical variations of temperature and pressure throughout. Air masses obtain these attributes from their areas of origin, known as source regions, and become less uniform with movement into different areas, and with the effects of *wind shear.

Air masses are classified by the source region: **continental air mass, c**, or **maritime air mass, m**; and by the latitude of the source area: the arctic, **A**, the poles, **P**, and the tropics, **T**. These two combine to distinguish most air masses, such as **cA— continental arctic, mP— maritime polar, mT—maritime tropical**, and so on. **Equatorial, E**, and **monsoon, M**, tags also exist. The **Spatial Synoptic (broad scale) Classification system** is an air mass typing system that uses ground-based measurements of atmospheric variables to determine the general makeup of an air mass at a particular observing station; see Lee and Sheridan (2012) *Int. J. Climat.* 32 for a six-step approach to developing future synoptic classifications.

air pollution An atmospheric condition in which substances (air pollutants) are present at concentrations higher than their normal ambient (clean atmosphere) levels that produce measurable adverse effects on humans, animals, vegetation, or materials. **Particulate air pollution** comprises solids and liquid droplets floating in the air, whereas **photochemical pollution** comprises ozone and other reactive chemical compounds formed by the action of sunlight on hydrocarbons, and oxides of nitrogen, largely from petrol-vehicle exhausts. Common pollutants include carbon dioxide, carbon monoxide, lead, nitrogen oxides, ozone, smoke, and sulphur dioxide. The most common meteorological conditions associated with air pollution are *inversions, which trap polluted air, and light winds. Fluctuations in emissions as well as changes in conditions due to sunlight, temperature, and wind affect the day-to-day levels of NO_2.

In 2020, the European Space Agency launched an online map showing averaged nitrogen dioxide concentrations using a fourteen-day moving average. The fourteen-day average helps to eliminate short-term effects such as weather conditions and cloud cover in order for users to see regional pollution level trends. NO_2 pollution can be traced to industrial activities, vehicle traffic, and power plants. NO_2 acts as an irritant and elevated levels can cause respiratory problems in humans by inflaming the linings of the lungs.

((()) SEE WEB LINKS
• Description of types of air pollution.
• EU emission standards.

air pressure *See* ATMOSPHERIC PRESSURE.

air–sea interaction *See* OCEAN–ATMOSPHERE INTERACTION.

Aitken nuclei *See* NUCLEI.

alas In a *periglacial landscape, a large, steep-sided, flat-bottomed depression which can be several kilometres across. Lakes often occur in alases, which may then develop into *pingos.

albedo That proportion of solar radiation reflected from a surface, such as clouds (low from thin stratus cloud; up to 80% from thick strato-cumulus), or bare rock. Lighter, whiter bodies have higher albedos than darker, blacker bodies. Earth's total albedo is about 35%. The **ice-albedo feedback** mechanism occurs when a reduction in greenhouse gases in the atmosphere, principally carbon dioxide and methane, make the global climate colder, creating larger areas of ice and snow. Ice and snow reflect more solar radiation than does bare ground or liquid water, which creates a 'positive feedback'. If the Earth ever became half-covered by ice or snow, the feedback would become self-sustaining and glacial ice would rapidly spread to the equator

Zhang et al. (2012) find that, over the last 300 years, surface albedo decreased due to conversions from grassland to cropland in the Northeast China Plain and increased due to conversions from forests to cropland in the surrounding mountains.

Surface albedo (percentage of incoming short-wave radiation which is reflected)	
Fresh, dry snow	80–95
Sea ice	30–40
Dry light sandy soils	35–45
Meadows	15–25
Dry steppe	20–30
Coniferous forest	10–15
Deciduous forest	15–20

alcohol, geographies of Alcohol is at once a social problem, a leisure activity, a pleasure, an accelerator of violence, and central to identity formation. In geographical studies of alcohol, drunkenness features heavily. It's good, therefore, to read that Eldridge and Roberts (2008) *Area* 40, 3 challenge the urban savage/social drinker binary, 'which wrongly presumes consumers cannot be both'. In Europe, a broad division was made between a northern dry area, where beer is the leading beverage, consumed at weekends and outside mealtimes; and a southern wet area, where wine is the main beverage, usually drunk at meals. Subsequently, European alcohol choices have started to converge.

P. Chatterton and R. Hollands (2003) see alcohol-related establishments as the driving force in recent city-centre regeneration; cities at night now attract vast numbers of people. See also D. Hobbs et al. (2003). However, Roberts et al. (2006) *Urb. Studs* 43, 7 conclude that British free market attitudes to extended licensing hours will undermine the government's professed aspirations for an 'urban renaissance' of cultural inclusion and animation. There is a link between alcohol and deprivation in the city—where alcohol use is more visible. Perhaps because of this, working-class cities are deemed to be places for excessive partying and wild stag or hen nights (Chatterton and Hollands (2001) *U. Newcastle on Tyne*), and the stereotype of a drinking culture remains firmly tied to the North-East's industrial heritage. Shaw (2000) *Australian Geogr. Studs* 38, 3 observes that when aboriginality was conflated with alcoholism, it tended to be exoticized and romanticized in its traditional 'bush' setting, but demonized in its urban setting. Magadan Oblast, in Russia's far north-east, has the country's highest alcohol consumption rates; Magadan is the location of a notorious gulag. There is also a clear link between the density of shops selling alcohol and alcohol-related hospital admissions in San Diego county.

alcove A semicircular, steep-sided cavity on a rock outcrop, caused by water erosion, especially spring-sapping.

Aleutian current A cold ocean current of the North Pacific. The **Aleutian trench** lies at the boundary of the North American and Pacific plates; try Choi (1990) *Pac. Geol.* 5. The **Aleutian low** is a semi-permanent 'centre of action' of persistently lower pressure (less than 1013 mb) and higher humidity and relatively higher temperatures for that latitude. In conjunction with its changing relationship with the North Pacific High in the eastern part of the Pacific, it is the major feature of the general circulation controlling the weather over parts of North America.

alfisol A soil order of the *US soil classification. Alfisols are young, acid soils, with a clay-rich B horizon, commonly occurring beneath deciduous forests. *See* BROWN EARTH.

algae A diverse group of simple plants that contain chlorophyll and can photosynthesize. An **algal bloom** is a dense spread of algae caused by changes in water chemistry and/or temperature; there is a positive correlation between increasing nutrients in water and algal density—leading to *eutrophication. See Chiara Faccaet al. (2014) *Hindawi* on harmful algae in the Venice Lagoon.

alienation The estrangement or separation of individuals from one another, said to include isolation, powerlessness, and meaninglessness. K. Marx (1844, 1975) saw this as a result of separating the workers from the benefits of industrialization as enjoyed by the owners of the means of production. L. Wirth (1938) saw alienation as a product of urban life. Others cite political causes; a major contemporary theme is the alienation of some Muslims from 'Western' lifestyles. See Martin 2017 *Ethnicities* 17, 3.

alkaline soil Any soil, such as a *rendzina, which has a pH above 7. Alkalinity usually reflects a high concentration of carbonates, notably of sodium and calcium.

Allerød A north-west European *interstadial, from *c.*12000 to 10800 years BP; in Britain not usually distinct from the *Bølling. Pollen records point to a cool temperate flora, with birch. See Willard et al. (2007) *Pal., Pal., & Pal.* 251.

allochthonous Of rock formed of materials brought from outside, for example from sediments transported by ice; *till is an example. *Compare with* AUTOCHTHONOUS.

allogenic Caused externally. Thus, an **allogenic stream** is one—for example, the Nile—fed from outside the local area, where precipitation and run-off generate the flow. **Allogenic succession** is an alteration in a plant *succession brought about by some external factor.

allopatric Describes species or populations of the same species that occur in habitats that are geographically separated and do not overlap. **Allopatry** is the occurrence of species or populations of a species in geographically separate, non-overlapping habitats. **Hard allopatry** is caused by physical barriers, and thus geographic isolation. **Soft allopatry** is caused by ecological processes that isolate populations geographically in allopatric refuges through niche conservatism/divergence. See Pyron and Burbrink (2010) *J. Biogeog.* 27.

alluvial Referring to *fluvial contexts, processes, and products. The form of an **alluvial channel** (cut in alluvium) reflects the *load and discharge of the river rather than the *bedrock. An **alluvial fan** is a triangular landform; its shape results from the river swinging back and forth during *aggradation while the apex is fixed at the point where the river emerges from the uplands. Alluvial fans develop their geometry mainly in response to different weathering intensities and of their catchment bedrock lithologies. See Pipaud and Lehmkuhl (2017) *Geom.* 293. An **alluvial cone**, similar in origin, tends to have a steeper slope.

Alluvial filling is *sedimentation from fluvial channels. During low water, **alluvial flood plains** are mostly dry, but when most of the *flood plain is swamped, coarse material is transported short distances, temporarily deposited, and picked up during the next flood. Major controls on alluvial flood plains include sediment, flow regime, and differing base levels. **Alluvium** is a general term for deposits laid down by present-day rivers, characterized by sorting (coarser alluvium is found in the upper course of rivers; finer in the lower courses), and *stratification (coarse material is overlain by finer). **Alluvial landforms** include *deltas, river beds, and flood plains.

alluviation The accumulation of sediment. See DeLong et al. (2007) *Geomorph.* 90, 1–2. See Blum and Törnqvist (2000) *Sedimentol.* 47, Suppl. on change in sediment yield and discharge, and alluviation during relative sea-level fall. **Cut-and-fill alluviation** results in a shifting channel eroding old surfaces and creating new surfaces.

Alonso model William Alonso's 1964 explanation of urban land use and land values is grounded on the concept of *bid rents. If the level of goods and services is constant, land prices should decrease with distance from the centre. The poor will live near the city centre, where high housing densities more than offset high land costs; the better-off will live at lower densities at the edge of the city (on more, but cheaper, land). Each household represents a balance between land, goods, and *accessibility to the workplace. However, commuting costs rise with distance from the centre, reducing disposable income, and higher-income groups may prefer the accessibility to the *CBD offered by the inner city to the space, quiet, and cheaper land of the suburbs, thus causing *gentrification. Hua (2001) *Int. Reg. Sci. Rev.* 24, 3 reviews and re-presents Alonso's model. See also Chang Su-Eun (2004) 'A Bid-Rent Network Equilibrium Model'.

Alps A chain of *fold mountains running from eastern France and northern Italy through Switzerland to Austria. An **alpine glacier** is a valley glacier, and **alpine topography** describes features of upland glaciation: *cirques, *glacial troughs, *hanging valleys, and *truncated spurs. The **Alpine orogeny** occurred in the *tertiary, creating fold mountains as far apart as the Andes, Japan, and East Indian Ocean *island arcs. See Rosenbaum (2002) *Tectonophys.* 359 on the movements of Africa and Europe during the Alpine orogeny.

alternative economies and spaces Spaces, activities, and economic geographies that are very different from those of mainstream *capitalism. These alternative practices include cooperatives, credit unions, ethical/fair trade, local exchange and trading schemes, and non-monetary currencies. These economies do not define success by growth, challenge unnecessary consumption, invest in localized systems of exchange, and minimize environmental damage.

alternative food networks Associations of food producers, distributors, and retailers committed to quality, *fair trade, and ecology. See Michel-Villarreal et al. (2019) *Sustainability* Feb, 7 for an assessment of alternative food networks.

alto- Prefix referring to *clouds between 3000 and 6000 m high.

ambient Surrounding, of the surroundings. **Ambient temperature** is the temperature surrounding a given point. The **ambient air standard** is a quality standard; see J. Stedman et al. (2005), DEFRA, on ambient air quality assessment in the UK.

ambivalence The condition of being in two minds. This may mean creating geographies and spaces in which tensions, contradictions, and paradoxes can be negotiated fruitfully and dynamically. Look for Bondi (2007) for a feminist geography of ambivalence.

amenity Pleasantness; those aspects of an area such as housing, space, and recreational and leisure activities which make it attractive to live in. **Amenity migration** is the purchase of first or second homes in rural areas prized for their aesthetic and/or recreational values It has been aided by rising incomes, the easing of limits on foreign property ownership in many countries, developments in transport and communications technologies, and increasing exposure to images of prized rural landscapes. Gosnell and Abrams (2009) *Geojournal* present a review of the social science literature related to the concept of amenity migration, focusing on the ways in which it has been conceptualized, theorized, and documented by different communities of scholars.

amphidromic point The point around which *tides oscillate; while there are tides along the coasts of East Anglia and the Netherlands, there is a point in the sea between the two where there is no change in the height of the water. High water rotates around the amphidromic point; anticlockwise in the Northern Hemisphere, clockwise in the Southern.

ana- Rising. An **anabatic wind** is an upslope wind driven by heating—usually daytime insolation—at the slope surface under fair-weather conditions (see AMS (1959) *Glossary of Meteorology*). These winds tend to be less frequent, and weaker, than their corresponding, descending, *katabatic winds.

anabranching channel A *distributary channel which leaves the main channel, sometimes running parallel to it for several kilometres, and then rejoins it; a channel separated by shorter-lived alluvial islands, cut into a pre-existing floodplain, or formed within a delta. Anabranching can sustain a stable river pattern where a single channel could not transport sediment. Anabranching channels differ from *anastomosing channels in that they are undivided.

anaerobic Describing any organism or process which can or must exist without free oxygen from the air, such as the anaerobic bacteria which are responsible for the process of *gleying.

ana-front A front at which the warm air is ascending. An ana-front is normally the initial state of a cold front, developing into a kata-front as the depression becomes more occluded.

anastomosis The division of a river into a stable multi-channel system with levées, backswamps, and large, stable islands. An **anastomosing channel**, unlike an *anabranching channel, has distributaries of its own. Valley incision starts at outflow points; up-valley incision begins where the flow rejoins the main channel. In time, the channel segments join, making a new anabranch of the anastomosing-channel system. Many anastomosed systems cannot be categorized; one common factor seems to be low specific stream powers, thus explaining the river's inability to maintain channel capacity via sediment transport and erosion.

anchor tenant A pioneer tenant, often a large innovative firm or a research university or public laboratory in a new development. The anchor may spin off new companies and attract others. A state may promote the 'anchor' with preferential treatments in terms of land prices, tax exemptions, infrastructure support, and administrative services, in order to persuade the firms along its supply chain to follow suit.

angle of repose The angle at which granular material comes to rest, varying inversely with fragment size in perfectly sorted materials, but directly in those imperfectly sorted; inversely with fragment density; directly with fragment angularity, roughness, and compaction; inversely with height of fall of material on free cones; and directly with moisture content up to saturation point, but inversely thereafter.

angular momentum The momentum of a body taking a curved path; for a rotating planet, angular momentum is defined as the moment of the linear momentum of a particle about a point, thus:

$$M = \underset{r}{\rightarrow} \, xm \, \underset{v}{\rightarrow}$$

where M is the angular momentum about a point 0, $\overrightarrow{r}$ the position vector from 0 to the particle, m the mass of the particle, and $\overrightarrow{v}$ the velocity. The Earth's angular momentum can easily be visualized by taking a marble on the end of a string and rotating it on a smooth table. If the string is shortened the rate increases; if lengthened it slows. The speed is inversely proportional to radius. Thus, part of the atmospheric envelope, rotating with the Earth at the equator, moves polewards, it will tend to speed up; see *conservation of angular momentum.

animal geographies Animals occupy human social life and space in many ways, yet, it is argued, humans have turned them into commodities or have degraded them by using them as pets. Major themes in animal geographies include: the relationships and interactions between animals and humans; the spaces and places occupied by animals in human culture; the environmental, social, and cultural implications of the representation and use of animals; the meanings and identities ascribed to animals and animal issues, and questions of animal welfare and rights. Sellick (2020) *Geography* 105, 1 traces the development of animal geographies using the example of cattle and their relationship with key geographical concepts of space, place, and landscape.

Geographers using *actor–network theory argue that the divisions between people and animals are subject to change and negotiation, while ecofeminists see the 'human chauvinism' towards animals as part of a hierarchy topped by European-American males. Wolch and

Emel (1998) argue that Western representations of animals such as wolves underpin the racism and sexism which label 'others' as savage. (Interestingly, in Russian representations, the wolf is the benevolent 'caretaker' of the forest, cleaning up the carrion.)

anistropic Of a phenomenon which has different values when measured in different directions; wood, for example, is easier to split along its grain than across it. Anistropic landforms include *drumlins and *dunes. See Houser et al. (2017) *Geom.* 297.

annular drainage *See* DRAINAGE PATTERNS.

annular modes Two hemisphere scale patterns of climate variability: a northern annular mode (NAM) and a *southern annular mode (SAM). Both annular modes explain more of the week-to-week, month-to-month, and year-to-year variance in the extratropical atmospheric flow than any other climate phenomenon. The annular modes describe variability in the atmospheric flow that is not associated with the seasonal cycle. These modes are characterized by north-south shifts in atmospheric pressure mass between the polar regions and the middle latitudes.

anomie The lack of traditional social patterns within a group; a lack of norms leading to conflict and confusion. In a study of second-generation British-Barbadians moving to Barbados, Potter and Phillips (2006) *AAAG* 96, 3 note that they are often thought by the locals to be mad. Anomie has been seen as the result of urbanism; see Pols (2003) *Osiris* 18.

Antarctic Denoting regions south of the Antarctic circle, 66° 32′ S. Within the **Antarctic circle**, the sun does not rise on

21 June (winter solstice in the Southern Hemisphere) or set on 22 December (summer solstice in that hemisphere). **Antarctic meteorology** is characterized by severe winters with double temperature minima, i.e. two separate occasions of minimum temperature, due to the absence of *insolation for several winter months, and to the frequent exchange of air with that of lower latitudes. *Blizzards are common. Temperatures rise in late summer as the long waves of the *westerlies bring incursions of warmer air. Nevertheless, precipitation is still almost always in the form of snow, since maximum temperatures, occurring at the summer (December) solstice, rarely exceed 0°C. See J. King and J. Turner (1997). Antarctica seems to be part of a single tectonic **Antarctic plate**, but there is clear palaeomagnetic evidence that in the past there has been a large relative rotation between different parts of the continent. Whittaker et al. (2007) *Science* 318, 83 detail major Australian and Antarctic plate reorganizations.

antecedent Prior to, before, as in **antecedent drainage** patterns: the Wind River, Wyoming, for example, probably pre-dates the dome it now cuts through, the river's erosion keeping pace with the uplift of the dome. **Antecedent moisture** is the amount of moisture already present in the soil before a specified rainstorm. An **antecedent boundary** is one drawn before an area is populated, as opposed to a superimposed boundary, fixed by colonial powers.

anthropocene era The current epoch of Earth's history when humankind has joined with the other environmental forces in shaping the planet. When did such an epoch commence? Crutzen (2002) *Nature* 415, 23 assumes the origin of Anthropocene to be synchronized with the appearance of the steam engine in the late eighteenth century, which prompted a significant upturn in the burning of fossil fuels, as shown by the increase in carbon dioxide and methane in the air trapped in the polar ice sheets and ice caps, corresponding to a sharp increase in carbon dioxide and methane concentrations in the atmosphere. However, Certini and Scalenghe (2011) *Holocene* 21, 1269 propose that the Anthropocene be defined as the last 2000 years of the late Holocene and characterized on the basis of anthropogenic soils.

During this era humankind has: increased its population tenfold; exhausted 40% of the known oil reserves; contributed to a 30% increase in atmospheric CO_2; transformed nearly 50% of the land surface (with significant consequences for biodiversity, nutrient cycling, soil structure, soil biology, and climate); dramatically altered coastal and marine habitats (50% of mangroves have been removed and wetlands have shrunk by one-half); and increased extinction rates. An entire issue of *The Holocene* (2011) 21 is devoted to early Anthropocene levels of CO_2 and CH_4.

The **early anthropogenic hypothesis** suggests that increased human agricultural production altered the natural interglacial trends in CO_2 and CH_4 during the latter part of the Holocene. Ruddiman et al. (2020) *Quat. Sci. Rev.* 240 review the early anthropogenic hypothesis.

anthropogenic Brought about by human agency.

anti-capitalism The clue is in the words, but essentially this is *anti-globalization, a global political project defined by notions of autonomy, ecology, democracy, self-organization, and direct action. David Harvey's

'Anti-Capitalist Politics in an Age of Covid-19' is accessible online.
See also CAPITALISM.

anticline See FOLD.

anticlinorium See FOLD.

anticyclone A region of relatively high *atmospheric pressure, frequently thousands of kilometres across, often formed as a response to *convergence in the upper *atmosphere, and also known as a high. As air near ground level flows into an anticyclone, its absolute *vorticity decreases, causing *divergence, and the descent of air. **Anticyclonic circulation** is clockwise in the Northern Hemisphere, and anticlockwise in the Southern Hemisphere.

Cold anticyclones (continental highs) form as continental interiors, such as Siberia, lose heat in winter through *terrestrial radiation, and cool the air above. They are marked by subsidence, which inhibits cloud formation and maximizes radiative cooling, making them self-sustaining.

Subtropical anticyclones are warm, and form due to subsidence below convergence resulting from the westerly sub-polar *jet stream at the northern limit of the *Hadley cell. The semi-permanent subtropical anticyclone over the North Atlantic (the *Azores high*) strongly influences weather and climate of much of North America, western Europe, and north-west Africa. Subtropical anticyclones bring stable atmospheric conditions, and fine, hot, dry weather. In *mid-latitudes, anticyclones often locate beneath the leading edge of ridges in the upper-air *westerlies, where they may be associated with *blocking weather patterns; see Wiedmann et al. (2002) *J. Climate* 15, 23.

anticyclonic gloom In winter, in north-west Europe, fog, or poor visibility caused by broad, persistent sheets of strato-cumulus cloud at the base of an *inversion trapping polluted air below. The inversion results from the arrival of a cold *anticyclone.

antidune A ripple on the bed of a *fluvial channel, travelling upstream. Antidunes are sediment bedforms, formed beneath *standing waves by fast, shallow water flows. A critical *Froude number of 0.84 may be taken as the discriminator between dunes and antidunes (Carling and Shvidchenko (2002) *Sedimentol.* 49, 6).

anti-globalization A colloquial term for the stance of those who, internationally, coordinate, organize, and mobilize opposition to *neoliberalism and the political power of *transnational corporations.

anti-humanism Broadly, anti-humanism criticizes humanism and looks to displace the human subject as the centre of philosophical and social inquiry. Within academic geography, it has emphasized the 'non-human' or 'more-than-human'.

anti-natalist Limiting population growth. The People's Republic of China has pursued anti-natalist policies, notably the now-suspended 'one-child' strategy. Countries with large populations are more likely to pursue anti-natalist policies in general, and to discourage single parenthood in particular, to justify controlling population.

antitrades Westerly winds in the upper atmosphere above, and in contrast to, the easterly *trade winds at ground level.

apartheid The system of racial segregation first promulgated by the largely Afrikaner National Party of South

Africa in 1948. **Petty apartheid** meant the separation of facilities such as lavatories, transport, parks, and theatres into white and non-white. On a much larger scale was the allocation of 12% of the land area into independent republics ('homelands, or bantustans') for the African population, which comprised 69% of the population when the policy began, in 1954. These 'homelands' were to be governed and developed separately from white South Africa, while allowing African workers strictly limited rights to live in the white areas, as and when their labour was required. J. Robinson (1996) suggests that the organization of urban space into racially segregated living areas was central to the persistence of the racial state. 'Without a gathering of the racially defined African population into spatially contained areas and the evolution of specific methods of administration and governance in these areas, the implementation of various racial policies would have been held hostage to the racial and physical "chaos" of the early twentieth-century city.'

With the election of South Africa's first democratic government in 1994, the last vestiges of apartheid were officially removed, but the policy will leave a mark on the South African landscape and its society for many years to come.

aphelion The point furthest from the sun of any body orbiting the sun; the Earth is at aphelion 152 000 000 km from the sun in July. *See* MILANKOVITCH CYCLES.

aphotic zone In any watery environment, the deeper zone, which is not penetrated by light.

applied climatology This describes, defines, interprets, and explains the relationships between climate conditions and countless weather-sensitive

activities. For example, in business, applied climatology embraces interactions with marketing, sales, customer services, research support, and the delivery arm for climatological data and information. See the journal *Theoretical and Applied Climatology*.

applied geography A type of problem-solving that uses geographical thinking. Four main areas of study are: the description and costing of contemporary environmental conditions; the evaluation of the value of particular environments for specified future uses; the identification and analysis of environmental impacts, mainly of human action, actual and proposed; and the prediction and design of environmental work.

Applied geography is, according to *Applied Geography*: 'research which utilizes geographic approaches (human, physical, nature-society and GIScience) to resolve human problems that have a spatial dimension. These problems may be related to the assessment, management and allocation of the world's physical and/or human resources. The underlying rationale . . . is that only through a clear understanding of the relevant societal, physical, and coupled natural-humans systems can we resolve such problems'.

(((•))) SEE WEB LINKS
• Website of the Applied Geography Specialty Group of the American Association of Geographers.

applied geomorphology The application of geomorphological insights to engineering, planning, and environmental/resource management problems.

applied meteorology The use of climate information in decision-making, impact assessments, seasonal climate forecast applications and verification, climate risk and vulnerability, development of climate monitoring tools, urban and local climates, and climate as it relates to the environment and society. It includes weather modification, satellite meteorology, radar meteorology, boundary layer processes, air pollution meteorology, agricultural and forest meteorology, and applied meteorological numerical models.

appropriate technology A technology which evolves or is developed in response to a particular set of needs, and in accordance with prevailing circumstances (Intermediate Techn. Dev. Group); 'a development strategy that would be appropriate to the material, technical, and organizational resources of nations poor but proud, and that at the same time would enhance local and national autonomy' (Brown (1979) *Dev. Econs.* 17, 1).

aquiclude A rock, such as London clay, which does not allow the passage of water through it; an impermeable rock. Such a rock will act as a boundary to an *aquifer.

aquifer A rock, such as chalk, which will hold water and let it through. Water runs into aquifers where the rock is exposed to the surface or lies below the *water-table. A **confined aquifer** is sandwiched between two impermeable rocks, and the water-table marks the upper limit of an **unconfined aquifer**. See A. Klimchouck, ed. (2000).

arch Along a coast, an arch is made when caves on either side of a headland meet. Arches are temporary; roof falls cut off the seaward end of the arch, which is then left as a *stack.

architecture, geographies of The geographical study of buildings. All geographies are concerned with space and place, and geographers of architecture are concerned with the built form. This includes the influences on the built form of power and social practices, of the relationships between the built form, and the signs, signifiers, and meanings of buildings (what is a skyscraper saying?). Kraftl (2010) *Geog. Compass* is good on this.

Arctic Denoting regions within the Arctic Circle, i.e. north of 66° 32′ N (often taken as 66½° N). Within these regions the sun does not set on 21 June (the summer solstice in the Northern Hemisphere) nor rise on 22 December (the winter solstice in the Northern Hemisphere). In climatology the Arctic is defined in terms of the treeless zone of *tundra and of the regions of *permafrost in the Northern Hemisphere. **Arctic air masses** are exceedingly cold, with the Arctic Ocean as their source region. During winter, they are very dry, and little different from a continental *polar (cP) air mass. In summer, the air is more *maritime. A **maritime Arctic (mA)** air mass is generally similar to maritime polar air. Initially it has continental characteristics, but when it crosses warmer water, it is warmed from below, and in consequence becomes unstable.

Arctic meteorology The Arctic regions experience an annual cycle of winter 'night' and summer 'day'. Most weather results from the intensely cold ground air which is chilled by contact with land, which loses heat from strong *terrestrial radiation—winter clouds are scarce. Only infrequently do depressions penetrate the *inversions so formed. Winter temperatures are close to −40 °C. Snowfall is slight, but winds cause frequent blizzards and drifting. In spring, days are longer and sunny, but temperatures remain low because of the high *albedo of the snow surface. In summer, some depressions bring thicker cloud and light rain, and snow- and ice-melt in June and July keep air temperatures low.

Since the early 1970s, sea-level pressure over the Arctic has decreased; the stratospheric polar vortex has become colder and has been persisting longer into spring; and water mass characteristics of the Arctic Ocean have changed. Ostermeier and Wallace (2003) *J. Climate* 16, 2 link these changes to a shift in the *North Atlantic Oscillation.

Arctic sea smoke A form of *steam fog. As very cold air passes over a warmer sea, it is rapidly heated. Convection lifts moisture up, which quickly recondenses to form fog.

areal scouring Landscapes of areal glacial *scouring have been described as comprising irregular depressions with intervening bosses scraped by ice and labelled 'knock and lochan' topography. An etched bedrock surface is a prerequisite for this type of landscape to develop.

Area of Outstanding Natural Beauty (AONB) In the UK, an area in which development is very carefully considered, so that the beauty of the landscape is not diminished. The Countryside and Rights of Way Act 2000 empowers local authorities to establish a Conservation Board for their own AONB, in order to conserve it, enhance it, and increase public understanding and enjoyment of its qualities.

area studies Interdisciplinary programmes specializing in research and teaching on specific regions of the world. Gibson-Graham (2004) *Env. & Plan. A* 2004, 36, 405 provides a handy table:

Thematic studies	Area studies
general	specific
universal	particular
abstract	concrete
nomothetic	ideographic
theoretical	empirical
deductive	inductive
social scientific	humanistic
quantitative	qualitative
hypothesis testing	case study
economic	cultural
contemporary	historical
the West	the rest
developed	underdeveloped
modern	premodern

areic Without surface drainage, that is, without streams or rivers. Areas of permeable rocks, such as limestone, often lack surface drainage.

arête A steep knife-edge ridge between *corries or *glacial troughs in a glacially eroded, mountainous region. Arêtes are possibly formed by the backward extension of corries into a mountain mass, or moulded by *nivation and *frost wedging.

argillaceous Clay-like in composition and texture, referring to rocks containing *clay minerals and clay-sized particles, for example, shale.

arid geomorphology The investigation and explanation of landforms and geomorphological processes in drylands. Tooth (2009) *PPG* 33, 2 identifies three emerging research themes that highlight the growing links between arid geomorphology and other disciplines: the role of fire in arid geomorphological systems; arid fluvial sedimentary systems, characterized by investigations that commonly focus on processes, landforms, and sedimentary products at larger spatial scales and longer

a

timescales; and arid geomorphology on Mars.

aridisol A soil order of the *US soil classification, found in arid environments. It is a *desert soil, predominantly composed of minerals, and often high in accumulations of water-soluble salts.

aridity Dryness; aridity is the most common criterion for the definition and classification of drylands, occurring when *evapotranspiration exceeds precipitation.

An **aridity index** is a numerical indicator of the degree of dryness of the climate at a given location, expressed as the ratio of annual potential evaporation to precipitation, and the UNEP index of aridity (1992 *World Atlas of Desertification*) is defined as:

$$AI_U = \frac{P}{PET}$$

where *PET* is the potential evapotranspiration and *P* is the average annual precipitation. *PET* and *P* must be expressed in millimetres.

armour The coarse surface material which protects the finer material below. The armouring of a river bed is caused by selective entrainment of smaller sediments, and reduces further entrainment of smaller sediments; see Berton and Friedrich (2018) *Geom.* 306. An **armour layer** is generally well sorted, and one or two grains in thickness. Migratory during floods, it will re-form when normal flow resumes.

arroyo A *gully, rectangular in cross-section, cut into debris-choked valleys, usually with a sandy floor, made of fine-grained cohesive sediments in arid environments. It results from the entrenchment of stream channels when run-off increases. *Badlands are associated with arroyos. Perroy et al. (2012) *AAAG* 102, 6, 1229 untangle the relative importance of climatic, tectonic, and anthropogenic drivers as triggers of arroyo formation in California.

art, geographies of Geographical understandings of art have focused on interpretations of art in urban spaces, and the role of art in urban regeneration and place promotion; where art—architecture, painting, and sculpture—is or was located, and how material and indigenous culture shaped the human environment and the nature of human civilization. This may include: paying close attention to landscape; to the layers of social and cultural meaning and power that could be found through visual interpretation of symbolic imagery; and to the variety of ways in which people's experiences, knowledge, and exploration of the environments which surround them may be understood through embodied–sensuous experience.

artesian basin A *syncline of permeable rocks, outcropping at the crest. Water from rain or streams seeps into this *aquifer; the rock becomes saturated, and the water is under pressure. If a hole is sunk to tap the water, an **artesian well** forms—water will initially flow upwards without pumping. Try B. Radke et al. (2000) on the Great Australian Artesian Basin.

artificial neural network (ANN) A computing methodology which simulates the human brain as it processes spatial data problems. Thurston has written an easily accessible summary, noting that 'in almost every instance where *GIS is being used, [artificial intelligence] applications could potentially be developed for the purpose of enhancing decision-making capabilities'.

SEE WEB LINKS
- Website that explains artificial neural networks in plain English.

ash *See* PYROCLAST.

ash cone *See* SCORIA.

ash fall The precipitation of volcanic ash (tephra) during and after a volcanic eruption.

ash flow The deposition of ash-sized fragments from an ash flow forms a deposit known as tuff. The flows known as *nuées ardentes* (glowing clouds) are mostly found in valleys, while *ignimbrites form plateaus. *See also* PYROCLAST.

aspect The direction in which a valley side or slope faces. In deeply cut east–west oriented valleys, the slopes facing the equator receive more sun and are more attractive to settlement than the shaded sides of the valley.

Aspect may be an important factor in the formation of landforms, since slopes facing away from the equator may be 6 °C colder than their opposites: gradational processes can be two to three times as active on northward-facing slopes in the Northern Hemisphere and in the Jotunheim of Norway, 70% of corries lie on the north side of the massif.

assarting In medieval Britain, the expansion of agriculture into the forested areas (Beaver in S. Wooldridge and S. H. Beaver 1950).

assemblage A process or set of processes whereby different entities meet, create relations, and work as 'wholes'. Assemblage theory argues that wholes, such as urban assemblages, are constructed by the coming together of separate and individual parts achieving some form of unity across differences.

Cities are wholes, and processes might include urban regeneration. You should find Anderson et al. (2012) *Dialogues in Hum. Geog.* clear and helpful; Kincaid (2019) *Prog. Hum. Geog.* provides an accessible critique.

assimilation Also known as *acculturation, or integration, this is the integration of an immigrant, outsider, or subordinate group into the dominant, host community, or *charter group; the minority group eventually takes on the values of the host, putting aside previous national identities and heritages. The degree to which immigrants remain spatially concentrated can be treated as a measure of their assimilation; see Cirenza (2015) *LSE Economic History Working Paper*, on Irish immigration to America. Sometimes the indigenous population is required to assimilate with a non-indigenous group; Lester (2006) *Geogr. Res.* 44, 3 explains how Australian aborigines' lands were sites where missionaries could prepare them for assimilation as black, Christian Britons.

associated number (König number) The associated number of a *node is the number of *edges from that node to the furthest node from it. A low number indicates a high degree of *connectivity; for an example of its use, see Wood (1975) *Geografiska B* 57, 109–19.

association (plant association) A *plant community unit; a floral assemblage with a characteristic dominant and persistent species.

asthenosphere That zone of the Earth's *mantle which lies beneath the relatively rigid *lithosphere, between 50 and 300 km below the surface, and representing a mechanical boundary between more rigid regions above and below. It is composed of hot, semi-molten, deformable rock, and may be plate- or density-driven, or both. It is, approximately, the zone of the mantle which transmits seismic waves at low velocity. The transition between a rigid lithosphere and a viscous asthenosphere is gradual.

asylum migration The international movement of *refugees and persons who, while having suffered generalized repression, violence, and poverty, do not qualify as refugees under the strict requirements of the 1976 UN protocol.

Economic hardship and economic discrimination against ethnic minorities lead to higher inflows of asylum seekers; political oppression, human rights abuse, violent conflict, and state failure are also important determinants, as are migration networks and geographical proximity. Attempts by the EU to reduce access to asylum systems or to curtail the rights of asylum seekers have generated an increase in illegal migration. See Hirsch Ballin et al. (2020) on variations in asylum, migration, and border control.

asymmetry Lacking symmetry. An **asymmetrical valley**, common in present or past *periglacial environments, has one side steeper than the other, since *aspect can be critical in determining where frost forms; see Meikeljohn (1992) *S. African Geogr. J.* 74.

at-a-station Investigations in hydraulic *geometry may take place at a particular station—at-a-station—or along the length of a stream.

A-tents Raised rock slabs or plates of varying thickness, mostly resulting from the buckling of laminae, plates, and slabs. These forms are generally cracked by tectonic activity, although decreases in lithostatic pressure through unloading, or, more rarely, from surficial expansion of plates caused by intense heat. Smaller plates and slabs may be changed in part through freeze-thaw.

Atlantic Period In north-west Europe, the period from around 7500 to 5000 years BP of wet, oceanic climate, when temperatures were warmer than at present. See Uścinowicz et al. (2007) *GSA Special Paper* 426 on sea-level oscillations in the Atlantic period.

Atlantic-type coast A coastline where the trend of ridges and valleys runs transverse to the coast. If the coastal lowlands are inundated by the sea, a *ria or *fiord coastline may result. The coast of south-west Eire is an example. *Compare with* PACIFIC COAST.

atmosphere A gaseous envelope gravitationally bound to a celestial body; in the case of the Earth, with an average composition, by volume, of 79% nitrogen, 20% oxygen, 0.03% carbon dioxide, and traces of rare gases. This surprisingly uniform composition is achieved by *convection in the *turbosphere and by *diffusion above it, especially above 100 km, where diffusion is rapid in the thin atmosphere, and stirring is weak. Also present are atmospheric moisture, ammonia, ozone, and salts and solid particles. The atmosphere is commonly divided into the *troposphere, the *stratosphere, and the *ionosphere. Since the troposphere contains the majority of the atmospheric mass, and virtually all of the atmospheric water vapour, most weather events occur within it. The **atmospheric energy balance** is the balance between incoming short-wave energy from the Sun and outgoing long-wave energy from the Earth. Outgoing energy loss is greater than incoming energy gain; this is equivalent to a cooling of 1°C or 2°C per day. However, this heat loss is mostly compensated for by the latent heat released by the condensation of atmospheric water vapour into *precipitation.

Typical annual global values for the energy balance of the atmosphere are:

- net radiative loss = -99.4 W m^{-2}
- sensible heat gain = 20.4 W m^{-2}
- latent heat gain = 79.0 W m^{-2}

Annual global precipitation is therefore linked closely to the atmospheric energy balance and can change only if there are associated changes in the global energy balance (see Lockwood in J. Holden 2012).

atmospheric boundary layer The lowest part of the atmosphere; as such, it is shaped through contact with the planetary surface. J. C. Kaimal and J. J. Finnigan (1994) define it as the 'lowest 1–2 km of the atmosphere, the region most directly influenced by the exchange of momentum, heat, and water vapour at the Earth's surface'. *Acta Geographica* 5, 2012 is devoted to the atmospheric boundary layer.

atmospheric cells Air may move, with a circular motion, northwards or southwards in a vertical cell. Atmospheric cells are major components in the transfer of heat and momentum in the atmosphere from the equator to the poles. Horizontal cells (winds) also fulfil this role.

atmospheric correction The retrieval of surface characteristics from

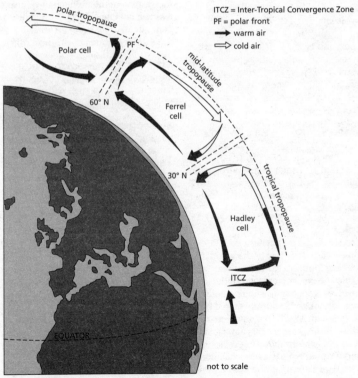

Atmospheric cells

remotely sensed imagery by removing the atmospheric effects. The quality of an image is degraded by *absorption, which causes haziness, and *scattering, which disrupts neighbouring pixels. For methods of atmospheric correction, see GISGeography.

atmospheric energetics The relations between internal, potential, and kinetic energy in the atmosphere. The bulk of the energy is heat, latent heat, and potential energy, but some of this can be converted into kinetic wind energy. Ackerman (*AIP Conf. Proc.* 247, pp. 1–23) reviews basic concepts.

atmospheric forcing *See* FORCING.

atmospheric heat engine The system of energy which drives and controls the nature of the pressure, winds, and climatic belts of the Earth's surface, with *solar radiation as the heat source, and *terrestrial radiation as the sink. Mechanical energy is expended by the 'engine' in the form of atmospheric processes.

atmospheric pressure The pressure exerted by the *atmosphere as a result of gravitational attraction exerted on the column of air lying above a particular point. Atmospheric pressure, measured in millibars, decreases logarithmically with height.

atoll A *coral reef, ring- or horseshoe-shaped, enclosing a tropical *lagoon. Most of the world's atolls are found in the Indian and Pacific oceans, and are said to be sensitive to fluctuations in relative sea level—although Woodruffe et al. (1999) *Marine Geol.* 160, 1, in a study of West Island, Cocos, dispute this assumption. Single colony, semicircular corals (**microatolls**) grow to a level constrained by exposure at low spring tides (Spencer and Viles (2002) *Geomorph.* 48, 1–3).

attentive geographies These consider writing geography as a research process, asking, for example, what does it mean to be a working geographer? What happens when you commit to deepening and developing skill? What difference does the practical 'doing' make? How does collaboration emerge through creative methods?

Atterberg limits Limits used to classify soils. The **liquid limit** is the minimum moisture content at which the soil can flow under its own weight; the **plastic limit** is the minimum moisture content at which the soil can be rolled into a 3 mm diameter thread without breaking; the **shrinkage limit** is the moisture content at which further water loss will not cause further shrinkage. See Fener et al. (2005) *Cana. Geotech. J.* on the Atterberg limits of cohesive soils.

attrition The wearing away or fragmentation of particles of debris by contact with other such particles, as with river pebbles.

aulacogen A long-lived, sediment-filled graben oriented at a high angle to either a neighbouring modern ocean or a neighbouring orogenic belt; an intracontinental graben (Umeji (1988) *Int. J. Earth Scis* 77, 2). An aulacogen can be seen as the failed arm of a triple-limbed spreading centre (Jansen (1975) *Trans. Geol. Soc. S. Afr.* 78).

aureole *See* METAMORPHIC AUREOLE.

autecology The ecology of particular species and individual organisms, particularly the relationships between species and their environments and within species, and especially the way organisms act within communities. See Sundberg et al. (2006) *J. Biogeog.* 33, 8.

autochthonous Referring to features and processes occurring within, rather than outside, an environment. An **autochthonous rock** has been formed *in situ*; coal is an example.

autogenic succession A succession which begins as a result of disturbance by an external physical factor, but whose sequence in community composition results primarily from interactions among the species. Sometimes the early species actually enhance the success of the later ones, perhaps by increasing the nutrient status or water-holding capacity of the soil, as in a psammosere (C. Townsend et al. 2003).

automobility The study of the economic, technological, social, and political consequences of the culture of the car. The analysis of automobility extends beyond congestion, commuting, suburbanization, and environmental impacts, to also consider the role of driving in the formation of cultural identity, social standing, and status; it shows that the automobile is more than just a technological object, it is in a very real sense 'a whole way of life' (A. Latham et al. 2008).

autonomous spaces (autonomous geographies) Self-managed spaces; those spaces where there is a desire to constitute 'non-capitalist, collective forms of politics, identity, and citizenship, which are created through a combination of resistance and creation, and the questioning and challenging of dominant laws and social norms' (in the words of Autonomous Geographies).

autotroph An organism which uses light energy to synthesize sugars and proteins from inorganic substances. Green plants are by far the most common autotrophs.

avalanche A rapidly descending mass, usually of snow, down a mountainside. **Powder avalanches** consist of a moving amorphous mass of snow. **Slab avalanches** occur when a large block of snow moves down a slope and can cut a swathe through the soil and sometimes erode the bedrock if the snow is wet. See A. V. Briuchanov (1967).

Avalanches of other substances are forms of *mass movement, and are distinguished by the type of material involved. **Debris avalanches** are typically triangular in shape, and are associated with morphological discontinuities, such as scarps and road cuts; see Guadagno et al. (2005) *Geomorph.* 66, 1–4 on man-made cuts and debris slides. **Rock avalanches** occur when a jointed rock loses internal cohesion, until some sections are held together only by the friction between blocks. If this frictional force is lessened through water seepage, or if lateral support is removed, failure will occur. See Mitchell et al. (2007) *Landslides* 4, 3.

avalanche wind The rush of air formed in front of an *avalanche. Its most destructive form, the **avalanche blast**, occurs when an avalanche stops abruptly.

avulsion The abandonment of all or part of a channel belt in favour of a new course. Bed *aggradation reduces channel carrying capacity, increases overbank flooding and sedimentation, and eventually results in avulsion.

azonal Referring to a soil without *soil horizons, such as a young soil developing on a bare rock surface. Alluvium and *sand dunes are examples of azonal soils.

back-arc basin A zone of tectonic enlargement, and substantial, and strongly asymmetric, sedimentation. A back-arc basin forms when an *island arc is split along the line of its magmatic axis, forming a remnant arc, which then glides away. Well-developed back-arc basins widen by **back-arc spreading**, which is thought to resemble normal oceanic crust formation, but in conjunction with a *subducting plate. See Tontini et al. (2019) *Nature Geoscience* 12 on the early evolution of a young back-arc basin in the Havre Trough.

backreef A sea area landward of a reef, usually including a lagoon. Antecedent topography probably plays an important role in the way backreefs develop during periods of rapid sea-level fall; see Fujita et al. (2020) *Geology* 48 1. Dahlgren and Marr (2004) *Bull. Marine Sci.* 75, 2 argue that backreefs are of critical ecological value to a coral reef ecosystem.

backshore The part of a beach between the ordinary limit of high tides and the point reached by the very highest tides. On cliffed coastlines, the backshore is the section of cliff foot and *shore platform affected only by storm tides; on low, shingle coasts it can take the form of a *berm. Waves are likely to have larger effects than winds on backshore sedimentation; see Udo and Yamawaki (2007) *Geomorph.* 60, 1–2.

backswamp On a *flood plain, a marshy area where floodwater may be confined between the river *levées and higher ground. See Takehiko et al. (2006) *Geog. Res.* 44, 1 on mangroves and backswamps.

backwash The return flow of water downslope to the sea after the *swash has moved upshore. Steep waves and long waves are linked with stronger backwash, and flatter beach gradients; see the great C. King (1959), and Hardisty et al. (1984) *Marine Geol.* 61, 1. When the path taken by the backwash differs from the path of the swash, *longshore drift occurs. See Bird et al. (2007) *Geogr. J.* 173, 2, on backwash erosion.

badlands Arid lands, generally bare of vegetation, dissected to form ravines and sharp-crested hills developed best in weakly consolidated sediments. This dissection is aided by the lack of vegetation, high *run-off, and heavy sedimentation, which increases *abrasion. Non-fluvial processes, such as mass wasting, *piping, and tunnel erosion, are also important. See D'Intino et al. (2020) *Journal of Maps* 16, 2 on badlands in the central Apennines of Italy.

bajada (bahada) A series of *alluvial fans, typically laid down by *ephemeral streams, which has coalesced along the foot of the mountains to form a gently sloping plain of unconsolidated sediments. The *weathering of the mountain front may also supply debris. Alluvium on the lower part of a *pediment may also make a bajada. See Leeder and Mack (2001) *J. Geol. Soc.* 158 for a clear review, plus a great title; and Wainwright et al. (2002) *J. Arid Envs.* 51.

bank erosion The erosion of material from the side of a river channel. The major processes are *corrasion and slumping, and these appear to be associated with high river flow levels and antecedent precipitation conditions, respectively. Rates of erosion vary with speed of flow (rates are highest on the outer bank of meander bends or where bars in the channel have diverted the *thalweg), bank vegetation and moisture content, and bank composition.

bankfull discharge The *discharge of a river which is just contained within the banks. This is the state of maximum velocity in the channel, and of maximum *competence.

bank storage Water held in the river bank which may contribute to stream flow, but, in arid conditions, may cause a decrease in discharge as water percolates from the river to the banks; see Burt et al. (2002) *J. Hydrol.* 262, 1. Chen et al. (2006) *J. Hydrol.* 327, 3–4 present a technical paper on bank storage and baseflow separation.

bar 1. On a gently sloping coastline, a submarine accumulation of marine sediment, which may be exposed at low tide. Most bars form where steep, *destructive waves break; such bars can be called **break-point bars**, with crests generally running parallel to the coast. **Bay bars** extend across an *estuary or a bay: Chesil Beach in the UK probably began as a bay bar, at a time of lower sea level. Some bay bars entirely enclose the inlet, and a *lagoon may then form on the landward side. A **mid-bay bar** is a spit which forms between the bay mouth and the head of the bay. **Offshore bars**, located further out to sea, are thought to

result from the breaking of larger waves, which erode the sea bed and throw up material ahead of them to form ridges: see Hoefel and Elgar (2003) *Science* 29.

2. Within a river, a deposit of alluvium, dropped where velocity and turbulence are low, and which may form temporary islands. **Alternating bars** develop as patches of alluvium, often regularly spaced, along opposite sides of a straight channel. **Braid bars**, roughly diamond shaped, are generally aligned along the channel course. **Point bars** form on the inner curves of a meandering river where *discharge is low; see Pyrce and Ashmore (2005) *Sedimentol.* 52, 4.

barchan *See* SAND DUNE.

baroclinic A term applied to sections of the *atmosphere where trends in pressure (pressure surfaces) are at an angle to trends in temperature. (The precise definition refers to the intersection of isobars and isopycnals—levels, or surfaces, of equal density—but, for most purposes, isotherms can replace isopycnals.) The number of intersecting isobars and isopycnals is a measure of **baroclinicity**. Polewards, temperatures fall very rapidly, while pressure remains constant in that direction, but falls with height. At some point, the slope for isotherms intersects with the slope for isobars, so that the two intersect in a **baroclinic zone**, most often in *mid-latitudes in winter.

In the mid-latitude baroclinic zones spontaneous generation of weather systems such as depressions and thunderstorms is common. These are **baroclinic disturbances**, characteristically, on *synoptic charts, with strong meridional pressure gradients in the constant-pressure surfaces and vertical *wind shear. When the temperature gradient along the *meridians is very steep, *atmospheric cells break down into cyclonic and anticyclonic *eddies. This failure is known as **baroclinic instability** which is characterized by the ascent of warmer, and the descent of colder, air.

barometric gradient See PRESSURE GRADIENT.

barotropic Describing *atmospheric conditions where trends in pressure align with trends in temperature, as in the ideal *air mass; the reverse of *baroclinic.

barrage A structure across a river or estuary, built in order to restrain or to use water; generally for irrigation, flood control, or to generate electricity. To guard against undue erosion, excess hydraulic energy downstream must be controlled. In addition, barrages obstruct the movement of sediment transported by the river, so that sediment may be deposited upstream of the barrage. This can be remedied by periodic flushing of the sediment. Tucked in the middle of their chapter, Rahman et al. in Nicholls et al. eds (2020) *Deltas in the Anthropocene* provide excellent information on barrages on the Ganges–Brahmaputra–Meghna Delta system.

barrier In *geomorphology, a general term for any offshore depositional form, usually running parallel to the coast or across an *estuary, and above water at normal tides. A **barrier coast** is a series of barrier islands, barrier spits, and *barrier beaches. **Transgressive barriers** are those that migrate landwards under the influence of rising sea level and/or a negative sediment budget. **Regressive barriers**, or strandplains, are large mounds of sediment that have developed along a shoreline with a surface exhibiting well-defined parallel or semi-parallel sand ridges separated by shallow swales. Landwards sediments are deposited on top of more seaward ones. Regressive barriers are linked to a positive sediment budget and/or falling sea level.

barrier beach A long, narrow beach, running parallel to the coastline, and unsubmerged by high tides. See Chadwick et al. (2005) *Procs Inst. Civil Engineer. Maritime Engineer.* 158, 4 on the barrier beach at Slapton Ley.

barrier island A long, narrow, offshore island, usually having beaches and *dunes on the seaward side, often with *lagoons on the landward side, such as the islands enclosing Pamlico Sound, off the coast of North Carolina. The formation of barrier islands has been explained as the drowning of beach ridges or the *progradation of *spits. Most barrier islands and spits are in a state of landwards retreat in response to a combination of *eustatic sea-level rise and local land subsidence. See Brantley et al. (2014) *PLOS One* 9.

barrier reef A *coral reef, stretching along a line parallel with the coastline but separated from it by a wide, deep *lagoon, the most famous of which is the Great Barrier Reef off north-west Australia (see D. Hopley et al. 2007).

barrow A communal burial mound built from the Stone Age until Saxon times. Long barrows, up to 100 m long and 20 m wide, were the earlier form, while round barrows were introduced during the Bronze Age; see P. Ashbee (1970 and 1960, respectively).

basal complex (basement complex) The ancient *igneous and metamorphic rocks which lie beneath Precambrian rocks and constitute the *shield area of the Earth's continents.

basal contact pressure In a glacier, the contact pressure between a *clast in basal transport and the underlying bed:

the greater the pressure, the more abrasion will occur. The pressure between a particle in contact with the glacier bed is related to the *effective normal pressure. The contact pressure between a clast and the glacier bed is a function of the rate at which ice flows towards the bed.

basal ice A layer of debris-rich ice at the bottom of a glacier. It is generally formed in layers, due to shearing within the mixture of debris and ice. Supercooling may be the principal mechanism responsible for creating debris-laden basal ice; see Lawson et al. (1998) *J. Glaciol.* Hall and Glasser (2003) *Boreas* 32, 1 model basal ice temperatures.

basal slipping (basal sliding) The advance of a glacier by comparatively rapid creep close to its bed, occurring where the ice at the base is at its *pressure melting point. The major types are enhanced basal creep, where ice, deformed by stresses created by jagged bedrock, bends round obstructions; *regelation (pressure melting); and slippage over a layer of water at the bed. Basal sliding, occurring in sporadic movements, accounts for the majority of

flow in warm glaciers, especially where the gradient is steep, the ice thin, or where *meltwater is present. Patterns of basal sliding strongly influence patterns of glacial erosion. See Cohen et al. (2000) *J. Glaciol.* 46, 155, and Cooke et al. (2006) *PPG* 30, 5.

basal thermal regime The general temperature pattern of ice at the base of a glacier. Glaciers are commonly divided into three types according to their temperature: cold-based, warm-based, and polythermal. Typically, the thicker, higher-level ice in the accumulation area is warm-based and the snout, lateral margins, and surface layer of the glacier are below the pressure-melting point.

base flow The usual, reliable, background level of a river, maintained by seepage from groundwater and by *throughflow, which means that the river can maintain the base flow during dry periods. The **base flow index** is the contribution of base flow to a stream. Total flow to stream systems consists of surface run-off, interflow, and base flow; streamflow partitioning methods are used for finding the relative contributions of each (try Morgan (1972) *TIBG* 56). With prolonged drought, base

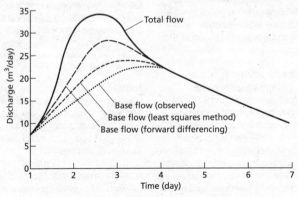

Base flow

flow itself will diminish, the rate of flow falling in a depletion curve.

base level The theoretically lowest level to which the course of a river can cut down. This level may be sea level, the junction between a tributary and the main river, or the level of a waterfall or lake, but streams rarely erode as far as base level. Base level may alter due to *eustatic or *isostatic change, and be termed **positive changes of base level** if the land sinks relative to the sea, or **negative changes of base level** if it rises. (Try Schumm (1992) *J. Geol.* 101 on river response to base level change; *see also* REJUVENATION.) **Marine base level** is the lowest point at which marine erosion occurs, perhaps as low as 180 m below the surface.

basement (crystalline basement) Rocks below a sedimentary platform or cover, that are metamorphic or igneous in origin.

base saturation The degree of saturation of the *adsorption complex of the soil; that is, its occupation by exchangeable basic cations, or cations other than aluminium and hydrogen, expressed as a percentage of the total cation-exchange capacity. Halvorson (1980) *EM2894* explains exchangeable cations, cation exchange capacity and base saturation, and the relationship to soil fertility.

base surge A type of *pyroclastic surge that forms at the base of volcano eruption columns and travels outward during some hydroclastic eruptions. As base surges move laterally away from the eruptive vent, they lose heat, sediment load, and velocity, which directly affects the resulting deposits. The sedimentary structures observed within surge deposits are also directly related to the amount of liquid water involved.

basin A major relief *depression, guided by structure, or formed by *erosion. **Basin and range** describes a landscape where ridges made of asymmetric *fault blocks alternate with lowland basins. In the USA, the basin and range country lies between the Sierra Nevada and the Wasatch Mountains; see W. S. Baldridge (2004).

bastide A planned, fortified strong point and centre of economic development created in the Middle Ages, mostly in France. The rectilinear street pattern contrasts strongly with the irregular, cramped layout of most medieval towns. This morphology was used by French colonizers in Canada (Louder (1979) *Cahiers de géo Québec* 23).

batholith A massive, frequently *discordant, intrusion of coarsely textured plutonic rocks, at least 100 km^2 in area and extending 20–30 km down into the layer of *magma, which may be composed of several plutons.

bathymetry Measuring water depth, mainly of seas and oceans but sometimes of deep lakes, such as the Aral Sea. GEBCO (General Bathymetric Chart of the Oceans) consists of an international group working on the development of a range of bathymetric data sets and data products.

(⊕) SEE WEB LINKS
• Website of GEBCO.

beach An accumulation of *sediment deposited by waves and *longshore drift along the coast. The upper limit is roughly the limit of high tides; the lower, of low tides. *See also* DRIFT-ALIGNED BEACH. Almeida et al. (2011 *ESPL* 36, 4, 523) show a clear relationship between offshore wave height, and the beach profile. They reveal three thresholds for morphological change: waves higher

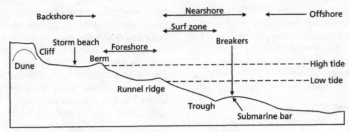

Beach

than 2.3 m are responsible for important changes to the berm and beach face; waves higher than 3.2 m are responsible for important changes to the sub-tidal terrace and long-shore bar; and waves over 4.1 m are needed to effect morphological change further seaward, near the depth of beach closure.

beach budget (beach sediment budget) The balance between the material deposited on, and eroded from, a beach; see Davidson-Arnott and Van Heyningen (2003) *Sedimentol.* 50, 6. During winter, waves tend to be destructive, and gradients steepen, but, with summer conditions, waves become constructive, and gradients slacken. See Martinez and Psuty (2004) *Ecol. Studs* 171 on *foredunes and beach budgets.

beach–dune interaction The interconnections between beaches and dunes. These involve the aeolian transport of sand from the beach to the erosion of the dune by storm waves. See Servera and Gelabert (2009) *Geomorph.* 110, 3–4 on the development and setting of a beach–dune system in Mallorca.

beach morphology Beach form. The occurrence of different types of beach morphology can be parameterized by the dimensionless fall velocity Ω. The equation is:

$$\Omega = \frac{H}{w_s T}$$

where H is the wave height, w_s is the speed at which a sediment particle falls through still water, and T is the wave period. *Reflective beaches tend to develop when $\Omega < 1.5$, *intermediate beaches are those where $\Omega = 1.5$–5.5, and *dissipative beaches form when $\Omega > 5.5$.

beach ridge A wave-deposited ridge, made of sand and reworked underlying beach material, running parallel to a shoreline, beyond the present limit of ordinary tides. Changes in beach ridge morphology and orientation reflect environmental factors such as changes in wave climate and wind regime.

bedding plane The surface separating distinct rock *strata.

bedforms (sedimentary bedforms)
1. In hydrology, accumulations of sediment on the river bed, formed by variations in flow competence, and ranging in size from ripples in the sand, a few centimetres apart, to 'dunes' tens of metres in length.

River bed form is related to energy level; as flow energy and sediment transport competence increase, bedforms change from ripples to dunes, which alters the ratio of inertial to gravitational forces in the flow field. This form-flow feedback is a reflection of river energy reorganizing the sediment mass in the most effective configuration for that flow condition.

2. Migrating megaripples are bedforms that appear in the surf zone of sandy coasts.

((⊕)) SEE WEB LINKS

• The excellent USGS bedform sedimentology site.

bed load Material moved along a river bed by traction and *saltation. Bed load usually comprises sands and pebbles, but when the water level is high and the current strong, large material boulders may be moved. Bed-load transport can fluctuate wildly even when the river flow is constant, due to stochasticity, bedform migration, grain sorting, *hysteresis, or variations in sediment supply.

bed profile See LONG PROFILE.

bed return flow Part of a sea-water circulation characterized by onshore flow in the upper part of the water column and a seaward flow near the bottom. This latter is the bed flow, sometimes termed the undertow.

bedrock The unweathered rock which underlies the soil and *regolith or which may be exposed at the land surface. **Bedrock gouges and cracks** are small cracks, gouges, and indentations created on bedrock surfaces. Both *chattermarks* and *crescentic gouges* are shallow crescentic furrows: in crescentic gouges the convexity is turned forward in the direction of ice flow, in chattermarks it is turned backwards. *Crescentic cracks* are vertical fractures of the rock without the removal of bedrock fragments, with a forward-turned concavity. **Bedrock channels** may be incised in a bedrock by *plucking along pre-existing joints or fractures, *cavitation, *abrasion, and dissolution. See Whipple et al. (2000) *Geol. Soc. Amer. Bull.* 112, 490.

bed roughness (bed-floor roughness) The frictional force of a river bed (also called *channel roughness*). A rough bed of boulders, pebbles, and potholes exerts more friction than a smooth, silky channel. *See* MANNING'S ROUGHNESS COEFFICIENT. The term is also applied to a glacier bed; see Rippin et al. (2004) *Ann. Glaciol.* 39.

behavioural geography An approach to human geography which looks for the processes underlying specific spatial actions. These processes include perception, learning, forming attitudes, memorizing, recalling, and using spatial thinking and reasoning to explain human activities in different environmental settings. The approach recognizes that people do not have complete information, do not always minimize distance, and are embedded in networks of social relations, and therefore may base decisions on factors other than sheer economic rationality. It specifically recognizes the role of cognition and the importance of social and cultural values and constraints, together with institutional, economic, and physical factors. Behavioural geography has been criticized for being inherently positivist (based entirely on natural things, their properties, and their relations), and considering individuals as separate from their social and cultural contexts. Nonetheless, behavioural approaches have been gaining ground in the field of economics; see Strauss (2008) *J. Econ. Geog.* 8, 2.

((⊕)) SEE WEB LINKS

• Blogspot on the Environmental Perception and Behavioral Geography Specialty Group of the American Association of Geographers website.

behaviourism The view that the actions of an individual occur as responses to stimuli. Through constant repetition, the individual learns to make the same response to a given stimulus. *Mental maps have especially been associated with the behavioural approach; see Soini (2001) *Landsc. & Urb. Plan.* 57, 3–4.

belonging The concept of belonging is intimately tied to place. Belonging is an attachment, to other people, places, or modes of being.

belonging, geographies of The geographical study of the practices, meanings, and processes of belonging: to the body, the house or home, the neighbourhood, the region, the nation, and the globe. Kustatscher (2017) *Children's Geographies* 15, 1 makes the concept clear as she describes the emotional geographies of belonging in primary school children.

Benguela current The eastern boundary current of the South Atlantic subtropical gyre, which begins as a northward flow off the Cape of Good Hope, then skirts the western African coast equatorwards until around 24–30 °S.

Benioff zone An inclined zone of earthquakes, plunging below the Earth's surface at an angle between 30° and 80°, but commonly at around 45°, and extending to a depth of 300–400 km. Benioff zones are associated with the downward movement of a lithospheric *plate at a destructive plate margin.

benthic Occurring at the base of bodies of water: lakes, oceans, and seas. **Benthos** is the life attached to the bottom, or moving in the bottom mud.

Bergeron–Findeisen theory A mechanism by which ice crystals grow from the vapour supplied by cloud droplets. In mixed clouds, where ice crystals and supercooled droplets coexist, the water vapour diffuses from supercooled droplets to ice crystals because the vapour pressure over ice is lower than over liquid water at the same temperature. See Korolev (2007) *J. Atmos. Scis* 64, 9 on the limitations of the Wegener–Bergeron–Findeisen mechanism in the evolution of mixed-phase clouds.

bergschrund A deep, tensional *crevasse formed around the head of a *cirque glacier. The crevasse forms as ice falls away downslope. Often, a sequence of bergschrunds forms.

berm 1. A low embankment or ridge on a sand beach, constructed by *swash or breaking waves. There are two models of berm development: either vertical growth at spring tides or 'following significant beach cut due to substantial swash overtopping'; or horizontal progradation at neap tides through the formation of a proto-berm, located lower and further seaward of the principal berm.
   2. A benchlike remnant of a surface resulting from the interruption of an erosion cycle. Berms may occur naturally in a river channel, or they may be constructed as part of a *river restoration.

beta index A simple measure of connectivity which can be derived from the formula

$$\beta = \frac{e}{v}$$

where e = number of edges (links), v = number of vertices (nodes). Values of the beta index range from 0.0 for network, which consists just of nodes without any arcs, through 1.0 and greater where networks are well connected. See J. Jenelius (2009) *Transport Geog.* 17, 234 for an illuminating account.

bid-rent theory At any point, holders of land might pay two prices: the cost of

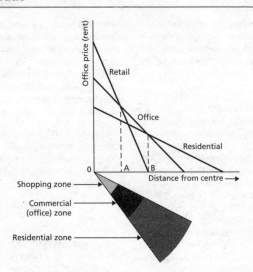

Bid-rent

the land, and the distance from that point to the place where they operate. Land is cheaper with distance from a city centre, but landholders need to pay travel costs. So there are two cost components, and the price of each depends on location. Thus, it is possible to trade off a quantity of land against location. A **bid-rent curve** is a graphical representation of the relationship between land or property rental costs and distance from the centre of a city or town. A **residential bid price curve** is 'the set of prices for land the individual could pay at various distances while deriving a constant level of satisfaction' (Alonso, 1964). The **bid price function** for the urban firm describes the prices which the firm is willing to pay at different distances from the city centre in order to achieve a given level of profits. See Egan and Nield (2000) *Urb. Studs* 37 on the intra-urban location decisions of hotels.

bifurcation ratio In a drainage basin, the ratio of the number of streams of a

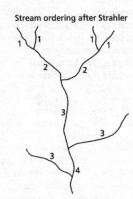

Bifurcation ratio

given order to the number of streams of the next, higher order. Generally speaking, small streams discharge into streams of the next highest order so that flow architecture develops from the lowest scale to the highest.

big data A collection of data sets so large and complex that it becomes difficult to process using general database management tools or traditional data processing applications. See Kitchin (2013) *Dialogues Hum. Geog.* 3, 3, 262 on 'big data and human geography: opportunities, challenges, and risks'.

binary distribution A *city-size distribution in which a number of settlements of similar size dominate the upper end of the hierarchy, said to be characteristic of nations with a federal political structure, such as Australia. *See also* RANK-SIZE RULE.

bioaccumulation The build-up of poisonous substances in an organism, occurring when that organism absorbs a toxic substance faster than it sheds it. The accumulation of toxins in specific parts of the ecosystem is usually at the higher levels of food chains due to the greater ability of some chemicals to accumulate in zones where they become bioavailable and are taken up and stored by producers and consumers. Bioaccumulation may result in severe adverse effects on an ecosystem: see, for example Peterson et al. (2015) *Procs Royal Soc. B* on mercury bioaccumulation in deep-diving northern elephant seals.

bioarchaeology The study of human remains within archaeological sites in order to reconstruct their economy and environment.

biocapacity The capacity of a given area to generate an enduring supply of renewable resources and to absorb its spillover wastes. Earth's biocapacity is determined by the biologically productive area available and the bioproductivity (yield) per hectare. The biocapacity of an area of land is affected by both physical and human factors: Iraq, for example has a low biocapacity, despite its large oil reserves and its position in the once 'fertile crescent'. See Guo et al. (2017), *Scientific Reports* 7, article 41150 on biocapacity optimization in regional planning.

bioclimatology The scientific study of the relationship between organisms and the long-term atmospheric characteristics of a specified area. See Di Filippo et al. (2007) *J. Biogeog.* 34, 11.

biocomplexity Those properties resulting from the interaction of the behavioural, biological, chemical, physical, and social factors that impact on living organisms. Cadenasso (2006) *Ecol. Complexity* 3, 1, 1 presents a thorough review of definitions.

biodiversity The number and variety of living organisms, from individual parts of communities to ecosystems, regions, and the entire biosphere, including: the genetic diversity of an individual species; the subpopulations of an individual species; the total number of species in a region; the number of endemic species in an area; and the distribution of different ecosystems. Greater plant diversity leads to greater primary productivity, because there is a greater chance that a more productive species would be present at higher diversity, and from the better 'coverage' of habitat heterogeneity caused by the broader range of species traits in a more diverse community. Biodiversity is inevitably damaged by human impact as the economic and ecological systems compete for space. See Rauscher et al. (2020) *Resource and Energy Econs* 32, 2.

The **biodiversity gradient** describes the greater biodiversity of living organisms at the tropics than at the poles in the biomass, the number of individuals, and, in many taxonomic groups, the number of species. Diversity gradients described for the Northern Hemisphere seem to be invalid for the Southern Hemisphere (see Boyero

(2006) *Ecology Info.* 32). Changes to biodiversity are most prevalent across oceans, particularly in the tropical marine regions, which are hotspots for species richness loss, but it has been surprisingly difficult and controversial to find signals of a **biodiversity crisis** in the context of local ecosystems.

SEE WEB LINKS

• Website of International Union for Conservation of Nature.

biodiversity hotspot Myers et al. (2000) *Nature* 403 identify 25 global hotspots, which together comprise only 1.4% of the Earth's terrestrial surface, but which, among them, contain 44% of all species of vascular plants, and 35% of all species of four vertebrate groups. Examples include the Mediterranean basin, Madagascar, the Western Ghats, and Sri Lanka.

biofuel Any liquid fuel derived from plant materials, including biodiesel which can be made from vegetable oils such as palm oil, soy, or rapeseed oil. A shift to biofuels means that less land is being used to produce food at a time when the world population continues to grow. Furthermore, as demand and prices have risen for crops such as palm oil, new land has been cleared to start plantations. The land that is being cleared in countries such as Indonesia and Malaysia is often rainforest, or other environments where plant and animal diversity is high and indigenous people live traditional lives in harmony with the delicate ecosystems.

biogenic Produced or brought about by living organisms. A **biogenous sediment** is one of biological origin, such as calcareous or siliceous ooze.

biogeochemical cycle The cyclical movement of energy and materials within *ecosystems; between their living organisms (the biotic phase) and their non-living (abiotic) components. The *carbon cycle is an example. See Slavekova (2019) *Environ. Sci.* on biogeochemical dynamics research in the Anthropocene.

biogeographic realm A continent- or subcontinent-sized area, separated from other biogeographical realms by physical barriers to plant and animal migration, such as major mountain ranges, and oceans: for example, the Afrotropical Biogeographic Realm, which contains one of the highest levels of biodiversity on Earth. Each biogeographic realm may include a number of different biomes. The use of *ecozone* for biogeographic realm is a fairly recent development.

biogeography The spatial analysis, and the search for patterns, in any biological feature: demographic, ecological, genetic, morphological, or physiological. 'Biogeography transcends classical subject areas . . . **ecological biogeography** is concerned with ecological processes occurring over short temporal and short spatial scales, whereas . . . **historical biogeography** is concerned with evolutionary processes over millions of years on a large, sometimes global, scale' (Crisci (2001) *J. Biogeog.* 28, 2). **Comparative biogeography** uses the naturally hierarchical phylogenetic relationships of *clades to discover the biotic area relationships among local and global biogeographic regions; **systematic biogeography** is concerned with the classification of biotic areas based on the naturally hierarchical phylogenetic relationships of clades; and **evolutionary biogeography** seeks to understand evolutionary mechanisms responsible for the distribution of its organisms. **Molecular biogeography** is the use of genetic data in the study of the biogeographic structure of lineages and biotas, together with the evolutionary and Earth history processes that have shaped them. Moodley and Bruford (2007) *PLOS One* use genetic data to explore the ecoregion connectivity, core habitats and ecological affinities of the African bushbuck.

biogeomorphology Any study linking plant and animal life with geomorphological form and process, for example bioweathering, bioerosion, bioconstruction, biotransformation, biostabilization, and bioprotection.

Many 'soft engineering' schemes for environmental management use biogeomorphological systems; see Cuong et al. (2015) *Ecol. Engin.* 81 on mangrove restoration.

biographical approach An approach which stresses that the behaviour of an individual is made within the context of their past: their culture, history, identity, and societies; the times and spaces that people pass through and the role of their biographies in their life journeys. Try Katz and Monk, eds (1993) on geographies of women over their life courses.

biological control The attempt to reduce numbers of pests by the use of predators, either from within the community, or by introduction from outside, rather than by chemicals. Harris et al. (2020) 151 review the causes and consequences of geographical variability in the biological control of weeds.

biological invasion The establishment of species in ecosystems to which they are not native. Invasive species cause significant ecological harm: they can alter ecosystem processes, act as vectors of disease, and reduce biodiversity. Around a quarter of the value of US agricultural output is lost to non-indigenous plant pests or the costs of controlling them. However, Hoffman and Courchamp (2016) *NeoBiota* 29 argue that human-mediated invasions are part of the spectrum of species movements, and not a unique phenomenon,

Case studies abound, among them the UK Environment Agency's material on American signal crayfish and Japanese knotweed, which is also of interest for its use of anthropomorphic phrasing: 'because [Japanese knotweed] does not originate in the UK, it does not compete *fairly* with our native species' (this author's italics).

biological magnification (biomagnification) The build-up of a toxin

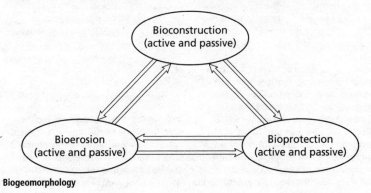

Biogeomorphology

as it moves up the *food chain. For example, pesticides wash into waterways, are eaten by aquatic organisms which are in turn eaten by large birds, animals, or humans, and become concentrated in tissues or internal organs. Contaminants include mercury, arsenic, pesticides, and polychlorinated biphenyls, which are then taken up by organisms because of the food they consume or the intoxication of their environment. Since aquatic food chains have more links, biological magnification in aquatic species is greater than in terrestrial species; see, for example, Sutherland et al. (2000) *Arch. Env. Contam. & Toxic.* 38, 4.

biological oxygen demand (BOD) An indicator of the polluting capacity of an effluent where pollution is caused by the take-up of dissolved oxygen by micro-organisms that decompose the organic material present in the effluent. Excess biological oxygen demand can kill oxygen-deprived fish.

biomass The total mass of all the organisms inhabiting a given area, or of a particular population or *trophic level.

biome An ecological zone whose uniformity is defined by the type of plant life that dominates within it, since plant life will often give a strong indication of other ecological features of a zone, such as animal life and soil type. Because biomes are defined by plant life rather than region, some biomes can stretch around the world. For example, the Boreal Forest biome (defined by the predominance of conifers) covers much of northern Europe, Russia, and Canada. Other major biomes are: *tundra, temperate (deciduous) forest, *tropical rain forest (selva), tropical grassland (*savanna), temperate grassland (steppe), and *hot deserts. A biome is an idealized type; local variations within a biome are sometimes more significant than variations between biomes.

biometeorology The description and understanding of the impacts of climate and weather on the biosphere and on humans. The International Society for Biometeorology defines it as 'an interdisciplinary science that considers the interactions between atmospheric processes and living organisms (plants, animals, and humans) with the central question within the field being "how does weather and climate impact the well-being of all living creatures"'. **Human biometeorology** comprises the study of the influence of weather and climate on both healthy and diseased persons.

(()) SEE WEB LINKS
• Website of International Society for Biometeorology defining biometeorology. The *International Journal of Biometeorology* is available online.

biophysical diversity The diversity of the natural environment, including soil characteristics and their productivity, the biodiversity of natural (or spontaneous) plant life, and of the soil biota. It takes account of physical and chemical aspects of the soil, surface and near-surface physical and biological processes, hydrology, microclimate, and also variability and variation in all these elements. Geertsima and Pojar (2007) *Geomorph.* 89, 1–2 recognize several types of biophysical/ecological diversity: site diversity, soil diversity, and derivative habitat or ecosystem (including aquatic ecosystems) diversity.

biopiracy The damaging exploitation of naturally occurring and economically valuable biological resources: animals, plants, and micro-organisms in nature. See Robinson and Raven (2017) *Australian Geographer* 48, 3 on identifying and preventing biopiracy in Australia. Other definitions add that individuals or companies then patent such resources for their own benefit. Also known as **ecopiracy**.

biopolitics An apparatus of control exerted over a population as a whole, citing the ratio of births to deaths, the rate of reproduction, the fertility of a population, and so on, in justification. Others characterize biopolitics as the political application of bioethics; the sociopolitical consequences of the biotech revolution; the administration and regulation of human and non-human life at the levels of both the population and the individual body: see Schlosser (2008) *Geog. Compass* 2, 5, 1621.

bioregion A region demarcated by the nature of the natural environment, not by man-made divisions. In *ecology, a bioregion is an area that constitutes a natural ecological community. Bioregions may be 'natural', but the contours of bioregions are drawn largely by humans; see Meredith (2005) *Ethics, Place & Environ.* 8, 1, 83. Bioregionalists argue that modern industrial societies have obscured or lost contact with the extent to which they are reliant on and embedded in ecosystems.

bioscience Any of the life sciences—biology, medicine, anthropology, or ecology—that deal with living organisms and their organization, life processes, and relationships to each other and their environment. **Geographies of bioscience** might analyse the differential impact of developments in relation to inequalities in gender, ethnicity, wealth, and location, and the role of international corporations in these developments; or the development of multinational regulatory endeavours and the distinctive take up of developments in biomedicine and biotechnology in different nations and regions.

biosecurity The protection of people and animals from pests and infectious diseases, notably by managing the movement of agricultural pests and diseases, reducing the effects of invasive species on supposedly indigenous flora and fauna, and preventing the purposeful and inadvertent spreading of biological agents into the human population. See Hinchliffe and Bingham (2008) *Envir. & Plan. A* 40, 1534.

biosphere The zone where life is found; the outer portion of the geosphere and the inner portion of the *atmosphere. This extends from 3 m below the ground to some 30 m above it. The biosphere also comprises that region of waters, some 200 m deep, where most marine and freshwater life is found. *Gaia theory postulates biosphere-scale evolution. See Phillips (2008) *PPG* 32, 1 on goal functions in ecosystem and biosphere evolution.

biotic Living or having sustained life. **Biotic interaction** is a mutual or reciprocal interaction between living organisms, in which the distribution of one species is influenced by the distribution of other species. Biotic interaction can be within the same group, for example, competition, facilitation, and parasitism in plants, or between groups, as best exemplified by relationships in food webs: herbivory, predation, and symbiosis. **Biotic potential** is the maximum population that an area can support. **Biotic homogenization** describes the process by which species invasions and extinctions increase genetic, taxonomic, or functional similarity; 'evidence is growing that changes in land cover are likely to be accompanied by the ongoing homogenization of biotas, and processes in which native species are replaced by a relatively small set of alien species' (Gaston and Fuller (2007) *PPG* 31, 2). See Cassey et al. (2008) *J. Biogeog.* 35, 5.

bioturbation The mixing of sediments and soils by animals or plants, for example, when earthworms turn over a soil. These sediments may be marine, lacustrine, or fluvial, where bed-sediment is engineered by plants and algae. Bioturbation is significant in downslope soil transport, particularly in soil formation, soil erosion, and soil creep; see Richards et al. (2012) *ESPL* 36, 9, 1240.

birth rate The crude birth rate is the number of live births per year, per 1000 population. This does not take the age structure of a population into account, and hence the numbers of women who are capable of giving birth in any year, making accurate comparisons difficult. Because of this, many demographers prefer to use a **standardized birth rate** which indicates what the crude birth rate would have been for a population if the age and sex composition of that population were the same as in a population selected as standard, allowing comparisons to be made between geographical areas, or across social groups within a society (see Fargues (1997) *Pop. & Dev. Rev.* 23, 1 for a demonstration of standardization).

black earth *See* CHERNOZEM.

black economy The part of the job market which is not reported to the tax authorities and does not appear in official statistics. 'Black' and 'grey' market activities encompass a very diverse set of practices that range from individual bribes and favours to get things done, to protection rackets, and to the illegal appropriation of Western aid payments and to human trafficking.

black geographies Black geographies study black agency in the production of space, and black geographic experiences in black geographical visions of society. The term is not a catch-all for all geographies of race but instead is a term encompassing works that emphasize black experiences. See Bledsoe et al. (2017) *Southeastern Geographer* 57, 1. Allen et al. (2019) *PHG* 43, 6 is a good place to begin.

black smoker A hydrothermal spring lying along the rift valley of a *mid-ocean ridge. Sea water seeps into fissures in the lava and becomes superheated, chemically interacting with the basaltic rock to create a black precipitate of metal sulphides. One such is located near the Carlsberg Ridge, north of the Seychelle Islands.

blanket bog A continuous covering of bog, needing high humidity and rainfall, mostly comprising peat; only steep slopes and rocky outcrops are dry. Human impact was the dynamic force responsible for the destruction of the woodland and the formation of blanket bog in upland Ireland; see Huang (2002) *J. Biogeog.* 29, 2. In the UK Peak District blanket bog has suffered a decline in species diversity as a result of air pollution, overgrazing, inappropriate or accidental burning, peat extraction and past drainage.

blind valley A steep-sided karst valley, abruptly terminating in a *streamsink.

blizzard As defined by the US National Weather Service, a wind over 16 m s^{-1} together with falling or blowing snow, causing visibility of less than 400 m and lasting for at least 3 hours. See Schmidlin (2002) *J. Climate*, 15, 1765 on the climatology of blizzards in the United States.

bloc An alliance of governments, groups, or parties with a common purpose, such as a trading bloc. Badinger and Breuss (2006) *Rev. Int. Econ.* 14, 4 investigate the gains for each member from enlarging a trade bloc, finding that small countries increase in competitiveness by attaining easier access to a larger market, while large countries may have advantages over small countries via relatively more group ties, high market power, and related terms-of-trade effects, advantages in competitiveness due to economies of scale and larger absolute endowments (particularly human capital), larger product varieties (reflecting the greater scope for specialization), and technological advantages.

block 1. A section of the Earth's crust, bounded by faults, that may be tectonically disrupted to form basins and block mountains or complex systems of troughs and ridges. The formation of a **tilt-block** may result in *tectonic deformation along plate margins; see Agarwal and Sharma (2011) *Zeitschrift. Geom.* 55, 2, 194.
    2. A generally angular boulder formed by frost-shattering *in situ* on a bedrock surface.

blockbusting A technique, used by estate agents in the USA, of inclining residents of a white neighbourhood to move out because they fear that the district is to be taken over by black families. See Hightower and Fraser (2020) *City and Community* 19, 1 on **reverse blockbusting**, the gentrification of black neighbourhoods in Nashville, USA.

block field (block stream) A sheet of *block spreads or lines in a *periglacial landscape. Block fields (*felsenmeer*) develop on slopes under 5° (A. L. Washburn (1979)), or under 10° (Caine

in White (1976) *Quat. Res.* 6). See van Steijn et al. (2002) *PPG* 26, 4. Thompson and Syverson (2006) *GSA Abstracts with Progs*, 38, 4 maintain that this debris is formed *in situ* by *freeze-thaw. Others suggest that the blocks have ridden down on the top of saturated debris during *gelifluction (Creemeens et al. (2005) *Geomorph.* 70, 1–2).

blocking A meteorological condition when the tropospheric circulation takes the form of large-amplitude stationary waves, possibly lasting for weeks. The upper air flow then guides depressions around the edge of the anticyclone(s). *Rossby wave propagation links anomalous convection in the tropics and blocking over the south-east Pacific Ocean. **Blocking anticyclones** develop when the Rossby pattern changes from zonal to strongly *meridional, often forming one or two high-level, closed anticyclonic circulations. These block the passage of *depressions, and may persist for a week or more. See Smith (1995) *Tellus* 47, 4.

block mountain An area of upland identified with an uplifted area bounded by *faults. The Sierra Nevada is a massive block mountain range that broke free along a bounding fault line and has been uplifted and tilted. See Huber (1987) *US Geol. Surv. Bull.* 1595.

blowhole A crack in the top of a cliff through which air and sea water blow. The blowhole is fed from the seaward end via joints and tunnels.

blow-out (blowout) A saucer-, cup-, or trough-shaped depression or hollow formed by *deflation on a pre-existing sand deposit. The adjoining accumulation of sand, the depositional lobe, derived from the depression, and possibly other sources, is normally considered part of the blow-out. Blow-outs usually result from the removal of vegetation; see Dech et al. (2005) *Catena* 60, 2.

bluff A steep, almost vertical, cliffed section of a river bank.

body The body may be seen as: the smallest unit of geography which can create or be moulded by place/space; a map of meaning and power; or a cultural representation of masculinity or femininity. It is also a form of reference by which supposedly 'disembodied' dominant cultures designate certain groups—the elderly, ethnic minorities, females, the obese, the disabled, and so on—as *other. **Sexed bodies** can create spaces, as in a gay pride parade (Johnston (2001) *Annal. Tour. Res.* 28, 1). The notion of the **raced body** tied to the land 'is not novel, having been used to discredit urban Aboriginals, claiming that once they leave the land, they cease to be Aboriginal' (Wazana (2004) *Refuge* 01-03-2004). Every researcher is situated in her or his own body—*embodied*.

body work Any work that involves intimate and often messy contact with the bodies of others through touch or close proximity. Body work has been, and continues to be, undertaken primarily by women, and is based on a sexual division of labour that assigns to women the care of bodies and the spaces they inhabit when doing so. Body work—for example, care work, childcare, and beauty therapy—is generally done by those on the lower rungs of the job ladder. C. Wolkowitz (2006) describes body work as involving the 'care, adornment, pleasure, discipline, or cure' of others' bodies.

bog-burst (bog-slide, bog flow) A peat slide, usually in temperate or cold and wet latitudes, triggered by water pressure beneath a peat surface which causes it to rupture.

bohemian Someone with creative or literary interests, who is socially unconventional. Florida (2002) *J. Econ. Geog.* 2 finds strong relationships between high concentrations of bohemians and concentrations of high human capital, with a particularly strong relationship between the concentrations of bohemians and concentrations of high-technology industry. See Faggian et al. (2013) *Regional Studies*, 47, 2 on disciplines and location determinants for entering creative careers.

bolide A meteoric fireball, or asteroid; see Dawson and Stewart (2007) *PPG* 31, 6.

Bølling The mild Bølling–Allerød interstadials began *c.* 13000, and ended between 12800 and 12300 years BP. In Britain the Bølling is not usually distinguished from the *Allerød.

Bora (fall-wind) A cold winter wind blowing down from the mountains onto the eastern Adriatic coast. See Pullen et al. (2007) *J. Geophys. Res.* 112.

border A boundary line established by a state, or a region, to define its spatial extent. The establishment and maintenance of borders are crucial to the integrity of the modern state. States use a variety of mechanisms to enforce their borders, including checking travel documents, limiting border crossings to select travellers, and in some cases building obstacles to prohibit unauthorized crossings. State borders are not only boundaries, but also symbolic social and cultural lines, based on collective historical narratives. Thus borders are not there only by tradition, wars, agreements, and high politics, but are also made and maintained by other cultural, economic, political, and social activities. A **hyper-border** goes beyond traditional understandings of a frontier; it might demarcate an intense mismatch between two adjacent states, such as Mexico and the USA, or be a porous border, where the inhabitants of one nation identify, or economically relate to, the nation next to them, as in the South Kiril Islands at the border of Russia and Japan; see Richardson (2016) *TIBG* 41, 2, and Cassidy et al. (2018) *Political Geography* 66. See also Dell'Agnese and Szary (2015) *Geopolitics* on **borderscapes**.

(⊕) SEE WEB LINKS

• A blog used as an archive for papers and sources regarding the China–India border dispute.

Boreal (Period) A division of the *Flandrian, *c.*10000–8000 years BP.

bornhardt A dome-shaped rock outcrop more than 30 m high, and sometimes several hundred metres in width, commonly rising above erosional plains in the tropics, but also found in unglaciated uplands in high latitudes. The origin of these features is problematic: faulting, lithological control, cross folding, and contrasts in fracture density have been cited. C. Twidale and J. R. Vidal Romaní (2005) give an excellent summary.

borral A soil order of the *US soil classification. *See* CHERNOZEM.

Böserup model E. Böserup's (1965) view that increases in population size stimulate agricultural change. According to Böserup's model, pre-industrial farmers were almost always technically capable of increasing the productivity of their land by applying methods (soil tillage, additional weeding, application of organic fertilisers and irrigation) which allowed it to be cropped more frequently and fallowed for shorter periods. But because such methods typically demanded higher labour inputs per unit of production, they tended to be adopted only when population pressure and the resulting land scarcity forced farmers to work harder in order to maintain existing levels of subsistence. Turner and Fischer-Kowalski (2010) *PNAS* take stock of Böserup's impact on research and practice and show how her ideas continue to shape and be reshaped by current research.

boulder field As distinct from a *block field, this is an area of boulders which is the result of *spheroidal weathering. See Sevon (1987) *N–E Section GSA* 5.

boundary 1. Of a river, the bed, and banks.
   2. A point or limit that distinguishes one social system or group from another and identifies and regulates who may participate in it. Boundaries are interesting to geographers because they mark change, for example at the edges of neighbourhoods, whether defined by socio-economic status, by employment, or by deprivation. Boundaries are rarely seen as natural; they are human creations imposed on a landscape. See Jaquez (2010) *Spat. & Spatiotemp. Epidemiol.* Dec 1; 1(4). On geographic boundary analysis, J. Nevins (2002) suggests that a boundary is a strict line of separation between two (at least theoretically) distinct territories, where a border is an area of interaction and gradual division between two separate political entities.
   Mostly in biogeography and plant ecology, **boundary detection** is the identification of the zones of greatest change in vegetation and associated abiotic/biotic factors; see Jaquez et al. in T. Poiker and N. Chrisman, eds (1998). Kent et al. (2006) *PPG* 30, 2 are a model of clarity in discussing boundary detection.

boundary layer Any layer of the *atmosphere significantly affected by the

Earth's surface. The **laminar boundary layer** is the few millimetres above the surface; the **turbulent boundary layer** denotes the conspicuously turbulent part of the atmosphere.

(((🌐))) SEE WEB LINKS
• NOAA pages on surface and planetary boundary layer processes.

bounded rationality A limited form of rationality, where individuals strive to be rational, after first greatly simplifying the choices available, accepting drawbacks such as computational difficulties, or the costs of obtaining information. In choosing between a small number, the decision-maker is a *satisficer. Salant (2011) *Amer. Econ. Rev.* 101, 2, 724 is clear on this.

Bowen ratio (*B*) The ratio of sensible heat (the heat exchanged by a body or thermodynamic system that has as its sole effect a change of temperature) to *latent heat. The Bowen ratio is the mathematical method generally used to calculate heat lost (or gained) in a substance. When the magnitude of *B* is less than one, a greater proportion of the available energy at the surface is passed to the atmosphere as latent heat than as sensible heat, and the converse is true for values of *B* greater than one.

braid bars *See* BAR.

braided channel A river channel in which *bars and islands have been deposited, differing from *anastomosis since the islands are less permanent. Braiding occurs with frequent variations in discharge: when the river cannot carry its load; where the river is wide and shallow; where banks are easily eroded; and where there is a copious bed load. The division of the flow into multiple threads is a prerequisite for bifurcation. Fotherby (*Geomorph.* 103, 4) argues for valley containment as a causal factor of

braided channels. See Bledsoe and Watson (2001) *Geomorph.* 38, 3–4 for an excellent summary.

braunerde *See* BROWN EARTH.

Brazil current A weak boundary current, flowing in the upper 600 m of the water, carrying warm subtropical water along the coast of Brazil—from about 9° S to about 38° S, where it joins the Malvinas Current. The two currents create the strong Brazil-Malvinas Confluence thermohaline frontal region (Bischof et al., no date, *CIMAS*).

break point (breaking point) The point at which the field of influence of one settlement ends, and that of another begins. See Haberkern (2002) *Issues in Shopping Centre Trade Areas* for W. Reilly's (1931) original equation, and a number of modifications to it.

break point bar *See* BAR.

bricolage The practice of creating things out of whatever materials come to hand, and thus describing an entity that is built out of whatever materials happen to be available, for example, a shack in a *squatter settlement. The term may be used in *postmodernist architecture.

brown earth (brown forest soil) A free-draining *zonal soil associated with deciduous woodland. Generally, brown earths have a thick litter layer and a *humus-rich A *horizon, containing iron and aluminium sesquioxides in small, crumb-like *peds. The weakly developed, lighter, B horizon contains blocky peds; there is little *leaching. See Storrier and Muir (1962) *Eur. J. Soil Sci.* 13, 2.

brownfield site Any previously used land and/or premises that is not currently fully in use, and may also be vacant, derelict, or contaminated. See Leger et al. (2016) *Planning Practice and*

Research 31, 2 on the planning challenges of brownfield development in coastal urban areas of England, and Gregorová et al. (2020) on transforming brownfields as tourism destinations, and their sustainability, in Slovakia.

buffer state A generally neutral state, lying between potentially belligerent neighbours; 'protective devices made obsolete by modern systems of communication' (Spykman (1947) *Geog. Rev.* 32). See V. Pholsena and R. Banomyong (2007) on Laos.

built environment The human-made surroundings that provide the setting for human activity, including buildings, parks, green spaces, neighbourhoods, and cities, and sometimes including their supporting *infrastructure. Very simply, it is defined as 'the human-made space in which people live, work, and recreate on a day-to-day basis' (Roof and Oleru (2008), *J. Environ Health* 71, 24).

buoyancy The quality of an air parcel allowing it to rise through, and remain suspended within, the *atmosphere, due to its relative warmth, and therefore lighter density. Should its water evaporate, the parcel will suffer an *adiabatic heat loss, and lose buoyancy. See Dupilka and Reuter (2006) *Weather & Fcst* 21, 3 on forecasting tornado potential.

bush encroachment The change in vegetation from open *savanna, or mixed grass and woodland, to scrubland. Occurring also in Australia and South Africa, bush encroachment affects the agricultural productivity and biodiversity of 10–20 million ha of South Africa (Ward (2005) *Afr. J. Range & Forage Sci.* 22, 2). Its probable causes are: increased

grazing levels; reduced use of burning to create grassland; changes in rainfall regimes; and the interactions between these factors. However, writing in 2005, Ward (op. cit.) writes, 'many people believe that we understand the causes of bush encroachment. We do not.'

bush fallowing A farming system whereby the fallow period is longer than the farming period. Hance (1969, in Omofonmwan and Kadiri (2007) *J. Hum. Ecol.* 22, 2) argues that bush fallowing is a major hindrance to food production in Nigeria, but Chokor (1993) *Env. Manage.* 17, 1 argues that 'modern' methods, which shorten the fallow period, disrupt 'the conservation principles inherent in the traditional man–nature relationship'.

bustee (basti) In India, a *shanty town; reading S. Mehta (2005) should be mandatory. N. Islam (1996) estimates that almost 50% of Dhaka's population live in bustees, and a major portion of this population is female. To combat negative views of bustees, see Mahmud (2003) *Cities* 20, 5.

butte A small, flat-topped, unvegetated, and very steep-sided hill of layered strata. It may be a volcanic neck or a cone that has resisted erosion, or the residue of a *mesa.

buzz (urban buzz) The joint potential to generate creative, innovative, and unconventional initiatives or activities in cities or specific urban districts; knowledge transfer via 'in-groups', socialization, and face-to-face contact. Buzz and socio-economic/cultural diversity are closely connected phenomena; see Arribas-Bel1 et al. (2016) on the socio-cultural sources of urban buzz.

C3 and C4 plants C3 plants use the C_3 carbon fixation pathway as the sole mechanism to convert CO_2 into an organic compound. They make up over 95% of Earth's plant species and flourish in cool, wet, and cloudy climates, where light levels may be low, because the metabolic pathway is more efficient there. With enough water, the stomata can stay open and let in more carbon dioxide. However, carbon losses through photorespiration are high.

C4 plants are able to raise the intercellular carbon dioxide concentration at the site of fixation, thus reducing and sometimes eliminating carbon losses by photorespiration. C4 plants first fix CO_2 into a compound containing four carbon atoms. They inhabit hot, dry environments, have very high water-use efficiency, and can double the C3 photosynthesis, but C4 metabolism is inefficient in shady/cool environments. Less than 1% of Earth's plant species can be classified as C4, but they account for around 20% of global gross primary productivity.

cadastre 1. A record of the area, boundaries, location, value, and ownership of land, achieved by a **cadastral survey**.

2. The term is also used in *geographic information systems (GIS). CARIS (Computer Aided Resource Information System) is a GIS system for hydrography (see Mioc, Anton, and Nickerson, no date, for a web-based GIS model for predicting flood events) and cadastral systems (see Steudler (2004) *GIS@development* for a worldwide comparison of cadastral systems).

Cainozoic *See* CENOZOIC.

calcrete A *duricrust made mostly of calcium carbonate, forming in arid climates via capillary action and prolonged evaporation. See Stokes et al. (2007) *Geomorph.* 85 on groundwater, pedagenic processes, and fan dynamics in calcrete development.

caldera A sunken crater at the centre of a volcano, formed either by an explosive volcanic eruption, or when the magma within a volcano founders, so that the centre of the volcano collapses. For magmatic processes and caldera collapse, see Kennedy and Stix (2007) *GSA Bull.* in New Hampshire; and Furuya (2003) *Earth Planets Space* 55 constructs an optimum source model of caldera formation.

calibration The way the elements which control a model are adjusted to make the model as close as possible to the reality that the model is designed to simulate. While validation may actually be impossible in some cases, calibration is obviously about the most essential component of modelling. Raimbault (2018) *PLOS One* demonstrates the calibration of a density-based model of urban morphogenesis.

calibre In rivers, the size of the sediment transported. With distance downstream, the calibre of the sediment decreases.

California current A cold *ocean current off the California coast, given to anomalous conditions, that can be termed a Californian *El Niño (Mullin (1998) *Glob. Change Biol.* 4, 1). Recent climate model simulations under a high-end greenhouse gas emissions scenario indicate California current waters will be 2 to 4 °C above the 1920–2016 average by 2100.

calving The breaking away of a mass of ice from an iceberg, ice front, or glacier, caused by: wind and/or wave action; over-extension of an ice shelf; or a collision with an older iceberg. Calving represents the major form of *ablation from a glacial system.

Cambrian The oldest period of *Palaeozoic time, approximately 570–500 million years BP; the time when most of the major groups of animals first appear in the fossil record.

(((⊕))) SEE WEB LINKS
• Provides an overview of the Cambrian period.

canyon A valley with very steep sides and no floor, such as the Grand Canyon of the Colorado River. A canyon differs from a *gorge in that the sides are stepped, reflecting alternating rock resistances. See Spamer (2006) *Annal. Imposs. Res.* 12, 2, who sounds off about the name 'Grand Canyon', but is quite sane in his book of 1984. **Submarine canyons** are deep troughs in the sea bed, sometimes as prolongations of river valleys on land. Canyons may also form via earth flows, turbidity currents, submarine springs, or the slipping of sediments.

capability constraint In *time-space geography, a limit to one's actions, due to biological needs, like sleep, and to restricted facilities, like access to transport, or low income; see Janusz

et al. (2018) *Tidschrift* 11, 4 on transport in Kampala, Uganda.

capacity The maximum amount of debris that a stream can move as *bed load. Capacity depends on *discharge, and on the nature of the load; for the same volume, a stream may be able to carry more gravel than boulders.

cap and trade A strategy for lowering the emission of greenhouse gases which aims to limit harmful pollutants and to establish a market-based price on emissions. Companies whose activities emit greenhouse gases are obliged to buy permits to cover their emissions; the permits ('allowances') can be bought or sold at prices determined by supply and demand. The 'cap' is the limit on permitted emissions, while the 'trade' is the market in emissions equal to or above this cap. The European Union 'cap and trade' scheme is known as the European Union Emission Trading Scheme.

capillary In a soil, the fine spaces between soil particles. **Capillary action** is the movement of a liquid in a very narrow space, such as a soil pore, due to surface tension, cohesion, and adhesion.

capital One of the four *factors of production, capital includes all the items designed by society to further the creation of wealth. Plant, machinery, and buildings are **fixed capital**—they earn profit without circulating further. **Circulating capital** (**floating capital**) includes raw materials, fuels, components, and labour inputs which are then sold—in the form of the product—at a profit. **Financial capital** is the money necessary for production.

The mobilization of **commercial capital** occurs through direct investments in companies or through syndicated loans, which allow risk sharing among the debt providers.

Commercial capital works for a fee in the circulation of commodities, but has no long-term control over it. Specialized commercial and financial capital drive down the costs of circulation and finance, speeding up capital turnover and reinvestment.

Knowledge is one of the key drivers of long-run economic growth, and **knowledge capital** covers know-how, know-what, relationships, skills, and talents. Amin and Thrift (2007) *PHG* 31, 2 link urbanism with **fast capitalism**— rapid information and communication technologies together with their impact on self, society, and culture. See P. Taylor (2004).

Capital can be seen as a social relation (link, exchange), symbolizing the economic exchanges between individuals and/or groups. In *Marxist geography, capitalism is seen as a process whereby one set of commodities, such as workers, is used to create other commodities, with the paramount aim of making a profit.

Geographers focus on capital as a contributor to *uneven development; Walker (1978) *Rev. Radical Polit. Econ.* 10 is still worth quoting: '[the] mobility of capital means that capital may use location as a strategy against labour; local development becomes more dependent on outside capital, and development comes and goes over time.' *See* SPATIO-TEMPORAL FIX.

Capital switching This concept is based on the work of David Harvey (1985), who argued that there are three circuits of capital: a primary circuit of capital accumulation; a secondary circuit, which is the capital that flows into the fixed assets, consumption funds, and the built environment; and tertiary circuit comprising the investments in science, technology, education, health, and other social expenditure. Capital switching is simply a change in investment from one circuit to another;

for example the switching of money into green infrastructures, with the aim of climate change mitigation. See Noel Castree (2015).

Another switch is the movement of investment from the production of goods to developing the urban built environment as a means to absorb surplus capital. Christophers (2011) *AAAG* 101, 6, 1347 finds a clear pattern of capital switching in the USA into the built environment since the turn of the millennium.

capital intensive With a high ratio of *capital to labour; wind farms are highly capital-intensive.

capitalism An economic system based on private property and private enterprise, with a major proportion of economic activity carried on by private profit-seeking individuals or organizations, and other material *means of production largely privately owned. Capitalism never stands still. Its central imperative—the search for profit and wealth creation—drives perpetual economic change, powered by innovation.

'Innovation and growth are at the heart of the capitalist system . . . As new large-scale technologies develop; new forms of firm and factory organisations result in increasingly global value chains producing for global market' (Kaplinsky (2008) *Geog. Compass* 2, 1). See also Metcalfe et al. (2006) *Camb. J. Econ.* 30, 7–32. *Accumulation is at the heart of capitalism. A feature of **advanced capitalism** is the possession of capital by fewer and fewer owners.

Capitalism is not uniform: nation-states have different trajectories of capitalist development and different national institutional frameworks support different forms of capitalism. See H. Wai-Chung Yeung (2006) and Buck (2007) *Antipode* 39, 4 on Chinese capitalism; E. M. Wood (2003) on English capitalism; and Clarke (2007) *TIBG* 32, 1 on British capitalism. Capital constantly searches for fresh markets. 'Spatial differentiation is the fundamental building block of capitalism' (Clark (1985) *Econ. Geog.* 61, 3).

See Peck and Theodor (2007) *PHG* 31, 6, and Faulconbridge (2008) *J. Econ. Geog.* 8, 4 on varieties of capitalism.

capitalist state The state provision of *infrastructure, *external economies, and a regulatory framework facilitates *accumulation. The state can limit or increase the freedom of market actors, raise or lower taxes, intervene in major financial crises, and defend and secure its territory—none of which can be accomplished by *transnational corporations. The **theory of the capitalist state** argues that such functions legitimize the state's existence. B. Jessop (2002) focuses upon the structural coupling and co-evolution of accumulation regimes and political regimes. It is argued that the state's capacity to manoeuvre the accumulation regime has been weakened by international finance, trading *blocs, and so on, leading to a loss of national sovereignty. The **environmental theory of the capitalist state** argues that the state is fundamental to capitalist development and a key part of what it does for capital is to manage nature.

capital–labour ratio The amount of capital per unit of work. In theory, as people move from lower real-waged city regions to higher-waged ones, the capital–labour ratio in stronger regions decreases, and rises in weaker regions. The process stops as soon as the capital–labour ratio is the same in all regions. In time, this development should improve the welfare of society as a whole.

capture (river capture) When a river is extending its channel upstream by *headward erosion, it may come into contact with the headwaters of a less vigorous river. The headwaters from the minor river may be diverted into the more rapidly eroding channel. There is often a sudden change of stream direction at the point of capture; this is the **elbow of capture**. See Mather (2000) *Geomorph.* 34, 3 for two examples of river capture.

carbonation A form of solution where carbonic acid, formed by the solution of atmospheric CO_2 in water, dissolves minerals. Carbonation is best seen in the solution of calcium-rich rocks such as limestone, but carbonic acid will also dissolve silicates.

carbon cycle Carbon is supplied to the *biosphere as carbon dioxide during volcanic eruptions. Most of this is dissolved in the sea or built into

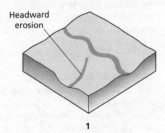

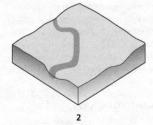

Headward erosion

1   2

River capture

calcareous sediments, which then form limestones and dolomites. As these rocks are folded and raised above sea level, they are subjected to solution by weak carbonic acid and form sediments once more. This is the largest and slowest of the carbon cycles. The shortest cycle involves respiration by plants and animals whereby carbon dioxide is expired, and photosynthesis by plants, changing carbon dioxide and water into organic compounds. Plant species affect ecosystem carbon uptake via biomass production, and carbon release via decomposition. Differences in the way plants respond to climate change may feed back to the atmospheric carbon balance, and the climate at a global scale.

Since 1957, only about half of the CO_2 emissions from fossil-fuel combustion have remained in the atmosphere; the rest being taken up by land and oceans. In the face of increasing fossil-fuel emissions, the rate of carbon absorption by the land and ocean has accelerated over time. There is no guarantee that the 50% discount will continue; if it disappears we will feel the full climatic brunt of CO_2 emissions from fossil fuels. Climate models that include descriptions of the carbon cycle predict that terrestrial uptake of carbon will decrease in the next century as the climate warms.

carbon footprint The amount of greenhouse gas (GHG) emissions emitted through human activity. It is expressed in terms of tonnes of carbon dioxide equivalent (the mass of carbon dioxide gas that has the same warming effect in the Earth's atmosphere as the given GHG emission) produced in a unit of time.

Carboniferous A period of *Palaeozoic time, stretching approximately from 345 to 280 million years BP, and subdivided into the Mississippian and the Pennsylvanian periods.

( ) SEE WEB LINKS
• An overview of the Carboniferous period.

carbon management The monitoring of carbon emissions and their removal through **carbon sequestration**; that is, the storage of carbon dioxide. Energy crops may have the greatest potential for carbon sequestration in Europe. Bourque et al. (2007) *Mitigation & Adapt. Stats. Glob. Change* 12, 7 provide a methodology for forestry and carbon sequestration.

CO_2 may also be stored underground or below the ocean *thermocline. Hoffert et al. (2000) *Science* 298 discuss CO_2 sequestration in deep seas ocean injections can substantially decrease peak atmospheric CO_2, although some CO_2 would return to the atmosphere. Harvey (2004) *Climatic Change* 63, 4 warns that the impact of carbon sequestering on atmospheric CO_2 rapidly decreases over time, as there is a reduction in the rate of absorption of atmospheric CO_2 by the terrestrial biosphere and the oceans.

carbon offset A reduction in emissions of carbon dioxide or greenhouse gases made in order to recompense or to offset for an emission made elsewhere. A **carbon tax** is a compulsory monetary payment imposed by governments. It is designed to reduce the *carbon footprint of commodity producers or consumers.

carbon sink A phenomenon, such as a forest or ocean, which can absorb atmospheric carbon dioxide. Between 2007 and 2016 the global carbon sink has removed around 33% of total anthropogenic emissions from industrial activity and land-use change. This sink constitutes a valuable ecosystem service, which has significantly slowed the rate of climate change. See Keenan and Williams (2018) *Annual Rev Climate & Resources* 43.

carceral geography Geographical perspectives on the spaces, practices, and experiences of incarceration (imprisonment). See D. Moran (2013) *Geografiska Annaler B* 94, 4.

care The provision of practical or emotional support. **Geographies of care** cover the topic from the global to the highly place-specific, examining the linkages between health and care-giving, the places in which 'caring' occurs, and those charged with its delivery. In many advanced capitalist countries, care-giving has shifted from large institutional settings to care in the community and at home, which affects who is responsible for that care. See Milligan et al. (2007) *New Zealand Geographer* 63, 2, and Morrison (2021) *Social & Cultural Geog.* 10. A **landscape of care** describes places beneficial to one's physical or mental health, such as clinics, gardens, hospitals, holy wells, temples, retreats, and spas.

carrying capacity The maximum potential number of inhabitants which can be supported in a given area. The upper limit is set at the point where the environment deteriorates. Carrying capacities are far from being universal constants, but alter with value judgements and objectives, and carrying capacities in the shorter term may differ greatly from those in the longer term. See Daily and Ehrlich (1996) *Ecol. Applics* 6, 4 on increasing Earth's carrying capacity.

Seidl and Tisdell (1999) *Ecol. Econ.* 31, 3 maintain that carrying capacity is a political concept 'generally highlighting that exponential growth, and thus environmental pressures, have to be curbed'.

cartel A system whereby producers divide up the market between themselves, avoiding direct competition and not encroaching on each other's share of the market.

cartography The production and study of maps and charts. Cartography is a system of information which is used to communicate something of the real world to other people; the map is a *model, to be decoded by the map reader. All maps are approximations; their clean lines and colours don't reflect the muddled nature of the reality they represent, and they can easily be used to support a point of view. So that the reader is not distracted by 'noise'— anything which stands in the way of understanding—the map has to be encoded using easily understandable signs, symbols, lettering, and lines. Perkins (2004) *Prog. Human Geog.* 28, 3 rewards careful reading, telling you as much about geography as about maps.

cascade 1. A sequence of small waterfalls.
2. A configuration in which the output of one component is the input of another. Thus, the catchment hydrological cascade is: bedrock, cascade, step-pool, plane-bed, riffle-pool, and dune-ripple. See Couthard and Van Der Wiel (2007) *Geomorph.* 91, 3–4: 'River systems can be viewed at distinctive hierarchical levels that represent a cascade of geomorphological interrelationships'.

(⊕) SEE WEB LINKS
• AUSRIVAS, Parsons et al. (2002).

cascading systems *See* SYSTEMS.

case hardening The production on a rock outcrop of a resistant *weathering rind mostly made of iron and magnesium hydroxides; sometimes of amorphous silica. Case hardening in limestone occurs when the calcium carbonate dissolved at the surface is re-precipitated internally, most commonly close to the surface from which it was just dissolved. Case hardening may be related to the production of visors over some *tafoni entrances.

cash cropping Growing a crop to produce goods for sale or for barter, rather than for the subsistence of the farmer and family. Lim and Douglas (1998) *Asia Pac. Viewpt* 39, 3 describe the results of pressure from the Malaysian government to switch from shifting cultivation to cash cropping: land shortages, increased use of agricultural chemicals, and the failure of sustainable agriculture.

caste A position in society inherited at birth, and from which there is no transfer throughout life. The system is at its strongest in the Indian subcontinent; in India's biggest cities, scheduled castes and scheduled tribes live an extremely segregated life. See, especially, Mosse (2018) *World Development* on caste and development. See also Müller-Böker (1988) *Mt. Res. & Dev.* 11, 2 on caste and spatial organization in Nepal.

catastrophic In geomorphology, any process which is rare, and acts briefly with great force and suddenness. See, for example, Clayton and Knox (2008) *Geomorph.* 93, 3–4 on the catastrophic drainage of Lake Wisconsin.
Catastrophism is a view that sees major changes in the physical environment as the result of catastrophic events. Although this view once fell out of favour, recently it has become accepted as valuable: for example, there have been many explanations of past mass extinctions invoking catastrophic events, and many geomorphological features can be explained by recourse to catastrophic ideas. See, for example,

Bintliff (2002) *World Archaeol.* 33 on catastrophism in the study of Mediterranean alluvial history.

catchment The area drained by a river. **Catchment shape** can be expressed by its circularity ratio:

$$Rc = \frac{A}{Ac}$$

where Rc is the circularity ratio, A is the area of the catchment, and Ac is the area of a circle with the same circumference as the catchment. Catchment shape is a major influence on hydrological relationships in a landscape:

$$L = 1.4\ A^{0.6}$$

where L = the length of the stream from its highest topographical point to the river mouth, and A = the area of the catchment. See K. Fryirs and G. Brierley (2012), for more detail. Sólyom (2011 *ESPL* 36, 4, 533) finds an intimate relationship between catchment shape and channel initiation. The initiation of a channel on an initially unchannelized surface sets up the initial flow-path structure, which consequently defines the boundaries of the channel's first-order catchment.

Catchment Abstraction Management Strategies (CAMS) A strategy by the *Environment Agency to license water abstraction in the UK.

(⊕) SEE WEB LINKS
• Sets out CAMS policy.

catchment area 1. The area served by a city, or some function such as retailers, airports, schools, etc. *See* URBAN FIELD.
   2. An alternative term for a *drainage basin/watershed.

categorical data analysis A range of techniques that may be used to analyse *nominal and *ordinal data. The key text is N. Wrigley (2002).

catena A sequence of soil types arising from the same parent rock, but distinct from each other because of the variations—such as *leaching and *mass movement—arising from topography. De Alba et al. (2004) *Catena* 58, 1 present a model of catena evolution.

cation An ion or group of ions having a positive charge such as the sodium ion in salt (Na⁺Cl⁻). **Cation exchange** is the interchange between a positively charged cation in the soil water and a negatively charged ion on a clay surface.

The most important ions held in this way are the four important plant nutrients sodium, potassium, calcium, and magnesium. **Cation exchange capacity** (CEC) is the overall net negative charge of clay minerals per unit mass of soil, usually expressed as milliequivalents (meq) per kg of oven-dried soil.

causation (causality) The relationship between two phenomena in which the first (the cause) brings about the second (the effect). A **causal relationship** usually requires: the existence of a spatial and temporal contiguity between two events, the occurrence of one event before the other, and the unlikelihood

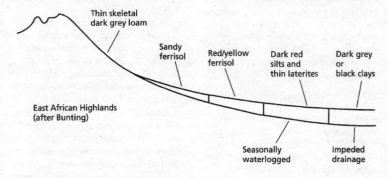

Thin skeletal dark grey loam

Sandy ferrisol

Red/yellow ferrisol

Dark red silts and thin laterites

Dark grey or black clays

East African Highlands (after Bunting)

Seasonally waterlogged

Impeded drainage

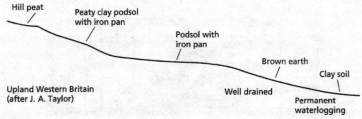

Hill peat

Peaty clay podsol with iron pan

Podsol with iron pan

Brown earth

Clay soil

Upland Western Britain (after J. A. Taylor)

Well drained

Permanent waterlogging

Catena

that the second event could have occurred unless the previous event occurred first. A causal relationship is also often thought to exist when a particular type of event always or almost always occurs in a particular way.

A **causal correlation** between two variables implies that one causes the other, but a correlation between two variables is not necessarily causal.

cave A large, natural, underground hollow, usually with a horizontal opening. **Karst caves** result from solution and *corrasion; see Miller (2006) *GSA Special Paper* 404.

cavitation The caving-in of bubbles in a liquid, close to a solid surface, causing shock waves. In a river, cavitation usually occurs downstream of an obstruction. It may aid the fluting and potholing of massive, unjointed rocks (Whipple et al. (2000) *GSA Bull.* 112, 3).

CBD *See* CENTRAL BUSINESS DISTRICT.

celerity In a river, the square root of the product of the acceleration due to gravity and the mean depth of the river flow. See Richardson and Carling (2006) *Geomorph.* 82, 1–2.

cellular automata An *agent-based model that works on a two-dimensional lattice grid, where the state of any cell is related to that of its neighbours. At its simplest, a grid of elementary cellular automata consists of a one-dimensional row of cells, where each cell can be in one of two states, and the rules for the transformation of a cell are based on the current state of the cell and its two closest neighbours. The neighbourhood thus consists of three cells [6, 7, 8].

In this example, the two states are called 'white' and 'black', and the rules might be that:

- If all three cells are white, the cell remains white
- If all three cells are black, the cell becomes white
- In any other case (that is, if there is a mixture of black and white cells in the neighbourhood), the cell becomes (or remains) black.

The diagram shows four successive generations of cellular automata conforming to these rules: an initial generation consisting of one black cell with all other cells white, and three subsequent generations created by three applications of the rules.

- *Parallelism*: The cells all change state independently and simultaneously (that is, in parallel)
- *Locality*: The new state of a cell is dependent on only the state of the cell itself and the states of its neighbours
- *Homogeneity*: All cells have an identical set of possible states and follow the same rules.

cellular clouds Roughly hexagonal patches of clouds. The open cell patterns

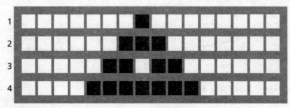

Cellular automata

typically found behind cold fronts can be determined by a large-scale sinking motion of a convectively unstable layer.

cellular convection *Convection occurring in semi-regular cells.

cellular modelling Cellular models in geomorphology can be defined as representing the modelled landscape with a grid of cells, over which the development of the landscape is determined by the interactions between cells (for example, fluxes of water and sediment) using rules based on simplifications of the governing physics. In fluvial geomorphology, cellular models use simplified or 'relaxed' versions of complex flow equations. See Coulthard et al. (2007) *Geomorph.* 90, 3–4; see also Nicholas (2005) *ESPL* 30, 5.

cementation The binding together of particles by adhesive materials, as in sedimentary rocks.

Cenozoic (Cainozoic) The most recent *era of Earth's history stretching approximately from 65 million years BP to the present day.

(()) SEE WEB LINKS
• An overview of the Cenozoic period.

census An investigation, usually into the size and nature of a population, but occasionally into other things, such as traffic. A **census tract**, known in the UK as an *enumeration district, is a small unit of area used in collecting, recording, and reporting census data. **Census geography** is the spatial organization of conducting a census and the geographical analysis and thematic mapping of census data (N. Castree, R. Kitchin, and A. Rogers 2013). In the UK, **Census Output Areas** are geographical areas consisting of about 125 households, with a population of about 300.

Census geography design is the delineation of appropriate geographical base units from which to create and use geography; see Vickers and Rees (2007) *JR Statist. Soc. A* 170, 2 and Rees and Martin in P. Rees et al., eds (2002). The UK Office of National Statistics (ONS) offers a range of 2011 Census geography products for England and Wales.

(()) SEE WEB LINKS
• For more detail on census output areas.

Center for Spatially Integrated Social Science (CSISS) A centre focusing on spatial analysis as an integrating force for the social sciences. Their website will lead you to their really useful *GIS cookbook: 'a collection of simple descriptions and illustrations of GIS methods, written with minimal GIS jargon'.

(()) SEE WEB LINKS
• The CSISS website.

central business district (CBD) The heart of an urban area, usually located at the meeting point of the city's transport systems, containing a high percentage of shops and offices. High accessibility leads to high land values, and therefore intensive land use. Consequently, development is often upwards. Within the CBD, specialist areas, such as a jewellery quarter, benefit from *external economies. Vertical land-use zoning is also common, so that retail outlets may be on the ground floor, with commercial users above them and residential users higher up. The CBD is under threat from traffic restrictions, out-of-town developments such as superstores, the Covid pandemic, and, possibly most of all, buying online.

Methods of delimiting the CBD include the *central business height index, recording the percentage of floor space given over to CBD functions, charting high level pedestrian flows, and surveying pavement chewing gum.

central business height index

(CBHI) A measure of the intensity of land use within the *CBD. The total floor area of all storeys of a building is compared with the ground-floor area of the building under consideration. The higher the index, the more intensive the urban land use. The index can also be used to delimit the CBD.

central flow theory An interlocking network model developed at GaWC. In central flow theory, flows generate a spatial network of urban settlements. City networks are essentially cities interlocked by agents of commerce. This suggests that urban policy-makers should focus on cooperation with other cities.

centrality The quality of being central, usually through *connectivity. Cities display the power of centrality economically, politically, and culturally as centres of a tributary region; and as centres or gateways for other distant regions. Lefebvre identifies 'urban' as a sociospatial form of centrality, but, as Kipfer et al. (2013), *PHG* 37, 1, 115 note, 'this is a tricky affair'.

centrally planned economy A synonym for a communist economy, also known as a **command economy**. The *factors of production are owned by the state, and their deployment is planned from the administrative capital. True central planning requires an enormous quantity of information, difficult to gather at the centre.

central place A settlement or nodal point which, by its functions, serves an area round about it for goods and services. **Central place theory** is an explanation, advanced by W. Christaller (1933, trans. 1966), of the spatial

arrangement, size, and number of settlements. Christaller modelled settlement locations in a hexagonal pattern to examine and define settlement structure and the size of the hinterland. His initial assumptions were of an *isotropic plain proportional to distance; and *economic man. Economic factors include demand density, economies of scope, localization economies, and agglomeration economies.

central vent eruption A volcanic eruption from a single vent or a cluster of centrally placed vents, fed by a single pipe-like supply channel. Lava from a central vent eruption accumulates to form a conical volcano; see Takada (1997) *Bull. Volcanol.* 58, 7.

centre of calculation A venue, usually an academic institution, in which knowledge is built up through repeated movements to and from other locations. Centres of calculation have been assisted by *capitalism and *imperialism. See Jöns (2011) in J. Agnew and D. Livingstone, eds.

centrifugal forces In *human geography, those forces which encourage a movement of people, business, and industry *away from* central urban areas; these include: intra-city location costs, improvements in transport, and systems the needs of clients dispersed over a wide area. Koch (2007) *UN Univ. Res. Paper* 2007/45, lists also the dispersion of demand, reputation (a new organization might take risks that, if successful, would raise its profile), competition effects (where organizations locate away from direct competitors), and high transport costs (when a business might locate near the ultimate users).

centripetal acceleration In meteorology, the force acting into the centre of a high- or low-pressure system, which makes winds blow along a curved path. It is crucial in *tropical cyclone and *tornado formation; see Wurman and Gill (2000) *Monthly Weather Rev.* 128, 7.

centripetal drainage *See* DRAINAGE PATTERNS.

centripetal forces The forces of concentration which move people, business, and industry towards a centre. These include: *accessibility, *functional linkages, *agglomeration economies, labour market externalities, intermediate services, technological and informational spillovers, and *external economies. Grote (2008) *J. Econ. Geog.* 8, 2 holds that, with ever-falling transaction costs, centripetal forces decline over time.

chain migration A process which depends on a small number of pioneer migrants, who settle in a new place. A feedback process is set up, from the destination to the origin, generating higher than expected levels of migration to these pioneers' neighbourhoods. A metropolis may then become both segmented and segregated.

channel A watercourse. The nature of any river channel is a result of the interaction of: the material properties of the bed and banks; the flow hydraulics; and sediment transport within the river. Any alteration in any one of them affects the other two. Narrow, stable channel stretches are linked to finer bed material and high levels of river vegetation; wider, unstable channels are associated with larger bed material and less riparian vegetation.

The Smith and Bretherton (1972, *Water Resources Res.* 8, 1506) explanation of **channel initiation** is based on the continuity equation for sediment transport. Its central result is a criterion for an infinitesimal

perturbation on the surface to begin to grow.

For more mature landscapes, channel extent is based on competition between diffusion and runoff processes: a channel is initiated once the magnitude of incising runoff processes exceeds the magnitude of infilling diffusion processes, such as creep, rainsplash, and shallow landsliding.

The **channel capacity** of a river is its cross-sectional area in square metres, but the borders of a channel are not easy to establish. **Channel resistance** slows or impedes the flow; friction with the bed is the major cause.

channel classification Howard et al. (1994) *J. Geophys. Res.* 99, primarily on the basis of bed morphology, define five channel types: live bed sand alluvial; live bed gravel alluvial; threshold gravel alluvial; mixed bedrock-alluvial; and bedrock. In a study of mountain drainage basins, Montgomery and Buffington (1997) *GSA Bull.* 109, 5 distinguish seven types: colluvial and bedrock, plus five alluvial channel types (cascade, step-pool, plane-bed, pool-riffle, and dune-ripple). Heritage et al. (2001) *J. Geol.* 109 use cluster and discriminant analyses to classify: bedrock anastomosed, mixed anastomosed, pool-rapid, braided, and alluvial single-thread.

channel flow Run-off within the confines (banks) of a channel. **Compound channel flow** occurs in *meandering channels, where the magnitude and directions of the main channel and the upper layer differ, causing sizeable internal *shear stress, and a consequent loss of energy. For meandering compound channels the important parameters affecting the boundary shear distribution are sinuosity (Sr), amplitude (a), relative depth (β), and the width ratio (α) and the aspect ratio (δ); see Khatua and Patra (2007) *Procs 5th Australian Stream Manag. Conf.* Compound channel flow also occurs during floods, when velocity differs between the main channel and the flooded area.

channelized flow At a glacier base, channels governed by a balance between viscous closure of the ice roof and melting due to turbulent eating. Such channels naturally tend to draw in surrounding water and therefore form branching arterial networks, aligned predominantly in the down-glacier direction. Two types of channels are thought to exist beneath glaciers: Nye (N) channels, which are incised into the bedrock, and Röthlisberger (R) channels, which are melted upwards into the glacier ice; see J. Holden (2012), pp. 475–6.

channel management Any proposals for changing a fluvial channel should: establish past and present morphology, stability, and responsiveness to change; determine the impact of the proposed change on the sedimentary regime; estimate the effects of change in the sedimentary regime on channel morphology, and on bank pore water pressure, and therefore on bank stability. Social and environmental conflicts of interest also arise: damming a river will strip out the finer silt, and increase the downstream concentration of coarser silt, leading to increased erosion, for example. Larsen et al. (2007) *Landscape & Urb. Plan.* 79, 3–4 demonstrate the need for taking a landscape-level, long-term view of managing a river channel.

channel order *See* STREAM ORDER.

channel roughness *See* BED ROUGHNESS.

chaos theory Chaos theory deals with simple, deterministic, non-linear, dynamic, closed systems that are extremely sensitive to initial conditions, so that the response to any tiny initial difference or disturbance will be chaotic and unpredictable. Chaotic behaviour is *not* random—current conditions are still linked to future conditions—and chaotic systems do not repeat themselves exactly, but may behave in a loosely recurrent fashion. See Phillips (2003) *PPG* 27, 1.

character displacement The accentuation of differences among similar species whose distributions overlap (or their minimization in regions where they do not). It results from natural selection; intensifying the aesthetic traits useful for species' discrimination, or for using different parts of a *niche, thus avoiding direct competition. The classic example is of finches on the Galapagos Islands. See Montoya and Burns, Russo et al., and Meiri et al., all in (2007) *J. Biogeog.* 34, 12.

charter group A group representing the most typical culture of a host community. The charter group creates a system that segregates it from other 'ethno-groups', who may be gradually assimilated, but in such a way that the charter group maintains its dominance. See Johnston et al. (2007) *AAAG* 97, 4.

chatter mark *See* PERIGLACIAL.

check dam A small, often temporary, dam constructed across a watercourse to reduce flow velocity and/or counter soil erosion. See Mustafa (2005) *AAAG* 95, 3 and Castillo et al. (2007) *Catena* 70, 3 on the effectiveness of check dams for soil erosion control.

chelate A chemical substance formed from the bonding of compounds to metallic ions. Intense podzolization in humic-iron podzols is linked to mineral chelation. The *leaching out of chelates is **cheluviation**. **Chelation**, the formation of chelates, is a process whereby relatively insoluble materials may become soluble, and be released into the soil, increasing nutrient availability to plants.

Chelford interstadial An interstadial during the Devensian glaciation, between 65000 and 60000 years BP. *See* PLEISTOCENE.

chemical weathering The breaking down *in situ* of rocks by *carbonation, *hydrolysis, *oxidation, *solution, and attack by organic acids. Hydrology could be the main factor controlling chemical weathering; see Bouchard and Jolicoeur (2000) *Geomorph.* 32, 3–4. See also Muhs et al. (2001) *Soil Sci. Soc. Amer. J.* 65 on the impact of climate and parent material on chemical weathering.

chernozem A *zonal soil (US soil classification mollisol, sub-order boroll) with a deep A *horizon, rich in *humus from decomposed grass, and dark in colour. The B horizon is lighter brown, but often absent. The lower horizons are often rich in calcium compounds.

    Chernozem development has been associated with temperate continental climates with marked wet and dry seasons, providing enough moisture to permit the decay of the grass litter into humus, but not enough for *leaching to be significant.

chestnut soil A *zonal soil (US soil classification mollisol, sub-order xeroll), found in more arid grasslands. The *xerophytic nature of much of the grassland under which chestnut soils develop retards the development of humus, and there is an accumulation of calcium carbonate in the B *horizon.

chevron A V-shaped, triangular, erosional microform, characteristically

developed on the shallow flanks of *cuestas in arid zones.

children, geographies of The study of the socially and spatially uneven impacts of economic, political, cultural, and social processes on children, and the ways in which these are experienced and responded to by individuals and groups. Matthews and Limb (1999) *PHG* 23, 61 emphasize the experiences of children and how they see the world around them.

choice modelling A group of statistical techniques to model the way people choose between alternatives, for example, where to retire. Duncombe et al. (2001) *Int. J. Pop. Geog.* 7 offer a mathematical but clear exposition of the method, together with some software options.

    (((●))) **SEE WEB LINKS**
- A full explanation of choice modelling on pdf at Rand Europe's website.

chorology The study of the variation from place to place on the surface of the Earth; as such it is the oldest form of geography.

choropleth A map showing the distribution of a phenomenon by graded shading to indicate the density per unit area of that phenomenon; the greater the density of shading, the greater the density in reality. Choropleths give a clear, but generalized, picture of distribution, but may mask finer details. The choice of values for the classes, and the areal units for which data are available, will affect the visual picture; *see* SCALE. See Goldsberry et al. (2004) *Procs Annual Meeting NACIS* on classification choice and animated choropleths.

chronosequence A group of soils whose properties vary primarily as a function of age variability. See Sauer et al. (2007) *Catena* 71, 3.

chronotope At its most simple, the chronotope is the 'setting' of a given narrative in time and space. Thus, the genre 'Western' in a movie is not solely a movie set in the 'Wild West', but also set in the late nineteenth century, with certain typical casts and locations, a plot following certain conventions, and a certain set of morals and mores: see Lawson (2011) *Antipode* 43, 2, 384.

c.i.f. pricing (cost, insurance, freight pricing) A form of uniform delivered pricing where prices are quoted with reference to a port.

cinder cones Also known as scoria cones, these are composite volcanoes, generally less than 300 m high, with straight, steep sides and very large summit craters. They are made of *tephra shot through with bubbles that were originally filled with volcanic gases. See Cascadden et al. (1997) *New Mexico Bureau of Mines, Bull.* 156.

circuit of capital Money *capital is needed to pay for *factors of production, such as raw materials, plant, energy, and labour. Commodities may then be produced which are more valuable than the cost of all the inputs; surplus value has thus been added. The commodities are sold (turned back into money) so that surplus value leads to reinvestment. *Accumulation occurs with each circuit of capital. D. Harvey (1996) discusses the different temporalities (time-scales) of short-term financial, medium-term industrial, and long-term infrastructural capital, and the pressures and distortions that these can create within the overall circuit of capital.

circulation 1. The movement of capital, labour, goods, and services through an economy.

2. In population geography, short-term, repetitive movements of individuals, where there is no intention to change residence permanently. Adepojou in P. Kok (2006) observes that, in southern Africa, poverty has drawn more people into circulation or temporary migration, and notes that 'brain drain' is being transformed into 'brain circulation'.

circumpolar vortex A vigorous, westerly flow of air in the upper and middle *troposphere over mid-latitudes. From summer to autumn, circumpolar flow in the Southern Hemisphere has contributed substantially to the observed warming over the Antarctic Peninsula and Patagonia, and to the cooling over eastern Antarctica and the Antarctic plateau. Frauenfeld and Davis (2003) *J. Geophys. Res.* 108 think the Northern Hemisphere circumpolar vortex could be a good indicator of climate change.

cirque Also known as a **corrie** or **cwm**, this is a circular, armchair-shaped, hollow cut into bedrock during glaciation, up to 2 km across. The steep side and back walls are subject to intense physical weathering; the front opens out downslope.

The formation of cirques remains unclear. Bedrock jointing determines the overall cirque morphology, and periglacial freeze–thaw weathering and glacial quarrying facilitate erosion of the headwall. Cirques can be initiated by large, deep-seated failures of rock slope, and fluvial erosion and large-scale slope failures can be important for the creation of hollows, which are then deepened during periods of glaciation. At these periods, ice is thought to undergo *rotational slipping, perhaps *overdeepening the cirque. Hooke (1991, *GSA Bull.* 103) suggests that glaciohydraulic supercooling is central to overdeepening. Cirques seem to grow by headward extension, biting back into the mountain mass until only *arêtes or *pyramidal peaks remain. See Oskin and Burbank (2005) *Geology* 33, 12 on cirque retreat.

Aspect plays an important role in the orientation of cirques: in the Fuegian Andes, for example, the cirques which are oriented towards the south-east are the most abundant. Glacial ice is preserved in these due to the limited exposure to sunlight, or the effects of humid air masses. Evans (2006) *Geomorph.* 80, 3–4 considers the effects of geology, relief, and region on the *allometric development of cirques in Wales.

citizenship The membership of a sovereign nation, and the enjoyment of the rights and norms prevailing in that country. Clearly, then, citizenship is geographical, in that it is attached to a specific location. The liberal/ individualist approach to citizenship stresses the importance of individual rights and the communitarian approach emphasizes the duties of a citizen towards fellow members of the community they belong to. Radical pluralists view the community critically; hence, they uphold policies that maintain differences. For the geographer, key concerns are: how notions of citizenship vary over time and space; why such variations occur; how the rights (or otherwise) and actions of the citizen structure the nation; and how the nation, in turn, structures the citizen. So citizenship is not solely a status; it is also a set of relationships. See Desforges et al. (2005) *Citizenship Studs* 9, 5 for a wide-ranging review of geographies of citizenship. *See* RIGHT TO THE CITY.

city A large urban centre, functioning as a central place, and providing specialized goods and services. There is no worldwide, or even European,

agreement over limiting figures of population size or areal extent for a city. Geographies of the city include: the politics of public space and how the city is experienced by different urban dwellers. Glaeser et al. (2001) *J. Econ. Geog.* 1, 1 argue that the critical urban amenities are: a rich variety of services and consumer goods; aesthetics and physical setting; weather; and ease of movement within the city. See Bell (2007) *PHG* 31, 1 on the **hospitable city**, and Hannigan (2007) *Geog. Compass* 1, 4 on **casino cities**, which 'elevate consumption and spectatorship at the expense of economic and social equality'.

A city may drive capital accumulation or it may be a creative field: R. Florida (2005) argues that urban fortunes increasingly turn on the capacity to 'attract, retain and even pamper a mobile and finicky class of "creatives"'. A city may be a multilingual and multi-ethnic community expressed through its infrastructure, public spaces, and festivals.

Painter (2005, *ICRRDS Background Paper*) sees access to public space as 'fundamental to the *right to the city'; the core elements of which are 'the promotion of equal access to the potential benefits of the city for all urban dwellers, democratic participation of all inhabitants in decision-making processes and [the] realization of inhabitants' fundamental rights and liberties'.

city marketing The promotion of a city, which aims to attract certain activities. City marketing starts with a thorough analysis of the city's current situation through extensive research on the city's assets, opportunities, and audiences. Then comes identifying and choosing a vision for the city that could be achieved with as broad a range of stakeholders as possible. After which, specific projects are identified, allocating clear roles for the participating bodies. Finally comes the phase of active implementation of city marketing measures: spatial/functional, financial, organizational, and promotional. The process ends with monitoring and evaluation of the results of all activities.

city network The connections a city has with other places, whether by canals, railways, air flights, or telecommunication networks. The network may then be analysed using graph theory. City networks also exist as cultural (including religious) links, whether of global finances, trade, markets, migration, or shared social spaces and/or histories. The city is the node where the different networks meet. See Zachary P. Neal (2012).

city of refuge San Francisco's City of Refuge ordinance prevents local police from collecting information about a person's immigration status if they have not been charged with any crime; see Ridgley (2007) *Urb. Geog.* 29, 1. Zenodo's *City of Refuge Toolkit* (2019) aims to explore the needs these actors have, such as the resources they need, and the obstacles they face when they try to build inclusive, safe, and equitable local and national communities and spaces.

city region The area around a city which serves, and is served by, the city (also called *umland or urban field).

city-size distribution The frequency with which the settlements of a country or region occur in certain, arbitrarily defined, population size groups. Generally, large settlements are less frequent than small ones, but when size is graphed against frequency, several patterns may be recognized. Iyer (2003) *Eurasian Geog. & Econ.* 44, 5 documents the change from a relatively uniform city-size distribution 'in accordance to

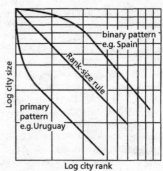

City-size distribution

Marxist ideals' to increasing evenness in post-Soviet Russia. See Overman and Ioannides (2001) *J. Urb. Econ.* 49, 3.

city/state relations J. Jacobs (1992) identifies two opposing 'moral syndromes' which give rise to different social spaces, spaces of flows, and spaces of places: cities are seen as commercial, honest, cooperative, enterprising, and conscientious; states are seen as guardians, supporting loyalty, tradition, ostentation, exclusion, and, when necessary, vengeance. Taylor (2007) *TIBG* 32, 2 concludes that the **city-state** is not appropriate in a globalizing world, since the guardian functions of the city-state impinge too greatly on economic expansion.

civil society A much-debated term to describe a mix of value-driven organizations—voluntary and community organizations, charities, social enterprises, cooperatives and mutual organizations—which aim to promote social, environmental, or cultural objectives; collective social interaction situated between the state, market, and household.

clade In biogeography, a grouping that includes a common ancestor and all the descendants (living and extinct) of that ancestor.

class Within a society, a set of people who are of the same economic position, and who may share the same tastes and social status. In capitalist societies, class is defined by socio-economic status, but *post-Fordism has created a new division: specialist, skilled workers, in highly paid positions with a large degree of self-determination, and low-paid workers with few skills, working in poor conditions with little or no security. This pattern is common in service employment, but is exacerbated among women.

The spatial mobility of middle-class lifestyles is taken for granted while the lives of those struggling to participate in British society are precarious. Working classness is seen as a simultaneously economic and cultural category. It rests not only on the material and labour market position of individuals, households, and communities, but also on symbolic values and cultural practices. See Stenning (2008) *Antipode* 40, 1.

classification It is important to remember that 'classification is not just a neutral activity' (R. Inkpen, 2005). The classifier looks for groups of like phenomena/objects that can be treated as a single unit for the purpose of making valid generalizations about aspects. Clearly, many different classifications are possible, depending on the purpose of the study being undertaken: streams, for example, form a single class, but they can be subdivided on a variety of criteria, such as order, depth, gradient, and so on. The purpose of classification procedures is to provide a grouping which is valid for the scientific activity being undertaken. See R. J. Johnston (1976).

Geography uses a range of classificatory systems: **intrinsic**
classification depends on natural differences or 'breaks' in the features studied; **extrinsic classification** uses arbitrarily defined class limits; **monothetic classification** uses one criterion; **polythetic classification** uses a number of criteria; **attribute-based classification** is based on 'present' or 'absent' evidence (a climate may or may not have a dry season); **variable-based classification** forms classes on the basis of the degree to which a variable is present.

class-monopoly rent The increases in rents that landlords, as a class, can charge when they have a monopoly over the supply of land. See Anderson (2019) *Antipode* 51, 4.

clast A rock fragment, eroded from a rock mass, and deposited in a new location. See Bourke et al. (2007) *Lunar & Plan. Sci. Conf.* 38 on fluvial clasts; and Clark (1995) *Geology* 19, 2 on **clast pavements**.

clay Mineral particles < 0.002 mm. When dry, clay is hard; when wet it swells and becomes pliable and sticky. **Clay colloids** are finely divided clays, dispersed in water, with a negative surface charge that attracts positively charged ions; they are among the most reactive constituents of a soil (W. van Olphen, 1991). **Clay micelles** are individual particles, platey in form, < 2 μm in diameter, having a negative charge, and therefore able to attract cations within a soil; try Murrmann and Koutz in S. Reed, ed. (1972). The **clay-humus complex** is a mixture of clay particles and decaying organic material that attracts and holds the *cations of soluble salts within the soil profile.

clay minerals A group of hydrous aluminium silicates, such as kaolinite, created by intense rock weathering. Clay minerals affect the structure of soils, as they expand when wet.

clearinghouse In *GIS, a network of the managers, producers, and users of data. The US National Geospatial Data Clearinghouse is a means of publishing descriptions of geospatial data from a variety of governmental and non-governmental sources, helping users determine their fitness for use, and providing tools to access, visualize, or order data as efficiently as possible. It is neither a central repository where data sets are stored nor a set of websites, but a federated system of compatible geospatial data catalogues that can be searched through a common interface—the geodata.gov portal.

(⊕) SEE WEB LINKS

• The Geospatial Platform.

cleavage 1. A division in society due to political or partisan allegiance.

2. The ability of a rock to split along a **cleavage plane**.

cliff A steep rock face, usually facing the sea; see Lee et al. (2001) *Geomorph.* 40, 3–4 on probabilistic models for different cliffs. While an **active cliff** is still subject to the forces of marine erosion, an **abandoned cliff** is protected from wave attack by a *wave-cut platform or by a *barrier beach. As the abandoned cliff is exposed to *subaerial denudation it becomes less steep, and its upper edge more indented. See Hutchinson (1984) *DTIC* ADA144040.

climate A summary of mean weather conditions over a time period, usually based on thirty years of records. Climates are largely determined by location with respect to land- and sea-masses, to large-scale patterns in the *general circulation of the atmosphere, *latitude, altitude, and to local geographical features.

climate sensitivity The relationship between the measure of climate *forcing

and the magnitude of the climate change.

(⊕) SEE WEB LINKS

• The Carbon Brief Explainer is a clear resource that shows how scientists estimate climate sensitivity.

climatic change During the last 55 million years the Earth has been cooling; during the last million years there have been alternating *glacials and *interglacials. *See* LITTLE ICE AGE.

External causes of climatic change include: changes in solar output (Haigh (2001) *Science* 294, 5549); changes in the number of sunspots, which seem to have an eleven-year cycle (Foukal et al. (2004) *Science* 306, 5693); changes in the ellipticity of the Earth's orbit, which follow a 100 000-year cycle (Goldsmith (March 2007) *Natural History*); and changes in the Earth's axis of rotation, which alters the season of perihelion, and which follow roughly a 100 000-year cycle (Zachos et al. (2001) *Science* 292, 5517). *See* MILANKOVITCH CYCLES.

Internal causes include changes in the distribution of land and sea, *continental drift, and changes in the atmosphere–surface–ocean system. *See also* GREENHOUSE EFFECT.

climatic climax community *See* CLIMAX COMMUNITY.

climatic geomorphology (clima-tomorphology) The association of landform types with different climates, such as the evolution of *periglacial landforms in *tundra climates. However, the geomorphic response to climatic changes, fluctuations, and episodes varies spatially and temporally. Some recent landforms result from the long-term and essentially continuous activity of particular processes. Others are surface manifestations of processes that hitherto were active but cryptic and yet others result from the impact of storm or catastrophic events, frequently, though not in every instance, on surfaces previously rendered vulnerable.

climatology The study of the origins and impacts of climates. For cultural climatology, *see* CULTURAL TURN.

climax community An *ecosystem which experiences a turnover of species, but with an overall steady state. This steady-state ecosystem is characterized by a complex, highly integrated community structure with high species diversity, but relatively low individual population densities not subject to serious fluctuation, and can be seen as the final stage of a *succession. An example is a mature oak woodland. 'Equilibrium or climax communities are rarely affected by intense natural disturbances and are characterised by a predominance of larger, long-lived and often deep-burrowing species. Such species can physically modify their surroundings.'

climograph A line graph of monthly average temperature against average humidity. The shape and location of line indicates the nature of climate. See R. G. Bailey (1998).

closed system A system marked by clear boundaries that prohibit the movement of energy across them. *Entropy in a closed system can never decrease.

closure The delimitation of, or setting boundaries to, an enquiry. Methods of closure commonly used by geographers include oversimplification and limiting an enquiry to a specified time period. 'The process of closing research must address the motives behind research topics and questions; [and] the reasons for preferring a particular methodology in addressing a given set of research questions, including the theoretical and practical issues raised by closure' (Brown (2004) *TIBG* 29, 3). The choice of

time–space model is also significant in closure; see Lane (2001) *TIBG* 26, 2.

cloud A visible, dense mass of water droplets and/or ice crystals, suspended in the air, and generally forming when air is forced to rise: at a *front, over mountains, or because of *convection. Clouds mirror atmospheric processes; the approach and passage of a warm front, for example, often follows the sequence: cirrus, cirro-stratus, alto-stratus, nimbo-stratus. At active *ana-fronts these clouds may take on a more cumulus form. Atmospheric *convection currents are generally indicated by the presence of cumulus or cumulo-nimbus clouds. A cumulus cloud will often form over a heated surface and then shift with the wind, so that further cumulus is formed over the same spot; *see* CLOUD STREET. *Turbulence is a common cause of stratus cloud, which is often trapped beneath an *inversion; it also creates a nearly continuous sheet of strato-cumulus cloud. See C. D. Ahrens (2000).

cloud classification Clouds may be classified by form and by height.
   1. Low-level clouds (0–2 km above sea level): stratus, S—extensive, shallow cloud sheet, often yielding *drizzle/light rain; strato-cumulus, Sc—shallow cloud sheet, in roughly recurring *cumuliform masses, often yielding drizzle/snow; cumulus, Cu—separate, hill-shaped clouds with flat, and often level, bases, often at the same height; cumulo-nimbus, Cb—large, high *cumulus, with tops often formed of ice crystals, dark bases, and often showery.
   2. Medium-level clouds (2–4 km above mean sea level): alto-cumulus, Ac—shallow cloud sheet broken into roughly regular, rounded clouds; alto-stratus, As—featureless, thin, translucent cloud sheet; nimbus, Ns—extensive, very dark cloud sheet, usually yielding precipitation.

3. High clouds (tropical regions 6–18 km high, temperate regions 5–14 km high, polar regions 3–8 km high): cirrus, Ci—separate, white, feather-like clouds; cirro-cumulus, Cc—shallow, more or less regular patches or ripples of cloud; cirro-stratus, Cs—shallow sheet of largely translucent cloud.

Suffixes may be added: *capillatus*—like a feather, or thread; *congest*—growing rapidly, in cauliflower form; *fractus*—broken or ragged; *humilis*—shallow; *lenticularis*—like a lens, especially of alto-cumulus, cirro-cumulus, and strato-cumulus; *radiates*—banded. See Li et al. (2007) *Remote Sens. Env.* 108 on cloud classification.

cloud droplet *See* PRECIPITATION.

cloud geographies The cloud is a bundle of experimental algorithmic techniques. Cloud geography is concerned with the identification and spatial location of data centres where the cloud is thought to materialize; see Amore (2018) *PHG* 42, 1.

cloud seeding Any technique of inducing rain from a cloud, usually by dropping on it crystals of dry ice (frozen CO_2), or silver nitrate, as *condensation nuclei.

cloud street Parallel lines of small cumulus: irregular over land, downwind from a sunny slope, and regular over warm seas. The cloud-free areas come from the mixing of saturated moist cloud air with overlying warmer, drier air: 'an individual cloud is capable of drying out a large area of the surrounding layer through cloud-top entrainment and thus is able to produce significant gaps between the clouds' (Chlond (1992) *Boundary-Layer Met.* 58, 1–2).

cluster A geographic concentration of: interconnected companies, specialized suppliers, service providers, associated institutions, and firms in related industries. The critical characteristics of clusters are: proximity—firms need to be sufficiently close in space to allow spillovers and the sharing of common resources; linkages; some level of active interaction; and critical mass—there need to be enough participants for the interactions to have a meaningful impact. Localization economies, specialization, and face-to-face contact also play a part. Places with higher levels of human capital are more innovative and grow more rapidly and robustly over time. Giuliani (2007) *J. Econ. Geog.* 7, 2 stresses the importance of embeddedness in local business networks. Clustering and globalization are closely intertwined; see M. Fujita et al. (2001).

Clusters begin as a group of firms with similar competencies, reducing expenditure on locational search costs, and enjoying *agglomeration economies. Further growth results from *cumulative causation, the presence of support institutions, and the power of the name of the locality as an attraction to a worker: the benefits of being physically located within, for example, Silicon Valley are perceived to be worth the costs of relocation. Huber (2012) *J. Econ. Geog.* 12, 1, 107 casts doubt on the benefits of clustering.

cluster analysis The assignment of a set of objects into groups so that the objects in the same cluster are more similar (in some sense or another) to each other than to those in other clusters. Cluster analysis is used when the researcher does not know the number of groups in advance but wishes to establish groups and then analyse group membership. For example, if the term 'geomorphological processes' is entered onto a search engine, there will be nearly 400 000 results. Careful study reveals that these could be clustered into (at least):

fluvial, aeolian, hillslope, glacial, tectonic, igneous, and biological processes.

coalescence theory *Bergeron–Findeisen's theory cannot explain the formation of all tropical rainfall, since ice crystals are often absent in tropical clouds. Langmuir's coalescence theory (1974) suggests that the small droplets in clouds grow larger by coalescence until they are heavy enough to fall. As they fall, they collide with other droplets, growing more. (Not every collision results in coalescence.) The deeper the cloud, the more the drops grow; up to about 5 mm in diameter. See Xue and Wang (2007) *J. Atmos. Scis* 31, 15. Urban geographers have identified coalescence as the growth of previously isolated clusters of settlement which grow together, clump, and form single entities; see Martellozzo and Clarke (2011) *Env. & Plan B*.

coastal classification Early coastal classification schemes were based on sea-level variations. Thus, submerged coasts are drowned river and glacial valleys (rias and fjords), while emerged coasts are characterized by coastal plains. A second type of classification distinguishes between primary coasts, which result mainly from non-marine processes and include drowned river valleys and deltas, and secondary coasts, mainly formed from marine processes and/or marine organisms, for example, barrier coasts, coral reefs, and mangrove coasts.

coastal dune A ridge or hill which forms when marine deposits of sand are blown to the back of the beach. 'Dune initiation is often site-specific and dependent on the mechanisms by which sand can be delivered from the shoreface to points where it will be reworked into available accommodation space' (Aagaard et al. (2007) *Geomorph.* 83, 1–2).

coastal flooding The inundation of coastal areas may be caused by *storm surges, *tsunamis, or linked to *El Niño. D. Sobien and C. Paxton (1998) SR/SSD 98-2 provide a primer on coastal flooding.

coastal geomorphology A major focus in coastal geomorphology is sediment transport by currents and waves, particularly at beach scale. Coastal geomorphology is currently concerned with the implied threats of climate change and sea-level rise, and by predicting morphodynamic behaviour at the 'engineering' scales. A major challenge is to create models revealing interactions between form and process.

cobble A stone, 60–200 mm, partly or wholly rounded by wave action or running water.

cockpit karst (kegelkarst) A landscape of star-shaped hollows surrounded by steep, rounded hills, found in tropical *karst. The cockpits, up to 100 m deep, usually containing a *streamsink, are the hollows (*dolines) formed by the solution of limestone, and now floored with alluvium. See Fleurant et al. (2006) *J. Geol. Soc.* special issue.

co-constitution The conjoint—joined together—development of two or more phenomena, such as the co-constitution between humans and non-humans. See Eriksson (2013) *Asian Geogr.* 30, 2 on the co-constitution of high technology and authoritarian politics in Indonesia.

coefficient of localization A measurement of the degree of concentration of a given phenomenon, such as industry, over a set of regions (also called the **index of concentration**). Values of the coefficient, L, lie between 0 (even distribution) and 1 (extreme concentration). Lafourcade and Mion

(2007) *Reg. Sci. & Urb. Econ.* 37, 1 use the coefficient to study the geographic distribution of manufacturing and the size of plants.

cognitive dissonance Dissonance is lack of agreement; between the beliefs one holds, and one's actions. It may be a mismatch between what is perceived and what is so that an individual may seem to act irrationally. **Cognitive dissonance theory** states that people experience conflict when deciding between alternatives, and try to reduce dissonance by increasing the supposed advantages of the alternative they chose, and downgrading the other. See Hobson (2006) *Area* 38, 3 on recycling, and Lundholm and Malmberg (2006) *Geografiska B* 88, 1.

cognitive geography Cognition is knowledge and knowing by sentient— responsive—entities (including humans, non-human animals, and artificially intelligent machines), and cognitive geography is the study of cognition, primarily human cognition, concerning space, place, and environment, using psychological theory to understand how people learn and remember spatial information and different environments. The comprehensive textbook is Montello ed. (2018) *Handbook of Behavioral and Cognitive Geography*, and the first pages of Butcher (2012) *PHG* 36, 1, 90 review the literature.

cognitive mapping The acquisition, coding, storage, manipulation, and recall of spatial information within the mind (also called mental mapping). A cognitive map helps break down complex research questions, establish priorities for follow-up research, and add clarity to abstract concepts. Cognitive mapping simplifies the complexity of the landscape, and a mental map influences behaviour.

See Sarah Gibbons (2019) Nielsen Norman Group.

cohesion Adhesion; the force holding materials together.

cohort A group of people who experience a significant event, such as birth or leaving school, during the same period of time. **Cohort analysis** traces the subsequent history of cohorts; see for example Frejka and Callow (2001) *Pop. & Dev. Rev.* 27, 1 on cohort reproductive patterns in low-fertility countries. A **cohort effect** is any effect associated with being a member of a group born at roughly the same time and bonded by common life experiences. The major problem of cohort analysis is to distinguish between the effects on the cohort of getting older (age effects), of common experiences like National Health orange juice (cohort effects), and particular historical events, like a war (period effects).

Cohort fertility is the total of live births born to a particular birth or marriage group.

col 1. In the landscape, a pass between two peaks or ridges. Landscape cols may have been formed by: the *headward erosion of a *cirque, *river capture, the beheading of *dip-slope valleys by *scarp retreat, or the localized *differential erosion of a ridge.

2. In meteorology, a narrow belt of relatively low pressure, but not a depression, between two *anticyclones. Here, isobars are few and therefore winds are slack.

cold glacier (cold-based glacier, polar glacier) A glacier with its base well below 0 °C, unlubricated by *meltwater, and therefore frozen to the bedrock. *Accumulation is slow; snow may take 150 years to turn to ice. Cold glaciers move very slowly; rates of 1–2 m per year are not uncommon, therefore they cause very little erosion through *abrasion, although *plucking may be reinforced. Where subglacial sediments are frozen to a cold glacier base, shear failure will occur at the bottom of the frozen zone, as high pore water pressures develop. See Glasser et al. (1998) *J. Glaciol.* 44 on Kongsvegen, Svalbard, and its role in landform genesis.

cold low (cold pool, polar low) An area of low pressure and temperature in the middle *troposphere. These lows don't appear on *synoptic charts, but are important in *Arctic and *Antarctic meteorology. **Cut-off low** systems are synoptic-scale low pressure systems, formed as a result of meridional shifts of jet streams in the upper troposphere, and playing a role in exchanges between troposphere and stratosphere.

collective farm A farming organization identified with socialist regimes. The farm is owned by the state, but leased to the members of the collective, and the workers are shareholders, not state employees. The collective is, in theory, self-governing, although the state may set production targets See Amelina (2000) *Post-Soviet Geog. & Econ.* 41, on why Russian peasants remain in collective farms.

collectivism A school of thought which maintains that the *factors of production and the means of distribution should be owned by all and not by individuals, who might pursue their self-interest at a cost to the state. See Vandello et al. (1999) *J. Personality & Social Psych.* 77 on patterns of individualism and collectivism across the United States.

collision margin The boundary between two continental *plates. These are hybrid margins since the two slabs of continental crust that eventually collide were initially separated by oceanic crust,

with passive or destructive margins; when the oceanic crust is consumed, the two continental crusts collide. The lower continental plate forms a double layer of crust together with the upper plate, as in the Himalayas; see D. S. N. Raju et al. (2005).

collision theory This states that raindrops grow by colliding and coalescing with each other, especially in tropical maritime *air masses. *See also* BERGERON–FINDEISEN THEORY and COALESCENCE THEORY.

colluvium The mixture of soil and unconsolidated rock fragments deposited on, or at the foot of, a slope. Colluvial boulder deposits are relatively stable colluvium, consisting of linear fields of well-varnished hillslope boulders, ubiquitous on desert hillslopes; see Liu and Broecker (2008) *Geomorph.* 93, 3–4.

colonial discourse theory European colonizers tended to construct the identities of colonized peoples and lands as *other: undeveloped, primitive, and immature; as homogeneous objects, rather than sources of knowledge; see Anand (2007) *New Polit. Sci.* 29, 1 on Western colonial representations of the (non-Western) other. The colonizer was represented as having a duty to the colonized which entailed both financial and emotional cost (and, for many, these costs were very real).

Colonial discourse analysis critically examines the role played by these *representations in colonialism and imperialism, 'rather than focusing on texts, systems of signification, and procedures of knowledge generation . . . a fuller understanding of colonial powers is achieved by explaining colonialism's basic geographical dispossessions of the colonized' (Harris (2004) *AAAG* 96, 1).

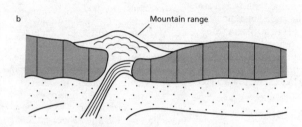

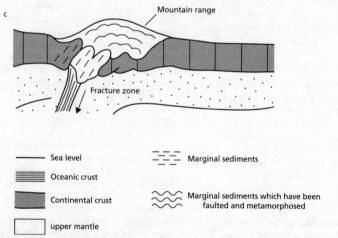

Collision margin (after J. F. Deaxy and J. M. Bird, 1970)

colonialism The acquisition and colonization, underpinned by a certain type of violence, by a nation of other territories and their peoples. This term took on a more specific meaning in the late 19th century when colonists saw it as the extension of 'civilization' from Europe to the 'inferior' peoples of 'backward' societies: white peoples were accorded the label of civilized in contrasting them from savage, non-white (read: non-European) peoples. Colonialism also entailed colonial philanthropy: 'a new doctrine of responsibility toward the unprivileged, underpinned by the Evangelical emphasis on the value of the human soul, and hence, of the individual.' Colonialism is also seen as a search for raw materials, new markets, and new fields of investment. Characteristics of colonialism include inequality between rulers and subjects; political and legal domination by the imperial power; and the exploitation of the subject people. Many commentators link colonialism with *uneven development.

Sometimes, but not always, colonialism was accompanied by **colonization**; that is, the physical settling of people from the imperial country. Although independence from former colonization has been achieved almost everywhere, most accept that it has been replaced by *neo-colonialism.

command economy A communist economic system where the state controls macroeconomic policy and entrepreneurial activity, but allows some local freedom for decisions on employment and consumption. Also called a **centrally planned economy**.

commercial agriculture The production of agricultural goods for sale.

commercial spaces (**commercial hospitality spaces**) Commercial hospitality is becoming increasingly important for the branding and promoting of cities, based around cafés, bars, and restaurants.

comminution Changing rock debris to fine powder via *abrasion and *attrition. *See also* ROCK FLOUR.

commodification The process of turning something—an artefact, symbol, or idea—into a *commodity that can be exchanged, usually for money.

commodity An object of economic value, intended for exchange in a *capitalist system, such as labour and money. **Commodity chains** (**commodity networks**) are the connections between the production, circulation, and consumption of goods; for example, the chains which bring cut flowers from Kenya to the UK.

commodity fetishism The collective belief that the value of things is measured by money and not by the human labour that created them; consumers perceive commodities as natural, rather than as socially produced. Kosoy and Corbera (2010) *Ecol. Econs.* 69, 1229 provide an excellent overview of this term.

Common Agricultural Policy (CAP) Since the early 1960s, the European Union's CAP has been highly protective, encouraging production through high price support, underpinned by import levies, state intervention and trading, and export subsidies. This helped to transform the European Union from a net importer of major temperate agricultural products to a major net exporter. EU exports needed heavy subsidization to be saleable internationally, and consequently depressed world prices. The EU has also maintained preferential access to regulated amounts of imports from members' former colonies and traditional trading partners.

SEE WEB LINKS
- European Commission Agriculture and Rural Development website for FAQs.
- European Commission Agriculture and Rural Development website with objectives of the CAP 2014–2020.

communication geography The study of the effect of communication on space and place; the geography of missives and messengers. Recently, geographers have been considering the way that information and communication technologies shape, and are shaped by, societies. Adams and Janssen (2012) *Communication Theory* 22, 3, 299 is helpful.

communism Historically, the principle of communal ownership of all property; basic economic resources are held in common. See Shaw and Oldfield (2008) *Pol. Geog.* 27, 1.

community 1. In ecology, a naturally occurring, non-random, collection of plant and animal life within a specified environment. Increases in habitat complexity due to the presence of ecosystem engineers (i.e. any animals that create, modify, maintain, or destroy a habitat) lead to higher **community** *diversity.

2. In human geography, the population and the interconnections of that population in a particular area, town, village, suburb, or *neighbourhood; a set of shared values, practices, and ways of being in the world. A community may be seen as a subject with its own construction of reality; unique needs, values, and assets; or as an organism for the development of social cohesion, human endeavour, empowerment, and place formation. See J. Peck and H. Yeung, eds (2003).

community of practice A group of people, not necessarily physically near to each other, who form common values, norms, and understanding through common training, experience, and expertise. See Amin and Roberts (2008) *Res. Policy* 37, 353.

community-supported agriculture (community-shared agriculture) (CSA) A CSA is organized around a contract between a farmer, commonly an organic producer, and a set of local residents, who share the risks of the farming enterprise by contributing money up front for a 'share' of the harvest prior to the farming season; see O'Hara and Stagl (2001) *Pop. & Env.* 22.

commuting The movement from suburban or rural locations to the place of work and back. Despite their shorter average overall commutes, women travel farther than men to reach jobs in the *CBD. While accessibility to job openings in surrounding regions significantly increases the likelihood of commuting, labour mobility decreases with access to job opportunities in neighbouring regions.

comparative advantage The advantage of some nations or regions to produce goods better and more cheaply than less favoured nations or regions; an important concept in understanding regional specialization.

comparative biogeography The study of biotic area relationships among local and global biogeographic regions: the comparative study of *biotic areas and their relationships through time.

comparative urbanism The systematic study of, and explanation for, similarities and differences among cities or urban processes. Every place is different, or even unique, in some ways, but separate places can be very similar in certain respects. See Nijman (2007) *Urb. Geog.* 28, 1.

compatibility 1. Being able to successfully accommodate more than one form of activity; for example, achieving a physical and economic coexistence of rural accommodation, tourism, and agriculture.

2. To harmonize existing land-use datasets in order to make comparisons within and between countries (Jensen (2006) *J. Land Use Sci.* 1, 2–4).

competence The largest size of particle that a river can carry. Competence varies with water depth, but is reduced with channel widening. The sixth power law suggests that a doubling, for example, of river velocity would raise competence by 2^6, that is, by a factor of 64. See Baker and Ritter (1975) *GSA Bull.* 86, 7 on shear stress as an estimate of competence. *See also* HJULSTRÖM DIAGRAM.

competition 1. Rivalry between suppliers providing goods or services for a market (Jonathan Law, ed. 2006). Essletzbichler and Rigby (2007) *J. Econ. Geog.* 7, 5 seek to show how competition may produce distinct economic spaces: 'as competition unfolds over time, new spaces of economic activity are created.'

Competitive advantage describes the relative economic gains a firm (or a group of similar firms) enjoys over its business rivals at home or overseas.

2. In biogeography, competition occurs when a necessary resource is sought by a number of organisms. See Case and Bolger (1991) *Trends Ecol. Evol.* 6 on the role of interspecific competition in the biogeography of island lizards.

complementarity A mutual dependency based on an ability to produce goods in one area which are needed in another, as with the import of Japanese manufactured goods to Australia, and the export of Australian agricultural goods to Japan. Nazara et al. (2006) *J. Geog. Sys.* 8, 3 explore the degree to which complementarity and competitive interaction at one level in the hierarchy persist at lower or higher levels.

complexity In *geomorphology, a **complex system** is characterized by nonlinear feedbacks, rather than straightforward relationships between *forcing and response. Complex systems spring from nonlinear dynamics and chaos theory and include: self-similarity and (multi) fractals; emergence and self-organization; network analysis; and self-organized criticality. A complex system when interfered with or modified is unable to adjust in a progressive and systematic fashion. Phillips (2003) *PPG* 27, 1, 3 is a good place to start.

complexity science (complexity theory) A study of complicated and chaotic systems and of the way that order, structure, and pattern can develop within these systems, and arise from them. The theory that agents which seem to be independent can unconsciously arrange themselves into an orderly system. The classic example is Zipf's *rank-size 'rule'.

See Trudghill (in S. Trudghill and A. Roy, eds 2003) for a wonderfully user-friendly discussion of models, narratives, and constructs in complexity theory. Martin and Sunley (2007) *J. Econ. Geog.* 7, 5, in a model of exegesis, write that 'to refer to complexity "theory" is perhaps to exaggerate the degree of conceptual coherence and explanatory power associated with the notion'. For **complexity in geomorphology** see (2007) *Geomorph.* 91, 3–4, special issue.

compositional data Population data based on percentages or proportions.

compressive flow (compressional flow) Where a glacier is slowing up, the flow is compressive; in zones of compression, the ice may erode the valley floor. Compressional flow on the *stoss side of hills causes till sheets to shear, and debris bands to stack. See also Swift et al. (2006) *Q. Sci. Revs* 25, 13–14.

concave slope A slope which declines in steepness with movement downslope, also known as a *waning slope. Where the scarp is not being undercut, the base and the crest make straight-line intersections with the mid-section to form a sharp basal concavity and crestal convexity. The basal concavity is poorly developed in humid temperate environments.

concealed unemployment A situation whereby individuals know that there are no jobs available and don't register as unemployed, and even though they would like a job; official figures thus under-represent the true situation.

concentration and centralization The tendency of economic activity to congregate in a restricted number of central places, encouraged by *functional linkages, and *external and *agglomeration economies. Industries become more concentrated with economic integration, and proximity appears to play a distinctive role in the geography of mergers and acquisitions.

concentration dynamics Changes in the concentrations of suspended sediment in a river. Concentrations usually rise with increased river velocity.

concentric zone theory R. Park and E. Burgess (1925) suggested that the struggle for scarce urban resources, especially land, led to competition for land and resources, which ultimately led to the spatial differentiation of urban space into zones. They predicted that

cities would take the form of five concentric rings with areas of social and physical deterioration concentrated near the city centre and more prosperous areas nearer the periphery. In the post-war period, Park and Burgess's model fell out of favour as critics suggested that the models were overly simplistic. However, M. Davis (1992) used the concentric rings model to describe Los Angeles as a city with an inner core of 'urban decay metastasizing in the heart of suburbia' (Brown, no date, *USBC Spatial Rev.*).

concordant Complying with. In geomorphology, relief and drainage may be concordant with geological structure. A **concordant coast** is parallel to the ridges and valleys of the country.

condensation The change of a vapour or gas into liquid form, accompanied by the release of *latent heat, altering the *adiabatic temperature change in rising air. See Wang and Wang (1999) *J. Atmos. Scis* 56, 18 on the classic turbulence condensation theory. Condensation in meteorology can be caused by the cooling of a constant volume of air to *dew point; the expansion of a parcel of air without heating; the evaporation of extra moisture into the air; the fall in the moisture-holding capacity of the air due to changes in volume and/or temperature; and contact with a colder material or *air mass. The likelihood of water vapour condensing will depend on the *saturation vapour pressures of water and ice at a given temperature, and/or the presence of *condensation nuclei, since water vapour can cool well below 0 °C before condensation occurs. *See* CONDENSATION NUCLEI; BERGERON–FINDEISEN THEORY. The **condensation level** is the height at which rising air will cool, condense, and form clouds.

condensation nuclei Microscopic *atmospheric particles that attract water droplets, that may then *coalesce to form a raindrop. Condensation nuclei come from pollen, salt from sea spray, dust from volcanic eruptions and soil erosion, and particulate air pollution. Many small condensation nuclei cause many small droplets to remain liquid well below normal freezing point, and small ice nuclei may increase the rainfall rate from single clouds, which is the principle of *cloud seeding. *See* HYGROSCOPIC NUCLEI.

conditional instability *See* INSTABILITY.

cone volcano A volcanic peak with a roughly circular base, tapering to a point. Cones may be built solely of *lava or of *scoria, or of both. Parasitic cones form round smaller vents on the flanks of the volcano. See Frazzetta et al. (1983) *J. Volc. & Geotherm. Res.* 17.

congelifluction (congelifluxion) *See* GELIFLUCTION.

congelifraction *See* FREEZE–THAW.

congeliturbation *See* GELITURBATION.

congestion Traffic jams. An overload of road vehicles results in increased costs, in terms of fuel and working time, and air pollution. Solutions to traffic congestion include *traffic management measures* such as traffic calming, park-and-ride sytems, and promoting cycling, and *traffic restraint measures* like traffic calming, parking controls, and road pricing. To achieve public/political acceptability for such policies, improved public transport, providing alternative means of travel, needs to be in place before the restraints are introduced. The danger is that once the situation improves, the cars will immediately return. 'All assessments of measures to reduce congestion should take account of the potential for induced traffic' (*House of Commons, Res. Paper* 98/16).

conglomerate. A *sedimentary rock composed of rounded, water-borne pebbles which have been naturally cemented together.

coniferous forest (boreal forest) Natural forest, found between 55° and 66°N, where winters are long and very cold, summers short and warm, and annual precipitation is around 600 mm. Pure stands are common and species relatively few. Trees are evergreen, and the needle-shaped leaves restrict surface area, thus preventing loss of water via transpiration. Undergrowth is sparse. Animal species are dominated by insects and seed-eating rodents; larger animals include deer, bear, and wolves. See Väisänen (1995) *IUFRO XX World Congr.*

connectivity In human geography, the degree to which the *nodes of a network are directly connected with each other. The higher the ratio of the *edges to the nodes in a network, the greater the connectivity. Connectivity in a network is said to increase as economic development proceeds. Taylor (2002) *Urb. Studs* 39 measures a city's connectivity by summing the products of every firm's service value in the city with their service values in all other cities. US cities are generally less globally connected than their European Union and Pacific Asian counterparts. See Urban and Keitt (2001) *Ecology* 82 on graph theory as applied to connectivity in heterogeneous landscapes.

In geomorphology, the transfer of matter, energy, or organisms between two landscape components.

Further reading: *Geomorphology* (2017) vol. 277.

consequent stream Any stream whose course is controlled by the initial slope of a land surface (A. Strahler and A. Strahler 1979).

conservation The protection of natural or man-made resources and landscapes for later use. A distinction is made between conservation and preservation; while a **conservationist** recognizes that people will use some of the fish in a lake, a preservationist would ban fishing entirely.

Conservation is a highly political enterprise: McKenna et al. (2005) *Area* 37 argue that a failure to conserve coastal dunes in Ireland results 'more from management deficiencies than from shortcomings in scientific understanding'. Duffy (2006) *Pol. Geog.* 25, 1 casts light on the struggles encountered in environmental politics over access to key natural resources; and Sundberg (2003) *Pol. Geog.* 22, 7 observes that the existence of democratic regimes and formal institutions does not guarantee that environmental projects will be implemented through demographic processes. C. Hambler (2004) favours establishment of an environmental fund into which polluters pay and from which societies can draw, such that those whose development opportunities are curtailed for global conservation objectives can access compensatory development funds.

Conservation biology is a sub-discipline of biology concerned with the impacts of people on the environment and the conservation of biological diversity See Murray et al. (2002) *Austral. Ecol.* 27; see also Klinkenberg (2001) *Canad. Geogr./Géog. canad.* 45, 3).

conservation of angular momentum The momentum of a body taking a curved path is *angular momentum*. A body's angular momentum will remain constant unless changed by some force. This is important in understanding *jet streams; a body of air at the equator will move at the same speed as the Earth. When this body of air moves polewards, it must increase its velocity to maintain its angular momentum, since Earth's radius decreases polewards. At 30°, for example, the air moves very much faster than the Earth below (which is much slower than at equatorial speed). This very fast-moving stream of air is a jet stream.

conservative margin A plate margin where the movement of the plates is parallel to the margin, such as the San Andreas Fault.

consolidation The reform and reorganization of land ownership in order to redress *fragmentation. See Gonzalez et al. (2004) *Ag. Systs.* 82, 1 on the practicalities of land consolidation.

constant slope A straight, sloping element of a hillslope, located either in the middle of a slope profile, or at the base of the *free face. A constant slope is characteristic of valley incision, but cannot survive at its original angle once downcutting stops. Calvache et al. (1997) *Geomorph.* 21, 1 note that particle size influences the longitudinal constant slope profiles of alluvial fans.

constructive margin A boundary where two *plates are spreading apart from one another and new oceanic crust is being created and added to the outer shell of the planet. See Jackson and Gunnarsson (1990) *Tectonophys.* 172 and Edmonds et al. (2003) *Nature* 421, 6920, on the slow-spreading Gakkel Ridge.

constructive wave A low-frequency (6–8 per minute) spilling wave, with a long wavelength and a low crest, running gently up the beach. *Swash greatly exceeds *backwash (which is reduced by *percolation), leading to deposition. Spilling waves usually occur on gently sloping beaches, and may form beach ridges and *berms.

constructivism The argument that each of us constructs her or his own understanding by relating new information to what they already know. It is the assertion that all forms of knowledge are shaped by the social and material cultures within which they are produced and consumed.

consumer 1. Organisms in all the *trophic levels, (herbivores, *carnivores, omnivores, and parasites) except for *producers. **Primary consumers** feed only on plant material; **secondary consumers** feed on primary consumers, and so on.
   2. A buyer of goods and services for her/his/others' personal satisfaction, as opposed to price. Jackson (2002) *PHG* 26, 1, sees garment labels that one is looking at as a confirmation of one's moral superiority, seeming to say to one 'you're an "alternative" person'; 'you'd like to help the Indian economy', and so on. Jackson sees these labels as rounds of *commodification in an all-encompassing consumer culture.

consumption The use of consumer goods and services to satisfy the present needs of individuals, organizations, and governments. 'Shopping is the medium by which the market has solidified its grip on our spaces, [and] buildings' (Leeong, 2001). **'Alternative' consumption** is consumption in the North that is underpinned by concern about social injustice and unfair labour practices in the South. See Coles and Crang in Lewis and Potter, eds (2011) on Borough Market, London, for an excellent real-world illustration. **Spaces of consumption** are purpose-built spaces, characterized by the provision of consumption-related services, visual consumption, and cultural products. They are the physical manifestation of consumerism as a way of life.

consumption, geographies of Major themes are: consumption as an arena in which governance, regulation, and citizenship are produced; relationships between consumption and forms of urban space; commodity chains; the globalization of commercial cultures; and the way in which consumption is embedded in social and spatial contexts which extend beyond simply buying. For example, in India, alcohol is prohibited in Gujarat, Manipur, Mizoram, Nagaland, and Lakshadweep, so the spatial pattern of alcohol consumption varies over India as it does by economic status and caste. Geographies of food consumption ask, among other questions, where do we eat? and what effect does our everyday environment have on what we eat, and vice-versa?

content analysis A methodology in social science for making sense of recorded human communication including news media, policy documents, letters, and video or novels—particularly written texts. Content in multimedia formats, from a variety of outlets, is collected and then coded for common themes and ideas. For an online, clear, step-by-step user's guide see Amy Luo *Scribbr*.

contested landscape A site of struggle between two groups, with each group claiming ownership of the area. It may be a common environment for two societies within which each group has its own separate cultural-symbolic landscape, or an environment containing a joint cultural landscape for both societies, but to which each society ascribes different purposes. Uluru,

Australia, is a major example. See Fernandez et al. (2019) *Landscape Research* 44, 1 on contested landscapes, territorial conflicts, and the production of different ruralities in Brazil.

context The setting of an event, statement, or idea. Geographical context is often thought about in terms of national or political territories, physical landscapes or exotic.

contextual effect The effect of their local community on the decisions that people make, so that they may be influenced by the views of others in the community; 'converted', in other words, by their neighbours. See Mancosu (2018) on contextual effects and voting behaviour. Pearce et al. (2008) *J. Epidemiol. & Comm. Health* 62 study the contextual effect on diet. *See also* EMBEDDEDNESS.

contiguous zone A zone of the sea beyond the *territorial seas of a nation, over which it claims exclusive rights.

(⊕) SEE WEB LINKS

• UN discussion of the Treaty on the Law of the Sea.

continental climate A climatic type associated with the interior of large land masses in mid-latitudes. Without the moderating influence of the sea, summer and winter temperatures are extreme. Precipitation is low, as the region is distant from moisture-bearing winds. The European Union PESETA project predicts that the greatest temperature rises related to global warming will be in continental climates.

continental crust The outer, rigid surface of the Earth which forms the continents. Continental crust probably existed on the Earth 4.4 gigayears ago. It is rich in feldspars and granitic rocks and about 35 km thick—thicker, but less

dense, on average, than *oceanic crust. Collins (2002) *Geology* 30, 6 explains the role of cycles of tectonic switching in the production of continental crust. See Hastie et al. (2016) *Geology* on the mechanisms responsible for generating the first continents.

continental drift The theory that continents which are now separate were united in a supercontinent, suggested by Alfred A. Wegener (1916). Wegener's ideas were vindicated in the mid-1960s when the development of plate-tectonic theory provided a new framework for planetary-scale tectonics, explicable only in terms of continental drift (Harper (2001) *Geol. Today* 17, 4): Hess (1962), Buddington memorial volume *Geol. Soc. Am.* 599–620, identified sea-floor spreading as a geologic mechanism to account for Wegener's moving continents. See Jackson (1993) *AAAG* 83, 2 on the role played by palaeomagnetic studies in continental drift research.

Dobson (1992) *AAAG* 82, 187–206 suggests that convection drives circular plate motion while gravity drives lateral motion. Convection currents well up at high-pressure centres, spiral outward, transfer to low-pressure cells, spiral inward, and descend at low-pressure centres. In most instances, upwelling and descent of the asthenosphere occur at opposing plate centres rather than at plate margins. Cells migrate laterally in a global pattern driven by gravity. Sea-floor spreading and subduction occur because of differential rates of lateral plate motion. Bokelmann (2001) *Geology* 30, 11 believes that movements of the mantle play an important role in driving the plates.

In 2006, NASA scientists released the first direct measurements of continental drift, showing that the Atlantic is gradually widening, and that Australia is receding from South America and heading for Hawaii (Anonymous (2006)

New Scientist 192, 2578; p. S24). Silver and Behn (Carnegie Institution) suggest that plate tectonics may have halted at least once, and may do so again.

continental shelf The gently sloping submarine fringe of a continent, ended by a steep continental slope which occurs at around 150 m below sea level. Most sediment supplied by rivers crosses continental shelves before being deposited offshore, making continental shelves dynamic environments, in geomorphological terms.

However, a continental shelf may be defined in political terms: the United Nations Convention on the Law of the Sea (1981) defines the continental shelf of a coastal state as 'the seabed and subsoil of the submarine areas that extend beyond its territorial sea throughout the natural prolongation of its land territory to the outer edge of the continental margin, or to a distance of 200 nautical miles from the baselines from which the breadth of the territorial sea is measured where the outer edge of the continental margin does not extend up to that distance.' This definition is of great significance in the right of a coastal state to exploit the resources of a continental shelf.

contingency An event or state of affairs dependent on another, uncertain event or occurrence. **Geographical contingency** can be described as the socio-spatial processes occurring at a location, or as the many conflicting predictions of the future which will influence current decision-making. Although it's jargon-heavy, try Nichols (2019) *Geoforum* 105 on contingency and global nutrition. **Contingence** in geomorphology is the influence of the history of a system in determining its response to an input. In the context of a fluvial system, for example, the response of a length of bank to a hydrological event will be influenced by the bank morphology resulting from a previous event. *Hysteresis—the dependence of the equilibrium position of a system on what happens during the process of dynamic adjustment—seems to be the human geographer's term for contingence.

continuity equation All landscapes must obey an equation for the conservation of mass: mass, energy, or momentum are conserved in a system, so that for any part of the system the net increase in storage is equal to the excess of inflow over outflow of the quantity conserved. In rectangular coordinates and in differential form, the continuity equation is:

$$\frac{\partial \rho}{\partial t} = -\left[\frac{\partial \rho u}{\partial x} + \frac{\partial \rho v}{\partial y} + \frac{\partial \rho w}{\partial z}\right]$$

where ρ is fluid density and u, v, and w are velocity components in the x, y, and z directions. Geomorphologists apply this equation to all cases of mass balance. Applied to mass transactions on a hillslope, the continuity equation states that if more material enters a slope section than leaves it, then the difference must be represented by aggradation; conversely, if less material enters than leaves, then the difference must represent net erosion. The same principle of continuity of sediment transport applies to other geomorphic systems including rivers; see Dietrich and Perron, 2006 *Nature* 439, 411.

contour A line on a map joining places of equal heights, and sometimes equal depths, above and below sea level. The **contour interval** is the vertical change between consecutive contours. **Contour ploughing** is a method of ploughing parallel to the contours rather than up or down a slope. It is used to check soil erosion and the formation of *gullies.

conurbation A group of towns forming a continuous built-up area as a result of *urban sprawl (also called **metropolitan area**). See Kassa (2013) *Ghana J. Geog.* 5 on conurbation and urban sprawl in Africa.

convection The process whereby heat is transferred from one part of a liquid or gas to another, by movement of the fluid itself. Convection carries excess heat from the Earth's surface and distributes it through the *troposphere (Takayabu et al. (2006) *J. Met. Soc. Japan* 84A). **Convective storms** include thunderstorms, hail, tornadoes, heavy rains, and high winds; Combardo and Colle (2011) *Weath. Forecasting*, 26, 940 document the convective storm structures over the northeastern United States.

   Thermal convection (free convection) is propelled by buoyancy (R. Barry and R. Chorley 2003). **Mechanical convection (forced convection)** is the upward movement of an air parcel over mountains, at fronts, or because of *turbulence. Kijazi and Reason (2005) *Theoret. & Appl. Climatol.* 82, 3–4 observe that wet conditions during *El Niño years are associated with **enhanced convection** (convection with a higher *Bowen ratio). See Ludlam and Scorer (2006) *Qly. J. Royal Met. Soc.* 79, 341.

convection rain When upward *convection occurs in a parcel of moist air, the rising air will cool. Further cooling will cause condensation of the water vapour in the air, and rain may result. If the air is very moist, the cooling results in condensation, and hence the release of *latent heat. This causes the rising air to accelerate, and very tall *cumulo-nimbus clouds form. See Diop (2018) Procs EUMETSAT Meteorological Satellite Conference on Sahelian convection rain.

convective condensation level The point on a thermodynamic diagram where a sounding curve (representing the vertical distribution of temperature in an atmospheric column) intersects the saturation mixing ratio line corresponding to the average mixing ratio in the surface layer (approximately the lowest 500 m). The dry adiabat through this point determines, approximately, the lowest temperature to which the surface air must be heated before a parcel can rise dry-adiabatically to its lifting condensation level without ever being colder than the environment. This temperature, the convective temperature, is a useful parameter in forecasting the onset of convection (AMS online).

convective instability Instability that is caused by the rising of very dry air over warm, moist air below. C.-G. Rossby (1932) argued that it exists when one of these conditions is met over a layer of atmosphere: the lapse rate of wet-bulb temperature exceeds the moist-adiabatic lapse rate; the equivalent potential temperature decreases with height; the wet-bulb potential temperature decreases with height. Convective instability of the terrestrial atmosphere is largely determined by water vapour. See Thual (2012) *Quart. J. Royal Meteorol. Soc.* on absolute or convective instability in the equatorial Pacific and implications for *El Niño–Southern Oscillation.

convective turbulence Turbulence generated when warm fluid parcels are accelerated upward by their buoyancy, normally initiated by daytime solar radiation. Chowdhuri (2020) *Physics of Fluids* 32 could be helpful.

convenience distance The ease, or otherwise, of travel. A town 50 km away may be well served by transport and thus 'nearer' than one 20 km away which is less accessible. **Convenience goods** are *low-order goods like milk, and occasional groceries, which are frequently bought locally, with little consideration of the prices.

convergence 1. The meeting of *tectonic plates. See Rosenbaum (2002) *Tectonophys.* 359, 1–2 on the convergence of Africa with Europe, and Gelabert et al. (2004) *Geologica Acta* 2, 3 on the convergence of the Indoaustralian and Eurasian plates.

2. In meteorology, air streams flowing to meet each other. An area of **convergence** is an area of rising air. Convergence can occur aloft over dense, cold air and is not necessarily confined to a layer bounded by the surface. Convective rainfall may result from the convergence of low-level winds, often restricted to well-defined lines of convergence with widths of the order of 1–2. In the upper troposphere, convergence causes air to subside, creating *anticyclonic conditions at ground level (Hastenrath (2007) *Dynam. Atmos. & Oceans* 43, 1–2).

3. Within human geography, the use of the same production methods or practices by firms operating in different national-institutional spaces. Economic convergence may be brought about by location, proximity, physical geography, economic structure, agglomeration, and economic potential; see Arvanitopoulos et al. (no date) *LSE Research Online* on drivers of convergence.

convergence space Spaces where geographically scattered, dispersed social groups meet (converge) from time to time temporarily but intensely, as on days of action, or in conferences. These temporary events can have lasting material effects, serving as important sites for making the relationships needed in forming and supporting

policies for action. Convergence spaces are made up of place-based, but not necessarily place-bound, movements, enacted both virtually through the internet, and in place- and face-to-face-based moments. See Temenos (2016) *Space & Polity* 20, on convergent spaces and drugs, law, people, place, and the state.

converging margin *See* DESTRUCTIVE MARGIN.

convex slope A slope, or slope element, that gets progressively steeper downhill. It may be determined by structure (*exfoliation domes), by lithology, or by landslides. Convex slopes may result from weathering and debris transport—see J. Wilson and J. Gallant (2000). See Lin and Oguchi (2004) *Geomorph.* 63, 3 on drainage density and convex slopes.

conveyor belt 1. An expansive upwards, polewards flow of air. As a mid-latitude depression develops, a warm conveyor belt rises at $c.20$ cm s^{-1}, ahead of the cold front, and a reciprocal cold conveyor belt descends from medium/upper levels, below the warm conveyor, well ahead of the surface warm front; see Pomroy and Thorpe (2000) *Monthly Weather Rev.* 128, for a useful diagram.

2. In the oceans of the world, a global system of currents. The **Atlantic conveyor** is a vast ocean current, which is part of the global ocean circulation. Warm water travels northwards close to the surface of the ocean. Beneath it there is a slow, deep-water flow carrying cold dense Arctic waters to Antarctica along the ocean bottom. As it travels north, water in the upper current becomes colder and denser. Winter sea ice forming in the Arctic Ocean leaves salt in the water making it denser still. At certain places in the North Atlantic the dense water sinks down towards the sea floor and begins to flow south. In this way the circulation 'overturns' to form the loop in the conveyor belt between the warm shallow current and the cold deep current (PEEP, *What Drives the Atlantic Conveyor?*). Broecker et al. (1985) *Nature* 315, 21 were the first to suggest a link between the Atlantic conveyor belt and possible climate change. *See also* THERMOHALINE CIRCULATION.

(⊕) SEE WEB LINKS
• Website of Physics & Ethics Education Project (PEEP).

co-occurrence An event or situation that happens at the same time as or in connection with another. For example, any map which shows the impact of having a number of resources in the same geographic area is a co-occurrence map. Co-occurrence used to be spotted by using actual transparent overlays, but with the development of computer mapping, a summary map can be generated showing the locations where different numbers of factors are co-located. Go to the excellent Natural Resources Inventory on the Natural Resources Conservation Service website.

(⊕) SEE WEB LINKS
• Natural Resources Conservation Service website.

cooperative A type of business organization owned by its employees or customers. **Agricultural cooperatives** are collective projects initiated by local producers which aim to organize the implementation, collection, and transformation of the members' production, and to ensure their reputation through collective actions. This means that they have to reconcile economic performance (marketing) with their social responsibilities towards their members.

Agricultural cooperatives have been encouraged in the developing countries, where several farms pool resources to jointly purchase and use agricultural machinery; see anon. (2005) *Rev. Int. Cooperation* 98, 2 on cooperatives in Colombia. Rice and Lavoie (2005) *Canad. Geogr.* 49, 4 evaluate **business cooperatives** in Canadian economic growth.

coral reef An offshore ridge, mainly of calcium carbonate, formed by the secretions of small marine animals. Corals flourish in shallow waters over 21 °C and need abundant sunlight, so the water must be mud free, and shallow. Fringing reefs lie close to the shore, while barrier reefs are found further from the shore, in deeper water. Reef health is mapped by live coral cover, diversity, and remote sensing. Coral reefs are the most biodiverse marine ecosystems on the planet, harbouring perhaps nearly one million species globally. However, the health of coral reefs is declining, due to, at the local scale, overfishing, nutrient enrichment, and coral diseases, and, at the global scale, ocean warming, acidification, and sea-level rise; see US NOAA Coral Reef Conservation Programs.

A **coral atoll** is a horseshoe-shaped ring of coral which almost encircles a calm lagoon. Many coral reefs are hundreds of metres deep; Charles Darwin's theory (1842) is that deep reefs formed during a long period of subsidence. Thus, coral forms in shallow waters and then sinks. Woodfine et al. (2008) *Sedimentol.* 52 seem to support the subsidence theory. Woodruffe et al. (1999) *Marine Geol.* 160, 1 dispute Darwin's theory. There is also the possibility of a positive feedback loop: sea level rises, providing more habitat for coral reefs, the reefs flourish and produce more CO_2, the additional CO_2

causes additional temperature increase, and sea levels rise even more; see G. Camoin and P. J. Davies (1998).

core 1. The central part of the Earth, the core is the point within the Earth where S-waves cannot penetrate, marked by an abrupt increase in pressure. It is believed to be composed primarily of a nickel-iron alloy (along with abundant platinum-group elements), with a liquid outer zone, 2900–5000 km deep, and a solid inner zone, 5000–6370 km deep.

2. *See* CORE REGION.

core–frame concept A model of the central area of the city that shows a core of intensive land use indicated by high-rise buildings (the *CBD), beyond which is the frame, with less intensive land use: warehousing, wholesaling, car parking, medical facilities, and so on.

core–periphery The core—a central region in an economy, with good communications and high population density, which conduce to its prosperity—is contrasted with the periphery—outlying regions with poor communications and sparse population (for examples, *see* UNEMPLOYMENT). The centre represents the locus of power and dominance and importantly, the source of prestige, while the periphery is subordinate.

Cores are associated with higher wages, high technology, and high profit inputs and outcomes. However, locations should never be seen as purely one or the other: so-called core countries encompass numerous, if minority, peripheral processes; and the opposite is so for 'peripheral countries'. When transport costs fall below a critical value, a core–periphery spontaneously forms, and nations that find themselves in the periphery suffer a decline in real income.

core region In a nation, a centre of power where innovation, technology,

and employment are at a high level, such as Madrid or south-east England.

(((●))) SEE WEB LINKS

• Brand and Bhatti (2006) on the core region of Madrid.

corestone *See* BOULDER FIELD.

CORINE The Coordination of Information on the Environment is an ecological commission of the European Communities located in Brussels, Belgium. Its main goals are to coordinate information and action within the EU, to define and protect biotopes, to combat air pollution, and to preserve the ecology of the Mediterranean region.

Coriolis force An apparent, rather than real, force which causes the deflection of moving objects, especially of air streams, through the rotation of the Earth on its axis. It shows up, for example, in the movement of an air stream, relative to the rotating Earth beneath it. The Coriolis force is proportional to the wind speed $U(z)$, and the **Coriolis parameter**, $f = 2\,\Omega\,\sin\Phi$—where Φ is the latitude and Ω is the Earth's rate of rotation—acts perpendicular to the wind direction (Orr et al. (2005) *Weather* 60, 10).

corporate governance The system by which companies are directed and controlled via systems, processes, controls, accountabilities, and decision-making.

corporate social responsibility The belief that a business or corporation is responsible for the economic, political, or social impacts of its business practices; for its wider contribution to society; for a form of ethical capitalism.

corrasion The erosive action of particles carried by ice, water, or wind (although the term is most often used in a fluvial context). Corrasion is another term for *abrasion.

corridor A limb of one state's territory cutting through another state, usually for access. The most famous is the Polish corridor (Howkins (2006) *J. Transp. Geog.* 4, 4). A **corridor of commerce** extends along a line of communication, such as the Mekong River (Furlong (2006) *Pol. Geog.* 25, 4), or the UK 'M4 corridor'.

corrie *See* CIRQUE.

cosmopolitanism Being open to, and respecting, very different ways of life to one's own. Fundamentally, it is the tenet that all human beings are to be valued, regardless of their location. It is an ideology that fosters respect for social and cultural differences and a wider community of caring. Many cosmopolitanisms exist, however, in different contexts or locations. Thus, Datta (2012) *Antipode* 44, 3, 745 describes and analyses cosmopolitanism in a Delhi squatter settlement, while Mith and Jenkyns (2012) *Antipode* 44, 3, 640 consider the relationship between cosmopolitanism, development, and civil society on Indian NGOs. In *biogeography, an organism with **cosmopolitan distribution** is found in a great variety of habitats (Smith (2007) *J. Biogeog.* 34, 10).

cost–benefit analysis The attempt to compare the total social costs and benefits of implementing a decision, usually expressed in money terms. Costs and benefits include money, externalities—for example pollution, noise, and disturbance to wildlife—and external benefits, such as reductions in travelling time or traffic accidents. For an evaluation of cost–benefit analysis, see Hockley (2014) *Env. & Planning C*.

cost–space convergence Especially in the last 50 years, innovations in air, rail, road, and sea transport technology, communications, and handling technology have made the distance between two points (the *friction of distance*) less significant. The effect of this reduced friction of distance is cost–space convergence. Buliung (2011) *AAAG* 101, 6 is a model of clarity. *See* CORE–PERIPHERY.

cost surface A three-dimensional 'contour model' representing the variation in costs over an area. See Martin et al. (2002) *Health & Place* 8 for a *GIS-based example.

Cost surface

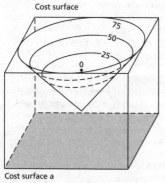

Cost surface a

Cost surface with revenue surface

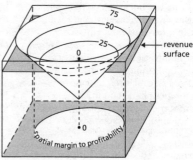

Cost surface b

Cost surface

cottage industry The production of finished goods by a worker at home, sometimes together with her or his family, such as the production of wool and woollen goods in the Himalayan district of Uttarkashi.

counter-mapping The mapping of phenomena that are not found on orthodox maps, such as safe river-crossings, sacred spaces, and places of safety, often with varying scales on the same map. Counter-maps exist to challenge predominant hierarchies, in order to, for example, protect indigenous land rights, demarcate and protect 'traditional' territories, manage community lands, protect biodiversity, raise awareness about conflict and mediate resolutions, empower communities, and promote cultural diversity. Counter-mapping is a cartography that lets the powerless speak. See Harris and Hazen (2006) *ACME* (*International Journal for Critical Geographies*) 4.

counter-radiation The long-wave radiation emitted from the Earth to the *atmosphere after it has absorbed the shorter-wave radiation of the sun.

countertrade A reciprocal trading system between two countries, useful for nations wishing to import high technology but lacking foreign exchange, or exporting goods for which world demand is low. See Onukwugha (2019) *Kent Student Law Review* 5, 43 for an evaluation of the utility of countertrade transactions in the Sino-Congolese barter deal.

counter-urbanization The movement of population and economic activity away from *urban areas—see Mitchell (2004) *J. Rur. Studs* 20, 1 for an exhaustive dissertation on the meaning of this term, and Escribano (2007) *Tijdschrift* 98, 1 for a more nuanced view.

Increased mobility and improved communications, higher incomes, changing household composition, changing technology in manufacturing, and growth of the service sector have led to increased commuting, counter-urbanization, and an urban-to-rural employment shift. See Bosworth (2010) *Env. & Planning A* 42, 4.

country rock A pre-existing rock which has suffered later igneous *intrusion.

coupled human-environment system (CHANS) Any system displaying interactions between humans and their environments. The study of these systems considers stability and change between human and environmental systems and the management of the interconnections between the two. Hua et al. (2018) *Sustainability* establish a new framework for a coupled human–environment system, based on the theory of event ecology, and illustrate it by presenting a case study of the Honghe Hani Rice Terraces. E. Sheppard and R. B. McMaster, eds (2004) note that social and natural systems are inseparable.

coupling (geomorphic coupling) The connectivity within the components of a system. For example, the physics, chemistry, and biology of the ocean are coupled since the oceanic CO_2 cycle is controlled by the ocean circulation, the supply of sunlight, and major and trace nutrients delivered by the oceanic and atmospheric circulation. Strong couplings link human dynamics, biology, biochemistry, geochemistry, geology, hydrology, and geomorphology. Coupling behaviour conditions the way a system responds to disturbance, and is therefore important in determining the response to human-induced, climatically induced, or tectonically induced environmental change. Savi et al. (2013) *Earth Surface Landform Procs* illustrate the connectivity between hillslopes and channels, and between sources and sinks.

coupling constraint In *time–space geography a limit to an individual's actions because of the necessity of being in the same space and time as other individuals. Fixed activities dictate strict coupling constraints, while flexible activities allow more fluid coupling in space and time.

coverage Used in GIS, coverage refers to the locations covered by a survey. These may be location names, various codes for localities, census map spots, and latitude and longitude ranges.

cover crop Any seasonal crop that is planted to protect and/or improve the soil. A cover crop can slow erosion, enhance water availability, smother weeds, help control pests and diseases, increase biodiversity, increase crop yields, add organic matter to the soil, improve crop diversity on farms, and attract pollinators. Its benefits are both economic and ecological.

crag and tail A large rock outcrop, with a rampart of *till on its lee side, formed by erosion on the up-glacier side of the outcrop and deposition of debris in the low-pressure lee area. Glasser and Bennett (2004) *Prog. Phys. Geog.* 28, 1 define **micro crag and tails** as small tails of rock that are preferentially protected from glacial abrasion in the lee of resistant grains or mineral crystals on the surface of a rock.

crater A circular depression around the vent of a volcano. Craters form the summit of most volcanoes. They occur where lava overflows and hardens or where the walls collapse as the *magma sinks down the vent after an eruption.

Funnel-shaped craters are typical of *stratovolcanoes, while kettle-shaped craters are characteristic of *shield volcanoes.

craton A core of stable continental crust within a continent and composed wholly or largely of *Precambrian rocks with complex structures. Two types are recognized: platforms, which are parts of cratons on which largely undeformed sedimentary rocks lie, and *shields. See Zheng et al. (2004) *Geology* 32, 3 on the North China Craton.

creative cluster An assembly, or group, of *creative industries, such as advertising, film-making, or music. Within the UK, creative clusters are found in, for example, Reading and Brighton; see Siepel (2020) on the UK's creative clusters and microclusters. Factors encouraging creative clusters include creative human capital; innovative activity associated with high-tech agglomerations; and the quality of a place as liberal, tolerant, and capable of attracting skilled people able to generate new ideas. Face-to-face contacts are fundamental, particularly in environments where information is imperfect, rapidly changing, and not easily codified.

creative industries Those industries that focus on creating and exploiting intellectual property—for example, advertising, architecture, arts and antique markets, crafts, design, designer fashion, film, interactive leisure software, music, performing arts, publishing, software, television, radio, and computer games—or providing creative services for business, such as advertising (also called the **creative economy**).

The creative industries are becoming more important in local economies across the UK. Between 2007 and 2014, creative employment rose by 28%. Creative communities in different parts of the UK work together across cluster and administrative boundaries; the geography of the UK creative industries is an interconnected system. In a commentary on the rise of the creative economy in the west of Ireland, Collins et al. (2018) *Creative Industries Journal* 11 suggest that western Ireland's approach to business could provide a sustainable development path for peripheral regions in Europe.

creep The slow, gradual movement downslope of soil, *scree, or glacier ice. Most creep involves a deformation of the material, i.e. *plastic flow. **Soil creep** is more frequent in slopes of calcareous rock with surface deposits and slope angles of approximately 25–35°. The rate of **ice creep** is a function of the *shear stress applied and is expressed in the Glen–Nye flow law:

$$\Sigma = k\tau^n$$

where:

Σ = shear strain (flow) rate

τ = stress

n = a constant between 2–4 (typically 3 for most glaciers) that increases with lower temperature

k = a temperature-dependent constant.

creolization Initially a term widely used to refer to processes of cultural change in the Caribbean (Lambert (2004) *PHG* 28). This term refers to the hybridization of a culture, as it absorbs and transforms forces from outside; the production of new local forms. This may be described as the development of cultural complexity, cosmopolitanism, hybridity, 'syncretism', and 'mixture', all in response to *globalization. In 2011, *Le Monde* wrote that 'Europe is getting creolized. . . . It has several languages

and very rich literatures that are interlinked and mutually influence each other. All the students learn them, they speak several of them and not only English . . . like the young Beurs or Antilleans of the [Parisian] suburbs'.

Cretaceous The youngest period of *Mesozoic time from *c.*136 to 65 million years BP.

((⊕)) SEE WEB LINKS
• An overview of the Cretaceous period.

crevasse A vertical or wedge-shaped crack in a glacier, varying in width from centimetres to tens of metres. Its maximum depth is about 40 m, as, at that depth, ice becomes plastic and any cracks merge together. **Transverse crevasses** occur when the ice extends down a steep slope. **Longitudinal crevasses** form parallel with the direction of flow as the ice extends laterally. **Marginal crevasses** occur across the sides of a glacier as friction occurs between the ice and the valley walls. **Radial crevasses** fan out when the ice spreads out into a lobe.

crime, geography of The analysis of crime—its effects, the offences, and the offenders—to understand the interactions between crime, society, and space. In the UK, Australia, New Zealand, and North America, the *inner city is a crime hot spot; Canada's Research and Statistics Division notes that 'several indicators were found to have a significant effect on crime levels [in Saskatoon] including higher proportions of single people and youth not attending school as well as lower average household incomes'. The residential pattern of the offenders resembles the pattern of offences. See Monmonier (2006) *PHG* 30, specifically 375–8, on mapping crime.

Feminist geographers stress that, while most girls are brought up to fear violence by strangers in public places (thus hugely limiting female spatial mobility), most violence against women is located within the home. It's difficult for many women to acknowledge that their homes are not a place of safety. Organizations devoted to reducing domestic violence are still rare in urban environments and virtually non-existent in rural ones. On the perception of crime, Stenning (2005) *TIBG* 30, 1 addresses the 'wider tendencies to characterize and stigmatize working-class communities'.

Herbert and Brown (2006) *Antipode* 38, 4 note that the onset of *neoliberalism in the United States coincided with 'an unprecedented expansion of punishment practices that intensify social divisions rooted in class and race'.

critical In a *geomorphological system, the magnitude of an input—object or process—required to cross a *threshold. Thus, **critical shear stress** in an overland flow is that point where the *shear stress is great enough for incision to occur; or within a channel, for mobilizing sediment (*entrainment). This point occurs at the **critical distance** from the *drainage divide, where the flow reaches the **critical depth**. When the velocity of a flow is less than critical, it is considered a **subcritical flow**. Subcritical flows are characterized by a water surface that is smooth and on which a wave produced at the water surface is quickly damped out; subcritical flows are also termed 'tranquil' (K. J. Gregory and A. S. Goudie, 2011). **Supercritical flows** are highly erosive, moving rapidly and efficiently through the channel, often over-shooting bends. Turbulent mixing is less intense. See Ono et al. (2019) *Sedimentology* on supercritical flow processes and sedimentary structures. For quantification of supercritical flows,

see Bagnold (1980, *Proc. R. Soc. Lond.* A 372, 453).

Critical stream power is the amount of stream power needed to transport the average sediment load. Ferguson (2005) *Geomorph.* 70, 33 amends Bagnold's equation, and Gob et al. (2010) *ESPL* 35 investigate the relationships between sediment size and critical stream power.

critical cartography A cartography grounded on *critical theory. Critical cartographers argue that maps are not neutral but have historically been produced to support and display the interests of the ruling classes: maps are social documents that need to be understood in their historical contexts. J. Crampton (2010) writes that 'critical cartography assumes that maps *make* reality as much as they represent it'.

The MIT SLAB website describes critical cartography as 'mapping the unmapped; [noticing] people and things that are overlooked or denied. This is *open-source mapping, where cartography is no longer in the hands of cartographers or GIScientists but the users.

(((•))) SEE WEB LINKS
• MIT SLAB website.

critical geography (critical human geography) The umbrella term for a group of geographical concepts and procedures that are centred on opposition to repressive and inequitable power relations in: *capitalism, *class, *colonialism, disability, ethnicity, gender, race, and sexuality. The keystone to all of these is the contention that 'knowledge can never come in an unpoliticized form'. Critical geographers stress the role of dominance and confrontation in the production and reproduction of *landscape, *place, and *space. *ACME* (an online international journal for critical and radical analyses of the social, the spatial, and the political) sees critical geographies as 'part of the praxis of social and political change aimed at challenging, dismantling, and transforming prevalent relations, systems, and structures of capitalist exploitation, oppression, imperialism, neo-liberalism, national aggression, and environmental destruction'.

To appreciate the hallmarks of critical geography, Unwin's words (2000, *TIBG* 25, 1) are inspirational: 'A critical geography needs to engage with the everyday practices of all of us who live in the places that we do; it needs to focus on the needs and interests of the poor and the underprivileged; it remains a very modern enterprise, retaining a belief that it is possible to make the world a "better" place.'

critical development geography

A geography characterized by the interweaving of political economy and *post-colonial theory; the deepening analysis of violence and development; and the internal social variegation of countries in the Global South. Critical development geography is dedicated to cultivating non-capitalist, and alternative capitalist, models, narratives, and practices. See Glassman (2011) *Prog. Hum. Geog.* 35, 5.

critical geopolitics *See* GEOPOLITICS.

critical legal geography The study of the connections between critical legal studies and *critical geography scholarship. Legal geography scholarship stresses that nearly every aspect of law occurs within a spatial frame of reference, and that social spaces, lived places, and landscapes are moulded by legal practices. Furthermore, spaces, lived places, and landscapes are not simply the inert sites of law, but are inextricably implicated in the way the law works.

critical race theory Critical race studies stress the socially constructed nature of race; race has no necessary epistemological status in itself, but depends on the spatial context and organization of its production for its political effects. The term 'race' is a complex social *construction*, and its meaning varies across time and space. The critical race approach is used to *deconstruct* racial ideologies, and to analyse the politics of 'race' and the links between nationalism/localism and racism For some geographers, critical race theory means a social commitment to activism within the academic world. See Anderson (2008) *Cult. Geogs* 15, 2.

critical zone The near-surface layer of the Earth, from the depths of the groundwater to the tops of the trees. Within this zone, atmosphere, rock and soil, and living organisms interact, providing the resources to sustain life. Xu and Lui (2017) *Geophys. Research Letters* provide a refreshingly accessible account.

critique A process of looking carefully at the assumptions made about any form of knowledge; a critique is not necessarily about finding fault; see J. W. Crampton (2010). The fundamental questions are: 'what is authority?' and 'who shall have it?', which means that a critique is a political project.

crop rotation The practice of planting a succession of crops in a field over a period of years. Rotations can maintain field fertility since different crops use different soil nutrients, so excessive demands are not made of one nutrient. In certain rotations, plants like legumes (peas and beans) are grown to restore fertility. See D. Stone (2005) on medieval agriculture, and Sainju et al. (2006) *J. Env. Qual.* 35 on contemporary practice.

(()) SEE WEB LINKS
- Wikipedia entry about crop rotation in India.

cross-bedding (current bedding) In a sedimentary rock, the arrangement of beds at an angle to the main bedding plane.

cross-correlation analysis A way of identifying any changes in land use that have occurred in a previously mapped area. See Yen (2015) *PLOS One* 10, 5 for a new methodology of spatial cross-correlation analysis.

crowd-sourcing (crowdsourcing) Getting data or ideas (user-generated content) by consulting large numbers of people, especially from the 'online' community; Wikipedia is a good example. The OpenStreetMap project is a crowd-sourced project, where detailed map data are produced by people pooling GPS and metadata. Matt Ball's blog of October 2010 explains the distinction between crowd-sourcing, volunteered geographic information, and authoritative data.

((⊕)) SEE WEB LINKS
• OpenStreetMap site.

crude rate A *vital rate which is not adjusted for the age and sex structure of a population.

crumb A spheroidal cluster of soil particles, i.e. a type of *ped.

crust The outer shell of the Earth including the continents and the ocean floor; the *lithosphere. *Sial overlies *sima in the continental crust; the oceanic crust is mostly sima.

cryofront *See* CRYOTIC.

cryogenic (cryergic) Processes carried out by ground ice. Needle ice and frost heaving are connected with the development of cryogenic phenomena (Grab et al. (2004) *Geografiska. B* 86, 2). *See* FROST HEAVING; FROST WEDGING; CRYOPLANATION.

cryopediment In *periglacial environments, a low-angle, concave, piedmont footslope developed by slope retreat, brought about through frost *weathering and the sapping of slopes, and by *nivation surfaces. Cryopediments may coalesce to form a **cryopediplain** (Tzudek (1993) *Permaf. & Perig. Procs* 4, 1—with useful photographs).

cryoplanation The lowering and smoothing of a landscape by *cryogenic processes. 'To deny the fairly widespread existence of features commonly called cryoplanation benches, terraces, and pediments would be foolish: to claim that there is anything approaching an adequate explanation of their origin(s) would be even more foolish . . . there is little to say about cryoplanation other than the concept needs to be challenged directly' (Thorn and Hall (2002) *PPG* 26, 4).

cryosphere The ice at or below the Earth's surface, including *glaciers, *ice-caps and *ice sheets, *pack ice and *permafrost. The mass balance of the cryosphere is most likely contributing to sea-level rise (J. Bamber and A. Payne, 2004).

cryostatic pressure The pressure exerted on rocks and soil when freezing occurs. As the *freezing front advances, the pressure of the trapped soil moisture increases. This pressure can separate individual grains of soil, forming a mass of fluid mud, which may dome up the ground, or form mud blisters.

cryotic Having temperatures below 0 °C. Van Everdingen (1976) *Can. J. Earth Scis* 13 proposes that 'cryotic' be used to describe permafrost on the basis of temperature, and that 'frozen' be restricted to ground in which part or all of the pore water is ice.

cryoturbation A general term describing all frost-based movements of the *regolith, including *frost heaving and *gelifluction. Grab (2005) *PPG* 29, 2 analyses cryogenic mounds.

cuesta A ridge with a *dip slope and a *scarp slope. *See* ESCARPMENT. For **cuesta topography**, see Pánek et al. (2008) *Geomorph.* 95, 3–4.

cultural capital The symbols, ideas, tastes, and preferences that can be strategically used as resources in social action, and serve as markers of collective identity and social difference. Cultural capital includes cultural goods: pictures, books, dictionaries, instruments, machines, and so on. See Hubbard (2002) *Env. & Plan. A* 34. The commodification of cultural capital is carried out by the **cultural industries** which convert cultural capital to economic capital. R. Florida (2002) argues that urban fortunes increasingly turn on the capacity to 'attract, retain and even pamper a mobile and finicky class of "creatives"'.

It's possible to be high in cultural capital, but low in economic capital: S. Fernandes (2006), for example, notes that Cuban arts and popular culture have been commercialized to entice foreign investment. Ottaviano and Peri (2006) *J. Econ. Geog.* 6, 1 find that US-born citizens living in cities of increasing **cultural diversity** earned significantly higher wages.

cultural distance A gap between the culture of two different groups, as between Vietnamese immigrants and the Czech majority in the Czech Republic (Drbohlav and Dzúrová (2007) *Int. Mig.* 45, 2).

cultural ecology An investigation that focuses on the dynamic interactions between human societies and their environments with an understanding of key ecological concepts such as resilience, stability, and biodiversity. Culture is seen as the primary adaptive mechanism used by human societies to deal with, understand, give meaning to, and generally cope with their environment.

cultural economy At its most simple, that part of the economy dealing with cultural goods and services: film, TV, radio, photography, advertising, and so on. However, with admirable clarity, Gibson and Kong (2005) *PHG* 29, 5, 54 explain why this term is so difficult to define.

cultural geography The study of the impact of human culture on the landscape; the ways in which place and identity are embedded in a range of cultural landscapes, and the ways in which those social and material landscapes have reflected and influenced various experiences. Cultural geographers emphasize the symbolic component of human activities and the relevance of historical societal processes, and are committed to a philosophy that investigates the origin, nature, methods, and limits of human knowledge as perceived and experienced by people and organizations, rather than as perceived only by social scientists. As things travel, their meanings and material nature can change; things themselves therefore have complex cultural geographies.

Mitchell (2004) *J. Cult. Geog.* 22, 1 argues that 'inequality, domination, oppression, exclusion, and power . . . at all scales from the household to the globe are the true object of cultural geographic study'.

cultural hearth The location in which a particular culture has evolved. Voigt-Graf (2004) *Glob. Networks* 4, 1 uses the term as 'the country, region or place of origin of migrants and their descendants'.

cultural landscape The impact of cultural groups in shaping and changing the natural landscape. This may be the three-dimensional patterns that cultures impress onto the land, such as field systems, transport networks, and urban forms. It may also be a way of interpreting these forms; looking to interpret cultural meanings embedded in landscapes. 'Landscape itself is a cultural image; a way of symbolizing, representing, and structuring our surroundings' (P. Cloke et al., eds 2006). Individuals interpret the landscape in many ways, 'all of them dependent on the viewer's mental ideas evoked by . . . their previous experiences' (Meinig, 1979).

cultural turn The increased focus, from the 1990s onwards, on culture as a geographical agent and product. This focus stresses the importance of culture within geography and demonstrates the significance of space in the production of cultural values and expressions. 'It is because all human realities are expressed through cultures that a cultural turn was needed in geography' (Claval (2007) *Tijdschrift* 98, 2). Thornes and McGregor in S. Trudgill and A. Roy (2003) propose a **cultural climatology**: 'the study of the processes of, and the interactions and feedbacks between, the physical and human components of the climate system at a variety of temporal and spatial scales' while Gregory (2006), *Geomorph.* 79, 172, argues for a **cultural geomorphology**; for example, 'with knowledge of processes and changes of river channels it seems expedient to consider cultural reactions and perceptions'.

cultures of work The attitudes, values, beliefs, modes of perception, and habits of thought and activity in the working environment. Spatial variations in corporate cultures may give rise to varieties of capitalism; for example, changing cultures of work in Canada, associated with the shift towards higher education, have led to increasing mobility and individualism in career choices; see Cowen (2005) *Antipode* 37, 4. Faulconbridge (2008) *J. Econ. Geog.* 8, 4 shows how managers in TNCs act as **cultural entrepreneurs**, driving change in institutionalized cultures of work through strategies that alter the cognitive frames of workers. See also Jarvis (2003) *Area* 34, 4.

cumec A measurement of *discharge. One cumec is one *cu*bic *me*tre of water per second.

cumulative causation The unfolding of events connected with a change in the economy, as a consequence of the *multiplier effect. Cumulative causation can be set in motion wherever the expansion of a cluster, through attraction of firms with complementary competencies, adds to its initial attractiveness. However, improvements through cumulative causation are made at a cost to some other part of the economy, and although underdeveloped regions offer the advantage of low-wage labour, these benefits tend to be offset by the agglomeration economies in the industrialized regions. Therefore, in order to overcome the negative effect of cumulative causation, intervention on the part of government may be necessary. Cumulative causation is not limitless, and can be reversed through negative externalities, such as congestion costs, or reductions in trade costs. Simulation models tend to show that large changes in trade costs may be required to cause deconcentration, and that at intermediate levels of trade costs

Cumulative causation

concentration remains high (Weiss (2005) *ADB Institute Discuss. Paper* 2005/04).

cumuliform In the shape of a *cumulus cloud. **Cumuliform convection** is vertical convection.

cumulo-nimbus A low-based, rain-bearing *cumulus cloud, dark grey at the base and white at the crown, which spreads into an anvil shape, as it is levelled by strong upper-air winds.

cumulus An immense, heaped cloud with a rounded, white crown and a low, flat, horizontal base, extending as high as 5000 m. Updraughts within the cloud are strong—up to 10 m s^{-1}, and cloud growth is often rapid, through *entrainment and the saturation of surrounding air. **Cumulus congestus** is a swelling, small cumulus which often becomes *cumulo-nimbus.

current A horizontal movement of water, as along a stream or through an ocean. The rate of flow of a river current varies with depth because friction operates along the bed and sides. **Tidal currents** are associated with the rise and fall of the sea, and the velocities of ebb and flow vary with the morphology of the coast and any outflow of fresh water; see

FitzGerald et al. (2002) *Geomorph.* 48, 1–2. **Rip currents** are generally strong, shore-normal, jet-like flows that originate within the surf zone and are directed seaward through the breakers. They form in the nearshore zone and balance the inflow of seashore currents, influence the morphology of the shoreline, and may be important for transporting fine sediments offshore (MacMahan et al. (2005) *Marine Geol.* 214, 1–4). *Ocean currents are driven by the planetary winds; see Hidaka (1952) *J. Oceanograph. Soc. Japan* (available online, captivating English, many equations).

current bedding In a sedimentary rock, bedding which is oblique to the 'lie' of the formation as a whole. The structure is not due to tilting or folding, but develops when sandbanks are built up in shallow water, or where sand dunes accumulate from wind-blown sands, where the pattern of bedding reproduces all or part of the outline of the dunes.

curvature In meteorology, the wind speed divided by the radius of curvature of the bending air stream. Conventionally, *cyclonic curvatures have positive, and anticyclonic curvatures negative, radii of curvature in the Northern Hemisphere. See W. Saucier (2003).

cusp A small hollow in a beach, U-shaped in plan; the arms pointing seawards. Beaches tend to have a series of cusps, formed when outgoing *rip currents and incoming waves set up nearly circular water movements. They form by self-organization, wherein positive feedback between swash flow and the developing morphology initiates the development of the pattern, and negative feedback—from the circulation of flow within the beach cusps—causes pattern stabilization. Ciriano et al. (2005) *J. Geophys. Res. Oceans* 110, C02018 suggest that beach cusp evolution might control low-mode edge wave dynamics.

cuspate foreland An accretion of sand and shingle, shaped by *longshore drift and constructive waves from two directions. See Roberts and Plater (2007) *Holocene* 17, 4.

customs union A common market encompassing two or more states within whose boundaries there is free trade; with no tariffs or barriers to the movement of goods.

cut-off low *See* COLD LOW.

cwm *See* CIRQUE.

cyberspace The information 'space' created by information technologies, notably the internet, the worldwide web, and virtual reality. Cyberspace is firmly embedded in the social spaces seen in the "real" world, and this relationship is reflected in how cyberspace is created, conceptualized, and studied (M. Dodge and R. Kitchin 2001). **Cyberspace geography** researches virtual, rather than real, spaces, considering a human–land–network relationship. CG research includes mapping relationships between cyberspace and real space, redefining the traditional geographic concepts of distance and regions for cyberspace, drawing maps of cyberspace, and researching the principles governing the evolution of cyberspace structures and behaviours. See Gao et al. (2019) *J. Geogr. Sci.* 2019, 29, 12.

cyborg An android whose artificial intelligence and sensitivity are comparable with those of its human creators; a machinic–organic hybrid. Thus, **cyborg geographies** enact hybrid ways of knowing, in contrast with disembodied, dualistic, masculinist, and teleological bodies of knowledge. See Wilson (2009) *Gender, Place and Culture* 16, 5.

cycle of poverty A vicious spiral of poverty and deprivation passing from one generation to the next. Poverty leads very often to inadequate schooling and then to poorly paid employment. As a result, the affordable housing is substandard—low housing costs in poor neighbourhoods attract migrants from rural areas—which leads to overcrowding, overuse of facilities and services, and can also contribute to the perpetuation of the cycle. Children growing up in such areas start off at a disadvantage, and so the cycle continues; see Chilton et al. (2007) *Indian J. Med. Res.* 126. J. Lin and C. Mele (2005) argue the need for class-specific policies in America, designed to raise educational levels, improve the quality of public schools, create employment, reduce crime, and strengthen the family: 'only a simultaneous attack along all fronts has any hope of breaking the cycle of poverty.'

cyclogenesis The formation of *cyclones, especially mid-latitude depressions. Cyclogenesis occurs in specific areas, such as the western North

Atlantic, western North Pacific, and the Mediterranean Sea, and is favoured where thermal contrasts between air masses are greatest. Cyclogenesis occurs when *divergence in the upper *troposphere removes air more quickly than it can be replaced by convergence at ground level. The net result is low pressure. The significance of upper-air movements in cyclogenesis is also indicated by the link with *Rossby waves; see Molinari et al. (forthcoming) *J. Atmos. Sci.* Surface depressions develop below the downstream, or eastern, limbs of Rossby waves, where the airflow is divergent. The routes of mid-latitude cyclones (*depression tracks) closely parallel the movements of the upper-air *jet stream.

cyclomatic number In *network analysis, the number of circuits in the network. It is given by:

$$\mu = e - v + p$$

where e = number of edges, v = number of vertices (nodes), p = number of graphs or sub-graphs. A high value of the cyclomatic number indicates a highly connected network. Gorman and Maleki (2002) *Telecomm. Policy* 2 note the relationship between the level of economic development of a region and the cyclomatic number of its major transport networks.

cyclone A *synoptic-scale area of low *atmospheric pressure with winds spiralling about a central *low. As air near ground level flows into a cyclone, its absolute *vorticity increases, and it is therefore subject to horizontal *convergence, causing air to ascend. This rising causes cooling, which often leads to *condensation, so that *precipitation is associated with cyclones. See McTaggart-Cowan et al. (2007) *Month. Weather Rev.* 135, 12 on Hurricane Katrina. See also Löptien et al. (2008) *Clim. Dynamic.*, online.

cyclone wave The wave-like distortion of flow in the middle and upper troposphere associated with a mid-latitude depression. See Orlanski and Katzfey (1991) *J. Atmos. Scis* 48, 17.

cyclostrophic Referring to the balance of forces in a horizontal, tightly circular flow of air. **Cyclostrophic flow** is a form of gradient flow parallel to the isobars where the centripetal acceleration exactly offsets the horizontal *pressure gradient. The **cyclostrophic wind** is the horizontal wind velocity producing such a centripetal acceleration; see J. Dutton (2002). It equals the real wind only where the *Coriolis force is small, or where wind speed and curvature, and hence centripetal acceleration, are great.

D

daily urban system The commuting area around a city. Bretagnolle and Pumain (2001) *Cybergeo* 335 show that this area varies according to the definition of 'city'.

dam A barrier for holding back water, usually in the form of a reservoir, which may be used to generate electricity and/or to supply irrigation water.

dambo A gently sloping, shallow *savanna wetland, with no permanent river channel, often wider upstream than downstream (perhaps through chemical sapping), and often infilled with sands and clays. Core boulders indicate *etchplanation; see von der Heyden (2004) *PPG* 28, 4.

Dansgaard–Oeschger (D–O) cycles Large-scale, but short-lived, climatic swings during the Quaternary period. See Boers et al. (2018) *PNAS*.

Darcy's law When the *Reynolds number is very low, the velocity of flow of a fluid through a saturated porous medium is directly proportional to the difference in water pressure at the two sites:

$$V = \frac{h}{P_l}$$

where h is the height difference between the highest point of the water-table and the point at which flow is being calculated, V the velocity of flow, P the coefficient of permeability for the medium in question, and l the length of flow. Darcy's law does not hold good for well-jointed limestone which has numerous channels and fissures.

Darcy's equation can be written in several ways: see GroundwaterSoftwater. com (March 2004). See also Buchan and Cameron in S. W. Trimble et al., eds (2003).

data acquisition The process of sampling signals that measure real-world physical conditions and convert the resulting samples into digitized data that can be used by a computer. Data acquisition systems (abbreviated to **DAS** or **DAQ**) typically convert analog waveforms into digital values for processing.

(((⊕))) SEE WEB LINKS
• Source of free data acquisition applications.

data mining A process of extracting information from a data set in order to transform it into an understandable structure. Multivariate statistical methods include multiple linear regression, *principal component analysis, *cluster analysis, and the parallel coordinate method; see Sun and Wang (2011) *IEEE International Conference* DOI: 10.1109/ICSDM.2011.5969006.

data model A GIS data model enables a computer to represent real geographical elements as graphical elements. The *vector data model uses points, lines, and polygons, the *raster data model uses cell matrices that store numeric values, and the *triangular irregular network represents geography as sets of contiguous, non-overlapping triangles. See K.-T. Chang (2014).

data structure In computer science, the way data are organized and stored in a computer.

datum In *geodesy and *GIS, a geographic coordinate system covering the Earth's surface, such as latitude and longitude, by which phenomena may be located. When loading a GIS, the correct datum for each map source should be used. When working with more than one map source at a time, all data must be converted to a common datum. The most frequently used datum is the Universal Transverse Mercator (UTM).

dead cliff A sea cliff no longer subject to wave attack, with an emerged strandline, often overlain with *shingle at its foot. Dead cliffs result either through a fall in sea level or because they are protected by a broad beach. See Mörner (2017) *J. Coastal Research* 33, 2 on coastal morphology and sea-level changes in Goa.

dead ice (stagnant ice) Static glacier or ice-sheet material. Ice may stagnate after a surging extension is abandoned, or at the end of a glacial period. As stagnant ice melts, the remnants are covered in *ablation moraine. Dead ice in valleys may generate *kame terraces.

death rate The number of deaths in a year per 1 000 of the population, as measured at mid-year. This is a crude rate—no allowance is made for different distributions of age and sex.
standardized death rate compares the rate with a real or assumed population which is chosen as standard.

debris Material such as scree, gravel, sand, or clay formed by *weathering.

Through the air, debris is transported by *saltation and *deflation; in water, debris moves by rolling, by saltation, in solution, and in suspension. Debris can be carried on a glacier—supraglacially—within a glacier—englacially—or subglacially. **Debris entrainment** is the process whereby ice picks up material: by the freezing of basal ice to the bed as the ice flows forward through the drag between ice and bedrock particles; via the closure of debris-filled basal cavities in the ice; by the refreezing of meltwater; and on the glacier surface. See Cuffy et al. (2000) *Geology* 28, 4; **Debris flow** is the very rapid downslope movement of saturated material, guided by stream channels, and both less deep seated, and rarer, than a landslide. **Debris flow fans** are used in inferring past flow characteristics, and, consequently, future hazards.

debt-for-nature swap A transaction whereby some of a developing nation's foreign debt is written off with the proviso that the debtor nation invests in environmental conservation measures. Go to Tristan Bove (2021) Earth.Org. Andrew Kessel (2006) provides a critical review of debt-for-nature swaps.

decalcification The *leaching of calcium carbonate from a soil *horizon by the downward movement of soil water. See van den Berg and Loch (2000) *Eur. J. Soil Sci.* 51, 1.

decarbonization The reduction or removal of the release of carbon dioxide (CO_2) into the atmosphere. It is achieved by switching to usage of low-carbon energy sources. See Cederlöf (2020) *TIBG* 45, 1.

decentralization (deconcentration) The devolution of decision-making powers to the lowest levels of government authority. National governments may use decentralization to regenerate declining regions. In 2003, the government of the Republic of Ireland began the transfer of complete government departments to provincial locations.

deciduous forest In the cold season of temperate latitudes, a tree's water supply is restricted when the temperature falls below 0 °C. To lessen water loss, deciduous trees shed their leaves until the spring brings more available moisture. See Bradshaw et al. (2005) *Ecography* 28, 2.

decision-making Geographical approaches to decision-making stress the spatial context in which the decision is made, and the reciprocal nature of a space–decision relationship. See Dengler (2007) *Pol. Geog.* 26, 4 on the formation and evolution of *ad hoc* organizational spaces of decision-making. Pickerill and Chatterton (2006) *PHG* 30, 6 value the **consensus-based decision-making** characteristic of alter-globalization (globalization which puts democracy, environmental justice, and environmental protection before economic considerations.

See Viehe et al. (2006) *Geografiska B* 88, 2 on the DaisyGIS decision-making model.

It is argued that 'satisficing' (as opposed to 'maximizing') is a much more faithful account of the way people make choices; people pick the alternative that satisfies a given level of aspirations without looking necessarily for the best possible option.

declining region A region suffering the economic decline associated with high unemployment, often because of factory closures or outmoded industry. Regions may be declining in terms of absolute or, more often, relative living standards and welfare indicators, but also in terms of an absolute or relative

fall in population, since the response to economic decline will be out-migration and reduced in-migration. This second kind of decline is often a symptom of the first. See Klosterman et al. (2006) *Int. J. Env. Tec. & Manage.* 6, 1–2, on the What if?™ *GIS system, as used in evaluating remedial strategies in declining regions.

decolonization 1. The achievement of self-governance by former colonies.
2. The freeing of minds from colonial ideology, especially from the idea that to be/have been colonized is to be inferior.
3. The dismantling of neocolonial relationships with former colonies.
4. A commitment to indigenous territory, indigenous sovereignty, and indigenous ways of thinking.
Clayton and Kumar (2019) *J. Hist. Geog.* 66 examine how ideas of decolonization are used within the discipline of geography. See also the entire issue of *TIBG* (2017) 3.

decommodification Given that a commodity is an artefact, symbol, or idea that can be exchanged, usually for money, *de*commodification would indicate that a commodity no longer has a status as such. This might result when it is thrown away, or when it is no longer deemed by ethical consumers to be exchangeable, for example if the commodity is produced by child labour, if its production or transportation pollutes the environment, and so on. As an example of this sort of decommodification, in 2006 the Dalai Lama stated that Tibetans should cease wearing clothing lined with skins of endangered animals (Yeh (2013) *J. Asian Studs* 72, 2, 319).

Decommodification is also any cultural, political, or social, process that reduces the scope and influence of the market in everyday life. In a decommodified system, welfare services such as education and healthcare are provided to all and are not linked to market processes because they are an entitlement, rather than as a commodity that must be paid for or traded.

decomposer *See* DETRITIVORE.

deconcentration *See* DECENTRALIZATION.

deconstruction Used for a text, a type of *critical analysis that considers the internal workings of the language and modes of thought used: a theory, with methodological implications, that emphasizes the incompleteness of categories and concepts used to make sense of the world. For the geographer, deconstruction should bring to light alternative layers of significance in any *discourse/discussion, pointing out contradictions and unspoken assumptions. Doel (2005) *Antipode* 37, 2 provides an entertaining, instructive account.

deepening A decrease in the central *atmospheric pressure, usually of a low pressure system. See Killer and Petty (1998) *Monthly Weather Rev.* 126, 9.

deep weathering The *weathering of rocks, up to hundreds of metres in depth, by air and water, forming *saprolite—see Smith et al. (2002) *Catena* 49, 1–2 on the premises of deep weathering. Deep weathering has played a major part in the geomorphological evolution of Europe.

defensible space The space managed and controlled by the inhabitants of a community. This involves physical and design changes to maximize the potential for surveillance by residents: if public space can be segmented into small, controllable areas, this would encourage residents to exercise control over these areas, which should discourage crime.

deflation The *denudation and lowering of a surface by wind action, for example, contemporary deflation from the Bodélé Depression, Chad, is controlled by topography and wind stress. See French and Demitroff (2001) *Permaf. Perig. Procs* 12 on deflation hollows in periglacial environments.

deforestation The complete clearance of forests by cutting and/or burning. Between 1995 and 2005, the world's top deforesters were Brazil, Indonesia, and Burma. Using high-resolution satellite images to study landscapes in 9-kilometre-wide blocks across the planet between 1992 and 2015, Stepinski and Nowosad found that deforestation occurs comparatively slowly in the blocks until about half of the forest is gone. Then the remaining forest disappears very quickly. The major drivers of deforestation include commercial and subsistence agriculture, illegal logging, urban expansion, and mining.

deglaciation The process by which glaciers thin and recede—through climatic change and/or by increased *calving when sea levels rise. Deglaciation is generally accompanied by the release of *meltwater and the formation of recessional *moraines. See Noormets et al. (2021) *Boreas* 50, 1 on glacial dynamics and deglaciation in Spitsbergen.

deglobalization The process, unwitting or deliberate, of reversing *globalization, of reducing the volume, scale, and velocity of social and environmental interactions. This might entail: producing for domestic markets only; protecting local industries through tariffs; and the replacement of global banks by regional ones.

deglomeration The movement of activity, usually industry, away from *agglomerations, perhaps when congestion makes further agglomeration in a region difficult and expensive, or as a result of economic decline; see Tazviona Richman Gambe (2019) *Resilience* 7, 1 on deglomeration in Harare, Zimbabwe. **Deglomeration economies**, such as lower transport costs, or longer commuting times and congestion externalities, lead to *decentralization.

degradation The lowering and flattening of a surface through erosion. **Land degradation** is the long-term loss of ecosystem function and productivity caused by disturbances from which the land cannot recover unaided. The relevant journal is *Land Degradation and Development*. See also Kriage (2013) *PPG* 37, 5, 664.

degree-day The concept of the degree-day is based on the assumption that any plant will require its own total heat input. Every day when temperatures would permit that plant's growth is a degree-day. Calculating a degree-day can be complicated; fortunately, plant degree-days are readily available online.

degrowth The opposite of growth. It entails a reduction of production and consumption—the contraction of local, national, and global economies. Devotees of degrowth argue that this does not require individual self-sacrifice or a decrease in well-being; instead, their aim is to maximize happiness and well-being through non-consumptive means—sharing work, consuming less, while devoting more time to art, music, family, culture, and community. See Jarvis (2019) *PHG* 43, 2.

((( ))) SEE WEB LINKS
• Degrowth magazine.

dehesa A system of agroforestry that creates a 'man-made savanna', where the trees are an integral part of the ecosystem, and, as such, are managed, planted, and pruned. See Joffre (1999) *Agroforestry Systems* 45.

deindustrialization The decreasing significance of the manufacturing sector, both in terms of employment, and national production. **Positive deindustrialization** is a normal feature of economic development, since, as manufacturing increases, workers are increasingly employed in services.

Negative deindustrialization is characterized by a vicious cycle linking falling incomes, rising unemployment, the erosion of competitiveness, low-wage economies, the casualization of employment, and increasing *social exclusion. Look for Schindler (2018) on the new international geography of deindustrialization.

delta A low-lying area at the mouth of a river, formed from *alluvial deposition, that occurs as the river's silt-carrying capacity is checked when it slows in the more tranquil waters of a lake or sea. Deltas develop through the progradation of river mouths and delta shorelines producing a subaerial deltaic plain over the seaward delta-front deposits. River deltas may be limited when sediment retention approaches zero, when the delta front advances into deep waters, or by the combination of sea-level rise and subsidence. See Bellotti et al. (2011) *Holocene* 21, 1105 for a detailed study of the Tiber Delta.

Inland deltas form in hot, arid inland drainage areas, where water dynamics and sediment transport are very different, especially deposition associated with expansion–contraction dynamics at the channel head. As a result, the mean slope of the delta decreases with distance from the apex.

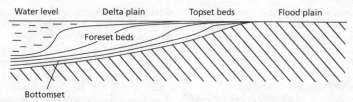

Water level   Delta plain   Topset beds   Flood plain

Foreset beds

Bottomset

Delta shore composition

demographic transition (DTM) An account, but not a complete explanation, of changing rates of fertility, mortality, and *natural increase. Four stages may be recognized: in the High Stationary Stage, birth and death rates are high and the death rate fluctuates from year to year; the Early Expanding Stage is characterized by high birth rates, falling death rates, and increases in population; in the Late Expanding Stage death rates are low and fertility is declining, but population is still increasing; and, lastly, in the Low Stationary Stage birth and death rates are low and the birth rate fluctuates. There seems to be a fifth stage where birth rates fall below death rates so population levels fall.

Death rates fall because of improved conditions and health care. The reasons for falling fertility are less clear, but include: anti-child labour laws; pensions (which remove the need for children as a support in old age); a higher proportion of women in work; and increasing costs of bringing up children, as standards of living rise. Everywhere, mortality decline appears to have played a central role for fertility decline. With falling child death rates, fertility falls rapidly in the developed world, and more slowly in more recent transitions. Because of their late start to the DTM, developing countries will not catch up demographically with more developed regions for over 200 years: 'in fact, less developed regions never seem fully to catch up' (UN Dept. Econ. & Soc. Affairs, Pop. Division).

The **second demographic transition** is marked by: a later age at marriage; an increase in the proportion of adults living alone or cohabiting; slower rates of remarriage; increased fertility outside of marriage; and delaying or forgoing childbearing. See Zaidi and Morgan (2017) *Annual Rev. Sociol.* 43 for a review and appraisal of the second demographic transition theory.

demography The observed, statistical, and mathematical study of human populations, concerned with the size, distribution, and composition of such populations.

dendritic drainage *See* DRAINAGE PATTERNS.

dendrochronology An absolute dating technique using the growth rings of trees. **Dendrogeomorphology** is the dendrochronological interpretation of geomorphic processes. See Šilhán et al. (2016) *Catena* 147 on an evaluation of dendrogeomorphic techniques.

density *See* URBAN DENSITY.

density dependent factors The checks to population growth, such as *competition, which result from overcrowding; factors which connect the dynamics of human-dominated landscapes, and 'natural' ecological processes. See Fagan et al. (2001) *Landsc. Ecol.* 16.

density gradient The rate at which the intensity of land use or the density of population falls with distance from a central point. See Chen (2009) *Int. J. Urban Sustainable Devt* 1, 1–2 for a new model of urban population density indicating latent fractal structure.

denudation A general name for the processes of weathering, transport, and erosion. See Twidale (2007) *Phys. Geog.* 28, 1 on variations in denudation.

dependency ratio The ratio between the number of people in a population between the ages of 15 and 64 and the **dependent population**: children (0–14) and elderly people (65 and over). It is used as a rough way of quantifying the ratio between the *economically active population and those they must support, but the age limits are somewhat arbitrary as, in the UK, for example, 30% of over-16-year-olds go on to higher education, and are therefore still dependent, either on the state or, increasingly, on their parents or bank managers.

dependency theory The view that the development of the advanced economies has caused the underdevelopment of less economically developed nations; see A. G. Frank (1967). Notions of what defines dependency are diverse and vague. It might be reliance on a set of international rules and institutions that are biased against developing nations, insufficiently nationalist and state-led development, the growing importance of international financial capital, and the ongoing reliance of developing countries on technological innovations in the *core. D. Slater (2004) finds that while dependency theory possesses a continuing relevance, owing to its concern with the marginalized and oppressed, a post-colonial version of dependency theory would introduce concerns with agency, discourse, and knowledge-making.

depopulation The decline, in absolute terms, of the total population of an area, more often brought about by out-migration than by a fall in fertility or excessive mortality. Depopulation typically occurs in more remote areas; it is initiated by loss of employment, effected through out-migration, and exacerbated by a vicious cycle of decline. See Zaslavsky (2007) *J. Reg. Sci.* 40, 4 on depopulation and Russia's demographic crisis.

depression An area of low pressure (roughly, below 1000 mb); *see* MID-LATITUDE DEPRESSION. **Depression tracks** are influenced by the courses of *jet streams, energy sources—such as warm seas—and mountain barriers.

deprivation Loss; lacking in desired objects or aims. Within the less developed countries deprivation can be acute: water, housing, or food may be lacking. Within the developed world basic provisions may be supplied but, in comparison with the better-off, the poor and the old may well feel a sense of deprivation. This is **relative deprivation**, which entails comparison, and is usually defined subjectively. In the UK, inequalities in health seem to be more marked in deprived areas than in more affluent ones. In more affluent Western societies relative deprivation seems to be more important than absolute deprivation in determining population health. The impetus for social change comes not so much from changes in absolute deprivation, but from relative deprivation.

The **cycle of deprivation** is the transmission of deprivation from one generation to the next: lack of knowledge and skills all contribute to a cycle of deprivation, leaving the poor with few resources and many barriers to overcome.

(⊕) SEE WEB LINKS
- Describes circle of deprivation.
- UK indices of deprivation.

deregulation Rolling back state 'interference' in social and environmental life.

desakota region A region characterized by an intense mix of agricultural and non-agricultural activities that often stretch along corridors between large city cores. Population density is high, stemming from agriculture, usually wet-rice production. See Kesteloot (2015) on desakota regions in the Mekong Delta.

desalination (desalinization, desalting) The removal of dissolved salts and minerals from saline water to produce fresh, drinkable water. Commercial desalination processes include: reverse osmosis, electrodialysis, multi-effect distillation, and vapour compression. The key issues associated with desalination are cost, energy use, and the environmental impacts of brine disposal and feedwater intake. See Trimble et al. (2005) *Encyc. Water Sci.*

desegregation The phasing out of the separation of humans, usually with reference to different ethnic, caste, or national groups. Policies prompting desegregation and social mix rarely meet policy expectations. Residential mixing cannot be assumed to enhance community cohesion or people's social capital. In fact, urban renewal policies have been found to disrupt communities. Displaced households experience difficulties in establishing new social ties. At least in the short term, dispersal often leads to a decline in social support. India has passed multiple laws to penalize the caste system present in the society, with little success.

desert An area of very low precipitation, sometimes specified as below 250 mm/yr^{-1} or where *evapotranspiration exceeds precipitation. A **desert pavement** is a continuous mantle of flat-lying, densely packed, partially overlapping pebbles, typically overlying a soft, silty layer. The pavement may be a lag deposit, made of rocks left behind after the wind has blown away all the fine-grained material. Alternatively, during the occasional rains, sheet flow washes away the silty layer. A third theory is that heaving moves stones to the top. See Quaid (2001) *Geology* 29, 9. **Desert soils** lack clear soil *horizons, as the climate is too dry for chemical *weathering or humus formation. *Leaching occurs only after occasional rain, and is soon reversed by evaporation. **Desert varnish**, also known as soil varnish, is a thin red to black glaze on exposed rock surfaces in arid regions. It is made of oxides and hydroxides of manganese and/or iron and clay minerals, together with particles of sand grains and trace elements, all bound together with silica gel. See Sarmast et al. (2017) *Catena* 149, part 1.

desertification The meaning of this word is much debated. It is defined by the UN as the degradation or destruction of land, which results in desert-like conditions. Thomas in J. Holden (2012) defines desertification as 'an often misunderstood and abused term used to describe land degradation in drylands. Misunderstanding often arises because natural environmental (especially plant system) responses to drought, from which recovery usually occurs, have been confused with longer-term and more persistent negative changes'. The causes and consequences of desertification are complex and little understood. 'Occasional droughts, due to the seasonal factors or inter-annual variations of rains, and severe droughts of long periods can be caused or aggravated by human influence on the environment, reduction of vegetation covering, change of the effect of albedo, local climate changes, greenhouse effect, etc.' (UNCCD 2004). Rasmussen et al. (2001) *Glob. Env. Change* 11, 4 conclude that 'broad generalisations on land degradation processes, based on local-scale studies, are risky. Significant variations exist at the landscape level, and trends in ecosystem dynamics are sometimes totally reversed even within small regions . . . conclusions concerning "irreversible" degradation, based on few years or decades of observations may be premature'.

SEE WEB LINKS

• The EU Science Hub 'desertification and drought' site is a fruitful source.

desilication (desilification) The removal of silica from a *soil profile by intense weathering and *leaching. It occurs when water that is undersaturated with silica percolates through a soil at the correct pH, and is facilitated by freely draining conditions in a hot, humid environment. See Puppe et al. (2020) *EGU General Assembly* on the anthropogenic desilication of agricultural soils.

deskilling Breaking down jobs into smaller units, each to be tackled separately, so that low levels of skill— and therefore cheaper labour—are required for restricted tasks. Hendrickson and Heffernan (2002) *Sociologia Ruralis* 42 claim that the deskilling of farmers is well advanced in parts of the USA. See Korzeniewska et al. (2021) *Migration Studies* 9, 1 on deskilling migrant nurses in Norway.

destructive margin The zone where two *plates meet, and oceanic crust is subducted beneath either oceanic or continental lithosphere, changing to the more ductile, fluid-like, denser rocks of the aesthenosphere via increases in temperature and pressure. Subduction creates strong compression at the plate

margins, piling up unconsolidated segments and pieces of rock slabs into a wedge.

destructive wave A plunging wave, with a short wavelength, a high frequency (13–15 per minute), and a high crest. *Backwash greatly exceeds *swash. Destructive waves comb beach material seawards (Joliffe (1978) *PPG* 2).

determinism The view that human actions are stimulated and governed by some outside agency. **Environmental determinism** assumes that the physical environment is the primary determinant of cultural forms. During the late 19th and early 20th century, some geographers claimed a link between imperialism and environmental determinism (see Smith in J. N. Entrikin et al., eds 1989)—a view that has long been rejected.

deterritorialization At its simplest, deterritorialization may mean the take-over of an area or place (territory) that is already established. G. Deleuze and F. Guattari (1972) use the term to represent the freeing of labour-power from specific means of production, as when the English Enclosure Acts (1709–1869) enclosed previously common land for private landlords. As such, deterritorialization is the severance of social, political, or cultural practices from their native places and populations. From this, we could see the decline in the role of the state as deterritorialization; 'certain *transnational corporations have more financial power, within a state, than the state itself' (Kapferer (2005) *Anthrop. Theory* 5, 3). Deterritorialization makes it increasingly difficult to maintain a convincing nationalistic discourse that involves cultural purity, but Sofia Voytiv (2020) argues that ethnicized armed conflict in a homeland can become deterritorialized; such conflicts can detach ideas, attitudes, and ethnicized narratives from a certain geographical location and 'settle them in the transnational space of interactions'. Connell and Gibson (2004) *PHG* 28, 3 claim that the expansion of world music exemplifies the deterritorialization of cultures. Globalization has been explicitly seen as deterritorialization by J. A. Scholte (2000).

detritivore An animal which feeds on fragments of dead and decaying plant and animal material. Detritivores have a vital part to play in *food webs and nutrient recycling.

detritus Fragments of weathered rock which have been transported from the place of origin. See Doyle (2006) *Geomorph.* 77, 3–4 on detritus loss rates; try Valdiya (2002) *PHG* 26, 3.

development The use of resources to relieve poverty and raise living standards; the means by which a traditional, low-technology society is changed into a modern, high-technology society, with a corresponding increase in incomes. This can be done through mechanization, improvements in infrastructure and financial systems, and the intensification of agriculture. This definition is based on the more obvious distinctions in living standards between developed and less developed countries. There is no definitive definition of what development should be for each society, and no blueprint for how to achieve it. See D. Gasper (2004).

Narrowly economic definitions of development have been criticized, and many believe that true development includes social justice. Smith (2002) *Area* 34 argues against development as 'some neutral concept of progress'. Potter

(2001) *Area* 33 argues that development is a non-linear process, and 'one should not mimic the experience of "developed" countries'. 'For many local communities, the maintenance of social and cultural practices is central to **participatory development** and just as important as income gains and poverty reduction' (Connell (2007) *Sing. J. Trop. Geog.* 28, 2). See O'Reilly (2007) *AAAG* 97, 3 on women's participation in development; see also Sharpe et al. (2003) *TIBG* 28, 3. **Development indicators** are measurements, such as life expectancy at birth, or per capita GDP , that are simply concerned with statistics, and do not indicate social structures and patterns of behaviour.

Ethnodevelopment is indigenous development in indigenous hands. See Hogue and Rau (2008) *Urb. Anthrop. & Studs Cultural Systs & World Econ. Devt* 37, 3/4 on the way ethnic identities and a framework for ethnodevelopment are emerging in Andean communities, in contrast and resistance to the neoliberal commodification of natural resources.

(((●))) SEE WEB LINKS
• World Bank development indicators.

developmental state A state that derives political legitimacy from its record in economic development, which results mainly from selective industrial policy. The classic model of the developmental state is derived from Japan between the 1950s and the 1980s. O. Edigheji ed. (2010), on South Africa as a developmental state is useful and comprehensive.

Devensian The last full glacial series in Britain, lasting from about 70000 to about 10000 years BP. During this time, sea levels were 130–160 m below current levels; Ireland was joined to Britain, and

Britain to northern Europe and Scandinavia by a broad dry plain. The later part of the Devensian embraces pollen zones 1–111 as defined by Godwin (1940).

devolution The transfer of powers from central government to regional and local governments. UK devolution has been based on the principle that decision-making powers be devolved to the appropriate level of *governance, creating new assemblies in Wales and Scotland for the purpose. In response to *nationalism, devolution may be seen as one way to avoid ethnic unrest. See Brownlow (2017). *Cambridge J. Regions, Econ. & Soc.* 10, 3, on the economic lessons of UK devolution. In the UK, devolution may also comprise delegating power from central government to city-regions; see the Liverpool City Region Devolution agreement.

(((♦))) SEE WEB LINKS
• Liverpool City Region Devolution Agreement.

Devonian A period of *Palaeozoic time stretching approximately from 395 to 345 million years BP.

(((♦))) SEE WEB LINKS
• An overview of the Devonian period.

de Vries cycle (Suess cycle) This is an approximately 200-year solar cycle, believed to be one of the most intense. Raspopov et al. (2008) *Pal., Pal., & Pal.* 259, 6 find that palaeoclimatic reconstructions correlate well with these variations in solar activity.

dew A type of condensation where water droplets form on the ground, or on objects close to it. Dew forms when strong night-time *terrestrial radiation cools the ground to chill the overlying air to *dew point.

dew point The temperature that a body must be chilled to for it to become *saturated with respect to water, so that condensation can begin; also seen as the temperature of a chilled surface just low enough to attract *dew from the *ambient air. The dew point of an *air mass varies with its initial temperature and humidity.

dialect A variant of a language in terms of vocabulary, grammar, syntax, and pronunciation. Dialect differences arise and persist because of geographical barriers (the Ohio River, for instance, helps to define the division between the dialect areas of the North and the South USA), and because of political boundaries, settlement patterns, migration and immigration routes, territorial conquest, and contact with other languages. The *Journal of Linguistic Geography* focuses on dialect geography and the spatial distribution of language relative to questions of variation and change. See also Langevelde and Pellenbar (2001) *Tijdschrift* 92, 3. **Dialectology** is the study of social and linguistic variations within a language; **dialect geography** is the study of local differentiations in a speech area: see Schneider (2002) in D. Kastovsky et al., eds *Anglistentag 2001 Wien.* on quantitative techniques in the analysis of dialect data.

dialectics Arriving at the truth by the exchange of opposing logical arguments. Especially associated with Hegel, this process states a thesis, then develops a contradictory antithesis, and resolves these into a coherent synthesis. Sheppard (2008) *Env. & Plan. A* (2008), 40, 2603 is not for the fainthearted, but rewards study.

diapir An upward-directed, dome-like intrusion of a lighter rock mass into a denser cover. See Navarro-Carrasco et al. (2020) *Journal of Maps* 16, 2 on the geology and evolution of the Cortes de Pallás diapir.

diaspora The dispersion of people from their homelands. Diaspora may be a descriptive typological term, or seen as a social condition, involving multiple allegiances and belongings, a recognition of hybridity, and the potential for creativity in diasporas; 'diaspora involves feeling "at home", in the area of settlement while retaining significant identification outside it' (B. Walter 2001). Laoire (2003) *Int. J. Popul. Geogr.* 9 is excellent. Drozdzewski (2007) *Soc. & Cult. Geog.* 8, 6 discusses the way that different groups of Polish people contextualize their diasporic identity, and attribute meaning to place. Writing on India's **diaspora strategy**, Dickinson (2012) *TIBG* 37, 4, 609 argues that 'specifically, incorporating the diaspora into a narrative of Indian sovereignty is an act that marks out India's own reinvention from that of a post-colonial developmentalist nation (and the concomitant excision of its colonial diaspora from its anti-colonial nationalist rhetoric) to an emergent power in the international economy'. New Zealand's diaspora strategy is to engage expatriates to extend international marketing opportunities, without requiring them to return home (Larner (2007) *TIBG* 32, 3).

diatoms Single-celled algae with a resistant siliceous outer shell that enables them to be preserved in marine and freshwater sediments. For the use of diatoms in physical geography, see Mannion (1982) *PPG* 6: 233.

diazotroph An organism which fixes atmospheric nitrogen to the soil.

difference Difference exists only with reference to another; in the case of the social sciences, this may be patriarchy, 'whiteness', 'colour', *feminism, *colonialism, or *post-colonialism, all of which may be defined by what they are *not*. 'The differentiation of an owning (capitalist) class from . . . the working class is foundational for capitalism in ways that gender, racial or sexual differences, however violent and life-destroying, are not' (Malden (2008) *PHG* 31, 1). Civil wars in Sri Lanka and sectarian fighting in Iraq are violent expressions of the difficulties of incorporating difference in the nation. I. Rivera-Bonilla (2000) analyses the way gating residential areas recreates spatial markers of class and racial differentiation. See also Valentine (2008) *PPG* 32, 3, 323.

differential erosion The selective erosion of surfaces, so that softer rocks, such as clays and shales, or lines of weakness, such as joints and faults, are eroded more rapidly than resistant, competent, and unjointed materials.

diffraction Bending and spreading. For diffraction in geography, see Sharp (2019) *PHG* 43, 5.

diffuse pollution A widespread pollution that cannot be attributed to a specific source. The OECD report *Diffuse Pollution, Degraded Waters: Emerging Policy Solutions* (2017) outlines the water quality challenges facing OECD countries today, presenting a range of policy instruments and innovative case studies of diffuse pollution control, concluding with an integrated policy framework to tackle this challenge.

diffusion The widespread dispersal of an innovation from a centre or centres. T. Hägerstrand's model of diffusion (1968) implies the existence of a *mean information field which regulates the flows of information around a regional system; flows that are moderated by barriers which can obstruct the evolution of information into innovation. 'To the extent that a high proportion of personal contacts are local, the diffusion of information must also be spatially constrained: the further people are from early adopters the later they are likely to be to adopt the innovation. To the extent that people's non-local contacts are hierarchical (more likely to be with people in big cities than in small), the diffusion of information must also be hierarchically constrained: the lower down the urban hierarchy people live, the later they are likely to be to adopt the innovation . . . the diffusion of innovations reflects supplier behaviour as well as adopter behaviour' (Webber (2006) *PHG* 30, 4).

digital divide The differential access to computers, information, the internet, and telecommunications, globally, regionally, nationally, and locally, that also comprises unequal access to job opportunities, resources, and training. It is not solely a technological divide, but results from the uneven spatial distribution of resources; advanced technologies have a pro-rich bias: they are essentially designed for developed countries. For less economically developed countries, strategies for closing the divide aim to leapfrog the earlier stages of economic development, bringing advanced communication technologies to areas of deprivation, in a way suited to local socio-economic conditions; see Vassilakopoulou et al. (2021) *Information Systems Frontiers* on bridging digital divides. See the World Bank's annual *Little Data Book* for data on technology and infrastructure from over 200 countries.

In the UK, there is a clear correlation between digital exclusion and social exclusion. This means that those already at a disadvantage and arguably with the most to gain from the internet are the least likely to be making use of it and become further disadvantaged by not using it. See Blank et al. (2018) *Soc. Sci. Computer Rev.* 36, 1 on local UK geographies of digital inequality.

digital elevation model (DEM) An array of regularly spaced elevation values referenced horizontally either to a Universal Transverse Mercator (UTM) projection, or to a geographic coordinate system. See USGS (2000) *Fact Sheet 040-00* for a thorough and helpful summary.

digital geographies The computer has drastically transformed both geography as an academic discipline and the geography of the world, and the **digital turn** in geography is the turn to the digital as both object and subject of geographical inquiry, and to the ways in which the digital has influenced geographic thought, scholarship, and practice. See Ash et al. (2018) *PHG* 42, 1.

digital terrain analysis (geomorphological analysis, landform parametrization, land surface analysis) The quanitification of terrain. A **digital terrain model** (*digital elevation model) is a digital relief map. See Deng (2007) *PPG* 31, 4; Lacroix et al. (2002) *Env. Model. & Software* 17, 2; and Thwaites (2002) *New Forests* 24, 2.

dike *See* DYKE.

dilatancy 1. The expansion of a mass of granular material—sand, for example—when its component grains are rearranged.
2. The solidification of a viscous substance under pressure.
Dilatancy theory emphasizes the role played by variations of stress in *till found at the base of moving ice masses. A local reduction in stress causes the till to settle and become more compact. See Minchew et al. (2020) *Proc. Royal. Soc. A*. Dilatancy theory has also been used in *earthquake prediction. See Main et al.

(2012) *Geological Soc. London Special Publication*. **Dilatation** is *pressure release.

diminishing returns, law of The principle that further inputs into a system produce ever lower increases in outputs. Radelet, Clemens, and Bhavnani (2005) *Finance and Dev.* 42, 3 find that *aid has a positive relationship with growth, but with diminishing returns.

dip The angle of inclination of a rock down its steepest slope, that is to say, the direction at right angles to the strike. Dip is the angle between the maximum slope and the horizontal. A **dip slope** occurs where the slope of land mirrors the slope of the underlying *strata.

direct cell *See* ATMOSPHERIC CELLS.

dirt cone A cone of ice, up to 2 m high, covered with a thin layer of debris, found in the *ablation zone of a glacier.

disability; geographies of disability, welfare, and social exclusion Hunt (1966) *Union of the Physically Impaired against Segregation* distinguishes between *impairment*, which relates to individually based bio-physical conditions, and *disability*, 'which is about the exclusion of disabled people from "normal" or mainstream society'. Geographies of disability consider the way space, place, and mobility have shaped, and continue to shape, the experiences of disabled people. Those physical barriers which bar or limit access for disabled people are not inevitable and unavoidable, but result from institutional and political processes that produce disabling spaces; it is argued that the biggest obstacle to disabled people's meaningful inclusion into mainstream community life is negative public attitudes. Place is influential in how disabled people feel about themselves. Wilton (2004) *TIBG* 29, 4 outlines the multiple strategies of control of the disabled in the workplace; notably the denial of access to accommodation. V. Chouinard, E. Hall, and R. Wilton, eds (2010) consider the 'spatial interpretation of disability' in a strongly recommended text. See Wilton and Schuèran (2006) *Area* 38, 2 for an examination of the interrelationships between disabled people and labour market opportunities and barriers.

disaster Usually, a sudden catastrophe with serious consequences. While the causes and nature of a disaster are largely the preserve of physical geographers, human geographers also consider the causes and impact of disasters. Le Billon and Waizenegger (2007) *TIBG* 32, 3, 411 note that 'disaster risk has "political roots", notably unequal power relations and under development processes'. Politics, rather than the strict needs of disaster victims, also influence responses to disasters; interpretations of disasters represent political choices with political impacts, particularly from a gender perspective, and for disaster recovery or future risk mitigation. See Hualou (2011) for a geographical perspective on the prevention and management of disaster, and Donovan (2016) *Prog. Hum. Geog.* 41, 1 on the critical geography of disasters.

discharge The quantity of water flowing through a cross-section of a stream or river in unit time. Discharge (Q) is usually measured in *cumecs, and calculated as $A \times V$ where A is the cross-sectional area of the channel and V is the mean velocity. See R. Charlton (2007) pp. 57–9 for instructions on measuring river discharge. Definitions of **dominant discharge** (also known as **effective discharge**) vary; at its simplest it is the flood discharge that achieves the greatest total geomorphic work, but see Ferro and Porto (2012) *Geomorph.* 139–40, 313 for a thorough review of definitions. For those rivers in which the relationships of bedload discharge to water discharge are not dominated by very large-sized particles, and where daily mean discharges are an adequate description of streamflow characteristics, dominant discharge is often considered to equate to bankfull discharge.

discordant Cutting across the geological grain, as in a stream cutting across an *anticline.

discourse Communication. Academic discourse is how we alter our communication in scholarly discussion, so it's a social process.

Discourses create their own tests for truth; that is, the acceptable formulation of problems, and solutions to those problems. See Nijman (2007) *Tijdschrift* 98, 2 on 'the prevailing political and ideological discourse [that] seeks to maintain the Haitian identity'.

discrete choice modelling *See* CHOICE MODELLING.

discrimination The exclusion of specific groups from certain sectors of society. Geographic discrimination is based on geographical location, where individuals receive disparate treatment from other people, or governments, solely because of where they live, or self-identify as home. See Rhee and Scott (2018) *W. Virginia Law Rev.* on place, space, hillbillies, and home. See also Galperin and Greppi (2017) *SSRN* on geographical discrimination in the gig economy.

diseconomies Financial drawbacks. **Diseconomies of scale** may occur when an enterprise becomes too large, sites become constricted, the flow of goods is congested, or the workforce is alienated. This might be due to the failure of management to operate effectively as a larger business, but there is little supporting evidence for this. *See* ECONOMIES OF SCALE.

disorder Why do we get lost? How do we orient ourselves? What are the cognitive and cultural instruments that we use to move through space? See Bissell and Gordon-Murray (2019) *TIBG* 44, 4, and Marcella Schmidt di Friedberg (2018).

dispersed settlement A settlement pattern of scattered, isolated dwellings. See Cruickshank (2006) *Norw. J. Geog.* 60, 3 on the fight in rural Norway for a dispersed settlement pattern.

dispossession Removal, withdrawal; in housing terms, eviction. See *urban dispossession.

dissection The incision of valleys by river erosion. A **dissected plateau** is a level surface which has been deeply cut into by rivers.

dissimilarity index A measurement of the overall difference between two percentage distributions. Dissimilarity indices are often used in studies of occupational, or racial, *segregation. Jones (2008) *Urb. Geog.* 29, 3 develops a typology to explain the different pathways by which the dissimilarity index may increase or decrease in a metropolis.

dissipative Describing a *non-linear system where energy is lost at positions far from equilibrium. A **dissipative beach** is gently sloping, running from the seaward edge of the forezone to the surf zone.

dissolved load Material carried in solution by a river; capable of passing through a 0.45-µm membrane filter.

dissolved oxygen Oxygen from the atmosphere and from photosynthesis, dissolved in the upper levels of bodies of water. Dissolved oxygen decreases with depth, rising temperatures, and the oxidation of organic matter.

distance Absolute distance is expressed in physical units such as kilometres and is unchangeable. The geography of the world as we experience it is being twisted and contorted so as to bear very little relation to the physical distances that are involved. The human aspect of this is sometimes called **social distance** or **cultural distance**—those gulfs that can exist in the understanding and in the history of experiences that we bring to the world. Physical proximity is not necessarily a good measure of social and cultural distance.

   Relative distance includes any other kind of distance such as **time-distance**, measured in hours and minutes, and changing with varying technology. Breschi and Lissoni (2006) *CESPRI W. Papers* 184 show that spatial distance and technological distance are both proxies of **social distance**. However, Carrere and Schiff (2004) *World Bank W. Paper Series* 3206 find that 'distance has become more important over time for a majority of countries'.

distance decay The lessening in force of a phenomenon or interaction with increasing distance from the location of maximum intensity; the **inverse distance effect**. Seidl et al. (2006) *BMC Health Serv. Rev.* 6, 13 find an inverse relationship between the geographical distance from a patient's home to a clinic and the likelihood of the patient actually turning up there.

distributary A branch of a river or glacier which flows away from the mainstream and does not return to it.

distributed mapping Sometimes called 'online GIS', this is the sharing of cartographic data and tools via the internet. *GIS software and spatial databases are on a server which is accessible online, enabling users to interact, manipulate, query, and map spatial data.

(((⊕))) SEE WEB LINKS
• A good website to start looking at distributed mapping.

disturbance In *ecology, any event, such as fire, flood, or earthquake, that disrupts the everyday running of an *ecosystem. A **disturbance regime** is marked by short, but recurring episodes of high magnitude, or at critical sites. **Disturbance-related landforms** tend to be polygenetic and heterogeneous, and are long-term/permanent adjustments to fires, volcanoes, etc. Church (2002) *Freshw. Biol.* 47, 4 argues for a disturbance-related interpretation of mountain rivers.

divergence The spreading out of an air mass into paths of different directions, linked with the vertical shrinking of the atmosphere. Upper-air divergence is associated with ground level, and closely related to vertical *vorticity.

diverging margin *See* CONSTRUCTIVE MARGIN.

diverse economies Non-capitalist economies, non-market economies; marginalized, hidden, and alternative economies. See Gibson-Graham (2008) *PHG*, 32, 5 and, better, Gritzas (2016) *European Urb. & Regional Studs* 23, 4.

diversification A spread of the activities of a firm or a country between different types of products or different markets, in order to reduce over-dependence. **Horizontal**

diversification is diversification between same-type investments, **vertical diversification** is adding different investments to a portfolio. Diversification for a small open economy is inevitably protracted, involving a change in economic structure. Supportive conditions include low inflation, stable exchange rates, and a labour-market environment with few worker–employer conflicts.

diversity The abundance of species within an ecosystem. **Alpha diversity** is the diversity of species within a particular area or ecosystem, expressed by its species richness. **Beta diversity** is the difference in diversity of species between two or more ecosystems, expressed as the total number of species unique to each of the ecosystems compared; see Vellend (2001) *J. Veg. Sci.* 12, 4 on beta diversity and species turnover.

divestment (divestiture) In business, the sale of an asset; the opposite of an investment. Geographical analyses of divestment focus on: the conduits through which surplus or excess effects are discarded; the practices of divestment; disinvestment as practice; and the connections between divestment practices. Surplus/excess things are routinely divested through specific conduits; throwing something *out* is actually throwing it *in* to another location—a bin, a skip, a charity shop. In international business, divestment is the closure or sell-off of units in foreign locations, or, conversely, units owned by foreign firms.

division of labour The partitioning of a production process into separate elements. **Gender divisions of labour** separate tasks into male and female. A **social division of labour** divides workers by product: Mitsui (2003) *30th*

ISBC, Singapore sees social divisions of labour as crucial to small and medium enterprises in manufacturing, unable to survive if they simply rely on own limited resources and specialized skills. A **spatial division of labour** develops when a firm actively increases profits by finding the least cost location for different phases of the production process, all linked via modern telecommunications. Capital and control functions are concentrated in the advanced economies or global core, while labour has been relocated to developing economies on the global periphery.

doldrums Regions of light, variable winds, low pressure, and high temperature and humidity, occurring over the East Pacific, the East Atlantic, and from the Indian Ocean to the West Pacific.

doline (dolina) A closed, steep-sided, and flat-floored depression in *karst country. The sides are 2–10 m deep; the floor 10–1000 m wide. **Solution dolines** form when solution enlarges a point of weakness in the rock into a hollow. **Subsidence dolines** form when limestone caves develop below insoluble deposits. The majority of dolines are corrosion forms.

dome An uplifted section of rocks, highest at the centre, from which the rocks dip all around. **Volcanic domes** may be formed from slow-moving, *viscous lava, rounded as the result of pressure from lava. A **plug dome** is a small, irregular dome within a crater, possibly with projecting, spiny extrusions.

domestic labour Work performed in the home. Domestic labour has been, and still is, essential to the reproduction of *capital, and has overwhelmingly been women's work. The COVID-19 pandemic has highlighted this state of affairs, as many women have struggled to combine domestic labour with home schooling and their own employment; see Ceuterick (2020) *Feminist Media Studies* 20, 6. Also operating within a home are paid domestic workers, often as migrants.

domestic space The home as a distinct, private, space; the household. It may be produced through the inhabitants' possessions or occupations or through *capitalist and patriarchal reproduction. Lau (2006) *Mod. Asian Studs* 40 argues that the division of domestic space within South Asian households reflects the social status of women. However, although the traditional model of female domestic space is still strong, there are some signs of the emergence of alternative domestic masculinity and femininity; see Rezeanu (2015) *J. Comp. Res. Anthrop. & Sociol.* 6, 2.

dominant discharge Usually, *bankfull discharge, but dominant discharge has also been equated to *effective discharge.

donga In South Africa, a steep-sided gully resulting from severe *soil erosion.

dormant volcano An inactive, but not extinct, volcano.

dormitory town A settlement made up largely of daily commuters, and with relatively few retail outlets, as the commuters will use city-centre services, or out-of-town shopping centres.

doughnut city A city with empty buildings, crime, and vandalism in the centre, and wealth, happiness, and family life in the suburbs. Look carefully at Musterd (2020) *Applied Geog.* 116.

downward transition region In John Friedmann's *core–periphery model, a peripheral region marked by

depleted resources, low agricultural productivity, or outdated industry.

downwearing, declining slope retreat A model of hillslope retreat where the slope angle decreases over time due to a combination of soil *creep, rain splash, and *sheet wash that causes slope convexities and concavities at the expense of straight hillslope segments.

drag The braking effect of friction, imparted by a fluid on bodies passing through it.

drainage The naturally occurring channelled flow formed by streams and rivers. A **drainage basin** is the area of land drained by a river and its tributaries (the term is synonymous with river basin, and termed *watershed* by North Americans). Stream channel length and equilibrium conditions within the system are revealed by the statistical relationships between the various parameters; see Gardiner (1978) *TIBG* 3, 4.

drainage basin geometry Drainage basin geometry postulates that, as the *stream order increases, (a) the number of streams decreases, (b) the average stream length increases, and (c) the average slope decreases. See Strahler (1952) *GSA Bull.* 63, and Giachetta and Willet (2018) *Nature, Scientific Data* 5.

drainage density The total length of streams per unit area. Any attempt to calculate drainage density is impeded by the difficulty of calculating total stream length, as the exact point at which a stream starts is problematical. See Lin and Oguchi (2004) *Geomorph.* 63, 3–4. Since water flow in channels is faster than *overland flow and *throughflow, the higher the drainage density, the faster the hydrograph rise, and the greater the peak discharge.

d

drainage network evolution The drainage basin can increase its area and extend its channels by landslides at the edge of the network; by *headward erosion; by the extension upslope of underground *pipes; and by rill formation. See Coulthard and Van De Wiel (2007) *Geomorph.* 91, 3–4 on *self-organization in drainage basin evolution.

drainage patterns The pattern of a drainage network is strongly influenced by geological structure. **Anastomotic drainage**, developing on nearly horizontal, coarse sediments, is the division of a river into several channels, and **annular** or **radial drainage**—where the major rivers radiate from a centre—develops on domes, particularly where belts of resistant rock are separated by weaker belts. **Centripetal drainage** moves into a centre created by a crater or depression. **Dendritic drainage**—a branching, tree-like network—is most common on horizontally bedded or crystalline rocks, with uniform geology. **Rectangular drainage** has tributaries running at right angles to the major river and occurs on rocks with intersecting, rectangular joints and faults; **contorted drainage** is rectangular drainage on complex metamorphosed rocks. **Trellised drainage** resembles a trained fruit tree, usually found on dipping or folded sedimentary or weakly metamorphosed rocks. See Riedel et al. (2007) *Geomorph.* 91, 1–2 on structural controls on drainage patterns.

dreikanter A stone with three clearly cut faces, like a Brazil nut, formed by sand-blasting in desert environments.

drift-aligned beach A beach which has developed parallel to the line of *longshore drift. If the wave direction becomes more right-angular, the beach realigns itself.

dripstone Secondary calcite deposits, precipitated as *vadose water drips from/flows over cave roofs, walls, and floors.

driving force A propulsive force; for example, the three main driving forces for the movement of the Earth's tectonic plates are mantle convection, gravity, and the Earth's rotation. The opposing forces are **resisting forces**.

drizzle Rain droplets, up to 0.2 mm across, and with a fall speed of around 0.8 m^{-1}.

drought A long, continuous period of dry weather; a significantly lower amount of water for a given climate. See Benjamin Cook (2019) for an explanation of drought from various perspectives, including hydroclimatology, climate change, land management, and groundwater.

drumlin A long hummock or hill deposited and shaped under an ice sheet or very broad glacier, while the ice was still moving. The classic view sees the drumlin longitudinal profile as asymmetric, with a steeper stoss (upstream) side and a gentler lee (downstream) side. However, in a study of 29 000 British drumlins, Spagnolo et al. (2011) *ESPL* 36, 6, 790 found that the average profile built from all mapped drumlins appears almost symmetric. Furthermore, they found that drumlin shape is not a good indicator of the palaeo ice flow direction. Drumlins have been formed via a variety of mechanisms; see Murray (in J. Holden, ed. (2012) pp. 496–7), for a very accessible summary of drumlin formation theories. Phillips et al. (2010) *Sediment. Geol.* 232, 98 highlight the role played by bedrock geology on the distribution of drumlins beneath an ice stream (this entire issue of *Sediment. Geol.* is devoted to drumlins). A large group of drumlins is a **drumlin swarm**, or **drumlin field**. Briner (2007) *Boreas* 36, 2 argues that the bedforms in the New York drumlin field indicate fast ice flow. **Rock drumlins** are more commonly known as roches moutonnées.

dry adiabatic lapse rate (DALR)
See ADIABAT.

Dryas Part of the characteristic threefold late-glacial sequence of climatic change and associated deposits following the last, Devensian, ice advance and prior to the current, Flandrian, interglacial. The colder Dryas phases mark times of cold, *tundra conditions throughout what is now temperate Europe (Britain did not experience the Older Dryas). The **Oldest Dryas** (Dryas I/Pollen Zone Ia) lasted from about 16000 to 15000 years BP, with tundra-grassland characterized by *Dryas octopetala* (mountain avens); the **Older Dryas** (Dryas II/Pollen Zone Ic) from 12300 to 11800 years BP; and the **Younger Dryas** (Dryas III/Pollen Zone III) from about 11000 to 10000 years BP. *See* PLEISTOCENE.

dry farming Farming without irrigation, using mulching, frequent fallowing, working the soil to a fine tilth, and frequent weeding to conserve water. See R. Shetto and M. Owenya (2007).

dry snow zone An area of a glacier where there is no surface melt, even in summer. Most alpine glaciers do not have a dry snow zone as some melting occurs, but much of Antarctica and the central region of the Greenland ice sheet are characterized by no melting.

dry valley A valley, usually in chalk or *karst, with no permanent watercourse. These valleys may have been cut during *periglacial phases, when the *permafrost would stop *meltwater from

soaking through chalk or limestone, and thus permit *dissection.

dual economy An economy consisting of a modern commercial sector alongside a traditional subsistence sector. Per capita output is lower in the subsistence sector than in the capitalist sector, through a lack of capital. With additional capital, more workers can move from the subsistence sector to the capitalist sector. The process of economic development then depends on the transfer of surplus labour from the subsistence to the capitalist sector. However, the nature of risk in a peasant-dominated subsistence sector tends to hinder the creation of surplus labour to feed the capitalist sector. In consequence, industrialization is not smooth and trouble-free.

dumping The off-loading of goods at below cost, usually as exports.

dune *See* SAND DUNE; COASTAL DUNE.

duricrust A hard surface layer on tropical upland soils. See Clarke (2006) *Geomorph.* 73, 1.

dust (atmospheric dust) The effect of dust aerosols on climate depends on both dust distribution in the atmosphere and the physical and chemical properties of the dust particles. Changes in atmospheric dust, derived mostly from deserts, can trigger large-scale climate responses; Yang et al. (2008) *Atmos. Chem. Phys.* 8 demonstrate the modulation of the monsoon climate by dust. Dust particles contribute to the *greenhouse effect by absorbing and emitting radiation; see K. Kohfeld and I. Tegen (2007).

Dust Bowl Parts of Colorado, Kansas, New Mexico, Texas, and Oklahoma, which were severely afflicted by *drought and dust storms in the 1930s. One preconception is that the Dust Bowl was created by farmers ploughing the prairies, but G. Cunfer (2005) thinks it was probably induced over many years by unusually low rainfall and unusually high air temperatures: 'the Dust Bowl would have happened even without farmers ever being there'.

dust storm Sand and dust storms are common meteorological hazards in arid and semi-arid regions, occurring when strong winds lift large amounts of sand and dust from bare, dry soils into the atmosphere.

(⊕) SEE WEB LINKS
• A comprehensive World Meteorological Organization site covering this topic.

duyoda In a *periglacial landscape, a steep-sided, shallow, and often circular depression formed as baydzharakhs collapse. A duyoda is smaller than, but can develop into, an *alas.

dyke (dike) A discordant vertical or semi-vertical wall-like igneous intrusion; see McDonnell et al. (2004) *Mineralogical Mag.* 68, 5 on the Slieve Gullion ring dyke, Ireland.

dynamic contributing areas When the soil is saturated, the water table rises to the surface, and the saturated area acts as an extension to the channel, transferring a significant volume of water in a short space of time. These saturated areas are known as variable source areas or dynamic contributing areas (R. Charlton 2008).

dynamic equilibrium In a landform, a state of balance, in spite of changes taking place within it. If this equilibrium is disrupted, the system will change to restore the balance. For example, dynamic equilibrium in an alluvial stream would be adjusted to provide just the velocity required for the transportation of the load supplied. See Biedenharn et al. (2000) *Geomorph.* 34, 3–4.

dynamic meteorology The study of those motions of the atmosphere that are associated with weather and climate. For all such motions, the discrete molecular nature of the atmosphere can be ignored, and the atmosphere can be regarded as a continuous fluid medium. See Weber and Névir (2008) *Tellus A* 60, 1 on DSI (Dynamic State Index) analysis in dynamic meteorology.

earth flow A form of *mass movement, where water-saturated, weak-slope material flows under the action of gravity at speeds between 10 cm s^{-1} to 10 cm day^{-1}. A rapid earth flow generally quickly follows the initial fall-slide phase. Di Crescenzo and Santo (2005) *Geomorph.* 66, 1–4 find relationships between: the height of the landslide crown zone and the slope relief energy; the slope relief energy and the sliding zone area; and the landslide area and the basin area.

earth pillar An upstanding column of soil that has been sheltered from erosion by a natural capstone on top, common where boulder-rich moraines have been subject to gully erosion.

Earth system science (ESS) A science concerned with how the Earth works today; its natural systems and how these have been perturbed by human actions; how the components of the system have evolved over time in response to changes in other parts of the system; and predicting what will happen in the future. See Reid et al. (2010) *Science* 330.

⊕ SEE WEB LINKS
• Online *Journal of Earth System Science.*

earthquake A sudden and violent movement, or fracture, within the Earth, followed by a resultant series of shocks. Earthquakes occur in narrow, continuous belts of activity which correspond with *plate junctions.

The scale of the shock of an earthquake is known as the **magnitude**; the most commonly used scale is the *Richter scale, while the **intensity** of an earthquake is measured by the *Mercalli scale.

Earthquake waves are of three basic types: P, primary, push waves travel from the focus by the displacement of surrounding particles and are transmitted through solids, liquids, and gases; S, secondary or shake waves travel through solids; and L, long or surface waves travel on the Earth's surface. Rhoades et al. (2011) *Acta Geophysica* 59, 4, 728 investigate the efficiency of earthquake forecasting models.

easterlies Winds *from* the east. Easterlies blow in low latitudes in both high and low *tropospheres, and in high latitudes in the lower troposphere. The downward propagation of easterly winds occurs throughout the equatorial zone; see Baldwin et al. (2001) *Rev. Geophys.* 39. **Easterly waves** are weak troughs of low pressure in *tropical areas. See Fuller and Stensrud (2000) *Monthly Weath. Rev.* 128, 8 on easterly waves and the North American monsoon, and Ross and Krishnamurti (2007) *Monthly Weath. Rev.* 135, 12 on easterly waves and tropical cyclones.

ecocompensation In China, ecocompensation describes any kind of payment that reflects the environmental impact from those who manage, maintain, or conserve a watershed. For example, Han et al. (2012) *Chin. Geogra.*

Sci. 22, 1, 119–26 calculate an economic value for the loss of wetlands using remote sensing, field investigations, and department visits to survey wetland types, assess wetland area changes, and calculate wetland economic value. Compensation is calculated using market valuations, environmental protection costs analysis, and outcomes. See Liu and Guo (2020) *Water Supply* 20, 8, on ecocompensation standards for water resource protection.

ecodevelopment Development which is based on self-reliance as opposed to cultural dependence; need-oriented, as opposed to a narrow productivistic approach and in harmony with the environment. It emphasizes economic equity, social harmony, individual fulfilment, and household and community self-reliance. See Lantitsou (2017) *Fresenius Environmental Bulletin* 26, 2.

ecofeminism A marriage of ecology and feminism. Ecofeminism cannot be easily defined, reflecting the fact that it draws on many feminisms, from New Age thinkers to socialists. It may be seen as the study of women's connections to nature and how these inspire particular forms of environmental activism, stewardship, and spiritual attachments to the Earth. Try Maria Mies and Vandana Sheva (1993).

ecogeomorphology An investigation of river systems that integrates hydrology, fluvial geomorphology, and ecology.

ecohydrology When vegetation types and patterns change, so do the ways in which water flows. Thus, ecohydrology is the study of the influence of dynamic hydrology, *hydraulics, and geomorphology on stream ecosystems, particularly nutrient cycling, primary productivity, and trophic interactions.

See Wilcox et al. (2011) *Geog. Compass* 5, 3 on dry-land ecohydrology, giving clear and practical insights into this field.

ecological balance The equilibrium between, and amicable coexistence of, organisms and their environment. Changes in air temperature, precipitation, and carbon dioxide levels, plus lesser ecological changes, may alter the ecological balance between communities: slight increases in individual mortality of one community, coupled with increased success rates of another, could eventually cause changes in the geographical distribution of forests, grasslands, and deserts.

(())) **SEE WEB LINKS**

• A useful gateway to exploring ecological balance in more detail.

ecological climatology (eclimatology) An interdisciplinary approach to the role of terrestrial ecosystems in the climate system. The central theme is that terrestrial ecosystems, through their cycling of energy, water, chemical elements, and trace gases, are important determinants of climate. See Beerling (2003) *Trends Ecol. Evol.* 18, 2.

ecological crisis A state of human-induced ecological disorder that could lead to the destruction of this planet's ecosystem, to the extent that human life will at least be seriously impaired for generations, if not destroyed. Evidence to suggest that this crisis has already been reached includes *deforestation, increasing levels of atmospheric carbon dioxide, and current rates of energy use.

ecological efficiency The ability of the organisms at one *trophic level to convert to their own use the potential energy supplied by their foodstuff at the trophic level directly beneath them; the energy content of prey consumed by a predator population, divided by the energy content of the food consumed by the prey population. Ecological efficiency tends to be inversely correlated with primary production.

(⊕) SEE WEB LINKS

- Article on how ecological efficiency is measured.

ecological energetics The study of the flow of energy from the sun through and up the *trophic levels, expressed in calories. With movement up each trophic level, there is a very great loss of the energy available as food. 'A more appropriate name for ecological energetics could refer to limitations by a single component of the resource' (Brooks et al. (1996) *Ecoscience* 3, 3).

ecological fallacy A wrong assumption about an individual based on aggregate data for a group; the presence of a relationship between two variables at an aggregated level that is due simply to aggregation, rather than to any real association. See Dark and Bram (2007) *PPG* 31, 5 on the *modifiable areal unit problem and the ecological fallacy; see G. King (1997) on a solution.

ecological footprint The demand of a society on biocapacity (Rees in J. Dewulf and H. Van Langenhove, eds 2006); the area of land functionally required to support a human society; the amount of the environment necessary to produce the goods and services necessary to support a particular lifestyle. In more economically developed countries, this will include land beyond the territory inhabited by that society; that is, the land from which its imports are sourced.

ecological imperialism A process of biological transformation brought about, intentionally or unintentionally, by colonizers in the developing world.

These 'newcomers' brought animals and plants with them, together with systems of land and resource use, which quickly altered the original natural landscapes. They also introduced microbes—diseases and germs—which profoundly reordered the lives of the indigenous populations. In addition, the newcomers brought with them radically different ideas about the ownership of the land, which clashed with the indigenous understandings of human–land relationships. The classic text is A. Crosby (2004).

ecological modernization The argument that the economy benefits from moves towards *environmentalism. Ecological modernization indicates the possibility of overcoming environmental crises without impeding modernization; see Jänicke (2020) in Mez, Okamura, and Weidner (eds). See also Gibbs (2006) *Econ. Geog.* 82, 2, 193 for a critique of this argument.

ecological overshoot The shortfall in Earth's biological capacity to meet the demands of its human population. Continued overshoot, although possible in the short term, means the global community is increasingly exposed to risks of environmental collapse. In 1961 humanity's load was 70% of the capacity of the global biosphere, rising to 120% in 1999.

ecological politics Advocacy for, or work towards, protecting the natural environment from degradation or pollution.

ecological restoration The re-creation of entire *communities of organisms, closely modelled on those occurring naturally, together with long-term maintenance and management to ensure integrity, and stability. See *Restoration Ecology* special issue (2019) 27 S1 on international principles and standards for the practice of ecological restoration.

ecological stoichiometry The study of the effects of the balance of energy and elements on organisms and their interactions. Try Ding et al. (2020) *Biog. and Forestry* 13.

ecology The study of the interrelationships between organisms and their surrounding, outer world; the study of animals and plants in relation to each other and to their habitats. **Production ecology** (**community ecology**) is the study of *communities in terms of the throughput of energy and chemical compounds; key concepts are primary *production, *trophic levels, and *nutrient cycles. See Shea and Chesson (2002) *Trends Ecol. & Evol.* 17, 4.

economically active population The total population between the ages of 15 and 65 in any country, used in the calculation of dependent population.

economic determinism The thesis, as advanced by Marx and Engels, that economic factors underlie all of society's decisions. The social relations specific to a particular *mode of production are said to structure social relationships between classes, underpinning the legal and political systems. This implies that all political, cultural, and social life can be predicted from the prevailing mode of production. This extreme view has been severely criticized, as it denies the existence of free will and individual independence. A more moderate view sees the means of production as a constraint on the ways in which individuals and superstructures can develop, 'but once the chains of causality anchoring consciousness and value to collective production of material goods are broken, as they are in "humanist" Marxism, the theory loses much of its effective explanatory edge' (Duncan (2005) *PHG* 29, 479).

economic development The transformation of a simple, low-income national economy into a modern industrial economy. The term may be used to denote simple growth in an economy, but generally it describes a change in the nature of a national economy. Taylor (2007) *GaWC Res. Bull.* 238 distinguishes between a growth in the economy, but with the same division of labour—*economic growth*—and the production of new commodities—*economic expansion*—where the division of labour is altered, creating a more complex economy.

economic distance The distance a commodity may travel before transport costs exceed the value of the freight. Economic distance can be reduced by better physical infrastructure, improvements in the institutional framework, better cost recovery and maintenance, simpler transit procedures, better information, and the reduction of unofficial payments/corruption (Crochet et al. (2004) *IBRD*, World Bank). See Blanc-Brude et al. (2014) *Int. Business Rev.* 23, 4 on economic distance in China.

economic dynamism Vigorous activity, progress, and strength in an economy. The Ewing Marion Kauffman Foundation measures economic dynamism by: the share of jobs in fast-growing firms; the degree of job turnover (a product of new business start-ups and existing business failures); and the value of companies' initial public offering (the event of a firm's first sale of stock shares). See Braunerhjelm and Henrekson (2013) *Indus. & Corporate Change*, 22, 1.

( ) SEE WEB LINKS

• Website of the Ewing Marion Kauffman Foundation.

economic geography The analysis of the spatial distribution of the transportation and consumption of resources, goods, and services, and their effects on the landscape.

One form of economic geography uses sophisticated spatial modelling to explain uneven development and the emergence of industrial clusters—by considering centripetal and centrifugal forces, especially *economies of scale and transport costs. See Krugman (1998) *Oxford Review of Economic Policy* 14, 2; for a critical review, see Martin (1999) *Camb. J. Econ.* 23.

The second form also attempts to explain the emergence of industrial clusters, but emphasizes relational, social, and contextual aspects of economic behaviour, particularly the importance of knowledge and learning— which takes place most effectively through personal contacts at the local-regional level. This economic geography emphasizes aspects of economic behaviour that are considered intangible by the first; see Perrons (2004) *Econ. Geog.* 80, 1.

Bathelt and Glückler (2003) *J. Econ. Geog.* 3 argue for a **relational economic geography** where economic actors operate within social and institutional relations. Here, the key words are evolution, innovation, and interaction. Crevoisier (2004) *Econ. Geog.* 80, 4 in an innovative milieux approach to economic geography, holds that space— or, more precisely, territory—is the matrix of economic development, and that economic mechanisms transform space. **Evolutionary economic geography** uses ideas from evolutionary economics to explain how an economic landscape changes over time. It also shows how the space in which an economy develops helps an understanding of the processes that drive economic evolution; in other words, how geography affects the nature

and development of an economic system. Boschma and Martin (2010) *Papers in Evolutionary Economic Geography (PEEG)* 1001 is clear and helpful. See also the *Journal of Evolutionary Economics*.

economic growth The growth in wealth of a nation, as measured by an increase in *gross national product, or in national income.

economic integration This includes regional cooperation and integration in the areas of infrastructure and software, trade and investment, money and finance, and regional public goods.

economic man A theoretical being, sometimes used in modelling, who has perfect knowledge of an economy and the ability to act in his or her interests to maximize profits; the concept of the *satisficer is more realistic.

economic nationalism There is no clear consensus on this, but economic nationalism might entail: a government policy to keep companies in domestic, rather than foreign, ownership; *protectionism; state-led development; or even Brexit policies (which stress freedom from foreign legislation and standards). See Helleiner (2021) *New Polit. Econ.* 26, 2, and Hsu (2017) *TIBG* 42, 2.

economic rent A measurement of land profitability, economic rent is a payment (rent) over and above that which is necessary to stay in business. Economic rent R, is given as:

$$R = Q(p - c)$$

where Q is the quantity of production (as in kilos of grain per hectare), p is market price per unit, and c is the cost per unit of production (E. Sheppard 2008). Economic rent is not synonymous with profit, since opportunity cost (the cost of choosing one alternative good or service in terms of the sacrifice of the next best alternative) is built into the concept.

economic system Economic systems mainly deal with the relationships between production/supply and consumption/demand: manufacturing, regulation, circulation, and distribution. Examples include traditional economies, command economies, and laissez-faire and market economies.

economies of scale The benefits of producing on a large scale: as the volume of production increases, cost per unit article decreases. After a certain volume, this fall in cost will be halted as *diseconomies arise, but only at very high levels of production, if at all. Parr (2002) *Env. & Plan. A* 34 has a classification of internal and external economies of scale.

economies of scope The economies or benefits that firms derive by producing particular combinations of output. These exist if one firm can produce two separate products more efficiently than two firms can independently and separately produce them. Economies of scope operate in a different manner from economies of scale, which refer to the lower costs involved in producing larger quantities of a single type of good. Economies of scope refer to the benefits generated from producing a mix of goods and are therefore common in multiproduct firms.

economism The belief that economic values are all that is real or important; that human beings are motivated mainly by economic drives.

economy For a country or region, the production and consumption of goods and services, and the supply of money.

ecoregion A large area (smaller than a *bioregion, but larger than an ecozone) with a distinctive assemblage of communities, environment, and species. A certain amount of physical variability exists in ecoregions, which makes them spatially diverse in their characteristics.

ecosystem A community of plants and animals within a particular physical environment that is linked by a flow of materials through the non-living (abiotic) as well as the living (biotic) sections of the system. **Ecosystem engineers** are organisms that, directly or indirectly, create, maintain, or modify habitats and resources.

ecosystem services The benefits to humans of one or more ecosystems. According to the United Nations 2005 Millennium Ecosystem Assessment (MEA), there are four broad categories of ecosystem services: *provisioning*—the production of food and water; *regulating*—the control of climate and disease; *supporting*—nutrient cycles and crop pollination; and *cultural*—spiritual and recreational benefits.

(((⊕))) SEE WEB LINKS
• UK National Ecosystem Assessment site.

ecotaxation (eco taxation) An economic instrument using taxes or subsidies to encourage behaviour that reduces pollution. Also known as pollution taxation, pigovian taxation, or green taxation, it has largely been associated with the 'polluter pays' principle. See Bazin et al. (2007) *Ecol. Econs.* 63, 4, 732.

ecotone A region of rapidly changing species between two *ecosystems. Within the ecotone, local factors, such as soil and groundwater conditions, determine the species types and distributions. See Goldblum (2010) *Geog. Compass* 4, 7 on the boreal forest ecotone.

ecotope A defined *niche or niche space within a habitat. Relatively homogeneous and spatially bound landscape functional units, ecotopes are the smallest ecologically distinct features in a landscape. Many ecotopes, either adjacent or overlapping, make up an *ecoregion.

ecotourism The development and management of tourism such that the environment is preserved. The income from tourism adds to the investment into landscape conservation. See Walter and Sen (2018) *Asia-Pacific J. Tourism Research* 23 on ecotourism in Cambodia. Klack (2007) *Geog. Compass* 1, 5 thinks that **sustainable ecotourism** should give equal weighting to ecological integrity, economic viability, and social justice.

ecozone *See* BIOGEOGRAPHIC REALM.

ecumene (oecumene) The inhabited areas of the world, as opposed to the non-ecumene which is sparsely or not at all inhabited. The ecumene of a nation is its more densely inhabited core. These very simplistic classifications pose difficulties of delimitation.

edaphic Of the soil; produced or influenced by the soil.

eddy A roughly circular movement within a current of air or water. An eddy may be a *vortex, show *helicoidal flow or the cylindrical motion of rollers, or appear as a surge phenomenon (a short-lived outbreak of greater velocity in a flow).

edge In *network analysis, another term for the *link between two *nodes.

edge city A self-contained employment, shopping, and entertainment *node, located at the periphery of a pre-existing city, that allows its inhabitants to live, work, and consume in the same place. The criteria for an edge city are: at least 465 000 m^2

office space; at least 56 000 m^2 retail space; more jobs than bedrooms; identification as a distinct 'place'; and a total dissimilarity to a 'city' of thirty years ago. Edge cities have developed with mass car ownership and supportive infrastructures, from highways to service stations, to drive-through fast food centres, out-of-town malls and auto-access leisure and retail complexes. The archetypal edge city is said to be Tysons Corner, Virginia, outside Washington, DC. It is located near the junctions of Interstate 495 (the DC beltway), Interstate 66, and Virginia 267.

edge effect The effect on an area of its boundary or edge. This underlines the notion of a *boundary as a transition zone, a porous area rather than a clear, linear demarcation. See Van Meter et al. (2010) *Int. J. Health Geographics* 9, 40.

education and learning, geographies of Geographies of education and learning stress the important influence of *spatiality in making and using education systems, whether formal—as in kindergarten and higher education—or informal - as in homes and work space; what is the role of space and place in educational experiences? Where you are affects what and how you learn. See Holloway and Jöns (2012) *TIBG* 37, 4.

effective discharge In a channel, the discharge that, on average, transports the largest proportion of the annual sediment load, calculated by multiplying the frequency of different discharges by the corresponding sediment transport rate. See Doyle et al. (2005) *Water Resour. Res.* 41, W11411 on effective discharge analysis.

effective normal pressure The force per unit area imposed vertically by a glacier on its bed. For a cold-based glacier, it is effectively equal to the weight of the overlying ice. This is expressed as:

$$N = pgh$$

where N is the effective water pressure, p is the density of the ice, g is the acceleration due to gravity, and h is the ice thickness. However, if water is present at the glacier bed, the effective normal pressure is reduced by an amount equal to the subglacial water pressure. The higher the subglacial water pressure, the more it can counteract the pressure of the ice. The equation then becomes:

$$N = pgh - wp$$

where N is the effective water pressure, p is the density of the ice, g is the acceleration due to gravity, h is the ice thickness, and wp is the subglacial water pressure. This is true for flat-bedded glaciers. Effective normal pressure is modified by the flow of ice over obstacles, increasing upstream of the obstacle and decreasing to its lee.

efficiency The ratio of the work done to the effort used; usually the relation of output to input. **Spatial efficiency**—the organization of space in order to minimize costs—is central to *location theory; see Halás et al. (2017) *PLOS One* 12, 11 on the spatial equity principle in the administrative division of the Central European countries.

EFTA Current members of the **European Free Trade Association**, established in 1960, are Iceland, Liechtenstein, Norway, and Switzerland.

(((●))) SEE WEB LINKS
• The EFTA website.

Ekman layer In the *atmosphere, the transition stratum between the surface *boundary layer and the free atmosphere. **Ekman layer rectification** describes the way the Ekman layer velocity profile differs with variations in

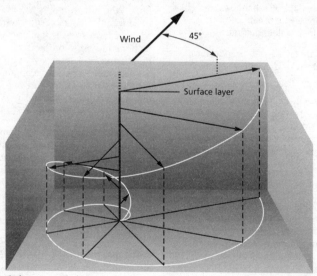

Ekman spiral

surface winds (McWilliams and Huckle (2006) *J. Phys. Oceanog.* 36, 8). The **Ekman spiral** describes how the horizontal wind sets surface waters in motion. As represented by horizontal vectors, the speed and direction of water motion change with increasing depth.

electoral geography The geographical analysis of elections; the study of the spatial patterns of voting and power (see Pattie and Johnston (1998) *Area* 30 on the *contextual effect). Other concerns include the influence of voting decisions upon the environment, and the drawing of constituency boundaries; in India, for example, constituency boundaries had to be redrawn frequently to take into account the latest census figures, reorganization of states, graduation of some of the union territories into states, and bifurcation or trifurcation of some states, to fulfil the demands and aspirations of the people within a new federal structure. See Singh (2000) *Pol. Geog.* 19, 4.

elevation The elevation of a geographic location is its height above a fixed reference point, most commonly a reference geoid (a mathematical model of the Earth's sea level as an equipotential gravitational surface) (K.-T. Chang 2014).

(⊕) SEE WEB LINKS
• An interface to query locations on the Earth for elevation data.

elites Powerful and/or very rich persons who disproportionately influence public and private policy. Ahmed (2009) *Hum. Geog.* 2, 3, 37 shows how the coercive power of global governance institutions has worked in tandem with the interests of local elites to produce neoliberal changes in India. Ahmed also points out that class elites are not a homogeneous group in India.

El Niño The atypical, sustained warming of sea surface temperature that occurs in the central-east equatorial Pacific every few years. The gradient between the low pressure of the warm, western tropical Pacific and the high pressure of the cold, eastern tropical Pacific decreases, weakening the easterly trade winds, and allowing warm surface water to move eastwards. The rising branch of the circulation cell, and therefore the *precipitation associated with it, moves with the water. An El Niño is only one element of a dual-phase oscillating ocean–atmosphere system; when the system reverts to its 'normal' phase, Pacific waters cool off the coast of Ecuador and Peru, and become warm again in the western Pacific. Sometimes the eastern Pacific becomes unusually cool; this is a *La Niña event. The 'normal' phase (sometimes developing into La Niña) is the **El Niño–Southern Oscillation**. Dawson and O'Hare (2000) *Geog.* 85, 3 give an exceptionally clear explanation. See also Yang et al. (2018) *National Science Review* 5, 6. For diagnostic discussions of El Niño, go to the International Research Institute for Climate and Society.

El Niño events have been *teleconnected to abnormally heavy rain in the southern USA and western South America, a decrease in tropical cyclones in the Atlantic, and droughts in Indonesia and Australia. See Nicholls (1988) *Bull. Am. Met. Soc.* 69, 2 on El Niño–Southern Oscillation Impact Prediction.

eluviation The movement of material in true solution or colloidal suspension from one place to another within the soil; soil horizons that have lost material through eluviation are **eluvial**.

embeddedness The state of being located or secured within a larger entity or *context. The economic life of a firm or market—infrastructure, operating environments, and conditions of production—is territorially embedded in its own social and cultural place. Yeung (2000) *TIBG* 23, 3 argues that *transnational corporations in the USA are highly constrained by dynamic and deep capital, while their Japanese counterparts are effectively bound by complex but reliable networks of domestic relationships.

There are interconnected societal, network, and territorial dimensions of embeddedness: **territorial embeddedness**—the extent to which an 'actor' is anchored in particular territories or places; **societal embeddedness**—the cultural, political, institutional, and regulatory framework the actor is located in; and **network embeddedness**—the structure of relationships among a set of individuals/ organizations. **Ecological embeddedness** is embedded in 'natural' or ecological processes.

embodiment A perspective/condition that routes our social existence and experience through the body, and the cognition of the self. As a symbol, the body is constructed; an individual takes on 'labels' and assumptions relating to the body. Intellectual production is always materialized through human bodies, for knowledge never arrives from pure brainpower. It is the outcome of embodied practice.

The human body is unique in playing a dual role both as the vehicle of perception and the object perceived. Bodies are relational and territorialized—they are woven together with space in intricate webs of social and spatial relations. S. Franklin and S. McKinnon (2001) list the 'vastly different scales of embodiment', from the gene, to the body, to the family or species, to the nation, to the commodity form, and to cyberspace.

emergence The creation of new phenomena, requiring new laws and principles, at each level of organization of a complex system. This concept from physics has been applied to geomorphology; see Harrison (2001) *TIBG* 26, 3.

emigration The movement of people from one place to another, usually from one country to another. Green et al. (2008) *NZ Geogr.* 64, 1 observe that the 'push–pull' model incompletely describes the motivations of those who migrate, and who are not necessarily either pushed or pulled, but motivated by things like a desire for a change, or for more adventure. Increasing gradients of difference between nations—in the workforce, in income and poverty levels, and in governance—have been important drivers of emigration.

emissions credits Through photosynthesis, green plants remove carbon dioxide from the atmosphere. Mindful of rising levels of atmospheric carbon dioxide and *global warming, the *Kyoto Protocol proposed that, if nations increase their forested areas, they should be granted emissions credits to count in their CO_2 **emission standards** (there are permissible, and legally binding, pollution maxima); see Depledge and Lamb (2005) *UNFCCC*. **Emissions trading** would allow a nation with *greenhouse gas emissions below its agreed target to trade its 'credit' to over-target nations.

emotional geographies The study of the location of emotions in both bodies and places; of the emotional relations between people and environments, for our emotional lives impact on our interactions with, and within, the places we move through every day, consciously and unconsciously. Emotional geography is concerned with the spatiality and temporality of emotions, with the way they coalesce around and within certain places. Look for the *TIBG* editorial (2019, vol. 2) for a summary on emotional geography. See L. L. Bondi et al. (2007) and Thien (2005) *Area* 37 for considerations of emotion in geography.

enclave A small area within one country administered by another country. Enclaves often become the regions of international tensions and wars: East Bengal, Cabinda, Northern Ireland, and the Palestinian territories. See E. Vinokurov (2007) for a theory of enclaves, and Rozhkov-Yuryevsky (2013) *Baltic Region*, 2, on the Kaliningrad region.

enclosure The conversion of common or open fields into private, exclusive parcels on which sole proprietorship gave rents and/or management decisions to individual owners.

encounter, geographies of Who— or what—we come across, and the nature of our responses to such encounters, depends on who and where we are. Putting this another way, encounters in public space carry with them a set of expectations about appropriate ways of behaving within that space, and the space itself serves as an implicit regulatory framework for our behaviour, as in a library. Piekut and Valentine (2017) *Soc. Sci. Research* 62 are excellent on this.

endemic Occurring within a specified locality; not introduced. **Endemism** describes the restriction of a species (or other taxonomic group) to a particular geographic region, and is due to factors such as isolation or response to soil or climatic conditions. See Kier et al. (2009) *Procs. National Acad. Sciences* 106, 23. Giokas et al. (2008) *J. Biogeog.* 35, 1 outline a technique to identify areas of endemism.

endogenetic (endogenic) From within. In geomorphology, this means those forces operating below the crust which are involved in the formation of surface features. In human geography, it is those forces acting from within, for example, a society. An **endogenous variable** in economics is a variable explained within the theory under consideration.

endogenous growth theory An economic theory claiming that economic growth is generated from within a system as a direct result of internal processes. (This theory treats as endogenous those factors—particularly technological change and human capital—which were classified as exogenous by neoclassical growth models.) Firms operating in places with pools of skilled professionals, for example, are more innovative, leading to faster growth in technology and productivity. Frenken and Boschma (2007) *J. Econ. Geog.* 7, 5 observe that the more product varieties are already present in a firm or city, the higher the probability that new product varieties can be generated through recombination of old routines. See Bond-Smith and McCann (2013) on incorporating space in the theory of endogenous growth.

energy The physical capacity for doing work. Nearly all our energy derives from the sun, and technical progress has reflected more and more sophisticated uses of energy, from wind and water, through *fossil fuels, to *nuclear power.

In the early stages of industrialization, energy consumption is closely related to levels of economic development, and per capita *GNP. Mature economies tend to be more energy efficient, perhaps because technology improves, and the emphasis shifts to service industries. Even so, the *advanced economies still account for most of the world's energy consumption.

World demand for energy has increased so much that an **energy crisis** (a potential shortage of energy) has been identified; see the Roosevelt Institution's 25 ways of solving the energy crisis. This crisis, together with the adverse environmental effects associated with the burning of fossil fuels (*greenhouse effect, *acid rain) has led to increased emphasis on energy conservation. **Energy intensity** is energy consumption per unit GDP. *AAAG* 101 (2011), issue 4 is devoted to new geographies of energy.

(((•))) SEE WEB LINKS
• Energy Administration Information.

englacial Within a glacier.

ENGO Environmental NGO.

ENSO *See* EL NIÑO–SOUTHERN OSCILLATION.

enterprise zones Geographically targeted areas chosen for development that have been designated on the basis of unemployment, poverty, population, age of housing stock, and other criteria. Firms that locate in the area and create jobs are given tax credits and infrastructure is improved. Firms and employees in the zone area benefit because of a reduction in the price of capital and/or labour. However, Paul Swinney (2019) has found that the number of jobs created in the first five years of the existing round of UK enterprise zones has not reached expectations; the jobs created have been overwhelmingly low skilled, meaning that the zones have done little to attract higher-skilled economic activity that would help to change the economic make-up of the economies into which they have been placed.

entisol In *US soil classification, young soils, high in mineral content and without developed soil *horizons. *See* RANKER.

entitlement A guarantee of access to benefits through rights, or through legal agreement. By entitlements, A. Sen (1981) means one's ability to achieve the commodities necessary for basic needs (including food and shelter), either through income-generating work or from direct production of these goods. The size of one's entitlement determines the amount of food (and other commodities) one can command. Thus Smith (2001) *PHG* 25, 140 writes that access to food in urban areas is affected by a wide range of entitlements and blockages, most of which revolve around the fact that unlike rural areas where self-sufficiency or exchange is possible, cities operate on the basis of a cash economy. Sen uses the **entitlement approach** to explain famines. See Watts (2000) *Zeitschrift*. Also try Fall (2020) *Political Geog.* 81.

entrainment 1. The picking up and setting into motion of particles, either by wind, water, or *ice. The main entrainment forces are provided by impact, *lift force, and *turbulence. Resistance to entrainment is provided by packing and by the submerged weight of the particle.
2. In *meteorology, the incorporation of buoyant air into a cloud; see Moeng (2000) *J. Atmos. Scis* 57, 21.

entrepôt A transhipment point where goods are held without incurring customs duties. See Dobbs (2002) *Sing. J. Trop. Geog.* 23, 3 on Singapore as an entrepôt, and Insoll et al. (2021) online, on the entrepôt of Harlaa as an Islamic gateway to eastern Ethiopia.

entrepreneur An organizer, singly or in partnership, who takes risks in creating, investing in, and developing a *firm. See Sorenson (2018) *Small Business*

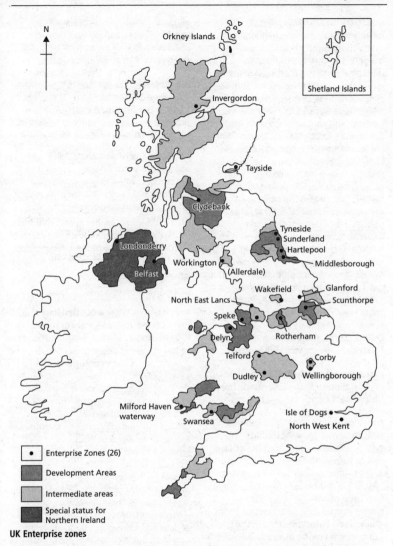

N

Orkney Islands

Shetland Islands

Invergordon

Tayside

Clydebank

Tyneside
Sunderland
Hartlepool
Middlesborough

Londonderry
Belfast

Workington

(Allerdale)

Wakefield

Glanford

North East Lancs

Scunthorpe

Speke

Delyn

Rotherham

Telford

Corby

Dudley

Wellingborough

Milford Haven
waterway

Swansea

Isle of Dogs

North West Kent

• Enterprise Zones (26)

Development Areas

Intermediate areas

Special status for
Northern Ireland

UK Enterprise zones

Economics 5 on social networks and the geography of entrepreneurship. A. Cronin and K. Hetherington (2008) see an **entrepreneurial city** as characterized by business-led urban development, technological innovation, social capital featuring skilled labour, and social and *environmental *sustainability, and supported by local authorities, educational and training institutions, and its own heritage and cultural industries. See Lauerman (2018) *PHG* 42, 2.

entropy maximization A framework for constructing spatial interaction and associated location models. Cities are complex, and it is difficult to analyse their interrelated issues, such as segregation, pollution, and access to infrastructure and services, all levels of complexity. They evolve dynamically, and predicting an exact future state is impossible. It should be possible, however, to produce models for likely configurations (structures), using 'big data', and increasing access to new geographic data. Entropy maximization models can be used to calibrate and test urban models, and to analyse large urban datasets. In other words, entropy maximization models can be used to derive the most probable state of a spatial interaction system, given information on total inflows, total outflows, and capacity constraints, which should help in decision-making. Now go to Purvis et al. (2019) *Entropy* 2, 1. Read slowly and carefully; it's all you need. See Wilson (2010), Entropy in Urban and Regional Modelling: Retrospect and Prospect.

enumeration district A unit of census survey; in the UK, about 150 households.

environment The surroundings. The **natural environment** includes the nature of the living space (sea or land, soil or water), the chemical constituents and physical properties of the living space, the climate, and the assortment of other organisms present. The **phenomenal environment** includes changes and modifications of the natural environment made by man. The effect of the environment on man is modified, in part, by the way the environment is perceived, and human geographers distinguish this—the **subjective environment**—from the **objective environment**—the real world as it is. But although the occurrences and elements of objective reality are, in essence, facts in space and time, the understanding of these facts is conditioned by the frame of reference they are related to. Nature and the environment are not only differently perceived and classified by different groups, but also used differently in policies and actions which greatly influence what is being perceived.

Environment Agency An independent public body, created to protect and improve the environment in England and Wales. Its remit includes tackling flooding and pollution incidents, reducing industry's impacts on the environment, cleaning up rivers, coastal waters, and contaminated land, and improving wildlife habitats.

(((•))) SEE WEB LINKS
• The Environment Agency website.

environmental accounting The use of economic and environmental information to measure the contribution of the environment to the economy, and the impact of the economy on the environment. Environmental accounts enable governments to set priorities, create and monitor economic policies and resource management strategies, and design more efficient market instruments for environmental policies. Go to the United Nations Handbook of National Accounting Series F, No. 78. Pulselli et al. (2006) *Ecol. Econ.* 60, 1 use environmental accounting in calculating an index of sustainable economic welfare.

An **environmental audit** is the practice of assessing, checking, testing, and verifying a process of environmental management, and/or the product(s) arising from it; see Ljubisavljević et al. (2017) *Economic Themes* 5, 4.

environmental archaeology The study of the relationship between climatic/environmental change and human cultural evolution during the *Holocene. The entire issue of *The Holocene* (2012) 22, 6 is devoted to multidisciplinary studies in environmental archaeology.

environmental determinism The view that human activities are governed by the environment, primarily the physical environment. J. M. Blaut (2005) challenges the central thesis of environmental determinism, while J. Sachs (2005) reconsiders the role of the environment in shaping human history. Judkins et al. (2008) *Geog. J.* 174, 1 think that Jared Diamond's books are pure environmental determinism: 'this resurgence threatens negative consequences if uncritically adopted by policymakers.'

environmental dynamics The behaviour of an environmental system in space and through time, considering the causes and consequences of past, current, and future changes to Earth's atmosphere, climate, and hydrological regimes, and examining paleo-environmental records and future projections of atmospheric composition and climate.

environmental economics Economic theory which both aims for sustainability and acknowledges the vital role played by environmental capital: identifying and understanding environmental issues exist, deciding if it is worth remedying the situation, and developing solutions. Hanley, Shogren, and White (2019) is easy to read.

environmental gradient A gradual and continuous change in communities and environmental condition over, for example, altitude, latitude, or longitude. The gradients can be related to environmental factors such as altitude, temperature, and moisture supply. See Riesch et al. (2018) *Current Zoology* 64, 2,

and Damgaard (2003) *Ecol. Modelling* 170 on modelling plant competition along an environmental gradient.

environmental impact A change in the make-up, working, or appearance of the environment. An **Environmental Impact Assessment** considers the probable consequences of human intervention in the environment, seeking to restrict environmental damage. In most developed countries, the methodology includes provision for the responsible authority to produce a draft, release this for public comment, and gather responses from a variety of perspectives: those of other agencies and levels of government, environmental and community groups, interested corporations, resource users, and ordinary citizens.

⊕ SEE WEB LINKS
- Interesting Friends of the Earth site on Environmental Impact Assessments.

environmentalism A concern for the environment, and especially with the bond between humans and the environment; not solely in terms of technology but also in ethical terms. 'If we are to understand modern conservation of nature, we need to see environmentalism not as a thing, but as a process' (L. Vivanco 2006). See Blanc (2019 *Ecology and Society* 24, 3. The doctrine of **market environmentalism** is to bring about positive environmental results through the use of markets and market-derived institutions and organizations, hopefully offering a fusion of economic growth, efficiency, and environmental conservation.

environmental justice The equitable treatment and involvement of everyone of whatever race, colour, national origin, culture, education, or income in developing, implementing, and enforcing environmental laws, regulations, and policies. See Ranco and Suagee (2007) *Antipode* 39, 4. It is also an interdisciplinary group of scholars interested in the linkages between social justice and environmental change, with a particular focus on the global dimensions of (in)justice. The **environmental justice foundation** campaigns on a number of issues, but is not an academic source. See Buckingham (2004) *Geog. J.* 170, 2 on environmental justice and ecofeminism.

environmental lapse rate The fall in temperature of stationary air with height, averaging 6 °C per 1000 m. This fall of temperature with height in stable air is due to a fall in the density of the air, varying greatly with the time of day, and with the nature of the *air mass concerned. See Schultz et al. (2000) *Monthly Weather Rev.* 128, 4143–8, and Houston and Niyogi (2007) *Monthly Weather Rev.* 135, 3013–32.

Environmentally Sensitive Area (ESA) A fragile *ecosystem area where the conservation or preservation of the natural environment is sustained by state controls and/or grants, such as the eastern South Downs, UK. Go to the England Rural Development Programme for comprehensive information.

environmental perception The way in which an individual perceives the environment; the process of evaluating and storing information received about the environment; the relationships of individuals and communities with the environment. Decision-makers cannot but base their judgements on the environment as they perceive it. See Marques et al. (2020) *Environmental Perception* 1.

environmental policy The statement by a supranational, national, or regional government of its approach to environmental protection. While

environmental policy internalizes externalized environmental values, sustainability policy achieves some form of long-lasting equity.

environmental science The study of environments; traditionally the physical environment, or, more widely, social and cultural environments. Major themes include: environmental accounting; environmental performance evaluation; reporting; participative processes in environmental policy-making; and sustainable management.

Environment Directorate An agency of the OECD, the environment directorate provides governments with the analytical basis to develop policies that are effective and economically efficient, including country performance reviews, data collection, policy analysis, projections and modelling, and the development of common approaches.

( ) SEE WEB LINKS

• The Environment Directorate on the OECD website.

eolian *See* AEOLIAN.

epeirogeny Broad, and generally large-scale, limited, or slow vertical movements of the Earth's crust which do not involve much alteration in the structure of the rock; hence **epeirogenic**—caused by the relatively gentle raising or lowering of the Earth's crust.

ephemeral Short-lived. R-strategist plants are ephemeral in that they grow and reproduce rapidly when conditions are favourable, dying within a short space of time. **Ephemeral streams** flow only during and after intense rain, and are typical of arid and semi-arid areas.

epicentre The point of the Earth's surface which is directly above the *focus of an *earthquake, usually 0–50 km below it.

epilimnion The upper layer of a body of water, penetrated by light, thus enabling photosynthesis. This zone is warmer, and contains more oxygen, than the layers below.

epiphyte A plant growing on another plant but using it only for support and not for food, most commonly found in *tropical rain forests. See Zotz et al. (1999) *J. Biogeog.* 26, 4 on the community composition, structure, and dynamics of epiphyte vegetation.

epistemic geographies 'Epistemic' refers to knowledge about knowledge, and particularly knowledge about academic disciplines; see Mahony and Hulme (2018) *PHG* 42, 3.

epistemology The philosophical theory of knowledge which considers how we know what we know, and establishes just what ought to be defined as knowledge. In *geography, the term is used to indicate the examination of geographical knowledge—how it is gained, sent, changed, and absorbed. O'Sullivan (2004) *TIBG* 29, 3 distinguishes between epistemology (1) 'experimenting on theories' and (2) epistemology 'learning from models'.

epoch An interval of geological time; several epochs form a period, several periods an era. An epoch is ranked as a third-order time unit (M. Allaby 2006).

equal area map A map so drawn that a square kilometre in one portion of the map is equal in size to a square kilometre in any other portion. Equal area maps of the whole globe tend to be elliptical in shape, and severely distort the shapes of regions far from the equator as in Peters' projection below.

equatorial rain forest *See* TROPICAL RAIN FOREST.

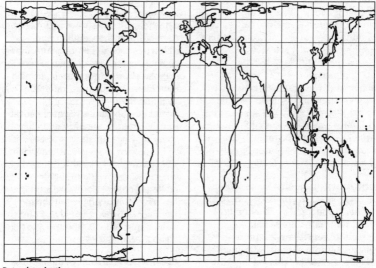

Peters' projection

Interrupted equal area

equatorial trough A narrow zone of low pressure, between the two belts of *trade winds, arising from high *insolation, especially in the centre of continents in summer. Also known as the *inter-tropical convergence zone, the equatorial trough is not constant in position, breadth, or intensity.

equifinality The principle that a given end state can be reached by many potential means. Thus, in geomorphology, this term indicates that similar landforms might arise as a result of quite different sets of processes. For example, Cruslock et al. (2010) *Geomorph.* 114 note that different erosion mechanisms—in this case, wave action and ice scour—have produced similar *shore platforms in Sweden and South Wales in the UK. Thus from the landform alone, without additional evidence, it might be difficult to identify the particular set of causes or to differentiate different feasible causes.

equilibrium In physical geography, the state of an open system where the inputs, throughputs, and outputs of energy or matter are in balance. Equilibrium relies on the presence of negative feedback loops within the system. J. Holden (2012) recognizes: '**static equilibrium** (no change over

time) or a **steady-state equilibrium** (short-term fluctuations about a longer-term mean value) while over longer time periods the equilibrium might be **dynamic** (shorter-term fluctuations with a longer-term mean value that is changing)'. Holden adds 'the concept of equilibrium has always been somewhat confusing because different people have chosen to identify different types of equilibrium and because the precise meaning is time dependent ... Often it depends on where and when you measure something as to whether it will show equilibrium'. Within a system, the **equilibrium time** is the time required for a system to establish a new steady state after the application of a stimulus or a change in external conditions. Within human geography **spatial equilibrium** is regional equality, brought about by the interaction of labour moving to high-wage areas and capital moving to regions where wage rates are low, and usually expressed in terms of wages: 'workers and firms are indifferent among alternative locations as they have eliminated any systematic difference in indirect utility and profits through migration' (Ottoviano and Perri (2005) *J. Econ. Geog.* 6, 1).

equilibrium line In glaciology, the point at which expansion of the glacier by accumulation is balanced by ice losses through *ablation.

equilibrium species Species that show characteristics consonant with a stable *niche. Dispersal is less important, perseverance is more significant than recovery from adverse conditions, and the survival of the young is more important than high fecundity.

equinox Equinoxes, those times when day and night are of equal length, occur twice a year. The **spring**, or **vernal**, **equinox** is on 21 March and the **autumn equinox** is on 22 September. On these dates, the sun is directly overhead at the equator. The changes in day length result from the changes in the tilt of the Earth with respect to its axis.

era The largest unit of geological time. The approximate datings of the eras are:

era	duration in millions of years before present
Precambrian	4600–570
Palaeozoic	570–225
Mesozoic	225–65
Cenozoic	65–0

erg Arid, sandy desert, particularly within the Sahara. The Rub Al Khali aeolian system of the Arabian Peninsula is the world's largest erg (Bray and Stokes (2004) *Geomorph.* 59, 1–4).

ergodic A form of reasoning whereby data are selected as examples of a landform at different stages of evolution, since we cannot travel back in time to see all the stages through which it has evolved. Ergodic reasoning is, in effect, a special case of the axiom that 'the present is the key to the past'.

erodibility The vulnerability of a material to erosion. This will vary according to the nature of the material:

degree of consolidation, organic content, particle size, hardness, and strength, among others; on the extent and type of ground vegetation; thus on *land use. See Darboux and Le Bissonnais (2007) *Eur. J. Soil Sci.* 58, 5 on measuring soil erodibility. Quantifying soil erodibility is crucial for forecasting dust events (Zender et al. (2003) *J. Geophys. Res.: Atmos.* 108, 4543).

erosion The removal of part of the land surface by wind, water, gravity, or ice. These agents can only transport matter if the material has first been broken up by weathering. Some writers use a very narrow interpretation of the word, claiming that erosion refers only to the transport of debris. If so, *denudation would be the correct term for the combination of weathering and the transport of rock fragments. Erosion may be measured with an **erosion pin**—a stake driven into the ground which is used as a benchmark to gauge the extent of erosion. See Kearing et al. (2018) *Catena* 163. **Primary erosion** is the initial *in situ* erosion of rock, *regolith, or soil.

erosion surface A relatively level surface produced by erosion. Much of Africa is composed of extensive plains, often cutting *discordantly across varied geological structures, and these have therefore been identified as erosion surfaces as such (Lister (1965) *Rec. Geol. Surv. Malawi* 7). See also Ojany (1978) *GeoJournal* 2, 5. The whole concept is bound up with the theory of the cycle of erosion.

erosivity The ability of an agent, such as rain or wind, to cause erosion, which depends on rainfall intensity and drop size, as well as the material it lands on. Rainfall intensity is used as the primary erosivity indicator at shorter times and smaller areal scales. Larger areas and greater time scales generally require use of run-off as the measure of erosivity. See Nearing et al. (2017) *Catena* 157 for a useful summary.

erratic A large boulder of rock which has been transported by a glacier so that it has come to rest on country rock of different *lithology; all erratic blocks in northern Germany, for example, originated in Scandinavia. Agassiz (1840–1) *Procs Geo. Soc.* 3, 2 used the presence of erratics as evidence of glaciation in the British Isles.

ESA *See* ENVIRONMENTALLY SENSITIVE AREA.

escarpment A more or less continuous line of steep slopes, facing in the same direction, and caused by the erosion of folded rock.

esker A long ridge of material deposited from *meltwater streams running *subglacially, roughly parallel to the direction of ice flow, and ranging from tens of metres to several hundred kilometres, as in Finland. Eskers wind up and down hill because subglacial streams are under great hydrostatic pressure, and can flow uphill. Brennan (2002) *Geomorph.* 32, 3–4 identifies and investigates five types of esker. Thomas and Montague (1997) *Quat. Sci. Rev.* 16 argue that glacial lake systems are crucial for the development of esker systems.

estuary That area of a river mouth which is affected by sea tides. In a **tide-dominated estuary**, tidal currents dominate geomorphologically; see Komatsu (1999) *Cretaceous Res.* 20, 3.

eta index (η) An expression of the relationship between a network as a whole and its *edges. For a worked example, see Alberto and Manly (2006) *J. Biogeog.* 33, 4.

etchplain A tropical *planation surface where deep weathering has etched into the bedrock. Fluvial denudation may lay the etchplain bare. See Stengel and Busche in X. Yang, ed. (2002) on a Late Proterozoic etchplain in Namibia.

ethical geography Ethics are the principles of right and wrong conduct; the basis for doing what is right, and the discernment of what is right, while morals are the accepted norms and standards of conduct of a society, community, or nation. Smith (2001) *PHG* 25, 2 seeks to clarify the difference between these two concepts. Evanoff (2007) *Eth. Place. Env.* 10, 2 argues that a single environmental ethic that can be universally applied in all geographic settings and across cultures cannot be formulated. Madge (2007) *PHG* 31, 5 provides a wide-ranging review of research ethics in geography.

ethical trade Trade based on social and environmental responsibility in existing large-scale businesses.

(⊕) SEE WEB LINKS
• Ethical Trading Initiative site.

ethnicity A cultural and geographic entity that emerges when a group shares a common ancestry, origin, and tradition. Ethnicity may relate to a geographical territory, world view, custom, ritual, and language. Ethnicity in a group may become pronounced as a result of migration: 'in Trinidad and Guyana, Hindus were at the bottom of the social structure . . . . When they migrated to Britain, they were still in an ethnic quandary, with the white British population thinking of them derogatorily as "Paki" (subcontinental Indian)' (Kong (2001) *PHG* 25, 2).

ethnic segregation The evolution of distinct neighbourhoods recognizable by their characteristic ethnic identity.

Phillips et al. (2007) *TIBG* 32, 2 discuss multiple readings of ethnic segregation.

External causes (imposed by the *charter group) of ethnic segregation include discrimination, low incomes—which direct them towards inner-city locations—and the need for minorities to locate near the *CBD since much of their employment is located there. Internal causes (springing from the ethnic group) include a desire to locate near facilities serving the group, such as specialized shops and places of worship, desire for proximity to kin, and protection against attack.

Ellis et al. (2004) *AAAG* 940 reveal that segregation by work tract is considerably lower than by residential tract. Lobo et al. (2007) *Urb. Geog.* 28, 7 argue that ethnicity itself is wielded as a potent force as a group moves into a neighbourhood and eventually dominates it. Phillips (2007) *Geog. Compass* 1, 5 explores the politics of data collection, categorization, and representation in ethnic segregation research.

ethnocentricity Making assumptions about other societies, based on the norms of one's own. This may lead to the development of global theories based, for example, on Western conditions. Inevitably geographers view other disciplines through their own cultural filter. See Ericka Diaz (2017) on overcoming ethnocentrism.

ethnography The study of the customs, habits, and behaviour of specific groups of people (usually people in non-literate societies). For geographers, ethnography provides insight into the processes and meanings that sustain and motivate social groups. These processes and meanings vary across space, and are central to the construction and transformation of landscapes; they are both place-bound

and place-making. See Herbert (2000) *Prog. Hum. Geog.* 24, and Megoran (2006) *Polit. Geog.* 25 on ethnography and political geography.

euphotic zone The upper layer of a body of water receiving light and thus where photosynthesis is possible. In a marine ecosystem, the euphotic zone may extend down as far as 200 metres below the surface. Most marine life inhabits this zone. Below the euphotic zone is the dysphotic zone.

European Economic Area A single, free-trade grouping, comprising the member states of the *European Union and *EFTA.

(⊕) SEE WEB LINKS
- The European Economic Area defined on the European Commission website.

European Neighbourhood Policy (ENP) Through its ENP, the EU works with its southern and eastern neighbours to achieve the closest possible political association and the greatest possible degree of economic integration. This goal builds on common interests and on values—democracy, the rule of law, respect for human rights, and social cohesion. Partner countries agree an ENP action plan with the EU, demonstrating their commitment to democracy, human rights, rule of law, good governance, market economy principles, and sustainable development. In early 2014, participating partners were: Armenia, Azerbaijan, Egypt, Georgia, Israel, Jordan, Lebanon, Moldova, Morocco, Palestine, Tunisia, and Ukraine.

European Union (EU) A *free trade area comprising Austria, Belgium, Bulgaria, Croatia, Cyprus, the Czech Republic, Denmark, Estonia, Finland, France, Germany, Greece, Hungary, Ireland, Italy, Latvia, Lithuania,

Luxembourg, Malta, the Netherlands, Poland, Portugal, Romania, Slovakia, Slovenia, Spain, and Sweden. Designed initially as an economic unit, the European Union now attempts uniformity in social policies.

(⊕) SEE WEB LINKS
• The European Union website.

European Water Framework This directive's key aims are: expanding the scope of water protection to all waters, surface waters, and groundwater; combining emission limit values and quality standards; getting the citizen involved more closely; and streamlining legislation. See Carter (2007) *Geog. J.* 173, 4 and Kay et al. (2007) *PPG* 31, 1.

(⊕) SEE WEB LINKS
• The EU Water Framework Directive website.

eustasy A worldwide change of sea level, which may be caused by the growth and decay of ice sheets, by the deposition of sediment, or by a change in the volume of the oceanic basins. It is commonly assumed that melting of polar ice sheets induced, for example, by climate warming would yield a nearly eustatic trend in sea level, but self-gravitation and loading effects would lead to significant departures from eustasy. See Rovere et al. (2016) *Curr. Clim. Change Rep.* 2 on eustatic and relative sea level changes.

eutrophication The process by which *ecosystems, usually lakes, become more fertile environments as detergents, sewage, and agricultural fertilizers flow in; the enrichment of nutrients. See Chambers et al. (2008) *Water Sci. Technol.* 58 on the eutrophication of agricultural streams and ecological protection.

evaporite A deposit formed when mineral-rich water evaporates; the most

common are of gypsum (hydrated calcium sulphate) and halite (sodium chloride). *Playas are **evaporite basins**; see T. M. Peryt, ed. (1988); see also Ayora et al. (2001) *Geology* 29, 3 on Mesozoic and Tertiary evaporite basins.

evapotranspiration The release of water vapour from the Earth's surface by evaporation and transpiration. Since evapotranspiration is so variable, physical geographers prefer to use the concept of **potential evapotranspiration** (**PE**). This is the greatest amount of water vapour which could be diffused into the atmosphere given unlimited supplies of water.

evolutionary economic geography Evolutionary economics is concerned with dynamics and change, and especially with entrepreneurship, innovation, industrial and institutional dynamics, and patterns of economic growth and development; all combined with the role of history in shaping the contemporary socio-economic landscape. This approach aims to understand the actions of participants in economic activity over time and space. It explains how behaviour of agents is situated and conditioned, but not determined, by structures, organizations, and the environment. There is a wide variety of approaches to evolutionary economic geography; key approaches include using the evolutionary principles of variety, selection, and retention. See Henning (2019) *Regional Studies* 53, 4, and Dieter Kogler (2016).

exclusion, geographies of The study of the way different social groups are geographically marginalized and discriminated against along lines of *gender, *race, *class, *sexuality, *age, and *disability; how some groups of people are deemed 'out of place' in some locales and are pushed to the margins of society. The seminal work is D. Sibley (1995). See also Mohan (2000) *Prog. Hum. Geog.* 24, 2.

See also SOCIAL EXCLUSION.

exfoliation The *sheeting of rocks and their disintegration, thought to be due to *thermal expansion, at least on small structures. See Collins et al. (2018) *Nature Commun.* 9 on thermal influences on spontaneous rock dome exfoliation. High rates of change in rock temperature may produce exfoliation sheets in *periglacial environments.

exhumation The removal of young deposits to reveal the underlying structure of older rocks.

exogenetic 1. Applying to processes which occur at or near the Earth's surface.
2. In human geography, as a result of outside, environmental influences.

expatriates, migrants, diaspora strategies New Zealand is a nation of 4 million people, plus 1 million expatriates: 'whereas early efforts involved encouraging highly skilled migrants to return to their home countries, today a range of new techniques [is] being used to engage expatriates in activities in their countries of origin without requiring them to return' (Larner (2007) *TIBG* 32, 3).

extended family A family unit of relatives by blood and by marriage as well as two parents and their children.

extending flow The extension and thinning of a glacier, often marked by an *ice fall. Extending flow occurs near the *equilibrium line, and where the velocity of the glacier increases, for example, down a rock *step. Extending flow can transmit material from the surface of a glacier to its base, thus increasing its powers of *abrasion. It is also responsible for *crevasses and is typical of the zone of *accumulation of ice. In zones of extending flow, ice reaches the bed tangentially, diminishing the ice's erosive capacity, and thus diminishing erosion.

external economies The cost-saving benefits of locating near factors which are external to a firm, such as locally available skilled labour, training, and research and development facilities. Moran et al. (2002) *PHG* 26, 4 write that regional specialization is sustained by external economies of scale that impart certain advantages and 'lock' regions into particular development paths. See also Crescenzi et al. (2007) *J. Econ. Geog.* 7, 6.

externality A side effect on others following from the actions of an individual or group. This effect is not brought by those affected and may be unwished for. Two types of externality are recognized: **public behaviour externalities** covering property, maintenance, crime, and public behaviour, and **status externalities** resulting from the social and ethnic standing of the household.

extinction The end of the existence of a species or group of taxa, or the end of their ability to reproduce. Gaston (2008) *PPG* 32, 1 observes that rare species are disproportionately vulnerable to extinction in the short term. Loss of a species to chance extinction leaves an area of resource space unoccupied; extinctions could trigger behavioural or evolutionary adaptations (Phillips (2008) *PPG* 32, 1). Van Kleunen and Richardson (2007) *PPG* 31, 4 consider the links between species traits and extinction risk and invasiveness. See Kiesling and Aberhan (2007) *J. Biogeog.* 34, 9 on geographical distribution and extinction risk. Losos and Schluter (2000) *Nature* 408 explain the relationship as the outcome of the effect of area on immigration and extinction rates. 'If the twentieth-century question was how to stop extinction, then perhaps the twenty-first-century challenge is how to avoid the total collapse of the biosphere, our life support system' (W. Adams 2004). See Sanderson et al. (2002) *Bioscience* 52, 10 on human activity and extinction rates. *See also* BIODIVERSITY CRISIS.

extra-tropical cyclone A cyclonic disturbance outside the tropics; for example, a *mid-latitude depression. See Knox et al. (2011) *Geog. Compass* DOI: 10.1111/j.1749–8198.2010.00395.x on the development of high winds in extra-tropical cyclones.

extrusion A formation of rock made of *magma which has erupted onto the Earth's surface as lava and has then solidified. The crystals in extrusive rocks are small, since the lava solidifies rapidly, giving little time for crystal growth. Extrusions emerge from *fissure eruptions and *volcanoes.

exurb American for *dormitory settlement. In many cases, amenity migration and the 'urbanization of the rural' it produces can be understood as the first signs of exurban development (McCarthy (2008) *PHG* 32, 1).

eye The calm area at the centre of a *tropical cyclone.

fabric The physical make-up of a rock or sediment. **Fabric analysis** can determine the *dip and orientation of particles in a sediment: see Carr and Goddard (2007) *Boreas* on fabric analysis and glacier dynamics.

facies The characteristics of a rock, such as fossil content, or chemical composition, which distinguish it from other formations and give some indication of its formation. See Torsvik et al. (2004) *J. Geol. Soc.* 161, 4.

factor analysis A way to get a small set of (preferably uncorrelated) variables from a large set of variables, most of which are correlated to each other. This will reduce the number of variables and detect structure in the relationships between them. For an example of its use in geography, see Brown and Raymond (2007) *Appl. Geog.* 27, 2.

⊕ SEE WEB LINKS
• Website with online tutorial on factor analysis using Microsoft Excel.

factorial ecology The investigation of urban spatial structure by *factor analysis. The classic study is Murdie's (1969) *U. Chicago Res. Paper* 116. For the problems associated with factorial ecology, see Martinez-Martin (2005) *ITC Diss. Series* 127, 44–5.

factors of production The resources required to produce economic goods. They are land (including all natural resources), labour (including all human work and skill), capital (including all money, assets, machinery, raw materials, etc.), and entrepreneurial ability (including organizational and management skills, inventiveness, and the willingness to take risks). For each of these factors there is a price, i.e. rent for land, wages for labour, interest for capital, and profit for the entrepreneur. Economic geography may be seen as the study of the location in space of factors of production. **Intangible factors of production** include trust, creativity, cooperation, and the sharing of knowledge.

factory farming A system of livestock farming in which animals—most commonly, pigs, laying hens, broiler chickens, and veal calves—are kept indoors, with very restricted mobility, for most of their lives. Factory farming has created its own set of environmental damage: raised output of greenhouse gases (farm animals account for 15–20% of methane emissions), and manure surpluses. In the US and Europe, factory farm animal welfare has become a matter of significant public concern; see Johnston (2013) *Geog. Compass* 7, 2.

factory system A concentration of the processes of manufacturing—fixed capital, raw material, and labour—under one roof, in order to provide the mass production of a standardized product or products, characterized by much more rigid, complex, and intricate divisions of labour, both within and between production processes—a system which may be described as industrial capitalism.

fact-value distinction The distinction between what *is*—the realm of facts, and what *ought to be*—the realm of values; the supposed *ontological difference between how the world is and how people evaluate the world. However, in many cases considerations of fact and value cannot be neatly separated; see Levine (2008) *J. of Philosophy & Geog.* 11, 1, 91.

failed state A *state may be said to have failed when it cannot carry out basic functions such as education, governance, or security; a state where the government is not effectively in a position to exert its authority over its entire territory. In many cases, the authority of the state is challenged by internal factions resulting in civil conflicts and a situation where the effective control of different parts of the state lies in the hands of different warring factions (see NUIM's *2013 Failed States Index*). See Orman (2015) *Int. J. Social Studies* 4, 1 for an analysis of the notion of a failed state.

fair trade A *commodity chain set up to promote worker welfare, minimize environmental damage, and allow producers to increase their living standards, and control within the supply chain, through *development projects.

((⊕)) SEE WEB LINKS
• Fairtrade Foundation website.

fall A form of *mass movement in which fractured rock and soil separates into blocks and falls away from the parent slope. A fall starts with the detachment of soil or rock from a steep slope along a surface on which little or no shear displacement takes place. The material then descends mainly through the air by falling, bouncing, or rolling.

fallowing An agricultural practice where the land is ploughed, but not seeded. Fallowing is based on the rotation of land between different uses rather than a single permanent use, to regenerate soils through a vegetation–soil nutrient cycling. See Awanyo in B. Warf, ed. (2010) on **bush fallowing**.

family In geolinguistics, a genetic group of languages with numerous cognates and regular correspondences, arising as a language diffuses and diverges over time. See Millar (2003) *J. Linguistics* 7, 1 and Sharma (2005) *J. Linguistics* 9, 2 on new varieties of English.

famine A relatively sudden flare-up of mass death by starvation, usually relatively localized, and usually associated with a sharp rise in food prices, the sale of household goods, begging, the consumption of wild foods, and out-migration. It is a socio-economic process which hits hardest the most vulnerable and marginal, and the least-powerful groups in a community. See Atkins et al. (2018) *Int. Encyclopedia Hum. Geog.* for a very useful bibliography on the geography of hunger and famine.

A. Sen (1982) sees famine as distinguishable from chronic hunger and deprivation, in that speedy intervention can prevent it. Sen developed the concept of *entitlement (the ability of people to get access to a resource), identifying declining wages, unemployment, rising food prices, and poor food-distribution systems as causes of starvation.

((⊕)) SEE WEB LINKS
• The famine early warning systems network.

farm fragmentation The division of a farmer's land into a collection of scattered lots. Fragmentation usually results from divisions on inheritance but may also reflect *bush fallowing. See Ciaian et al. (2018) *Land Use Policy* 76 for an Albanian case study.

fascism A political system characterized by dictatorship, strong socioeconomic controls, suppression of the opposition through terror and censorship, and often militant nationalism and racism. It is a difficult and contentious term, but Heffernan (2005) *Pol. Geog.* 24, 731 provides a useful analysis.

fault A fractured surface in the Earth's crust along which rocks have travelled relative to each other. Usually, faults occur together in large numbers, parallel to each other or crossing each other at different angles; these are then described as a **fault system**. The slope of the fault is known as the *dip. Where rocks have moved down the dip there is a **normal fault**; where rocks have moved up the dip, there is a **reverse fault**. A **thrust fault** is a reverse fault where the angle of dip is very shallow and an **overthrust fault** has a nearly horizontal dip. A **fault plane** is the surface against which the movement takes place. A **tear fault** is where movement along the fault plane is lateral. This latter type of fault may be termed a **strike-slip fault**. Regions, such as the Harz of Germany, that are split by faults into upland *horsts or depressed *rift valleys are said to be **block faulted**.

SEE WEB LINKS

• The USGS Visual Glossary provides definitions and outstanding photographs.

fault block A section of country rock demarcated by faults, which has usually been affected by *tectonic movement. See Young et al. (2002) *Basin Res.* 4, 1 on the Hammam Faraun Fault Block, Egypt.

fault breccia A zone of angular rock fragments, located along a fault-line, and formed by the grinding action associated with movement either side of the fault. Killick (2003) *S. African J. Geol.* 106 outlines a system of fault rock classification.

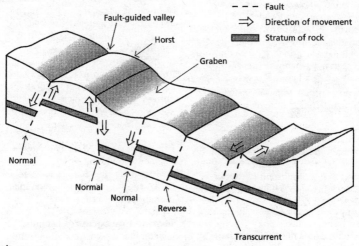

Fault

fault scarp A steep slope resulting from the movement of rock strata down the *dip of a normal fault; Locke et al. (1992) *Bull. Seismological Soc. America* 82 describe the Teton Range fault scarps. A **fault-line scarp** is formed when faulting juxtaposes stronger and weaker rocks, and the latter are eroded.

faunal realms The simplest groupings of the world's animals. The *holarctic* realm covers the *nearctic* (most of North America, plus Greenland) and the *palearctic* (extra-tropical Asia, Europe, and North Africa). The *neotropical* realm covers Central and South America; the *Ethiopian* Africa south of the Sahara and Arabia. The *oriental* realm is tropical Asia, with an ill-defined boundary between it and the *Australian* realm (New Zealand, Australia, Oceania, and some of South-East Asia).

fear, geographies of The study of those places that people indicate and understand as fear-filled or dangerous (Modly (2000) *PHG* 24, 3, 365 is worth reading). Fear itself is complexly tied to our individual expectations and messages, our gender/class and our capacity, or belief in our capacity, to manage fearful and/or dangerous situations. See also R. Pain and S. Smith, eds (2008). Paul (2011) *Gender Tech. & Develop.* 15, 3 411 argues that women's restricted access to public space in India is a manifestation of socially produced fear.

fecundity The potential of a woman/women to bear live children. Fecundity in a population is, of course, closely linked to the proportion of women of childbearing age. See Bhalotra and van Soest (2006) *IZA Dis. Paper* 2163 on fertility and neonatal mortality in India.

federalism A two-tier system of government. The higher, central government is usually concerned with matters which affect the whole nation such as defence and foreign policy, while a lower, regional authority generally takes responsibility for local concerns such as education, housing, and planning, although, of course, the division of responsibilities varies from case to case.

Federations have been designed to preserve regional characteristics within a united nation, or to constrain *nationalist elements. Thus, the Nigerian federation was substantially amended after the 1967 Biafran war; see Philips (2005) *African Studs Rev.* 48, 2. Crispin et al. (2012) *Conflict Management and Peace Science* 29, 1 look at the way the geographic distribution of groups across a country affects the ways in which federalism contributes to conflict resolution.

feedback The response within a *system to an action or process. **Negative feedback** causes the situation to revert to the original; in a river system, erosion is a negative feedback mechanism that restores stream stability by lowering channel gradient and increasing bed material size (Nunally (1985) *Env. Manage.* 9, 5). **Positive feedback** causes further change: positive feedbacks involve a move away from an equilibrium state. Perry and Enright (2002) *J. Biogeog.* 29 note that positive feedback loops between *fire and vegetation have degraded tropical and humid forest ecosystems.

felsenmeer A surface of broken rock fragments found in *periglacial environments. See Ebert and Kleman (2004) *Geomorph.* 62, 3–4 on circular moraines on felsenmeer.

feminist geography Feminist geographies are concerned with the relationships between gender relations and space—how space impacts upon gender relations, and how gender relations express themselves spatially. Dixon and Jones in S. Aitken and G. Valentine, eds (2006) are more vehement: 'feminist geography is concerned, first and foremost, with improving women's lives, by understanding the sources, dynamics, and spatiality of women's oppression, and with documenting strategies of resistance'. The defining characteristic of feminist geography is a concern with *patriarchy—feminist geographies question the patriarchal and hierarchical assumptions on which geography is based, and emphasize the oppression of women and the gender inequality between men and women. G.-W. Falah and C. Nagel (2005) examine gender relations as they are mediated by Islamic practices: 'Muslim women are not only subject to patriarchal forces within their own communities and societies, but are also subjected to patriarchal structures of the west.' Gender can be understood as an organizing principle of societies; a social relation that shapes the forms, functions, structures, and governance of cities (Bondi (2006) *estudos feministas/ études feministes*).

Women's subjectivities, relationships, and symbolic productions also vary with *economic development. M. Domosh and J. Seager (2001) discuss the gendering of space, and the ways that geographies restrict women's access to, and movements within, space. Nagar et al. (2002) *Econ. Geog.* 78 examine the 'double marginalization' of subjects and space characteristic of dominant accounts of globalization: women are sidelined and women in southern countries are feminized differently from those in advanced economies. Hyndman (2003) *ACME* 2, 1 argues that **feminist geopolitics** provides more accountable, embodied ways of seeing and understanding the intersection of power

and space; that it refers to an analysis that is dependent on context, place, and time, rather than it being a new theory of geopolitics or a new ordering of space. See also Evans and Maddrell (2019) *Gender, Place and Culture* 26, 7–9.

L. Staeheli and E. Kofman, eds (2004) argue that a **feminist political geography** recognizes a different outline of the political from other forms of *political geography.

Dias and Blecha (2007) *Prof. Geogr.* 59, 1 believe that feminist geographies should be seen not solely as a separate sub-discipline in geography, but as a critical perspective useful to all sub disciplines. L. Nelson and J. Seager (2004) offer a useful, concise, and densely referenced overview of the history of feminist geography.

feminist standpoint theory A theory concerned with the way that women's perceptions are shaped by their daily activities and realities. The central tenet is the belief that members of marginalized groups may, in certain contexts, be likely to generate knowledge that is less privileged and less distorted than knowledge generated by the socially privileged; all knowledge is *situated knowledge. See Silvestre Cabrera et al. (2020) *Investigaciones Feministas*. Feminist standpoint theory has proven controversial because of its implication that women form a coherent, homogeneous group—they are not all alike. See Thien in R. Kitchin and N. Thrift (2009).

feminization The social process which has resulted in certain occupations, such as nursing, being regarded as women's work. Watson and Stratford (2008) *Soc. & Cult. Geog.* 9, 4 suggest that the feminization of HIV/ AIDS involves both hazard and vulnerability for women and their dependants.

This term is also used to describe the increasing presence and influence of women in previously male-dominated occupations or institutions. Gallo (2006) *Glob. Netwks* 6, 4 explores the relationship between a 'feminization of migration' and the construction of masculine identities among migrants from Kerala.

feminization of poverty Women represent a disproportionate percentage of the world's poor. This trend is deepening, women's increasing share of poverty is linked with a rising incidence of female household headship (single parent families). See Chant (2006) *Gender Institute Working Paper* Series, 18, London School of Economics). Chant has a long and thoughtful piece for UNRISD, 2015. Skip the abstract.

ferrallitization The effect on a soil of strong, tropical *leaching and intense weathering. Organic matter is rapidly destroyed (so there is little *humus), and bases and silica are leached. Sesquioxides in the B2 *horizon form a hardpan that may be revealed by the erosion of the upper horizons. In *US soil classification, ferrallitic soils are oxisols.

Ferrel cell A weak *atmospheric cell lying between the Polar cell and the Hadley cell, and travelling in the direction opposite to that of the Hadley cell. It transfers warm air to high latitudes and shifts cold air back to the subtropics, where it is warmed. The Ferrel cell is a consequence of the stationary and transient eddy circulations in the mid-latitudes.

ferricrete A soil *horizon made from the *cementation of iron oxides at or near the land surface. See Phillips (2000) *Geografiska A.* 82, 1 on groundwater and ferricrete formation.

ferruginous Of or containing iron or iron rust. A **ferruginous soil** is a very deep, *zonal soil found in warm temperate climates without a dry season, or in tropical *savannas/bushlands. The A *horizon is dark red-brown with a weak crumbling structure; the B horizon is stained red by ferruginous gravel. In *US soil classification, a ferruginous soil is an ultisol.

fertility The level of childbearing; in an individual, but more often in a society or nation. Crude *birth rate is the simplest measure of fertility, but does not relate the number of births to the number of women of childbearing age, while the **general fertility rate**, or **fertility ratio**, shows the number of births in a year per 1000 women of reproductive age (generally 15–45, sometimes 15–49). **Cohort fertility rates** show the number of births to women grouped according to either their year of birth or their year of marriage. The **total fertility rate** is the average number of children that would be born per woman, if women experienced the age-specific fertility rates of the year in question throughout their childbearing lifespan. Globally, fertility rates vary widely; in 2006, the rate for the UK was 1.84; for Malawi, 5.7. Abernethy (1999) *Pop. & Env.* 21, 2 argues for the **fertility opportunity hypothesis**—that humans are alert to environmental signs that indicate whether conditions for childbearing and nurture are more or less optimal.

There are very strong global correlations between fertility rates and per capita *GNP, fertility rates and women's education, and fertility and infant mortality rates. Fertility has been declining in industrial societies since the late nineteenth century; a decline which preceded easily available artificial contraception.

(⊕) SEE WEB LINKS
- UN website with tables on estimated and projected total fertility for the world, development groups, and major areas (p. 11).

fetch The distance that a sea wave has travelled, from its initiation to the coast where it breaks. The fetch controls the energy and height of a wave, and the longest fetch, and hence the dominant wave direction, will affect beach orientation and the direction of *longshore drift. See Stephenson and Brander (2004) *PPG* 28 on the effect of fetch on a beach, and Davidson-Arnott et al. (2005) *Geomorph.* 68, 1–2 on sediment transport and fetch.

feudalism A system, common in Europe in the Middle Ages, where access to farm land was gained by service to the owner: the **feudal lord**. Initially, no money was involved in transactions between the serf and the lord, although the payment of cash in lieu of service became common in the later Middle Ages; see T. Aston, ed. (1987). In India, the feudal *zamindari* system divided rural society into three broad classes: landlords (*zamindars*), tenant farmers, and landless labourers. Women, Untouchables, and tribal peoples were excluded from decision-making at both regional and national levels. See N. Mohammad (1992).

fiard (fjärd) An inlet of the sea with low banks on either side, common along the Gulf of Finland, and formed by the post-glacial drowning of the Fenno-Scandian shield.

fictitious capital Money that is thrown into circulation as capital without any material backing in the form of commodities or productive activity. D. Harvey (2006) explains: 'consider the case of a producer who receives credit against the collateral of an unsold commodity. The money equivalent of the commodity is acquired *before an actual sale*. This money can then be used to purchase fresh means of production and labour power. The lender, however,

holds a piece of paper, the value of which is backed by an unsold commodity. This piece of paper may be characterized as fictitious value'. A **fictitious commodity** is one with an ecological and/or social value exceeding its monetary value functions.

field capacity The volume of water which is the maximum that a soil can hold in its *pores after excess water has been drained away. Field capacity generally increases with finer soil textures. Field capacity hardly applies for sandy soil, because of its high hydraulic conductivity.

field system The layout and use of fields. The extent and use of fields varies with the natural environment, the nature of the crops and livestock produced, and aspects of the culture of the farming community such as inheritance rights and available technology. See M. Aston (1983) for medieval field systems in Britain; M. Widgren and J. Sutton, eds (1999) on the Engaruka field system, Tanzania; and Xiong (1999) *Toung Pao* 85, 4–5 on the highly political equal-field system in China and Japan.

filière *See* GLOBAL COMMODITY CHAIN.

film Cinema may be seen as 'an essentially geographic art, a way of "writing the world"', according to D. Strauch (Geography of Film, Geog. 425 syllabus) for film contributes to the production of meaning, culture, economy, and identity. Film is important to geography because it mediates social knowledge, reinforces ideological constructions of the status quo, and is an active agent of the governing powers.

filtering down The movement of progressively poorer people into housing stock. It is suggested that when the rich move away from the city to newly built houses, the 'next down' social and occupational class moves in. There are reservations to this theory: it is not only the rich who get new homes, many higher-status housing areas withstand infiltration, and the well-off *gentrify run-down areas.

financial ecology The shaping of the financial system so that it produces economic, social, and environmental value in a way that leads to greater stability and a more balanced economy.

financial exclusion The barring of the disadvantaged and the poor from financial services. Financially excluded people typically: lack a bank account; rely on alternative forms of credit (such as doorstep and online lenders and pawnbrokers); and lack key financial products such as savings, insurance, and pensions.

The socio-spatial characteristics commonly associated with financial exclusion include low-income households, single- and lone-parent households, social housing tenants, and residence within a deprived urban area. Financial exclusion often reinforces other aspects of social exclusion: see Chakravarty (2006) *Reg. Studs* 40.

financial services, geography of
Mergers, acquisitions, and takeovers, online and phone banking databases of customers' credit rating, and the central processing of cheques have radically altered the spatial distribution of providers of financial services. The 1990s saw a 28% fall in the number of UK bank branches. The spatial implications of these changes have been: disproportionately more bank closures in less affluent areas; a shift in *central business district land use from financial services to restaurants and bars; and the development of call centres. Call centres locate close to existing concentrations of allied activity, with preferences for

densely populated areas mediated by needs to maintain employee access and avoid staff turnover problems: this has important implications for the spatial division of labour, with call centre growth likely to reinforce existing spatial unevenness in employment in key service activities. See Henegham and Hall (2020) *European and Urban Studs.* on the emerging geography of financial centres in the European Union and the UK.

London has four distinct clusters of financial services: a very cohesive City of London cluster featuring banks, insurance, auxiliary finance, law, and recruitment firms; a less cohesive West End cluster, with distinctive cluster zones, such as banks near Mayfair, and advertising in Soho; an incipient general cluster north of the City of London featuring architecture, and business support; and the law cluster that straddles the City of London and the West End (Corporation of London 2003).

fines Small stone particles, whose poor thermal conductivity may aid the development of *patterned ground in *periglacial environments. For fines in desert landforms, see A. Goudie and N. Middleton (2006).

fiord (fjord) A long, narrow arm of the sea which is the result of the 'drowning' of a glaciated valley; the classic text is J. P. Syvitski et al. (1987).

Fiords are distinctive because of their great depth, and the overdeepening of their middle sections which are deeper than the water at the mouth. Søgnefjord, for example, is 1 200 m deep, but its mouth is only 150 m below sea level. The shallow bar at the seaward end of the fiord is thought to represent the spreading and thinning of ice as it was released from its narrow valley and spread out over the lowland. Norway's 2004 nomination for its fjords to be added to the UNESCO World Heritage List makes interesting reading: the geomorphology is outlined, together with issues of management, developmental pressures, and environmental pressures. Glaciation may not be the only fiord-forming factor; see Nesje (1992) *Geomorph.* 5, 6 on fracture systems and Søgnefjord.

fire ecology The study of the impacts of fire on ecosystems, fire ecology can be considered a sub-discipline of landscape ecology as both span the temporal, spatial, and social dimensions of landscapes. The primary focus has been to understand the direct and indirect effects of fire disturbance on plants and, to a lesser extent, animals. See McLauchlan et al. (2020) *J. Ecology* 108, 5 on fire as a fundamental ecological process. See also Bowman (2007) *PPG* 32, 2; and Roberts (2001) *PPG* 25, 2 on tropical fire ecology.

firm In *neoclassical economics, an ordered, autonomous, rational independent unit which utilizes the factors of production to produce goods and services. Revenue is kept high enough to cover costs and to generate profit. In Marxist analysis, a firm is a complex organization embodied in a logic of *accumulation, driven to increase profits by avoiding the costs of community or environmental degradation, exploiting labour, and manipulating interactions with governments and unions.

Economic geographers are interested in the nature of a firm: its dimensions of gender, sexuality, corporate culture, and cultural embeddedness, and its locating factors, including *institutional thickness. See Taylor and Asheim (2001) *Econ. Geog.* 77, 4 on the concept of the firm in economic geography.

firn Ice which has formed when falls of snow fail to melt from one season to another. As further snow accumulates, its weight presses on earlier snow, compacting and melting it to a mass of globular particles of ice with interconnecting air spaces. Where temperatures are around 0 °C, snow can turn to firn within five years, but the process takes much longer in very cold conditions. See Spötl et al. (2014) *Holocene* 24, 2 on perennial firn and ice in an Alpine cave. The **firn line** is the line, close to the *equilibrium line, at which firn forms, and varies with *aspect.

First World A misleading term (since 'first' does not mean better) for western Europe, Japan, Australia, New Zealand, and North America; the first areas to *industrialize. Widely used synonyms include 'the developed world', 'the North', 'the more economically developed countries' (MEDCs), and 'the advanced economies'.

fish farming The rearing of fish in pools or tanks. In China, manuring has traditionally taken place by raising ducks and allowing their droppings to fall into the water. In a somewhat left-field approach to the topic, Tejeda and Townsend (2006) *Gender, Tech. & Dev.* 10, 1 discuss the different meanings of fish farming for different family members as a key issue in understanding relative successes and failures in fish farming; look, too, for Ho's atmospheric account of the Chinese dyke-pond system (2006, *Institute of Science in Society*). See Ruiz-Zarzuela (2009) *Water Sci. Technol.* 60, 3 on the effects of fish farming on the water quality of rivers in north-east Spain.

fissure eruption A volcanic eruption where *lava—usually very fluid and basic—wells up through fissures in the Earth's crust and spreads over a large area. See Walker (1995) *Geol. Soc. Memoirs* 16 on the Antrim plateau, Northern Ireland. See also Witt et al. (2018) *Front. Earth Sci.* 18 on fissure venting in Iceland.

Flandrian The present *interglacial, during which there has been a global rise in sea level—the **Flandrian transgression**, caused by the melting of *ice sheets and *glaciers. The Flandrian is divided into the *Pre-Boreal, *Boreal, and *Atlantic. See Smith et al. (2006) *ESPL* 8, 5.

flashy In hydrology, applied to a natural watercourse which responds rapidly to a storm event, hence **flash flood**; a very sudden, brief, and dramatic flood event—rising and falling limbs are steep, and the period of peak flow is short. Flash floods are major hazards in deserts, for, although rainfall is rare, the hard-baked ground may be impermeable, and there is little or no vegetation to intercept the rainfall or slow the floodwater. Flash flooding is the major process in the formation of *wadis—which, incidentally, should therefore be avoided as camp sites. See Chhopel (2006) *Procs Int. Workshop Flash Flood Forecasting* on flash floods and glacial lake outbursts, and Chan (1997) *Disaster Prevention & Management* 6, 2 on urbanization and flash flooding in Kuala Lumpur.

(((⊕))) SEE WEB LINKS

- USGS description and depiction of effects of flash flooding precipitation on desert geomorphology during El Niño years.

flattening 1. Of a data set, a reduction in the divergence of statistical data over a geographic area; for example, trade union membership.
  2. In *cartography and *GIS, representing the shape of the Earth as a spheroid—that is to say, not a perfect sphere—rather than a sphere.

(The Earth is slightly flattened in the direction of its axis.)

flexible accumulation The use of innovative industrial technologies, adaptable inter-firm relations, variable organizational structures, and flexible consumption, in response to *competition from *newly industrializing and *less developed countries, and to the saturation and fragmentation of markets within *more economically developed countries.

flexible city Across the globe, change is happening faster and faster. In response, cities need to be flexible: making buildings that can be retrofitted to change their use, for example, and creating spaces that can utilize extra capacity. Thus, a space that is used as a café in the morning might become a co-working space in the day, and a bar or community events space in the evening. Increasingly, individuals, too, are flexible: students, and precarious, part-time and self-employed workers, especially in the creative sector. These low-income 'flexible' individuals are able to live inexpensively in both unusual sites and central urban locations. See Ferreri et al. (2017) *TIBG* 42, 2.

flexible space Space which has multiple uses and meanings, which overlap in time and space. It may be envisaged as fluid, 'folded', and eventful. ICT and the developing knowledge-based economy are driving forces for a more flexible space, and cyberspace creates and sustains a virtual space that is still flexible enough for individuals and communities to take part in, especially in the opportunities to acquire technologies. Ardeshiri (2016) *Researchgate* looks useful.

flexible specialization A strategy at company level of permanent innovation, and accommodation to, rather than control of, change. This strategy is based on innovation, multi-use equipment, flexible, skilled workers, or on subcontracting to major firms in sweatshops and homeworking. Suppliers that adopt flexible specialization, through multi-skilled workers operating in groups, working with incentive bonuses and empowered to stop the production line to ensure quality, seem to have better outcomes for their workers—in terms of wages and working conditions—and for themselves—in terms of productivity gains, efficiency, and compliance. See Ylinenpää (2016) *Eur. J. Spatial Devt*, 63 'from flexible to smart specialization'.

flocculation The process whereby very small soil particles, usually clays, aggregate to form *crumbs. Flocculation plays an important role in the transport and behaviour of fine sediment in aquatic systems. See Guo (2018) PEARL on flocculation dynamics on an intertidal mudflat. In some subsoils of arid areas, downward translocation of soluble salts leads to the breakdown of these crumbs in the process of **deflocculation**.

flood Floods occur when peak discharge exceeds channel capacity, and this may be brought about naturally by intense *precipitation; snow- and ice-melt; the rifting of barriers, such as ice dams; the failure of man-made structures; deforestation and urbanization, (which reduces *infiltration and *interception); and by land drainage and the straightening and embankment of rivers. Extreme upland flooding appears to be associated with negative *North Atlantic Oscillation index values.

  Flood prevention and **flood control measures** include afforestation, the construction of relief channels and reservoirs, water meadow areas in which to divert flood water, and a ban on building in flood-prone environments, such as *flood plains. These measures may be increasingly adopted with higher per capita income, individual preparedness, and/or experience with flooding, but may decrease with distance from a river, acceptability of flood risk, and provision of environmental information. In *more economically developed countries, there is a move for flood costs to be borne by the private citizen. Chen and Hou (2004) *J. Am. Water Resources Ass.* 40, 1 develop a multi-criterion, fuzzy recognition model for flood control.

flood frequency analysis (FFA) The calculation of the statistical probability that a flood of a certain magnitude for a given river will occur in a certain period of time. Each flood of the river is recorded and ranked in order of magnitude with the highest rank being assigned to the largest flood. The *return period here is the likely time interval between floods of a given magnitude and can be calculated as:

$$\frac{\text{number of years of river record} + 1}{\text{rank of a given flood}}$$

All methods of FFA are methods of extrapolation, which requires the fitting of a model, and this, in turn, needs an assumption about the underlying distribution generating flood events: 'not only is this not known for extreme hydrological events beyond the observed record, but it is untestable within human timescales' (Kidson and Richards (2005) *PPG* 29, 3).

flood plain The relatively flat land stretching from either side of a river to the bottom of the valley walls. Flood plains are periodically inundated by the river water; hence the name. Flood plains are often ill drained and marshy, and characteristic *fluvial features include meanders, levées, and ox-bow lakes.

The **flood-plain hydrology system** is dependent on interactions between dynamic, non-linear physical and biological processes, which link water, heat, and materials such as biota, sediment, and nutrients. The key processes include flood-caused scour and sedimentation (cut and fill alluviation), routing of river water and nutrients above and below ground, and channel movement (avulsion). The movement of groundwater through the flood plain and welling up back to the surface involves penetration of river water into subsurface palaeochannels. Bed sediments are created by channel scour and subsequent filling with sorted gravel and cobbles. Strong interactions between short-lived, powerful floods, channel and sediment movement, increased roughness due to presence of vegetation and dead wood, and upwelling of groundwater creates a complex, dynamic landscape.

floral (floristic) realms Areas characterized by the indigenous plant species and not by the *biome. Moreira-Muñoz (2007) *J. Biogeog.* 34, 10 presents a detailed revision of 19th- and 20th-century floralistic realm classifications.

flow 1. A 'river' of rock, earth, and other debris saturated with water; differing from *slides in that no shear plane exists. Flows are classified by the size of the particles: **debris flow** refers to coarse material; earth flow to soil; and mud flow to clay. See Quinta-Ferreira (2007) online *Bull. Engineer. Geol. & Env.* 66, 1 on the natural and man-made causes of an **earth flow** in Portugal, 2000. Earthflows commonly form in steep river canyons, and initiate from rapid incision that destabilizes the toes of hill slopes. They are linked with

seasonal precipitation and resulting changes in stress. See Finnegan and Nereson (2017) *NASA Astrophysics Data System*.

2. The movement of goods, people, services, and information along a *network. All places are connected to other places, and traversed by all sorts of flows, like migrants, money, goods, germs, satellite images, and digital data. P. Dicken (2003) observes that flows of goods, people, and information are rising, based on changes in scale, and driven by innovations in information and communication technologies. Try Bertaut et al. (2018) on globalization and the geography of capital flows. Carr et al. (2005) *J. World Business* 40, 4 write about **talent flow**—whereby economically valuable individuals migrate between countries. A **flow dynamic** is the set of forces that operate within a flow of, for example, water or air, producing change. **Spaces of flows** are the networks that bind a world system of cities.

flowage The movement of solids such as ice or rock without fracturing. The term is also used in tectonics (Dickinson (2004) *PPP* 213, 3–4).

flow regime The 'general behaviour' of a river, which is defined by average flow conditions over a year. Characteristics of the flow regime include: seasonal variations in *discharge; the size and frequency of floods; and frequency and duration of droughts. See Martínez-Fernández et al. (2013) *Prog. Phys. Geog.* 37, 5 on recent trends in rivers with near-natural flow regime in Spain.

fluid stressing The erosion of weak, cohesive rocks by the force of water in a river. The effect of this force depends, among other factors, on the strength of the bed, the percentage of clay in the bed, and the velocity and *turbulence of the water. It is effective for weakly cohesive muds. Evorsion is an alternative term.

flume In hydrology, an apparatus placed across a watercourse to measure *discharge.

fluvial Of or referring to a river, including the organisms within a river or the landforms produced by river action. **Fluvial processes** include erosion, flow processes, and sediment and solute transport in rivers. **Fluvial erosion** is the destruction of bedrock on the sides and bottom of the river; the erosion of channel banks; and the breaking down of rock fragments into smaller fragments. **Fluvial flux** refers to the material carried by a river. **Fluvial deposition** dumps material worked or deposited by rivers. At current sea levels, the locus of fluvial deposition is not necessarily the ocean, estuary, or delta, but floodplains in and upstream of the fluvial-estuarine transition zone. **Fluvial dunes** are trains of highly ordered sediment waves along the sediment on a river bed. Coleman and Nikora (2011) *ESPL* 36, 1, 39 review current understanding of these bedforms.

fluvial geomorphology The study of the processes and pressures operating on river systems. Changes in the independent variables of *discharge, *sediment *load supplied to reach, and valley slope, give rise to adjustments in the dependent variables of sediment load and particle size, hydraulic characteristics, and morphologies, all of which interact with each other. See Gregory (2008) *Geomorphology* 98, 1 on applying fluvial geomorphology to river channel management.

fluvial system A river can be viewed as an *open system; see G. M. Kondolf and H. Piégay, eds (2003). The fluvial system is a process-response system that includes the morphology of channels, floodplains, hill slopes, and the cascading component of water and sediment. The system changes progressively as a result of normal erosional and depositional processes, and it responds to changes in climate, base level, and tectonics.

A fluvial system can be divided into three zones:

Zone 1   is the drainage basin, watershed, or *sediment-source* area. This is the area from which water and sediment are derived. It is primarily a zone of sediment production, although sediment storage does occur there in important ways.

Zone 2   is the transfer zone, where, for a stable channel, input of sediment can equal output.

Zone 3   is the *sediment sink* or area of deposition (delta, alluvial fan).

These three subdivisions of the fluvial system may appear artificial because sediments are stored, eroded, and transported in all the zones; nevertheless, within each zone one process is dominant.

fluvio-glacial Of, or concerned with, watercourses from melting glaciers. The website for the Littlemill Site of Special Scientific Interest (Inverness) describes a suite of fluvio-glacial landforms.

flux In geomorphology, the amount of matter that flows through a unit area per unit time. **Solute flux** (sometimes called the **solute load**) is the solute concentration of the water at a point in time multiplied by the discharge occurring at the same time:

solute flux, mgs^{-1} = solute concentration
$$(mgL^{-1}) \times \text{discharge} (Ls^{-1})$$

See Heal in J. Holden (2012) pp. 372–3 for techniques in estimating solute flux.

flying geese theory An East Asian variety of *modernization theory, which argues that East Asian economies have followed Japan's economic development pathway of import substitution-cum-export promotion and the progression from crude and simple goods to complex and refined goods. See Ginnsberg and Schimanazzi (2005) *J. Asian Economics* 15, 6.

SEE WEB LINKS
• Ozawa (2010) provides a thorough critique of the theory at Columbia University website.

focal area In geolinguistics, an area of relative uniformity, as indicated by sets of shared linguistic features, which acts as the central area of a *dialect: for example, the central area of industrial Scotland. See J. Chambers et al. (2008). Britain (2005) *Linguistics* 43, 5 on Estuary English looks interesting.

focus The point of origin of an earthquake. Akira (2000) *J. Geog.* 109, 6 explains how multi-beam echo sounding and seafloor imaging provides information on an earthquake focus.

fog A cloud of water droplets suspended in the air, limiting visibility to less than 1000 m. Fog forms when a layer of air close to a surface becomes slightly supersaturated and produces a layer of cloud, that is, when vapour-laden air is cooled below *dew point. In **advection fog**, this cooling is brought about as warm, moist air passes over cold sea currents.

Radiation fog forms during cloudless autumn nights when strong *terrestrial radiation causes ground temperatures to fall. Moist air is chilled by contact with the ground surface. See Underwood et al. (2004) *J. Appl. Met.* 43, 2 on radiation fog in California. Where cold air streams across warm waters, **steam fog** forms. This is common when relatively warm surface air over lakes in *frost hollows convects into the cold *katabatic airflow above it, and is also the mechanism behind *Arctic sea smoke; see Walker (2003) *Weather* 58, 5 on radiation and steam fog. **Frontal fog** forms when fine rain falling at a warm front is chilled to dew point as it falls through cold air at ground level.

föhn When moist air rises over a mountain barrier, it cools at the slow saturated *adiabatic lapse rate; precipitation is common. Once past the mountains, the air, now much drier, descends, warming at the dry adiabatic lapse rate, higher than the saturated rate by some 3 °C/1000 m. A dry, warm, gusty wind, which can reach gale force, results. In summer, desiccation brings a serious risk of bush fires; in winter, snow melt can be rapid. See Chan (2005) *Croatian Met. J.* 40 on föhn winds in Hong Kong, and the British Antarctic Survey report on the characteristics and variability of föhn winds in South Georgia.

fold A buckled, bent, or contorted rock. Folds result from complex processes including fracture, sliding, *shearing, and *flowage. Delcaillau et al. (2006) *Geomorph.* 76, 3–4 report on recent fold growth in India. An arch-like upfold is an **anticline**, a downfold is a **syncline**. A complex anticline is an **anticlinorium** (see J. Conley 1973 on the Blue Ridge anticlinorium) and a complex syncline is a **synclinorium** (see Röhlich (2007) *Bull. Geosci.* 82, 2 on the Prague synclinorium). In an **overturned fold** the upper limb of the syncline and the lower limb of the anticline dip in the same direction. In **recumbent folds** the beds in the lower limb of the anticline and the upper limb of the syncline are upside down. G. Bennison and K. Moseley (2003) have excellent block diagrams of various fold types.

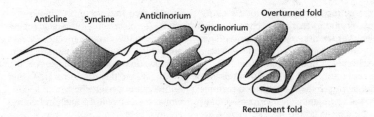

Anticline   Syncline   Anticlinorium   Overturned fold

Synclinorium

Recumbent fold

Fold

fold mountain An upland area, such as the Alps or Andes, formed by the buckling of the Earth's crust. Many fold mountains are associated with destructive or collision margins of *plates. **Young fold mountains**, such as the Caucasus and Alps, were formed by the Alpine orogeny of 65 million years BP, and reach elevations of 10000 m. **Old fold mountains**, such as the Grampian mountains of Scotland, were created by earth movements pre-dating the Alpine orogeny but have been extensively eroded. Some old fold mountains have been uplifted and re-eroded.

food chain Plants (primary *producers) and *consumers at various *trophic levels are interconnected in their diet and in their role as sources of food. A food chain is a linear sequence representing the nutrition of various species from the simplest plant through to top *carnivores, as in: rose → greenfly → ladybird → sparrow → sparrowhawk. (This direct pathway is too simplified, and plants and animals are usually linked together in a *food web.) Schoener (1989) *Ecology* 70 shows that **food-chain length** is governed by the total energy available to a given trophic level. See Steinberg et al. (1995) *Procs Nat. Acad. Sci. USA* 92 on the evolutionary consequences of food-chain length.

food conversion ratio The ratio of the number of calories of a prey required to produce one calorie for a predator.

food production, industrialization of The increasing intensification and capitalization of the production, transport, storage, and retailing of foods, associated with *agribusiness and *transnational corporations. The Indian broiler industry, for example, grew from 31 million birds a year in 1981 to 800 million two decades later. Very powerful corporations dominate many sectors. Primary producers are locked into tight specifications and contracts. Consumers may benefit from cheaper food but there are implications for quality and health. In consequence, new quality agencies have been created. Food policy is thus torn between increased productivity, reduced prices, and the demand for higher quality and higher production standards. See Lang (2003) *Dev. Policy Rev.* 21, 5/6.

food security Living without hunger, fear of hunger, or starvation. Key components of food security are building local capacity to produce and distribute food and control food supplies, and possibly keeping decision-making power within the community rather than losing it through dependence on external sources of food. See Anderson and Cook in J. Harris (2000).

Nally (2010) *TIBG* 36, 1, 37 explores the way a moral economy of hunger (viewing hunger as immoral and requiring intervention by governments and NGOs) is gradually being replaced by a political economy of food security that promotes market mechanisms as a better protection against scarcity. 'The spectre of hunger in a world of plenty seems set to continue into the 21st century. It bears repeating that this is *not* the failure of the modern food regime, but of its reliance on over-production in some places and under-production in others. To think seriously about global hunger means addressing the legal, institutional, and biotechnical mechanisms—including trade tariffs, agricultural subsidies, enforcement of intellectual property rights, and the privatization of public provisioning systems—that directly restrict certain people's ability to subsist.

foodshed The flow of food from the area where it is grown into the place where it is consumed. See Schreiber et al. (2021) *Environ. Res. Lett.* 16.

food web A series of interconnected and overlapping *food chains in an *ecosystem. A small change in the number of species in a food web can have consequences both for community structure and ecosystem processes.

footloose industry An industry whose location is not influenced strongly by access either to materials or markets, and which can therefore operate within a very wide range of locations. Any form of 'direct line' business, operated almost entirely through telephone and fax lines, would be an example. Such industries are also liberated from locational constraints by **footloose capital**, that is, capital that is freely mobile between countries. After the global financial crisis in 2008, the G20 supported efforts to improve global capital flows; see Bertaut et al. (2019) *FEDS* notes.

foraminifera Usually marine micro-organisms of plankton and *benthic animals with calcite skeletons, found over much of Earth's ocean beds. Foraminifera are very sensitive to temperature, and their fossils may be used to reconstruct past environments; see, for example, the journal *Holocene* (2011) 21, 4 on foraminiferal faunal evidence of twentieth-century Barents Sea warming.

forcing A causal change or perturbation of the climate system.

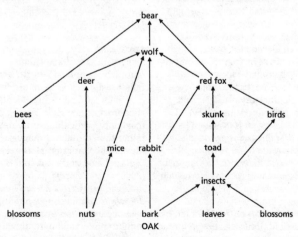

Food web

Radiative forcing is the net flux of *radiation in or out of a system. The five major greenhouse gases account for about 97% of the direct radiative forcing by long-lived gases. Interannual variations in the growth rate of radiative forcing due to CO_2 are large, and probably related to natural phenomena such as volcanic eruptions and ENSO events, as well as to *anthropogenic activity. The same forcing factors can produce hugely divergent evolutions of the climate system given infinitesimally small differences in initial conditions. **Orbital forcing** (also known as **astronomical forcing**) is the climatic effect of slow changes in the tilt of the Earth's axis and shape of the orbit.

A **forcing factor** would therefore be a factor that affects forcing. Tooth (2008) *PPG* 32, 1 lists climate, tectonics, and human activity as forcing factors in an arid geomorphology, and Burningham (2008) *Geomorph.* 97, 3–4 writes of sea-level rise as a forcing factor on estuaries.

förde An elongated bay in an area of glacial deposition, formed when glacial tongue basins are drowned by rising water levels. The type locations are northern Germany and Denmark.

Fordism Henry Ford (1863–1947) paid workers high wages in return for intensive work. The term Fordism was coined by Antonio Gramsci (1971) to describe a form of production characterized by an assembly line (conveyor belt) and standardized outputs, linked with the stimulation of demand brought about by low prices, advertising, and credit. Fordism is associated with the spatial separation of the development of the product, at the centre of research and development (usually in a more developed country), and the actual sites of the production of a standardized product (often in a developing country).

foredune A ridge of irregular sand dunes, typically found adjacent to beaches on low-lying coasts, and partially covered with vegetation. Hesp (2002) *Geomorph.* 48, 1 reviews the initiation, dynamics, geomorphology, and evolution of foredunes.

foreign direct investment (FDI) The purchase of real assets abroad; through acquiring land, buying existing foreign businesses, or constructing buildings, or mines. **Inward foreign direct investment** is the acquisition of real assets within a country by non-residents.

T. Moran et al. (2005) argue that FDI almost inevitably delivers economic development benefits: FDI from other developing countries can add to inflows of other external financial resources, including commercial bank lending, and portfolio investment. For poorer developing countries, it can be significant, accounting for over half of total FDI inflows into several LDCs (UNCTAD 2006).

However, attempts to attract FDI may reduce the benefits to host countries, as managers may make their investments more 'footloose', and readily relocate to somewhere that has become cheaper, and FDI tends to be skewed towards the interests of *TNCs, rather than the interests of local and national communities and governments.

foreign exchange The foreign capital earned by a country's exports. Since the currency of many less developed countries is not accepted by international markets, it often becomes necessary to earn foreign exchange in order to buy imports. See Wojcik (2014) *SSRN Electronic Journal* on the geography of foreign exchange trading; of currencies and international financial centres. See also P. Wang (2009) on the

economics of foreign exchange and global finance.

formal game An exercise in *game theory. Buttyán and Hubaux (2001) outline a formal model of rational exchange. See Padon (1999) *J. Urb. Affairs* 21, 2 on a game theoretic model of local public good provision.

formal sector All those types of employment which offer regular wages and hours, which carry with them employment rights, and on which income tax is paid—compare with the *informal sector—the 'gig economy'. However, these two sectors are by no means entirely separate; see Arimah (2001) *Afr. Dev. Rev.* 13, 1 on informal and formal sector linkages in Nigeria.

fossil fuel Any fuel found underground, buried within sedimentary rock: coal, oil, and natural gas.

foundationalism The use of some secure foundation of certainty (justified belief) in theories of knowledge; a methodological claim made for the existence of an underlying physical structure or an integrated theory that is designed to explain individual events. Livingstone *TIBG* 15, 3 is still worth reading.

fractal A geometric figure, each part of which has the same statistical character as the whole. In a landscape, **self-similar fractals** occur regardless of the scale at which it is viewed; topographic contours are described as self-similar, and a numerical sequence is statistically self-similar if its statistical moments are the same upon resampling. However, in **self-affine fractals**, different geometrical directions are scaled differently to preserve shape or statistical moments; see Sung and Cheng (2004)

Geomorph. 62, 181; see also Pelletier (2007) *Geomorph.* 91, 3–4.

fractus *See* CLOUD CLASSIFICATION.

fragmentation *See* FARM FRAGMENTATION.

FRAGSTATS A computer software program designed to compute a wide variety of landscape metrics for categorical map patterns. The original FRAGSTATS future patch updates, and releases can be downloaded from the internet.

framing A way of organizing our everyday experience; for example, we need the frame of play-fighting in order to know not to punch someone really hard! Schmueli (2008) *Geoforum* 39 describes framing as a cognitive process whereby individuals and groups sort out their perceptions, interpretations, and understandings of complex situations in ways that fit their own socio-political, economic, and cultural world views and experiences. See Demeritt (2009) *TIBG* 34, 1 on framing environmental research in Geography.

free atmosphere The portion of the Earth's *atmosphere above the *planetary boundary layer in which the effect of surface friction on the air motion is weak.

free face (fall face) An outcrop of rock which is too steep for the accumulation of soil and rock debris.

free market Essentially, the free market is the price mechanism through which consumers choose goods or services on the basis of their willingness to pay for these. **Free market environmentalism** is an ideology that argues that the free market is the best tool to preserve the health and sustainability of the environment. See Kolstad et al. (2011) *PERC* 29, 2 on the promise and problems of free market environmentalism.

free trade Trade between countries which takes place completely free of restrictions; there are no *tariffs and *quotas. This allows specialization in member states of free trade areas, and should lower costs because both competition and markets are increased. See Martin, Paris, Thoenig, and Mayer (2010) *RePEc* on the geography of free trade agreements.

free trade zone A designated area, often within a *less economically developed country, where normal tariffs and quotas do not apply. It is common for the conditions of employment there to be more repressive; such zones are often mandatorily union-free, and working conditions can be harsh. See Hu et al. (2020) *Int. J. Financial Studs* on the comparative advantages of free trade port construction in Shanghai.

freeze–thaw The weathering of rock said to occur when water, which has penetrated joints and cracks, freezes. In a bright and breezy paper, Murton (2007, *Planet Earth* 4) finds that the long-held explanation that freezing, expanding water shatters rocks is 'probably not very significant in nature because it requires some pretty unusual conditions. The rock must essentially be water-saturated and frozen from all sides, to prevent the piston-like effect of freezing water driving the remaining liquid water into empty spaces or out of the rock through an unfrozen side or crack. So we need to look for another explanation.'

freezing front The edge of frozen or partially frozen ground. In areas of seasonal freezing with no *permafrost, the freezing front moves downward through the earth. In areas of permafrost, the front can also move upwards: 'during frost heaving of soil

when the soil is freezing from the top down, water in the soil pores flows upward to the freezing front because of a gradient in the soil moisture pressure (or tension). This occurs even when the soil pores are not saturated. Ice lenses form and grow at or slightly above the freezing front and cause great uplifting forces' (Henry (2000) *US Army Corps of Engineers*).

freezing nucleus A *nucleus on which a water droplet will freeze to form an ice crystal. Ice nuclei thus formed are much less common than *condensation nuclei, but their effectiveness rises as the temperature falls below 0 °C. They grow by *sublimation if the *ambient air is saturated with respect to water.

freight rate This cost of transporting goods reflects a number of factors besides basic transport costs, such as the nature of the commodity. Non-breakable, non-perishable items, like coal, are carried most cheaply as they can be carried in bulk on open wagons. The more careful the handling required, the more expensive is the freight rate.

Distance is an important factor. Many freight rates are tapered; that is, the rate per tonne-mile or tonne-kilometre drops as the distance increases, but this change in rates is expressed in a series of distance 'bands' so that, on a graph, the relationship between cost over unit distance and distance would appear as a series of downward steps rather than a smooth diagonal line.

friction The force which resists the movement of one surface over another. Friction between the surfaces of two mineral grains is related to the hardness of the mineral, the roughness of the surface, and the number and area of the points of contact between the grains. It is of major significance in any study of the movement of sediment since the forces moving the sediment must be greater than the resistance provided by friction. In meteorology, the **frictional force** is the roughness and irregularity of the Earth's surface that reduces wind speeds. The **friction layer**, where this effect is strongest, roughly comprises the lowest 100 m of the atmosphere.

friction coefficient The ratio of the force that maintains contact between an object and a surface and the frictional force that resists the motion of the object; the ratio of two forces acting, respectively, perpendicular and parallel to an interface, between two bodies under relative motion (Blau (2001) *Tribology Int.* 34, 9). Under the same friction coefficient different normal stresses cause different friction levels. In a rockfall, the friction coefficient is the strongest factor in determining velocity.

friction of distance As the distance from a point increases, the interactions with that point decrease, usually because the time and costs involved increase with distance. Distance need not be reckoned solely in spatial terms; the frictional effect of distance 'on the ground' is far less in a lowland area with good communications than in an upland area of difficult terrain, but has slackened with improvements in transport and communications. Human activities tend to organize with respect to geographic location due to the friction of distance and the consequent competition for advantageous location, but technologies to mitigate the friction of distance influence the relationships between people, place, and activity; see (Miller (2007) *Geog. Compass* 1, 3). Ellegård and Vilhelmson (2004) *Geografiska A* 86, 4 argue that geographical immobility and proximity in everyday life indicate the continuing and often neglected importance of the friction of distance.

friendship, geographies of Geographers of friendship are concerned with: geographies of affect/emotion, children and young peoples' geographies, and the (re)production of social ordering; and geographies of mobility and transnationalism in a world of increased human spatial movement and virtual social relations. See Korkiamäki and Kallio (2018) *Area* 50, 1 on experiencing and practising inclusion through friendships.

fringe belt At the edge of a town or built-up area, a zone of varied land use. Typical features in the UK are: Victorian hospitals and cemeteries (located beyond the city for reasons of public health); recreation facilities such as playing fields, riding stables, and golf courses; and utilities such as water and sewage works. Many of the functions of the fringe belt have been squeezed out from the town centre due to congestion, high land prices, the need for a special site, or disturbances in the central area. Sometimes further urban expansion leapfrogs the fringe belt.

front The border zone between two *air masses which contrast, usually in temperature. A **warm front** marks the leading edge of a sector of warm air; a **cold front** denotes the influx of cold air. Fronts are intensely *baroclinic zones, about 2000 km long and 2000 km wide, moving at around 14 km per day. In *mid-latitude depressions, fronts develop as part of a horizontal wave of warm air enclosed on two sides by cold air. These **frontal wave forms** move from west to east in groups known as **frontal wave families**.

The basic classification into warm fronts, with a slope of 1 in 100, and steeper cold fronts is further divided by the type of air movement at the front. In **ana-fronts**, the warm sector air is rising,

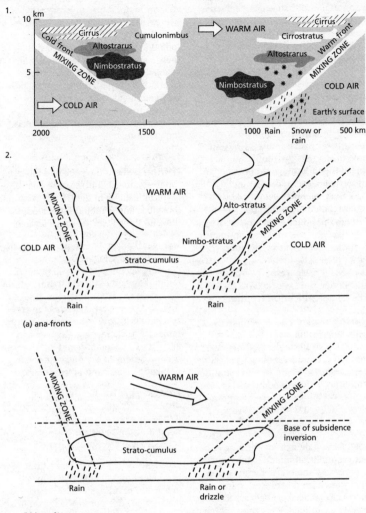

Fronts

and a succession of cloud types and precipitation results. In **kata-fronts**, the warm sector air is descending, clouds are few, and precipitation is reduced to a drizzle.

frontier That part of a country which lies on the limit of the settled area; the term indicates outward expansion into an area previously unsettled by a particular state. 'The frontier arrives and passes, in a historical sense, replacing one set of social relations with another' (Simmons et al. (2007) *AAAG* 97, 3). Some frontiers have occurred where two nations advance from different directions, leading to boundary disputes. A **settlement frontier** marks the furthest advance of settlement within a state while the **political frontier** is where the limit of the state coincides with the limit of settlement. The **agricultural frontier** is the fringe of market-oriented agriculture and ranching, which advances on subsistence farming or uncultivated wilderness, as the case may be.

Pallot (2005) *TIBG* 30, 1 sees the Soviet use of the peripheries as a place of exile as driven by a powerful state intent on expanding the **resource frontier**; see Barney (2009), *Geographical J.* 175, 10 on the concept of the resource frontier in Laos. Butler and Lees (2006) *TIBG* 31, 4 write on **gentrification frontiers**.

frontogenesis The development of *fronts and frontal wave forms. Frontogenesis occurs in well-defined areas; for example, off the east coast of North America. Fronts are less common in the tropics, where contrasts between *air masses are less marked, but see Chen et al. (2007) *Monthly Weather Rev.* 136, 1.

frost Frozen dew or fog forming at or near ground level. **Black frost** is a thin sheet of frost without the white colour usually associated with frost, occurring when few or no ice crystals are formed, because air in the lower atmosphere is too dry. Air below 0 °C is **air frost**. **Radiation frost** (sometimes called **ground frost**) is a ground-level fog formed by nocturnal radiational cooling of a humid air layer so that its relative humidity approaches 100%, characterized by a low vertical ceiling of the cold air layer. The stratification of cold air temperature on spring nights with radiation frost is much stronger than that on nights with advective-radiation frost. **Hoar frost**, or rime, is a thick coating of white ice crystals on vegetation and other surfaces; Karlsson (2001) *Met. Appls* 8 finds a correlation between the amount of hoar frost, average wind speed, and difference between dew point and road surface temperature.

frost cracking When the frozen ground reaches very low temperatures, it contracts, splitting up to form a pattern of polygonal cracks. Liquid water content has an important impact on freezing-induced rock damage; sustained freezing can yield much stronger damage than repeated freeze-thaw cycling; and that frost cracking occurs over the full range of temperatures measured extending from 0°C down to –15°C. Anderson (1998) *Arct. & Alp. Res.* 30, 4 defines the **frost cracking window**—the optimal thermal range for frost cracking—as between –3° and –10° C. Vegetation and accumulations of organic matter may reduce the likelihood of frost cracking.

frost creep The net downslope displacement that occurs when a soil, during a *freeze–thaw cycle, expands perpendicular to the ground surface and settles in a nearly vertical direction.

frost heaving The upward dislocation of soil and rocks by the freezing and expansion of soil water. See Matsuoka et al. (2003) *Geomorph* 52, 1–2 on frost heaving and sorting.

frost hollow A concentration of cold air in a hollow or valley floor when nighttime *terrestrial radiation is greatest on valley slopes. Air over these slopes becomes colder, and hence denser, and flows downslope. Temperatures in frost hollows can be tens of degrees colder than the surroundings. See Payette (2015) *J. Ecol.* 103, 3.

frost pocket *See* FROST HOLLOW.

frost shattering The fracturing of rock by the expansionary pressure associated with the freezing of water in planes of weakness or in pore spaces; the major weathering process in cold areas.

frost thrusting The lateral dislocation of soil and rock by the freezing and expansion of water. See Zhu (1996) *Perig. & Permaf. Procs* 7, 1.

frost wedging *See* FREEZE–THAW.

Froude number The ratio of the velocity (v) of a river to its *celerity where celerity is the product of the acceleration due to gravity (g) and the mean depth of flow (d). The Froude number (Fe) is calculated from the equation:

$$F_e = \frac{v}{\sqrt{gd}}$$

Where F_e is less than 1, deeper flow is tranquil. Where F_e exceeds 1, the flow is turbulent. See Lewin and Brewer (2001) *Geomorph.* 40 on Froude numbers and channel patterns.

fugitive species *See* OPPORTUNIST SPECIES.

fumarole A vent in a volcano through which steam and volcanic gases are emitted. See Walker et al. (2006) *GSA Spec. Paper* 412.

fumigation 1. The trapping of pollutants beneath a stable layer or *inversion, and their transport down towards the ground. See Kim et al. (2005) *J. Atmos. Scis* 62, 6.
   2. Soil fumigation is a chemical or physical process that kills viable weeds, seeds, soil-borne pathogens, and nematodes.

( ) SEE WEB LINKS
• Cockx and Simonne article on University of Florida website.

functional linkage The link between industries, including information, components, raw materials, finished goods, and transport links. Functional linkages are at the heart of *agglomeration economies. A firm's linkages are classified by their direction of movement: **backward** or **input linkages** are received by the firm—producer services have large backward linkages (Besser (2003) *Rural Sociol.* 68, 4). **Forward linkages** are supplied by a firm to another undertaking: consumers and/or firms like to locate close to their suppliers, as this decreases their living/production costs. See Crozet (2004) *J. Econ. Geog.* 4 on the forward linkages that relate labour migrations to the geography of production.

functional region A type of region characterized by its function, such as a city-region or a drainage basin. Coombes et al. in D. Herbert and R. Johnston, eds (1982) define functional regions for the UK Census.

Gaia hypothesis J. Lovelock (1988) argues that planet Earth—*atmosphere, ecosphere, geosphere, and *hydrosphere—is a single ecosystem/organism, regulating itself by feedback between its *abiotic and *biotic components. Lovelock stresses the overriding importance of Gaia, rather than any individual species.

Gaia tends to equilibrium, but human agency seems to be overriding its regulatory mechanism. 'We have spread thousands of toxic chemicals worldwide, appropriated 40% of the solar energy available for photosynthesis, converted almost all of the easily arable land, dammed most of the rivers, raised the planet sea level, and now . . . are close to running out of fresh water. A collateral effect of all this genetic activity is the continuing extinction of wild ecosystems, along with the species that compose them. This also happened to be the only human impact that is irreversible' (E. O. Wilson 2006).

gaining stream Also known as an effluent stream, this is a stream that is fed, wholly or in part, from groundwater from the underlying *aquifer. As long as the level of the water table is higher than the channel, the flow will be maintained, even during rainless periods. A stream losing water through its bed and banks is, obviously, a **losing stream**.

game theory Game theory models ideas of rational choice in conditions of interdependent decision-making, that is, situations in which the decision of one

actor is dependent on the decisions of the other players, and vice versa. The fates of the players are mutually implicated in this way. In geography, game theory is often used to overcome or outwit the environment. A. Dinar et al. (2008) apply game theory to real-life issues in natural resources and the environment.

gamma index (γ) In a network, a measure of connectivity that reflects the relationship between the number of observed links and the number of possible links. The value of gamma is between 0 and 1 where 1 indicates a completely connected network (extremely unlikely in reality). Where e is edges, and v for vertices (nodes),

$$\gamma = \frac{e}{3(v-2)}$$

See Xie and Levinson (2007) *Geog. Analysis* 39, 3.

garden city A planned settlement, as conceived by Ebenezer Howard (1850–1928), with low housing densities, and many parks, open spaces, and allotments; the maximum city size to be about 30000. The first UK garden city was Letchworth, 1903. Howard's ideas were echoed in the construction of *new towns in the UK. A **garden suburb** is a planned suburban development inspired by Howard's ideas. Garden suburbs were built in the late 19th and early 20th centuries, for example London's Hampstead Garden Suburb,

1907. See P. Hall and C. Ward (1999) on Howard's legacy.

gatekeeper An individual—or possibly a group—able to control access to goods and/or services. This may be a body that vets research; see Schurr et al. (2020) *Professional Geographer* 72, 3. In 1994, Operation Gatekeeper was set up by the US administration to 'regain control' of its border with Mexico; see Joseph Nevins (2002). For estate agents as gatekeepers, *see* REDLINING.

gay geography This is concerned with all the relationships and interactions between human sexuality and space and place. Brown (2009) *Env. & Plan. A* 41 examines a number of gay spaces and practices in order to consider the different forms of enterprise, transactions, and labour that take place within them. See Hubbard (2018) *Sexualities* 21, 8 on geography and sexuality, and Alldred and Fox (2015).

gaze The way we look at things, for the way we look at things influences what we see; see R. Dubbini (2002). Different people will gaze at a landscape and 'see' different features according to—among others—their gender, race, class, age, and sexuality. Thus, the **male gaze** looks at things from the perspective of a heterosexual man. D. Arnold (2006) outlines the ways in which India's material environment became increasingly subject to the **colonial gaze** at Indian landscapes.

GDP (gross domestic product) The total value of the production of goods and services in a nation over one year. Components for a finished product are not taken into account; only the finished articles (but the definition of 'finished product' varies). GDP is an imperfect measurement of a nation's economy because certain forms of production,

especially *subsistence production, are not recorded.

gelifluction The downslope flow of soil in association with ground ice, occurring in *periglacial environments, where water cannot percolate through the *permafrost. Gelifluction is promoted by abundant groundwater, fine-grained sediments, and slopes of 10°–30°. See Harris et al. (2000) *Arctic, Antarctic, and Alpine Research* 32, 2.

gelifraction Synonymous with *freeze–thaw, an umbrella term for those physical weathering processes resulting from the expansion of water as it freezes into ice. The main influential factors are altitude, aspect, temperature and precipitation, lithological characters, and structure and relief (Liu and Xiong (1992) *J. Glaciol. & Geocryol.* 14, 4: 332).

geliturbation Any frost-based movement of the *regolith, including *frost heaving and *gelifluction.

gendarme On an *arête, an abrupt rock pinnacle which has resisted *frost shattering.

gender While it is generally accepted that sex is biologically determined, societies construct appropriate behaviour for each gender. The relevant journal for geography and gender is *Gender, Place & Culture*. Padmanabhan (2007) *Sing. J. Trop. Geog.* 28, 1 notes that in rural West Africa the **gendered division of labour** extends to labelling certain crops as 'male' or 'female'; with the introduction of new crop varieties, these gendered plant constructions are renegotiated. See Guyat (2005) *NZ Geogr.* 61, 3, writing on bar staff, and masculinities and femininities.

The ordering of space is strongly gendered, and may also reinforce gender stereotypes; when spaces (secluded woodlands, dark alleyways, ill-lit multi-storey car parks) make them feel unsafe, women feel vulnerable, and this will further constrain their movements, so fulfilling the stereotype that women are less adventurous than men. 'While patriarchal structures of inequality often result in the spatial entrapment of women, the spatial boundedness of women's lives can be both enabling and constraining' (J. Lin and C. Mele 2005). Hmm.

See Dua (2007) *Gender Place Cult.* 14, 4 on the importance of gender in constituting the racialized practices of Canada. See also Heynen (2008) *ACME* 7, 1.

The study of **gender and development** considers the importance of gender in development issues: the global diversity of gender roles, relations, and identities such as nation, race, class, age, and sexuality; and the need to integrate gender into research and policy agendas in less advanced economies. See V. Kinnaird and J. Momsen, eds (2002).

genealogy 1. The study of family history; see Nash (2002) *Env. & Plan. D* 20 (1) 27, who observes that 'genealogy can serve to anchor and protect exclusive national cultures'.

2. The history of systems of thought; the history of those structures within society that 'have produced and shaped the boundaries of knowledge, ideas, truths, representations, and discursive formations in different historical periods' (Crowley in R. Kitchin and N. Thrift 2009). S. Aitken and G. Valentine (2006), p.15 throw light on this concept: in geography 'professional organizations may privilege or reinforce particular fashionable ways of thinking, but there are always dissenting voices. In reality, most ways of knowing are partial or in flux'.

general circulation of the atmosphere The world-scale system of pressure and winds that transports heat from tropical to polar latitudes, thus maintaining the present patterns of world temperatures. This global circulation, influenced by the *Coriolis force, is driven by intense differences in *insolation between the tropics and the poles. Air moves vertically along the *meridians and horizontally with the wind systems, both at ground level and in the upper atmosphere. See B. Haurwitz (2007). See Yu (2000) *Math. Comput. Simul.* 52, 5–6 for a general circulation model.

general linear models (GLM) A class of statistical models that aims to quantify any correlation between independent variables and dependent variables. R. F. Haase (2011) explains how to use GLM.

generator cell A localized cyclonic cell, caused by fluctuations in the polar front *jet stream, bringing precipitation. See Trapp et al. (2002) *Monthly Weather Rev.* 129, 3.

genetic modification The modification of the characteristics of animals through very precise ways, most recently through the transference of a gene. Animals altered in this way are called *transgenic*. Holloway (Economic and Social Research Council web pages for funding and guidance) provides a useful summary, noting that, to the layperson, the adoption of genetic techniques is portrayed as being 'straightforward and inevitable', which it is not.

((●)) SEE WEB LINKS
• BBSRC web pages for guidance.

genocide A collective, organized attempt systematically to destroy a politically or ethnically defined group. Ideas of 'difference' and '*other' underpin genocide: 'to truly understand how . . . contemporary genocides and camps remain legitimate within liberal democratic regimes, as well as to

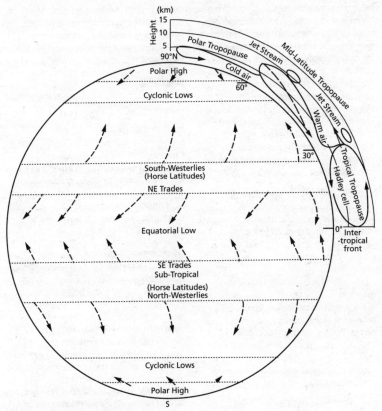

General circulation of the atmosphere
Source: Managing Water for Peace in the Middle East

understand how sovereign states discriminate between an authentic life and a life lacking political value, we need to theorize biological constructions of difference and their connections to citizenship' (Mitchell (2006) *PHG* 30, 1). Territoriality is a further component: late 20th-century genocides exhibited a marked spatial pattern, attacks being more frequent in the peripheral areas of the supposed 'ethnically pure' unit: in eastern Croatia in the former Yugoslavia, and in north-west Rwanda, for example.

Wood (2001) *TIBG* 26, 1 uses the concepts of Lebensraum, territorial nationalism, forced migration, and ethnic cleansing to explain the production of genocide in Rwanda, and J. Anderson (2010) gives a clear and chilling account of that genocide.

'Ethnic cleansing' (the foul euphemism for genocide) is not achieved solely through mass murder, but through forced expulsions and systematic rape—to 'cleanse' the lineage. 'After a decade of displacement, the legacy of ethnic cleansing endures, forming limits to returns and persistent insecurity for returning communities, thus permanently altering Bosnia's human geography and political future' (Dahlman and Ó Tuathail (2005) *Pol. Geog.* 24, 5). Typically, the overall ploy moves from segregation to isolation and elimination—which explains the vital role that 'safe areas' can play in countering genocide.

Do see Joe Sacco's graphic novel *The Fixer* (2003) on Bosnia.

gentrification The rebuilding, renewing, and rehabilitation of depressed areas of the inner city as more affluent families seek to live near to the city centre, trading space and quiet for access to the goods and services of the city centre.

Gentrification initially emerged as a quaint and local anomaly in some cities, but has become a sustainable urban form, characterized by street-level spectacles, trendy bars and cafes, social diversity, and funky clothing outlets. Lees, Slater, and Wyly (2008) is a fine introduction to the topic. See Phillips (2004) *PHG* 28, 1 on rural gentrification.

geochronology The chronology of the Earth generated from geological information; the science of the absolute and relative dating of geological formations and events.

geocoding (georeferencing) The process used to convert location codes, such as street addresses or postal codes, into geographic (or other) coordinates. Geocoding software is available online, and can offer services such as converting postcode to longitude/latitude, or longitude/latitude to postcode.

geocomputation The art and science of solving complex spatial problems with computers, focusing on the research,

development, and application of spatial information technologies to address social, economic, and environmental problems. See Xue et al. (2004) *Future Generation Computer Sys.* 20, 7.

(⊕) SEE WEB LINKS
• A survey of geocomputational techniques in spatial data analysis.

geo-constitution The concept of geo-constitution provides a framework for considering the importance of the geography of political institutions, the way they shape political relationships and decision-making, and their impact on political outcomes. It's an approach to geopolitics, and not a real entity. See Wills (2019) *Prog. Hum. Geog.* 43, 3.

geodemographics A form of information technology that enables marketers to predict, through the use of statistical models, the behavioural responses of consumers, based on identity and residential location. See Singleton and Spielman (2014) *Prof. Geogr.* 66, 4 on the past, present, and future of geodemographic research in the USA and the UK.

geodesy The science of the shape and size of the Earth. **Geodetic** is the relevant adjective, and a **geodetic datum** is a set of reference points on the Earth's surface against which position measurements are made, and (often) an associated model of the shape of Earth to define a geographic coordinate system (K.-T. Chang 2014).

geodiversity The variety of rocks, minerals, fossils, landforms, sediments, soils, and associated processes which are used as assets.

geodynamics A subfield of geophysics dealing with dynamics of the Earth's lithosphere and mantle and, using data from geodetic GPS, InSAR, seismology, and numerical models. For example, White and Lister (2012) *J. Geodynamics* 7, review the relative motion of India and Asia for the last 100 million years.

geoecology (landscape ecology) An interdisciplinary science which studies the interactions and interrelations in the environment; a view of geo-ecosystems as dynamic entities constantly responding to changes within themselves, their near-surface environments, and external influences, both geographical and cosmic. See *Catena* 2017 149 (3).

geoeconomics A branch of *geopolitics concerned with the study of the spatial, temporal, and political aspects of economies and resources. Primary concerns and objectives of geoeconomics comprise building international partnerships that promote harmonization, efficiency, economic leverage, and growth, as against the supposed threats of political radicalism, anachronism, and anarchy. See Sparke and Lawson in J. Agnew, K. Mitchell, and G. Cairns, eds (2007) 316.

geogenic Relating to the history of the Earth; mostly resulting from geographic and/or geological processes.

geographic information systems (GIS) Essentially, the merging of cartography, statistical analysis, and database technology. The critical functionalities of any GIS package are: data capture, storage, management, retrieval, analysis, and display.

   Qualitative GIS aims to look at how individuals understand space and what the impacts of these understandings are for the production of socio-spatial relations. For an example, see Verd and Porcel (2011) *Forum Qualitative Sozialforschung/Forum: Qualitative Social Research* 13, 2 on the way qualitative GIS was used to study the social production of urban space. A **geographic resources analysis support system** is a free, open source GIS capable of handling *raster, topological vector, image processing, and graphic data (K.-T. Chang 2014).

geographies A term encompassing the multiple states of the world and the way we understand and are *situated in them. 'Geographies' is not simply the plural of geography—it is argued that the complex world can be seen through many different eyes. There are many geographies and many ways to understand them.

geography The study of the landscape—both physical and human; the study of place and space—where things are located, why they are located where they are, how places differ from one another, and how people interact with the environment.

geohistory 1. A study of the role of environmental features in history, and the way these features are interpreted and reinterpreted over time.
   2. For F. Braudel (1949), the historical understanding of the spatial and environmental contexts of human activities. Look for *Geohistory and Connected History*, Cairn International Edition (2018).

geoinformatics The integrated study and examination of hardware, software, tools, techniques, and pedagogy related to developments in geographical science, including geographic information systems, remote sensing, photogrammetry, and global positioning systems.

(((●))) SEE WEB LINKS
• Geo Informatics website.

geolegal Just as *geoeconomic refers to a worldwide commercial and economic logic, geolegal describes a global legal system and logic. Thus, the geolegal power of the European Union is the influence it wields, as a bloc, over legal rights and duties external to its borders. See Jorgensen (2018) *KFG Working Paper Series, No. 21, Berlin Potsdam Research Group*, on the rising Chinese geolegal order.

Geolegality has been described as 'the indissolvable relations between law, space and the workings of power across intimate and global scales'—see Gupta et al. (2020) *Political Geog.* 83. See also Brickell and Cuomo (2019) *PHG* 43, 1.

geological column Also known as the **stratigraphical column**, this is the separation of geological time into *eras and periods, as in the table below.

geomagnetism (terrestrial magnetism) The magnetic field of the Earth. The axis runs through the *magnetic poles, whose position varies over time; sometimes the north and south

magnetic poles switch places. The pattern of the magnetic field at any one time will be preserved in contemporary *extrusions of volcanic rock. The study of past magnetic fields, **palaeomagnetism**, can yield information about the creation of new material at the oceanic ridges, and continental drift.

geomatics The complex topics and techniques that can be assembled under Geospatial Information, namely, geodesy, cartography, photogrammetry, remote sensing, informatics, acquisition systems, global positioning systems, digital image processing, geographic information systems, decision support systems, and WebGIS. *Geomatics* is an international and interdisciplinary peer-reviewed open-access journal of geomatic science, published quarterly online by MDPI.

geometry In geomorphology, the shape, size, and relative position of a landform.

geomorphic effectiveness Comparing the magnitude of an event,

Era	Period	Epoch	End date, million years
Quaternary		Holocene (recent)	
		Pleistocene (glacial)	
Tertiary (Cenozoic)		Pliocene	2
		Miocene	
		Oligocene	
		Eocene	
Secondary (Mesozoic)	Cretaceous		65
	Jurassic		135
	Triassic		190
Primary (Palaeozoic)	Permian		225
	Carboniferous		280
	Devonian		345
	Silurian		395
	Ordovician		440
	Cambrian		500
	Precambrian		570

such as an increase in river volume, with its effects, such as fluvial erosion. See Lisenby et al. (2018) *Earth Surf.* 43, 1 for a detailed review of this concept.

geomorphic signature The distinctive characteristics of a landform which indicate the processes giving rise to it.

geomorphic systems Open systems where the energy of operation comes from outside, and moves across, the system boundary. Any change in energy or mass causes both the processes and their related landforms to adjust in order to maintain equilibrium in the system. The stability of a geomorphic system will depend on whether the changes are linear, non-linear, or threshold-related, and whether the effects of change are locally damped or propagate through the system; see Harvey (2007) *Geomorph.* 84, 3–4.

geomorphic threshold *See* THRESHOLD.

geomorphological sensitivity In a geomorphological context, the degree of response of a landform or system to anthropogeomorpholgical impact. Social factors and natural conditions will affect the degree of sensitivity to anthropological impact; see Tullos et al. (2014) on geomorphic and ecological disturbance and recovery, *PLOS One* 9, 9.

geomorphological threshold A point at which a landform switches, as in the change from an incision stage to an accumulation stage on the slopes of semiarid areas (Gutiérrez-Elorza and Peña-Monné (1998) *Geomorph.* 23, 2–4). An **intrinsic threshold** indicates possibly independent changes within the system; see Hugenholtz and Wolfe (2005) *Geomorph.* 70, 1–2 on the activation of a dune field, while an

extrinsic threshold indicates a response to change in an external factor; see Brunsden (2001) *Catena* 42, 2–4.

geomorphology The science concerned with understanding the form of the Earth's land surface and the processes by which it is shaped, both at the present day as well as in the past (British Soc. Geomorphology). Since many landforms cannot fully be explained by present-day geomorphic processes, geomorphologists also consider the impacts of past events on the present-day landscape; the landscape is a physical system with a history. The **process geomorphology** approach is concerned with the processes of surface mechanics.

Landscape change is dominated by uncertainties and probabilities, and is place-dependent.

With humans becoming one of the Earth's major modifiers, land-surface processes pose a suite of threats to human resources, through soil erosion, slope instability, and river and coastal flooding.

For cultural geomorphology, *see* CULTURAL TURN.

geomorphometry (morphometry, terrain modelling, digital terrain analysis) The quantitative analysis of the morphology of landforms. It uses general geomorphometric measures to capture the basic characteristics of landscapes, like slope, convexity, concavity, aspect, linearity, etc., and **specific geomorphometry**: the use of quantitative measures designed for analysis of specific landscape features. See Pike (2000) *PPG* 24, 1.

Geomorphometry uses input digital elevation models (DEMs, also known as digital land surface models). See Pike et al., eds in T. Hengl and H. Reuter, eds (2007) for a 'brief guide'.

geopolitics The study of the relationship between geographical features, whether natural or constructed, and international politics; the influence of geographical factors on political action. Geopolitics resembles International Relations, with more emphasis on geographical factors such as location, resources, and accessibility. The go-to textbook is Geoffrey Sloane (2017).

Critical geopolitics holds that all truth claims are political: that they are made on behalf of vested political interests and often in the pursuit of political economic imperatives. Thus Dalby (1996) *Pol. Geog.* 15 investigates 'the use of geographical reasoning in the service of state power', exploring the way the production of geopolitical knowledge about the relationship between states both exercises political power and affirms identity. But beware: 'after two decades of scholarship on "critical geopolitics", the question of whether it is largely a discursive critique of prevailing knowledge production and geopolitical texts or critique with an implicit, normative politics of its own, remains open' (Hyndman (2010) *Pol. Geog.* 26, 247).

geopolitik A view of *geopolitics developed in 1920s Germany. Individuals are subordinate to the state which must expand with population growth, claiming more territory—*Lebensraum—to fulfil its destiny. See Herwig (1999) *J. Strategic Studs* 22, 2.

geopower There is no clear consensus on the meaning of this term, but geopower has been described as the power of the Earth as a political personage, or as the forces of nature; 'earthly and inhuman powers of chaos, differentiation and creativity'; see Ingram et al. (2016) *Cultural Geog.* 23, 4.

geosciences *See* EARTH SYSTEM SCIENCE.

geosophy The study of geographical knowledge; 'the study of the world as people conceive of and imagine it' (John Kirtland Wright, quoted in Keighren (2005) *J. Hist. Geog.* 31).

geospatial The combination of spatial software and analytical methods with terrestrial or geographic datasets. **Geospatial object-based image analysis** (GEOBIA) is a technique used to analyse remote-sensed images (K.-T. Chang 2014). Hay and Castilla in T. Blaschke, S. Lang, and G. J. Hay (2008) provide a formal definition of GEOBIA, conduct a SWOT analysis of its potential, and discuss its main tenets and plausible future.

geospatial intelligence Earth-referenced information about natural and man-made objects or events with national security implications; the exploitation and analysis of imagery and geospatial information to describe, assess, and visually depict physical features and geo-referenced activities that have national security implications. See Clarke in R. Kitchin and N. Thrift (2009).

geospatial technologies Technologies that enable an ever-expanding range of individuals and social groups to create and disseminate maps and spatial data; Google Earth, Microsoft's Virtual Earth, and Wikimapia allow users to add their own geographic information to web-based displays or modify the contributions of others. Other technologies include geotagging, geoblogging, GPS-enabled devices, photographs, narrative text, and video clips that may now be embedded with details about their geographic location.

geostatistical modelling The use of statistical methods to analyse and predict spatial phenomena.

geostrophic wind A theoretical wind, occurring when the *pressure gradient force equals the opposing *Coriolis force (assuming straight/nearly straight isobars; when the isobars are strongly curved, the effect of centrifugal force should be added). The net result is a wind blowing parallel to the isobars. Except in low latitudes, where the Coriolis force is minimal, the actual wind has the same direction as the geostrophic wind.

   Supergeostrophic flow describes winds faster than the expected geostrophic wind, and occurs at a jet entry; see Cunningham and Keyser (1999) *Qly. J. R. Meteorol. Soc.* 125. **Subgeostrophic flow** describes winds slower than the expected geostrophic wind, occurring when air rotates counterclockwise around a low.

geosurveillance The monitoring of activity, often through the use of drones. See Swanlund and Schuurman (2019) *PHG* 43, 4.

geosyncline A thick, rapidly accumulating sediment body, usually parallel to a plate margin, formed within a subsiding sea basin. See Kashfi (1976) *GSA Bull.* 87, 10.

geotechnology The use of engineering techniques and scientific methods to exploit and utilize natural resources. See Gallagher et al. (2017) *PHG* 41, 5.

geothermal flux (geothermal heat) Heat from the Earth's interior generated by early gravitational collapse, and later radioactive decay. **Geothermal energy** is a potential energy source in a volcanically active area; see J. Marti and G. Ernst, eds (2005).

geovisualization An abbreviation of geographic visualization, using cartography, cognitive sciences, computer science, data visualization, *GIS, graphical statistics, human–computer interaction design, and information visualization. Geovisualization differs from *cartography in that it affords direct interaction with the content and presentation of what is seen. Gilbert and Masucci (2006) *Trans. GIS* 10, 5 find that the geovisualization of information in ways relevant to people's experiences may be more important to them than 'geographic accuracy'.

geo-web The body of geographical information that is increasingly used to organize and deliver content over the Web; examples include Google Earth, Google Maps, Live Search Maps, Yahoo Maps, and NASA World Wind, which are raising awareness of geography and location as a means to index information. Much of this locational-based data is supplied by users with no formal background in geography or cartography who 'voluntarily' contribute spatial information over the Web. This is the 'crowdsourcing' of geographical information—that is to say the practice of harnessing the 'power of the crowd' to create collectively represents a pronounced shift in what constitutes 'geographic information' at all levels of data, technologies, and practices. The impetus for this shift is not geographical *per se*. Its emergence and development have been overwhelmingly driven outside academic geography by the Web business and computing communities. Leszczynski (2012) *PHG* 36, 1, 72 explains it all.

gerrymandering Redrawing constituency boundaries for political gain. It involves 'careful drawing of constituency boundaries by a party so that either it wins a particular seat or, more generally, it wins more seats than its opponents' (Johnston (2002) *Pol. Geog.* 21, 1).

Gestalt theory A theory of perception and social understanding that emphasizes the degree to which individuals understand phenomena as wholes, greater than the sum of their individual parts, as when we perceive all of the elements of a map as a unified whole.

geyser A jet of hot water and steam, usually as a result of the *geothermal heating of underground water, connected to the surface by a narrow outlet.

ghetto A part of a city, not necessarily a slum area, occupied by a minority group (a **gilded ghetto** is simply a more affluent ghetto).

giant nuclei *See* NUCLEI.

gilgai Undulations of the soil surface, characteristic of many vertisol soils, and resulting from the shrinking and wetting of soils. See Joeckel and Howard (2018) Univ. Nebraska.

Gini coefficient In a *Lorenz curve, a measure of the difference between a given distribution of some variable, like population or income, and a perfectly even distribution. More simply, it tells us how evenly the variable is spread; the lower the Gini coefficient, the more evenly spread the variable.

GIS *See* GEOGRAPHIC INFORMATION SYSTEMS.

GIS analysis When spatial data are keyed into GIS software, the location of each point is recorded, together with the way in which that point relates to its real position on the surface of the Earth. When this database is loaded and the

software turned on, most programs will report on: the number of *points, lines, and *polygons for the theme loaded; the individual polygon areas and the total area for the theme loaded; and the total lengths of linear objects of a particular category.

glacial 1. Of or relating to a glacier.
2. An extended length of time during which Earth's glaciers expanded widely. The most recent glacial epoch, better known as the Pleistocene glacial epoch, previously the Quaternary, covered at least the last 3 000 000 years. A. Penck and E. Brückner (1909) established the classical sequence of four glacial periods in the Alps: the First Glaciation, at its peak some 550 000 years ago, the Second Glaciation, at its peak 450 000 years ago, the Third Glacial Period (Riss) with a climax 185 000 years ago, and the Fourth Glaciation (Wurm) at its height 72 000 years ago. Ice-sheets grew only during parts of these so-called 'glacials', each of which was followed by an *interglacial period.

glacial abrasion The grinding away of bedrock by fragments of rock incorporated in ice. It can be achieved by bodies of subglacial sediment sliding over bedrock, or by individual clasts within ice (Glasser and Bennett (2004) *PPG* 28, 3). Abrasion is favoured where effective basal pressures are greater than 1 MPa and where there are low sliding velocities. Landforms of glacial abrasion include striae, grooves, micro-crag and tails, bedrock gouges and cracks. Ice ceases to be an effective agent for abrasion when the weight of the ice is thick enough to bring about plastic flow.

The Boulton model (1996, *J. Glaciol.* 42) sees abrasion as controlled by effective normal pressure, and the Hallet model (1979, *J. Glaciol.* 23) argues that abrasion is highest where basal melting is greatest.

glacial breaching The erosion of a breach between two adjacent valleys when a transfluent glacier flows across the *watershed (divide) between them. See Hall and Jarman in S. Lukas et al. (2004).

glacial debuttressing The exposure of glacially steepened rockwalls, due to the down wastage and retreat of glacier ice. Ballantyne (2002) *Quat. Sci. Revs* 21, 1935, sees glacial debuttressing as 'one of the most important geomorphological consequences of deglaciation in mountain environments'.

glacial deposition The laying down of *sediments which have been removed and transported by a *glacier. The sediments—known as *till, or drift—are deposited when the ice melts; when *ablation is dominant. Glacial deposition is predominant in marginal areas of present and past ice sheets, such as the Eden Valley in the English Midlands, but also occurs in uplands, especially in the form of *moraines. See Escutia et al. in Cooper and Raymond et al. (2007) *Online Procs 10th ISAES* on meltwater production and high sediment discharge in the East Antarctic. *See also* DRUMLINS; ERRATICS; KETTLE HOLES.

glacial erosion Glacially eroded landscapes are moulded by abrasion, *debris entrainment, the 'conveyor belt' transport of *moraine on top of a glacier, rock fracturing, *pressure release, and the action of subglacial water pressure. Glacial erosion can modify the bed of an ice sheet and alter large-scale ice dynamics and mass balance.

The extent and nature of glacial erosion depend on ice characteristics. *Glen's Flow Law explains glacier behaviour with respect to strain rates and shear stress (Tarasov and Peltier (2000) *Annal. Glaciol.* 30, 1). If a glacier

digs too deep, strong thermodynamic feedbacks prevent further deepening. Olymo and Johansson (2002) *ESPL* 27 provide a useful summary of the roles of rock structure, lithology, and pre-glacial relief. For the response of glaciated landscapes to rapid rock uplift, see Brocklehurst (2007) *J. Geophys. Res.* 112.

In glacially eroded lowlands, relief is confused, weathered rock is stripped away, and *abrasion and *quarrying of the bedrock are active. Drainage is rambling, as earlier patterns are disrupted by erosion and the deposition of *moraines. Lakes and ponds abound, and perched blocks and erratics may be common. Jansson et al. (2011) *ESPL* 36, 3, 408 give a detailed and helpful account of using a GIS filter to replicate patterns of glacial erosion.

((⊕)) SEE WEB LINKS

• Glaciers online's thumbnails of glacially eroded landforms.

glacial–interglacial cycle The most recent third of the Quaternary is most clearly characterized by the alternation between cold glacial periods and warm interglacials. The dominant period of these climate cycles is 100 ka. The terms 'glacial' and 'interglacial' emphasize the changes in ice volume and sea level associated with the appearance and disappearance of large Northern Hemisphere ice sheets. About eight major glacial–interglacial cycles have occurred during the past 0.8 million years. Ice core records indicate that greenhouse gases co-varied with Antarctic temperature over glacial–interglacial cycles, suggesting a close link between natural atmospheric greenhouse gas variations and temperature (S. Solomons, ed. 2007).

glacial meltwater The effectiveness of glacial meltwater as an agent of erosion depends on: the susceptibility of the bedrock involved (particularly its structural weaknesses or proclivity to chemical attack); the water velocity and the level of turbulent flow; and the quantity of sediment in transport.

A **glacial margin channel** is a stream running between the side of the glacier and the valley sides. Landforms of glacial meltwater erosion include both subglacial and ice-marginal meltwater channels.

glacial surge The swift and dramatic movement of a glacier, associated with the growth of ice up-glacier to unstable proportions and with severe *crevassing. See Russell et al. (2001) *Glob. & Planet. Change* 28, 1 on glacier surging as a control on the development of proglacial, fluvial landforms and deposits.

glacial trough A deep linear feature, carved into bedrock by glacial erosion that is channelled along a trough or valley. Troughs may be cut beneath ice sheets by a combination of glacial *abrasion and *quarrying. The effects of glacial abrasion are more obvious on the smoothed and polished trough walls whilst the effects of glacial quarrying— such as roches moutonnées and rock basins—are most evident on the valley floors.

Hebdon et al. (1997) *ESPL* 22 calculate glacial erosion rates as between 1 and 2 mm a^{-1} in Scotland; Harbor et al. (1988) *Nature* 333 cite rates of 10^{-3} ma^{-1} landslide material in a single glacial cycle.

Although the cross-sectional form of a glacial trough is traditionally described as 'U-shaped', their true cross-sectional morphology is better described by empirical power-law functions and by second-order polynomials. Furthermore, this cross-sectional morphology varies with lithology rock

mass strength and the degree of alluvial modification.

Other characteristics of glacial troughs include classic features of glacial erosion such as hanging valleys and divide elimination.

glacier A mass of ice which may be moving, or has moved, overland: when enough ice has accumulated, a glacier will start to move forwards. A glacier may be seen to be the result of a balance between *accumulation and *ablation. Glaciers are classified by their location (*cirque glacier, expanded-foot glacier, valley glacier, *niche glacier, *piedmont glacier), by their function (diffluent glacier, *outlet glacier), or by their basal temperature (*cold glacier, warm glacier, *polythermal glacier). The **glacier balance velocity** is the velocity required to maintain the glacier in equilibrium without change to its geometry. Glaciers where the actual and balance velocities are similar are said to be in balance with the current climate, and are unlikely to experience major changes in flow rate unless the climate changes.

glacier budget The balance in a glacier between the input of snow and *firn, that is, *accumulation, and the loss of ice due to melting, evaporation, *sublimation, and *calving, that is, *ablation. A glacier grows where the budget is positive and retreats when it is negative.

glacier drainage systems In a warm glacier, meltwater will drain into the ice through a network of veins, *moulins, channels, and conduits, which can connect to an *englacial system, discharging at the snout, or to a *subglacial system of channels, linked cavities, subglacial lakes, and water films. These englacial and subglacial systems may become connected during the melt season; in cold glaciers, however, the superglacial and subglacial drainage systems are always separate. Hydrological features related to the evolution of the glacier drainage system include supraglacial routing, moulin types, and chasms (Yde and Knudsen (2005) *Geografiska A* 87, 3).

glacier hydrology The study of the deposition, distribution, and depletion of snow; the evaluation of hydrologic water balance of basins where appreciable snow occurs; the physics of snow and glacier melt; the storage and transit of liquid water in snow and glaciers, and of possible methods for estimating the rate of stream flow, and the volumes of run-off in basins where snow affects those quantities. See Cooke et al. (2006) *PPG* 30, 5 on glacier hydrology and supercooling; Hock in U. Huber et al. (2005) on melt modelling; and Hock (2005) *PPG* 29, 3 for a review of glacier melt.

glacier mass balance The balance between accumulation and ablation, which is largely dependent on climate. On a valley glacier or ice sheet, mass is added at the top in the accumulation zone and taken away through melting and calving at the terminus in the ablation zone. Consequently the surface profile of the glacier will steepen with increasing accumulation (M. Bennet and N. F. Glasser 2009). Thus, the mass balance gradient is the glacial driving mechanism.

glacier response time The duration of the morphometric adjustments following change in net mass balance. This is a vague concept, because the reaction time may depend on the glacier history in an untransparent way. A distinction should be made between response times for glacier *volume* and glacier *length*. When the specific balance changes, the volume will respond

immediately. For glacier length it may take a bit more time, in particular when the changes are more apparent at higher parts.

glacier retreat (glacier recession) Between 1980 and 2018, glaciers tracked by the World Glacier Monitoring Service have lost ice equivalent to 21.7 metres of liquid water—the equivalent of cutting a 24-metre- (79-foot-)thick slice off the top of each glacier; and the pace of glacier loss has accelerated from 228 millimetres (9 inches) per year in the 1980s, to 443 millimetres (17 inches) per year in the 1990s, to 676 millimetres (2.2 feet) per year in the 2000s, to 921 millimetres (3 feet) per year for 2010–18. Fahrenkamp-Upprnbrink (2015, September) *Science* explains recent glacier retreat.

glacier sliding *See* BASAL SLIPPING.

glacio-epoch A multimillion-year period of glaciation. Eyles (2008) *PPP*, 258, 89 reviews the relationship between glacio-epochs over the last 3 Ga of Earth history and phases of supercontinent breakup and assembly.

glaciohydraulic cooling A process that allows water to remain liquid below 0 °C. It is associated with basal water flow through subglacial overdeepenings; as the subglacial melting point changes in response to changing pressure, water may become supercooled, and freeze, but if the adverse bed slope is steep enough, the heat generated will not be enough to match the changed pressure melting point, and it will become supercooled. Alley et al. (2003) *Nature* 424 view supercooling as a key link between glacier dynamics, basal ice formation, and subglacial landform evolution.

Glaciohydraulic cooling is associated with subglacial deformation, and ice creep at crystal boundaries. Cooke et al.

(2006) *PPG* 30, 5 review recent work on glacial supercooling.

Glen's Flow Law The relationship between glacier creep and glacier stress can be given by Glen's Flow Law:

$$\varepsilon = A\tau^n$$

where ε is the strain rate, A and n are constants, and τ is the basal shear stress. The constant $n = 3$, and the value of A are dependent on ice temperature, crystal orientation, debris content, and other factors (Davis at Antarctic glaciers.org website). In using this law to model real ice masses, remember that many simplifications may have been assumed (B. Hubbard and N. Glasser 2005).

(((()))) SEE WEB LINKS
• Website of Antarctic Glaciers organization.

gley soils Soils with mottled grey/yellow patches. **Gleying** occurs in waterlogged soils since oxygen supply is limited there. Under these conditions, *anaerobic micro-organisms extract oxygen from chemical compounds, particularly reducing ferric oxide to ferrous oxide. See Vladychenskii (2011) *Eurasian Soil Sci.* 44, 7, 834.

global care chain A global network of person-to-person contacts between people working as carers, whether paid or unpaid. See Hochschild in W. Hutton, W. and A. Giddens, eds (2000). A common care chain typically entails an older daughter from a poor family who cares for her siblings while her mother works as a nanny caring for the children of a migrating nanny who, in turn, cares for the child of a family in a rich country. Yeates (2004) *Feminist Rev.* 77, 79 is excellent on this.

global circulation *See* GENERAL CIRCULATION OF THE ATMOSPHERE.

global city To rank as a 'global city', a city must have high scores in the following features: a concentration of advanced producer services (accountancy, advertising, banking/finance, law); a number of major TNC headquarters; the major provision of global financial services; and high levels of business activity, human capital, information exchange, cultural experience, and political engagement.

(((()))) SEE WEB LINKS
• Yeung and Olds on the global city.

global civil society The totality of the multitude of groups operating across borders, beyond the control of governments.

global commodity chain A network of organizations and production processes resulting in a finished commodity; a chain of nodes from raw material exploitation, primary processing, through different stages of trade, services, and manufacturing processes to final consumption and waste disposal. See Suwandi et al. (2019) *Monthly Review* 70, 10 for an admittedly left-wing view. Some have argued that the more inclusive language of *global value* chains should replace the more specific concept of *commodity* chain.

global commons Goods and resources—the oceans, Antarctica, outer space, and the atmosphere—that no person or state may own or control and which are central to life; as such they can be thought of as belonging to the whole of humanity.

global energy balance The difference between the total influx of *solar radiation to the Earth's surface and the loss of this energy via *terrestrial radiation, evaporation, and the dissipation of sensible heat into the ground.

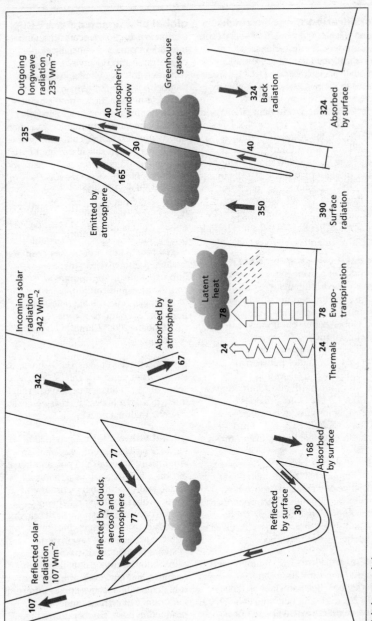

Global energy balance

globalization The increase in the volume, scale, and velocity of social (and environmental) interactions. Globalization is not new, pre-dating *colonialism. Ash (2004) *TIBG* 29, 2 describes globalization as a politically driven project, led by the US government, the World Bank, the World Trade Organization, and the G8. Jackson (2004) *TIBG* 29, 2 prefers to think of 'globalizing' rather than 'globalized', suggesting that globalization 'might be better thought of as a site of struggle rather than as a foregone conclusion'. See Broz et al. (2020) *International Organization* 75, 2 on the economic geography of the globalization backlash.

global justice movements A global assembly of movements focused on poverty, environmental degradation, racism, sexism, and neo-imperialism. It is often, wrongly, viewed as anti-globalization, but it does criticize the current mode of corporate neo-liberal globalization. It also criticizes inequalities between elites and ordinary citizens within nations of the north and south. King (2018) *J. Global Ethics* provides interesting insights, and Graness (2014) *J. Global Ethics* 11, 1 introduces concepts of global justice from African scholars.

But P. Routledge and A. Cumbers (2009) argue that, rather than being indicative of a coherent 'movement', such forms of political agency contain many political and geographical fissures and fault-lines, and are best conceived of as 'global justice networks'.

global pipeline The channels used in distant communications and/or interactions. See Bathelt et al. (2004) *PHG* 28, 1, 31, and Bahlman et al. (2009) *VU Research Portal*, who are not impressed.

global production network (GPN, **Global Production Network**) A global network of producers, usually of complicated goods, spread over several countries; in other words, the spatial fragmentation and dispersion of production. As manufactured products are becoming more technologically sophisticated, their production has often involved an increasing number of stages. These production stages can be decoupled and farmed out to a network of geographically dispersed locations to take advantage of lower costs . . . International trade then entails flows of goods belonging to a single industry but at different stages of production. This cross-border multi-staged production process has been facilitated immensely by major improvements in transportation, coordination and communication technologies, and the liberalisation of trade and investment regimes' (G. Nandan (2006) Commonwealth of Australia).

global space Global space is not a surface of sameness. It can be seen as a space in which finance capital can range freely in search of ever-increasing profit; a public multinational business space.

global value chain A production network which can link households, enterprises, and states to one another; the specific industrial, geographical, social, and institutional environment within which production networks operate; or the entire process of bringing a product from its beginnings to its end use: the extraction of raw materials, design, manufacturing, marketing, distribution, warehousing, delivery, retailing, and customer support. The value chain perspective stresses the importance of activities other than production itself, notably design, logistics, and marketing. Gereffi and

	Global commodity chains (GCCs)	Global value chains (GVCs)	Global production networks (GPNs)
Disciplinary background	Economic Sociology	Development Economics	Relational Economic Geography
Object of enquiry	Inter-firm networks in global industries	Sectoral logics of global industries	Global network configurations and regional development
Orienting concepts	Industry structure Governance Organizational learning Industrial upgrading	Value-added chains Governance models Transaction costs Industrial upgrading and rents	Value creation, enhancement, and capture Corporate, collective, and institutional power Societal, network, and territorial embeddedness
Intellectual influences	MNC literature Comparative development	International business Trade economics	GCC/GVC analyses Actor-network theory Varieties of capitalism

Lee (2012) *J. Supply Chain Managt* is wonderfully clear.

SEE WEB LINKS
• Global value chains.

global village A term coined by H. Marshall McLuhan (1962 and 1964) who correctly foresaw that continued improvement in electrified communication would eventually lead to instantaneous interaction across the globe. Some use this term to describe the *internet and its effects on human interactions.

global warming The increase in global temperatures brought about by the increased emission of greenhouse gases into the atmosphere.

glocal A combination of the local and global, made possible by travel, business and communications, and the willingness and ability to think globally and act locally. Thus, glocal financial institutions can use relatively local banks with sufficient global experience to operate in major financial markets, and

to so-called soft factors, such as the initiation of business deals, and an in-depth analysis of the company that sets out to issue bonds or shares.

GNP (gross national product) The *GDP of a nation together with any money earned from investment abroad, less the income earned within the nation by non-nationals. *Compare with* GDP. **GNP per capita** is calculated as GNP/population and is usually expressed in US dollars. It may be used as an indicator of *development.

goal function A function that can describe the succession of an ecosystem. 'Ecosystems do not, of course, have goals; but their successional development may be described by a goal function' (Jørgensen and Nielsen (2007) *Energy* 32, 5). Phillips (2008) *PPG* 32, 1 reviews theories of evolution and ecosystem development based on goal functions.

Gondwanaland A 'supercontinent': a continuous land surface formed of the now separate units of Africa, Madagascar, Antarctica, Australia, and India. Gondwanaland started in the earliest Paleozoic at the break-up of a Late Proterozoic supercontinent and assembled by 650–570 Ma.

gorge A deep, narrow cleft in an upland area, usually containing a river. Gorges occur in *karst scenery partly as a result of the collapse of caves, or when the incisive power of a river is greater than the processes of valley-wall erosion.

governance The action, fact, or manner of governing (not solely by governments); the various means used to shape society to a desired end, which include:

- the processes by which governments are selected, scrutinized, and modified
- the capacity of government to generate, establish, and apply policy and legal procedures
- the systems and methods of interaction between the executive, the legislature, and the judiciary
- the relationship between the state, economy, and civil society, and the regulation of economic regimes, and of social authority and practice.

Griffin (2012) *Pol. Studs Rev.* 10, 2, 208 clarifies the terms **multi-level governance**, which is concerned with continuous negotiation among nested governments at several territorial tiers, and **network governance**, emphasizing horizontal relations between and across institutions. **Environmental governance** is concerned with: the formal and informal institutions that influence how power is exercised; how public decisions are taken; how citizens become engaged or disaffected; and who gains legitimacy and influence in environmental policy.

governmentality A concept of M. Foucault (2003) to combine the mentalities, rationalities, and techniques used by governments, within their territory, to structure their subjects (the people they govern), and the social, economic, and political ways their policy can best be implemented. In other words, governments try to produce the citizen best suited to fulfil their policies. It encompasses the techniques and rationalities involved in both governing the self and governing others, and also includes ways in which spaces and places are created, and used, in order to pursue policies. Thus, **colonial governmentality** was aimed to construct a wage-earning, property-owning, taxpaying, and literate subject population; see Legg (2006) *Soc. & Cult. Geog.* 7, 5.

government in exile An exiled administration, such as the government in exile of Tibet that 'claims legitimacy as the official representative of the Tibetan population, performs a number of state-like functions in relation to its diasporic "citizenry" and attempts to make its voice heard within the international community', McConnell (2009) *Pol. Geog.* 28, 343.

GPR Ground-penetrating radar.

grade In geomorphology, a state of *equilibrium in a system. A river is said to be **graded** when it can: transport the 1.5 year water discharge within the bankfull channel; transport all grain sizes supplied by the watershed without net erosion or deposition; and balance any bank erosion with an equal amount of bank sedimentation; that is to say that it has sufficient *discharge to transport its *load, but not enough energy for erosion; if any of the factors controlling discharge changes, the river will react to re-establish grade. Grade over one section of a river course is a **graded reach**. A **graded slope** has an angle at which output, throughput, and input remain in equilibrium.

gradualism Change over a long period of time. Within a species, gradualism contrasts with *stasis.

granular disintegration (granular disaggregation) A form of weathering where the grains of a rock become loosened and fall out, to leave a pitted, uneven surface.

graph theory The term '**graph**' can refer to the structure of a network. Thus, graph theory is the mathematical analysis of the way a network is organized, considering the way all the *nodes in a graph are connected to each other, in order to understand its functions. Cantwell and Forman (1993) *Landscape Ecol.* 8, 4, 239 take you through the procedure, step by step.

gravity anomaly The difference between the actual gravitational force and the calculated force. For example, when a plumb line is set up near a mountain range it will be pulled from the vertical towards, and by the gravitational pull of, the mountains—but by far less than expected from calculations. This is a **negative gravity anomaly**. See Nakada et al. (2002) *Technophysics* 351, 263 on the Eastern Kyoto gravity anomaly.

gravity model In human geography, gravity models evaluate, or forecast, the spatial interactions of goods, people, and so on, between origins and destinations. 'The gravity model assumes that the probability of patronizing a facility (facility utility) is proportional to the attractiveness of the facility and inversely proportional to a function of the distance' (Drezner and Drezner (2006)

Geog. Analys. 38, 4). These interactions are expressed as:

$$I_{ij} = kM_iM_jDist_{ij}^{-\beta}$$

where I_{ij} denotes the interaction between two locations i and j; M_i and M_j represent the 'masses' measuring the strength of i and j (usually the population numbers of two settlements); $Dist_{ij}$ stands for the distance between i and j; and k and β are constants (G. Robinson 1998). 'While the gravity model is now accepted by economics as a useful tool (it is said to be all over the place at the International Monetary Fund), it is considered only in an incidental manner in economic analysis that employs it' (Isard (2001) *Int. Reg. Sci. Rev.* 24, 3). See Rodríguez-Pose and Zademach (2006) *Tijdschrift* 97, 3 on the gravity model, and mergers and acquisitions.

great circle Any circle on a sphere formed by a plane that passes through the centre of the sphere. The equator and the meridians of longitude are all great circles on the Earth's surface. Any great circle route between two points will represent the shortest line between the two; great circles are therefore relevant to studies of migration and other interactions. Try Zar (1989) *J. Field Ornithol.* 60 for the calculations.

great soil groups The primary classification of global soils into groups; a classification similar to the formations of vegetation. *See* SOIL CLASSIFICATION.

green belt An area of undeveloped land encircling a town, restricting development in the semi-rural areas beyond the built-up zone. The amount of new building is restricted although by no means completely banned. An early green belt was set up in the 1950s around Greater London, in order to limit the spread of suburbs. Other cities have followed this example: in Beijing, all the land in the 'green belt' was forced to keep its status in 2004 without any changes until 2020. 'Under this policy, the trend of central city to form a big pancake would be stopped, while the sub-cities in the region would develop very fast with many small and medium "pancakes" forming outside the 6th Ring Road' (He et al. (2006) *Appl. Geog.* 26, 3–4).

greenfield site Areas beyond the city where development can take place unfettered by earlier building and where low-density, high-amenity buildings can be constructed.

greenhouse effect The warming of the atmosphere as some of its gases absorb the heat given out by the Earth. Short-wave radiation from the sun warms the Earth during daylight hours, but this heat is balanced by outgoing long-wave radiation over the entire 24-hour period. Much of this radiation is absorbed by atmospheric gases, most notably water vapour, carbon dioxide, and *ozone, but also by methane and chloro-fluorocarbons. All of these may be called **greenhouse gases**. Without this absorption, which is also known as *counter-radiation, the temperature of the atmosphere would fall by 30–40 °C.

This is the **natural greenhouse effect**. Through human agency, such as the clearance of rain forest, or the increased rearing of livestock, the concentration of greenhouse gases in the atmosphere is increasing. This is the **enhanced greenhouse effect**. Measurements taken at Mauna Loa, Hawaii, show that the concentration of atmospheric CO_2, for example, increased by 19.4% between 1959 and 2004. It would follow, therefore, that increased concentrations of such greenhouse gases would lead to a rise in global temperatures; *see* GLOBAL WARMING.

green manure A leguminous crop ploughed into the fields after it matures, since it fixes nitrogen from the air and thus improves soil fertility.

green politics (greening) A political creed that aspires to the creation of a sustainable society founded on *environmentalism, grassroots democracy, and *social justice. See Hermelin and Andersson (2018) *Scottish Geograph. J.* 134, 3–4 on green growth in Sweden.

green revolution The development of high-yielding crop varieties (HYVs) for developing countries began in a concerted fashion in the late 1950s. In the mid-1960s, scientists developed HYVs of rice and wheat that were subsequently released to farmers in Latin America and Asia. The success of these HYVs was characterized as a 'Green Revolution'.

Early rice and wheat HYVs were rapidly adopted in tropical and subtropical regions with good irrigation systems or reliable rainfall. Over the following years, the Green Revolution achieved broader and deeper impacts, extending far beyond the original successes of rice and wheat in Latin America and Asia. Increased food production has contributed to lower food prices globally. Average caloric intake has risen as a result of lower food prices—with corresponding gains in health and life. See Howlett (2008) London School of Economics, for an assessment of the Indian Green Revolution.

grèzes litées Deposits down a hillslope of *imbricated rock fragments bedded parallel to the slope. It may be that the debris is shattered by freeze–thaw, with larger fragments rolling downwards; wind might be the depositional process.

gross domestic product *See* GDP.

gross national income (GNI) A broad measure of national prosperity constructed by adding *GDP (Gross Domestic Product) to income accruing from investments outside the country. It is usually expressed in $US.

gross national product *See* GNP.

gross reproduction rate The number of female babies born per thousand women of reproductive age. The net reproduction rate also takes into account the number of women who cannot or do not wish to have children; in the UK, the cohort net reproduction rate in 2020 was 0.85.

gross value added (GVA) The additional value at a particular stage of production, or through image and marketing. It shows the contribution of the factors of production, in increasing the value of a product. GVA measures the contribution to the economy of each individual producer, industry, or sector in a nation.

grounded theory This involves the collection, coding, and categorization of qualitative data from interviews, focus groups, and researchers' notes, photographs, and other images. The purpose of grounded theory is to build theories from data about the social world grounded in people's everyday experiences and actions. Knigge and Cope (2005) *Env. & Plan. A*, 38 (11), 2021 take you through the process step by step. Hypotheses, *models, or theories might then emerge gradually, with careful thought.

ground failure *See* AVALANCHE; DEBRIS; LANDSLIDE; ROCK FALL.

ground frost *See* FROST.

ground moraine *See* MORAINE.

ground truth In *remote sensing, ground truth refers to the comparison of remotely sensed data with real features and materials on the ground. Search for Art Kalinski, *Geospatial Solutions*, for a fascinating and wholly relevant account of the use of ground truth websites.

groundwater All water found under the surface of the ground which is not chemically combined with any minerals present, but not including underground streams. The GroundWaterSoftware.com newsletter is really useful.

growth pole A point of economic growth. Poles are usually urban locations, benefiting from *agglomeration economies, and should spread prosperity from the core to the periphery. See Ionescu-Heroiu et al. (2013) *World Bank Report* on the development of growth poles in Romania. Other growth pole developments have been less successful; see G.-J. Hospers (2004) on the Mezzogiorno.

groyne A breakwater running seawards from the land, constructed to stop the flow of beach material moved by *longshore drift.

guanxi In Chinese society, the clan ties and relationship obligations formed on the basis of kinship, and which are an autonomous unit within the whole of society. See Gold, Guthrie, and Wank (2009).

Guinea current A warm *ocean current off the coast of West Africa, which transports low-saline, warm waters from the Gulf of Biafra south-west along the north Gabon coast.

gully A 'catch all' definition would be 'a water-made cutting, usually steep-sided with a flattened floor', but Bergonese and Reis (2011) *ESPL* 36, 11, 1554 provide six definitions currently in use. An **ephemeral gully** is a small channel eroded by concentrated overland flow that can easily be filled by normal tillage, only to reform in the same location with further runoff. Gullying usually occurs in unconsolidated rock and rarely cuts through bedrock. It can be stopped by restoring a vegetation cover, by contour ploughing, and by making terraces and small dams.

A **badass gully** is defined as one which does not conform to any existing gully model, and generates 'disproportionate results' (Marden et al. (2018) *Geom.* 307).

gust A temporary increase in wind speed, lasting for a few seconds. A typical ratio of gust speed to wind speed in rural areas is 1.6:1, increasing to 2:1 in urban areas due to the effects of high buildings and narrow streets.

guyot A truncated sea-floor volcano occurring as a flat-topped mountain which does not reach the sea surface. Guyots are thought to be associated with *hot spots.

habitat The area in which an organism can live and which affords it relatively favourable conditions for existence. **Habitat generality** means being able to survive in a multitude of environments. See Brown (2007) *J. Biogeog.* 34, 12.

habitat rehabilitation The management of a degraded habitat with the aim of restoring it to an ecologically superior state. An understanding of the recovery and reassembly of plant communities is imperative for habitat rehabilitation. Rehabilitation projects should provide the appropriate elements for a healthy system; one would then let these elements interact to produce some target state. The problem lies in determining what that target state should be, since habitat changes that are beneficial to one species may be detrimental to another; thus, in some ways, habitat rehabilitation is based on explicit or implicit species values. See Clough (2003) *Rept. Fish Habitat & Species Recovery Wkshp.*

habitus Bodily movements, tastes, and judgements characteristic of a social class. Our relationship with the world is shown through our embodied expressions, understandings, and actions. So habitus helps in understanding and explaining the way people act in specific situations while not making fully conscious decisions. Habitus might be described as a 'frame of mind'. It's a useful tool for thinking about how social relations are internalized and experienced as 'natural'.

hactivism Any operation using computer hacking against an internet site, aiming to disrupt normal operations, but not to cause major damage. It may be argued that hacktivism is motivated by a libertarian ethos that promotes free flows of information, securely accessible to all. Curran and Gibson (2013) *Antipode* 45, 2, 294 are really good on wikileaks, anarchism, and technologies of dissent. See also Denning (2001) *Special Reports, December 31* on activism, hacktivism, cyberterrorism, and the internet as a tool for influencing foreign policy.

Hadley cell A simple, vertical, thermally direct, tropical, *atmospheric cell. Air rises at the *inter-tropical convergence zone (ITCZ) and drifts polewards, cooling a little through a net loss of radiation, and thereby transferring heat. As the circumference of the Earth decreases with distance from the equator, the moving air has to converge. This combination of cooling and convergence, together with deflection by the *Coriolis force, causes the air to sink around latitude 32°. The air then returns to the equator as surface winds. The basic components of this cell are: weak upper-air easterlies above the ITCZ, subtropical anticyclones at the descending limb, and easterly *trade winds associated with the return of air to the equator, together with *synoptic scale disturbances, such as easterly waves or *tropical cyclones. The westerly subtropical *jet is located at the poleward limit of the Hadley cell.

In the tropics, the rotationally confined, thermally direct Hadley circulations are important for maintaining the observed small meridional temperature gradients in low latitudes. The intensification and poleward expansion of the cross-equatorial Hadley cell can lead to westerly acceleration in the winter subtropics, and enhanced vertical shear of the zonal wind in the subtropics and mid-latitudes. *See* GENERAL CIRCULATION OF THE ATMOSPHERE.

hail A form of snow made of roughly spherical lumps of ice, 5 mm or more in diameter, formed when a frozen raindrop is caught in the violent updraughts found in warm, wet *cumulo-nimbus clouds. As the drops rise, they attract ice, and as they fall, the outer layer melts, but refreezes when the droplet is again lifted by updraughts, often causing the hailstones to show an onion-like pattern of alternating clear ice (glaze) and opaque ice (rime). A hailstone will descend when its fall speed is enough to overcome the updraughts in the cloud. **Soft hail** is white and low-density as it contains air. See Billett et al. (1997) *Weather & Fcst* 12, 1 on the prediction of hail size and Gavrilov et al. (2010) *Phys. Geog.* 31, 5 on hail suppression.

halo A ring of light around the sun or, more rarely, the moon, caused by the refraction of light by ice crystals. A **cloud halo** is a zone of increased humidity around cumulus clouds; see Lu et al. (2002) *J. Appl. Met.* 41, 8.

halocline Within a sea or ocean, the layer where salinity concentration changes rapidly with depth. See

Bourgain and Gascard (2011) *Deep Sea Research* 1, 58.

halophyte A plant which can grow in saline conditions: salt marshes, estuarine environments, and the lower parts of sea cliffs. Controversy exists over whether halophytes require saline conditions for their existence—or merely tolerate them. See Glenn et al. (1999) *Crit. Rev. Plant Sci.* 18.

hamada A desert surface of pebbles/gravel/boulders, formed by the washing or blowing away of the finer material. See Dłużewsk and Krzemień (2008) *Prace Geograficzne*, zeszyt. 118 (it is in English).

hamlet A small settlement without services or shops, and usually without a place of worship.

hanging valley A high-level tributary valley from which the ground falls sharply to the level of the lower, main valley. The depth of the lower valley may be attributed to more severe glaciation. The level of a hanging valley may be used as an indicator of glacier ice thickness.

haptic Relating to the sense of touch, especially when relating to the perception and manipulation of objects. Paterson (2009) *PHG* 3, 6, 766 provides a commendably accessible overview of haptic knowledges in geography that respond to bodily sensations and the reactions of the embodied researcher. See also Bonner-Thompson (2021) *TIBG* 46, 2.

hardpan A cemented layer in the B *horizon of a soil, formed by the *illuviation and precipitation of material from the A horizon, such as clay (forming a **clay pan**), humus (a **moor pan**), or iron (an **iron pan**). See Lee and Gilkes (2005) *Geoderma* 126, 1–2.

Harmattan *See* LOCAL WINDS.

Hawaiian eruption A volcanic *fissure eruption where large quantities of basic lava spill out with very little explosive activity. Hawaiian eruptions are typically gentle because their lava is highly fluid and thus tends to flow freely both beneath the surface and upon eruption. Search the USGS site.

hazard An unexpected threat to humans and/or their property; a hazard includes both the event and its consequences. By this definition, the Indian monsoon is not a hazard, but its failure is. A natural event becomes a hazard through: the social processes which cause some people to be much more at risk from the effects of a hazard; a lack of options to minimize, or escape from, the effects of the hazard; the location of human settlement on potentially dangerous physical sites; the location of human settlement near potentially dangerous economic activity; and activities that aggravate events with natural causes.

Many highly vulnerable communities may deliberately choose to inhabit a hazard-prone environment if this reduces other risks, such as poverty.

Hazard perception is the view which an individual has of a hazard. Robertson (1999) *Geog. Rev.* 89, 4 compares the reality of the tornado hazard with the portrayal in movies, finding the latter seriously misleading. *See* COGNITIVE DISSONANCE.

haze A suspension of particles in the air, slightly obscuring visibility. These particles may be naturally occurring— sea salt or desert dust—or man-made, like smoke.

headcut Very rapid change in the height of river water, most often a waterfall cascade eroding upstream; a *knick point.

headward erosion The lengthening of a river's course, by erosion back towards its source. A *knick point on a stream profile suggests that the fluvial system has reacted to a lower base level by headward erosion. See Hill et al. (2008) *Geomorph.* 95, 3–4 on the Grand Canyon, USA.

headwater stream A somewhat subjective term for a small upland tributary; typically a first- or second-order stream.

health, geographies of The studies of health from geographical perspectives, including the mapping and interpretation of the geography of health; inequalities in health outcomes and provision/utilization of health services; the impact of migration on health; the impact of health status on migration; health and human modification of the environment; the impact of air/water quality, air/water contamination or pollution, global climate change; and other global environmental changes on health and disease, such as infectious diseases, toxicity, neoplasms, and heat stress. Pearce and Barnett (2012) *PHG* 36, 1, 3 have a very interesting, clear, and nuanced study of the impact of the local economic and social environment upon smoking. Try also Andrews and Evans (2008) *PHG* 32, 6, 759. Issue 5 of *AAAG* (2012) is entirely devoted to geographies of health.

heat island In a city, air temperatures are often as much as 3–4 °C higher than over open country. Causes include: the presence of large vertical faces; a reduced sky-view factor; surface materials of high heat capacities and conductivities; large areas of impermeable surfaces; anthropogenic heat fluxes; and atmospheric pollution.

For more detail, see Yow (2007) *Geog. Compass* 1, 6.

hedonic pricing A hedonic model of prices is one that decomposes the price of an item into those separate components that determine the price. For example, the wages paid to baseball players might depend on the basis of hitting and pitching performance, the player's experience, and so on. A hedonic model would not necessarily separate all the factors that could be separated, only those that affect the usefulness to a buyer of what is being sold. This is entirely separate from the use of the word 'hedonic' as 'happy'.

hegemony In terms of power and politics, leadership (especially by one state of a federation); more widely, the way in which one social group will represent its interests as being everyone's interests. This ruling group may keep its grip on society either by **social hegemony**—the use of force to maintain order—or, more commonly, by **cultural hegemony**—producing ways of thinking, especially by subtly eliminating alternative views to reinforce the status quo (Laurie and Bennett (2002) *Antipode* 34, 1; Hoelscher (2003) *AAAG* 93, 3).

New approaches to hegemony examine how social identities (of race, gender, and so on) become dominant and how their particular interests are defined and accepted as universal interests; see Gallaher in C. Gallaher et al. (2009).

helicoidal flow A continuous corkscrew motion of water along a river channel.

henge In the UK, a Neolithic circular earthwork with an encircling bank and inside ditch.

heritage Inherited circumstances or benefits. Heritage can formulate certain national identities with which individuals may come to identify themselves in their everyday lives. The heritage industry is any industry concerned with the management of historic buildings, museums, and sites, as tourist attractions. See While and Short (2011) *Area*, 43, 1, 4 on the production of local heritage and design discourses, and the regulation of conservation and change in the built environment in Manchester, England. See J. Scarpaci (2005) on heritage tourism in Latin America.

hermeneutics For geographers, the interpretation of the relations between humans and place. There has been more than one description of the nature of hermeneutic geography; see Lisa Renning (no date) on hermeneutics in qualitative research.

heterogeneous nucleation The freezing of water droplets around a *freezing nucleus. See Khvorostyanov and Curry (2004) *J. Atmos. Scis* 61, 22.

heteronormativity The societal privileging of heterosexuality; that is, the assumption that heterosexuality is the 'better' option. Hubbard (2008) *Geog. Compass* 2, 3, 640 explores the theorization of heteronormativity. See the special issue of *Antipode* 34, 5 (2002).

heterotopia Created by Michel Foucault—a concept of other spaces; the coexistence of a large number of fragmentary possible worlds. Cyberspace is a good example. See Johnson (2013) *Geog. Compass* 7, 11, 790.

heterotroph An organism which has to acquire its energy by digesting food that has been manufactured by other organisms. Thus all organisms are heterotrophs except the primary *producers which can manufacture their own food, usually by photosynthesis.

heuristic Describing a technique that promotes and uses discovery. A **heuristic model** does not directly answer a question , but rather helps to create a *way* to answer the question. So, for example, someone with a housing problem might search for a new house. The searcher can choose a *rational* approach to the problem, either by improving the old home, or by finding a new, less stress-inducing home. Or using a *heuristic* model, the searcher uses the information built up from each house visited to update his/her perceptions of the types of available houses within a price range. See N. Xiao in B. Warf (2010) on heuristic methods in spatial analysis.

hierarchical diffusion *See* DIFFUSION.

hierarchy Ordering of phenomena with grades or classes ranked in sequence; *central place theory, for example, posits a hierarchy of settlements from regional capitals to hamlets. In a **nested hierarchy** each level contains/is composed of the level below, with the individuals in that level being smaller-scale sub-systems of the level above. The metasystem (highest level) does not respond to the lowest sub-system directly, as the dynamics of the sub-system are on too small a scale to be of any consequence; in other words, information across any three hierarchical levels is *buffered*. This suggests that the sensitivity of a system to change takes a different form, depending on where in the hierarchy the drivers of change originate. Whatever your own focus in geography, you will never find hierarchy expressed more clearly than in Couper (2007) *TIBG* 32, 3.

high A region of high atmospheric pressure—in Britain, generally over 1000 mb.

high-order goods and services Goods and services with a high *threshold population and a large *range. Examples include furniture, electrical goods, and financial expertise. These goods are usually *shopping goods.

high-technology (high-tech) Any industry concerned with advanced technology, such as biotechnology, computers and microprocessors, fibre optics, aerospace, electronics-telecommunication, and pharmaceuticals. In theory such industries are *footloose, since the products have a high value to weight ratio. High-tech hubs in India are Bangalore, Chennai, Delhi, Hyderabad, Mumbai, and Pune. Florida (2002) *J. Econ. Geog.* 2 finds a significant, positive correlation between his 'bohemian index' and concentrations of high-tech industry.

hill farming The *extensive animal farming of an upland area. European Parliament's resolution, A6-0327/2008/ P6_TA-PROV(2008)0438, highlights the importance of hill and mountain farming in the European Union, particularly stressing the importance of environmental protection, the need to produce quality products, and the need to maintain the cultural landscape as part of the European heritage.

hill fort A fortified site on a hilltop, usually with a ditch and ramparts. The earliest date from the Iron Age but some British examples were created as late as the Dark Ages. Favoured sites for such structures in southern England seem to be chalk uplands.

hillslope Any geometric element of the Earth's solid surface at an angle from the horizontal. Slope profiles are generally surveyed along a straight line (on a map) from the divide (watershed), following a line of steepest descent, to the nearest point at their base. The four possible elements of a hillslope profile are: an upper convexity, a free face, a straight slope of almost constant gradient, and a basal concavity. Not all of these features are found in every slope profile, and some may be repeated more than once within a profile.

The amount of the hillside that is convex in profile is usually expressed as a proportion of the total slope length, and most slopes are mainly convex. This proportion of convexity results from the processes acting on the slope, and the effects of the river at the base of the slope; rapid undercutting results in steeper and more convex slopes. With *aggrading streams, more of the slope profile will be concave.

Hillslopes evolve over time in response to the redistribution of their surface sediments, usually with some net removal of material to rivers or the ocean. Where there is a plentiful supply of material, and the removal processes can only move a limited amount for a short distance, the rate of transport is limited by the transporting capacity of the process, which is defined as the maximum amount of material that the process can carry. Hence **transport-limited hillslopes**, where the rate of erosion/deposition is related to the divergence of shear stress, power, or velocity; see Pelletier (2012) *ESPL* 37, 1, 37.

hill wave *See* LEE WAVE.

hinge line A line either side of which *isostatic readjustment is uneven; the line separating uplift from subsidence.

In glaciology, it is the point where a glacier, reaching the coast, starts to float.

hinterland The area serving, and being served by, a settlement. Kaushik (2013) *Int. J. Sci. Engineering & Res.* (*IJSER*) 1, 3 cites the example of Agra (India) and its fertile hinterland. An **information hinterland** is the region which provides the best access to information flows for a city. See Wang et al. (2007) *Tijdschrift* 98, 1.

historical geographical materialism Essentially *Marxist geography. See Cox (2013) *Dialogues Hum. Geog.* 3, 1.

historical geography The study of past human geographies; of past landscapes; the study of the organization of space over time geographic perspectives on the past (*Historical Geography* website). 'Historical geography may be viewed as being concerned with the historical dimension in geography, and geographical history with the geographical dimension in history. Prominent themes are: the historical geographies of geopolitics; the state and territory; landscape and environmental change; and matters of memory and commemoration. McGeachan (2018) *PHG* 42 highlights the range of different ways in which historical geographers have explored lives, deaths, and their fleeting traces through varied biographies. See Offen (2012) *PHG* 36, 4, 527 for a useful summary of the concerns of historical geography.

historical materialism *Materialism* asserts that matter is the primary substance of all living and non-living things. *Historical* materialism differs from broader materialism by stressing that humans create their own realities and worlds; see Mitchell (2004) in J. S. Duncan, N. C. Johnson, and

R. H. Schein. It is an analysis of history, most closely associated with Marx, which points out that economic systems underlie the development of history and ideas. See D. Harvey (1996).

historicism A concept that holds that the aim of social and historical inquiry is to formulate and verify general laws of historical development. J. M. Blaut (1993), for example, argues that nations follow similar paths of economic development, so that different parts of the world are at different stages along a general path of economic development. This view has been vigorously countered; see Lowe and Barnett (2000) *Env. & Plan. D* 18, 1. Rowe (2010) *Western Political Science Association* has a different view of historicism, seeing it as a form of nostalgia, prominent in postmodernist urbanism.

historiography The study and writing of history; see Trimble (2008) *PPG* 32, 1.

histosol *See* US SOIL CLASSIFICATION.

Hjulström diagram A diagram showing the relationship in a channel between particle size and the mean fluid velocity required for *entrainment. It shows that an entrained particle can be transported in suspension at a lower velocity than that required to lift the particle initially. When the stream velocity slows to a critical speed, the particle is deposited. Note that higher velocities are needed for the entrainment of clay-sized particles because of the electrostatic forces which bind them together.

holism The view that everything is linked to everything else, and should be acted upon as such, and that the whole is more than its parts. In earlier geographies, for example, the *region has been seen as having a distinct identity which does not come solely from its separate parts. See Antrop (1998) *Landsc. & Urb. Plan.* 41, 3.

hollowing out The reduction of political power; for example, of a member state on joining a supranational body, such as the *European Union; of a local authority by national government; or of a state by global organizations such as the *World Bank or the *International Monetary Fund. The term may also

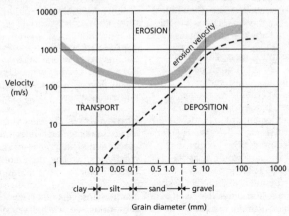

Hjulström diagram

apply to the loss of local government power when national government takes more control; see Skelcher (2000) *Pub. Policy & Admin.* 15, 3.

Holocene The most recent geological *epoch, stretching from 11 000 years ago to the present day. The Holocene may be divided into:

Early Holocene	Preboreal	$10,300 \pm 200$ years BP
	Boreal	9,000—7,000 years BP
	Atlantic	7,000–5,000 years BP
Mid-Holocene	Sub-Boreal	5,000–2,500 ka) years BP
Late Holocene	Sub-Atlantic	2,500 years BP–present

See K. Alverson (2004) on global change in the Holocene.

During the early Holocene, atmospheric methane (CH_4) concentrations gradually decreased and then reversed direction around 5 000 years ago. This latter is the late **Holocene methane anomaly**; see Singarayer et al. (2011) *Nature*, 470, 82 for the orbital explanation, and Zhou (2012) *Holocene* 22, 731, who ascribes it to the expansion of early rice paddy fields.

homelessness, geography of
Homelessness is commonly perceived as an urban problem. Lee and Price-Spratlen (2004) *City & Commy* 3, 1 observe that, in the USA, the homeless, and services for the urban homeless, are mostly located relatively close together, usually in more deprived central city areas, and almost forming a service ghetto. Fagan (2003, *San Francisco Chronicle*) identifies Main Street as the centre of welfare and medical services in San Francisco. For the homeless, such a concentration renders services more

accessible, but has a number of disadvantages: For many, simply getting to the local night shelter is frightening, and this fear is gendered. Informal areas develop in streets, parks, and public car parks, where people find places to set up 'home', sleep, eat, make friends, and look for security, money, and entertainment. These niches are colonized by the homeless at particular times: 'at dusk, the street families lay out boxes, build plastic houses, warm themselves, and cook on small fires. The children rarely have anything. Public space is *their* space only at night' (Morelle and Fournet-Guérin (2006) *Cybergeo, politique, culture, représentations* 342). See also P. Cloke, J. May, and S. Johnsen (2010), which is both informative and well-reviewed.

Differentiation is not merely temporal; it is also social. Homeless Research record a division between 'straightheads', 'smackheads', and 'pissheads': what for one homeless person might be experienced as a space of care, might for another become a space of fear because of the dominant presence of a particular homeless sub-group responding to a specific spatial logic. See Cloke, Milbourne, and Widdowfield (2003) *Geoforum* 34, 1 on rural homelessness.

homeorhesis While *homeostasis is the maintenance of a steady state, homeorhesis is the preservation of a steady flow. This flow is maintained by positive feedback.

homeostasis The maintenance of a steady state regulated by self-adjusting mechanisms, under ever-changing conditions. See Lau and Lane (2001) *PPG* 25, 2.

homocline One of a regular series of hills made from a large area of rock strata of uniform thickness and dip.

homogeneous nucleation The spontaneous freezing of water droplets at around −40 °C, as clusters of water molecules within a droplet settle, by chance, into lattice formation of ice, making the entire droplet freeze. See Easter and Peters (1994) *J. Appl. Met.* 33, 7.

homonormativity
*Heteronormativity 'privileges' the straight world. Homonormativity does not contest dominant *heteronormative assumptions and institutions, but *promises* the possibility of a demobilized gay constituency and a privatized, depoliticized gay culture anchored in domesticity and consumption. 'In other words, many gays and lesbians want nothing more than to be considered normal so that they might go on with their day-to-day lives as part of the status quo' (Oswin (2008) *PHG* 32, 1). See Brown, Browne, and Lim in V. J. Del Casino Jr et al. (2011), p. 299.

honeycomb weathering The removal of material from sandstone while the cement is preserved, leaving a lacy network; see Bruthans et al. (2018) *Geom.* 303.

honey-pot A location, such as Shakespeare's birthplace, which is particularly appealing to tourists. Planners in National Parks often develop honey-pot sites with car parks, shops, cafés, picnic sites, and toilets, so that other parts of the Park will remain unspoilt; see Pickering and Buckley (2003) *Mt. Res. & Dev.* 23, 3 on Mt Kosciuszko, Australia.

horizon (soil horizon) A distinctive layer within a soil which differs chemically or physically from the layers below or above. The **A horizon** or topsoil contains *humus. Often soil minerals are washed downwards from this layer. This material then tends to

accumulate in the **B horizon** or subsoil. The **C horizon** is the unconsolidated rock below the soil. These three basic horizons may be further subdivided. Thus, **Ah horizons** are found under uncultivated land, **Ahp horizons** are under cultivated land, and **Apg horizons** are on *gleyed land. The B horizons are also subdivided by means of suffixes: **Bf horizons** have a thin iron pan, **Bg horizons** are gleyed, **Bh horizons** have humic accumulations, **Box horizons** have a residual accumulation of sesquioxides, and **Bs horizons** are areas of sesquioxide accumulation. **Bt horizons** contain clay minerals and **Bw horizons** do not qualify as any of the above. **Bx horizons** or fragipans contain a dense but brittle layer caused by compaction. **C horizons** are also subdivided: **Cu horizons** show little evidence of gleying, salt accumulation, or fragipan; **Cr horizons** are too dense for root penetration; and **Cg horizons** are gleyed. Additional suffixes may be used. Some soil scientists use the term **D horizon** for the consolidated parent rock.

In addition to these soil horizons, other layers are distinguished. Thus, the layer of plant material on the soil surface is classified as: the **L horizon** (fresh litter); the **F horizon** (decomposing litter); the **H horizon** (well decomposed litter); and the **O horizon** (peaty). A *leached A horizon is termed an **E horizon** or *eluviated horizon.

horst A block of high ground which stands out because it is flanked by normal faults on each side. It may be that the block has been elevated or that the land on either side of the horst has sunk.

Hortonian overland flow An overland flow of water occurring more or less simultaneously over a *drainage basin when rainfall intensity exceeds the rate of water infiltration in the soil (Horton (1933) *Trans. Am. Geophys. Union* 14; Kirkby (1988) *J. Hydrol.* 100).

hot desert Located on the west coasts of tropical and subtropical climes, these have average temperatures over 20 °C and annual rainfall under 250 mm. Deserts are too dry for most plant species except for *xerophytes. Desert insects, reptiles, mammals, and birds are all adapted to drought. See M. Evenari et al., eds (1985). See also Nash (2000) *PPG* 24, 3 on geomorphological processes in arid areas.

Hotelling model A model of the effect of competition on locational decisions, showing that locational decisions are not made independently but are influenced by the actions of others (Hotelling (1929) *Econ. J.* 39, 1). Kim (2007) *MPRA Paper* 2742, U. Munich, develops the Hotelling model.

hot spot An area of localized swelling and cracking of the Earth's crust caused by upwelling *magma; deep-seated thermal anomalies fixed with respect to the surrounding mantle. Plumes from the lower mantle rise to the surface, forming volcanoes. As plates move over the Earth's surface, the volcanoes delineate and record the history of plate motion, and the radiometric age of seamounts should show a simple progression away from the present-day hot spot locations. See Wessel and Kroenke (1998) *Earth & Planet Sci. Letts* 158, 1–2 on the relationship between a hot spot and its seamounts.

hot towers Immensely tall, tropical *cumulo-nimbus, at the rising branch of a *Hadley cell, that siphons sensible and latent heat upwards, to the equatorial *tropopause. See Montgomery et al. (2006) *J. Atmos. Scis* 63, 1 on vertical hot towers and cyclogenesis.

hub-and-spoke network Hubs are special *nodes that are part of a network linked by spokes, and located in such a way as to facilitate connectivity between interacting places. Song and Ma (2006) *Chinese Geogr. Sci.* 16, 3, 211 show how commercial and operational freedoms have led most of the larger airlines to develop hub-and-spoke networks.

hukou A system established in China in 1951 to regulate the movement of people within China, enabling the state to decide where people should live and work, and thus acting as a type of domestic passport system to living and working in a named location. See Ho (2011) *Citizenship Studs* 15, 6–7, 643, and Bosker et al. (2012) *J. Urb. Econs* 72, 2–3, 252, on relaxing hukou.

human capital Talent. The economic geography of talent is strongly associated with high-technology industries. 'The presence and concentration of *bohemians (creatives) in an area creates an environment or milieu that attracts other types of talented or high human capital individuals' (Florida (2002) *J. Econ. Geog.* 2, 1). **Human capital externalities** are workers whose skills have been provided by bodies outside the firm. Education is a major human capital externality. **Human capital spillovers** occur when an individual's productivity is increased by interacting with and learning from high-skilled workers (Winters (2013) *J. Econ. Geog.* 13, 5, 799).

Human Development Index (HDI) A compound indicator of *economic development, established by the United Nations, which attempts to get away from purely monetary measurements by combining per capita GNP with *life expectancy and literacy in a weighted average. 'Human Development is a development paradigm that is about

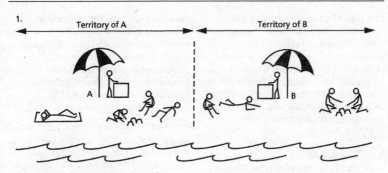

1. Territory of A — Territory of B

First pattern of market share

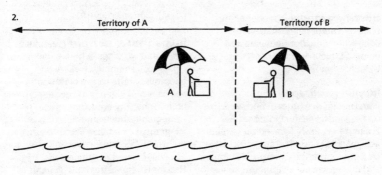

2. Territory of A — Territory of B

Second pattern of market share

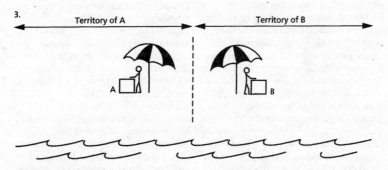

3. Territory of A — Territory of B

Final pattern of market share

Hotelling model. Two vendors, A and B, are positioned in the centre of their own half of the beach. If A moves closer to the centre, he increases his market share, which leads B to also move to the centre, ensuring that market share is equal again.

much more than the rise or fall of national incomes. It is about creating an environment in which people can develop their full potential and lead productive, creative lives in accord with their needs and interests. People are the real wealth of nations. Development is thus about expanding the choices people have to lead lives that they value' (UNDP).

SEE WEB LINKS
• The Human Development Index database.

human ecology An approach whereby the human social system and the ecosystem of the planet are seen as a mesh of reinforcing connections, each influencing and being influenced by the others. See G. G. Marten (2001).

human geography The study of the patterns and dynamics of human activity on the landscape, including studies that are more narrowly focused on human activity, and those that are particularly concerned with the dynamics of humans and environment, or of nature and society; part of the discipline of geography concerned with the spatial differentiation and organization of human activity and its interrelationships with the physical environment. 'More than just analysing and taking note of where people and their possessions are, however, geography studies the interactions and social habits of humans *within* and *across* all spaces. Basically, geographers critically examine how humans organize and identify themselves in space—how we create "places". Thus, the science of human geography does not look at location per se, as much as it does at the mobility of, access to, and barriers against human processes' (Oas (no date) *Implications* 1, 7). For the student, it may be helpful to know that the different human geographies

arise from the way any geographer believes that the world works, be it structuralist, performative, materialist, queer, humanist, transactional, and so on. Academic papers on human geography are notoriously difficult to understand; ask for advice.

human influence index An index of human influence on natural environments: based on population density, land transformation (leading to loss and fragmentation of habitats), human access, and power infrastructure. See Sanderson et al. (2002) *Bioscience* 52, 10.

humanism A belief in the unmatched worth and workings of human beings. As such it places human welfare and happiness at the centre of ethical decision-making; it is a 'people-centred philosophy concerned with meanings, values, and understanding our taken-for-granted experiences and "being in the world"' (S. Aitken and G. Valentine, eds 2009). *See* POST-HUMANISM.

Humanistic geography The focus of humanistic geography is on people and their condition. This is a geography centred on human perception, capability, creativity, experience, and values. The Nepalese geographer Kanhaiya Saptoka (2017) explains: 'for a long time Nepalese geographers followed the Western Eurocentric view and appear to be content in following western notions and ignored understanding our own social and cultural aspects/landscapes that enrich our knowledge of geography.' Humanistic geographers also maintain that any investigation will be subjective, reflecting the attitudes and perceptions of the researcher, who may also be an influence on the very field of his study; see the originator of the concept, Y.-F. Tuan (1974).

Human Suffering Index A compound indicator of distress, compiled by 'adding together ten measures of human welfare related to economics, demography, health, and governance: income, inflation, demand for new jobs, urban population pressures, infant mortality, nutrition, access to clean water, energy use, adult literacy, and personal freedom' (Population Crisis Committee, Washington, DC).

humanitarianism The concern for the welfare and happiness of human beings. **Geographies of humanitarianism** consider the webs of connection that can be mobilized to nurture caring relationships. See Lynn-Ee Ho (2017) *Trans. Inst. Brit. Geog.* 42.

humic acid A complex acid formed when water passes slowly through humus. Humic acid is an example of an organic acid in that it is formed from carbon-based compounds. It is significant in chemical weathering and in the formation of soil. See Yadav and Malenson (2007) *PPG* 31, 2. *See also* MODER; MOR; MULL.

humidity The amount of water vapour in the atmosphere. It is more exactly defined as the mass of water vapour per unit volume of air, usually expressed in $kg\ m^{-3}$. This is **absolute humidity**. **Relative humidity** is the moisture content of air expressed as the percentage of the maximum possible moisture content of that air at the same temperature and pressure. The **humidity mixing ratio** is the ratio of the mass of water vapour in a sample of air to the mass of dry air associated with that water vapour.

humilis *See* CLOUD CLASSIFICATION.

humus A naturally occurring complex combination of substances resultant of the extensive biochemical breakdown of plant and animal residues. This breakdown is **humification**—the

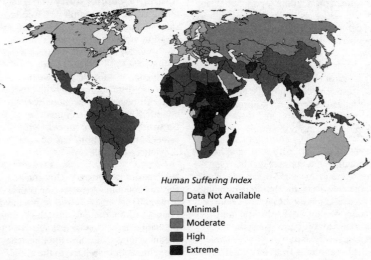

Human Suffering Index

Data Not Available
Minimal
Moderate
High
Extreme

Human Suffering Index

process whereby the simple mineral compounds released by weathering combine with the organic residues to form large, stable organic molecules which bond the soil. Humus can form relatively stable complexes and chelates with polyvalent cations, thus increasing assimilation by plants. See D. Alexander and R. W. Fairbridge (1999).

hunger Hunger is not simply a function of hardship; *see* *ACCESS.

hunting and gathering A form of society with no settled agriculture, or domestication of animals, and which has little impact on the environment. The hunting of animals and the collection of edible plants depend on adapting to the environment, rather than changing it. Try R. K. Guatam (2011) on the Baigas hunter-gatherers of Central India.

hurricane *See* TROPICAL CYCLONE.

hybrid capitalism A form of capitalism that combines elements of culturally specific imprints—such as employment relations—and globalizing norms—such as international standards of corporate governance. The concept of hybrid capitalism offers a way of combining the focus of the global convergence literature on systemic 'drivers' of change with the emphasis on domestic institutional logic found in the 'varieties-of-capitalism' approach (Dyson and Padgett (2005) Anglo-German Foundation).

hybrid geography A geography that recognizes the connections between the human, the non-human, and the more-than-human. Hybrid geographies can overcome the boundaries between binary divisions like 'physical' and 'human' and produce something ontologically new.

Have a look at Hull (2017) *LSU Master's Theses, 4534* on the hybrid geographies of Baton Rouge urban gardens. The go-to author is S. Whatmore (2002).

hybridity The mixture of meanings that emerges when two cultures interact; the new forms that are created when cultures merge. Acculturation engenders new forms of cultural negotiation which materialize as hybridity, creoleness, and inbetweenness. Identities are not stable, but ambivalent. H. Bhabha (1994) claims that subjects at the margins of cultures, or in a *third space, are best placed to resist a domineering culture/narrative, and to create new hybrid forms.

hydration The incorporation of water by minerals. Hydration is believed to be a major force in weathering coarse-grained igneous rocks; see Hughes (1998) *Geoarchaeol.* 3, 4.

hydraulic action (hydraulic force) The force of water, including *cavitation and fluvial *plucking.

hydraulic conductivity The ability of soil or rock to transmit water under saturated or nearly saturated conditions. The conductivity of dry soil or rock is low, as water must form a film of water around the soil particles before conduction occurs. Pore geometry and pore-size distribution determine the relations between saturation, hydraulic conductivity, and air conductivity. **Saturated hydraulic conductivity** refers to the maximum rate of water movement in a soil.

hydraulic geometry The study of the interrelationships along the course of a river. *Discharge is linked to the mean channel width, the mean channel depth and slope, the suspended *load, and the mean water velocity, and the wavelength of the meander is related to the radius of curvature.

hydraulic gradient The rate of change in *hydraulic head with distance. Within a drainage network, topography and valley morphology will accentuate or diffuse the local distribution of flow energy through changes in the hydraulic gradient and constrictions or widening of the river channel.

hydraulic head The pressure exerted by the weight of water above a given point.

hydraulic radius Also known as **hydraulic mean**, this is the ratio of the cross-sectional area of a stream to the length of the *wetted perimeter (cross-sectional length of a river bed). The hydraulic radius is a measure of the efficiency of the river in conveying water—if the hydraulic radius is large, there is little friction.

hydraulic resistance A measure of the resistance offered by a material, such as vegetation, to a flow of water. The calculation of flow velocities within stream channels and over hillslopes cannot be made without accounting for the hydraulic resistance experienced by flow over different morphologies. Moore and Burch characterize hydraulic resistance by choosing an appropriate value of the *Manning roughness coefficient, and it can also be characterized using the Darcy–Weisbach friction factor, expressed as a function of the Reynolds number of the flow; see Prosser and Rustomji (2000) *PPG* 24, 2 for the equations.

hydraulics The study of the mechanical properties of liquids. See Boyer et al. (2006) *J. Geophys. Res.* 111, F04007 on flow turbulence and bed morphology, and Wilcox and Wohl (2007) *Geomorph.* 83, 3–4 on three-dimensional hydraulics.

hydraulic sorting The process by which particles of river bed material are sorted into sections of near-uniform particle size due to the change in river *competence throughout its journey from source to mouth; see Attal and Lave (2006) *Spec. Pap. Geol. Soc. Am.* 398, 143. Hydraulic sorting also occurs on beaches; try Hughes et al. (2000) *J. Sed. Res.* 70, 5, 994.

hydric soils Soils that formed under conditions of saturation, flooding, or ponding long enough during the growing season to develop anaerobic conditions in the upper part. Under natural conditions, these soils are either saturated or inundated long enough during the growing season to support the growth and reproduction of hydrophytic vegetation. **Hydric erosion** is the breakdown and transport of soil particles by raindrop erosion and superficial draining. The triggering factor is precipitation—both intensity and duration. However, the relationships between rainfall characteristics, infiltration, drainage, tillage, and soil loss are very complex; see Ballais et al. (2012) *Zeitschr. Geom.*

hydroclimatology The study of the effects of the climate system on the hydrologic cycle over time and space, both globally and locally. Changes in the relationship between the climate system and the hydrologic cycle underlie floods, drought, and possible future influences of global warming on water resources. See M. L. Shelton (2009). Simultaneous variations in weather and climate over widely separated regions are commonly known as **hydroclimatic teleconnections** (Kashida et al. (2010) *J. Hydrol.* 395, 23).

hydro ecology *See* ECOHYDROLOGY.

hydrogeomorphology An investigation in geomorphology focusing on the interaction and linkage of hydraulic processes, whether surface or sub-surface, on earth materials or landforms. See Sidle and Onda (2004) *Hydrol Procs* 18.

hydrograph A graph of *discharge, or of the level of water in a river throughout a period of time. The latter, known as a **stage hydrograph**, can be converted into a discharge hydrograph by the use of a stage-discharge rating curve. Hydrographs can be plotted for hours, days, or even months. A storm hydrograph is plotted after a rainstorm to record the effect on the river of the storm event.

hydrography The measurement, description, and depiction of the nature and form of water's topographic features.

hydrological connectivity The transport via watercourses of material, energy, and/or organisms. Hydrological connectivity refers to the passage of water from one part of the landscape to another, and is expected to generate some response in the catchment runoff. See Gooseff (2011) *Geog. Compass*,

DOI: 10.1111/j.1749-8198.2011.00445.x, for a worked example.

hydrological cycle (water cycle) This is the movement of water and its transformation between the gaseous (vapour), liquid, and solid forms. The major processes are *condensation by which *precipitation is formed, movement and storage of water overland or underground, *evaporation, and the horizontal transport of moisture. The length of time any water stays in the *atmosphere is about eleven days.

hydrological pathways The routes that water takes through the catchment from precipitation inputs to river channel outputs.

hydrology The study of the Earth's water, particularly of water on and under the ground before it reaches the ocean or before it evaporates into the air. This science has many important applications such as flood control, irrigation, domestic and industrial water supply, and the generation of hydroelectric power.

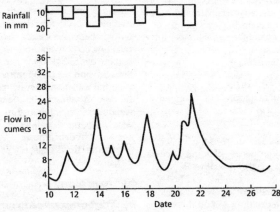

Hydrograph

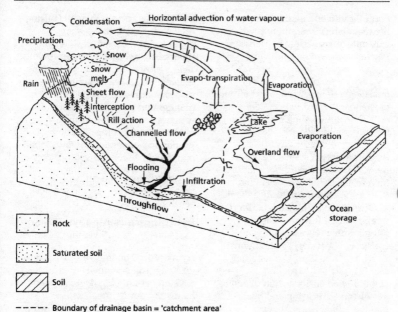

- - - - - Boundary of drainage basin = 'catchment area'
Hydrological cycle

hydrolysis The chemical reaction of a compound with water. Hydrolysis is an important component of soil formation, and of *chemical weathering—for example, as feldspars in granite decompose to make china clay.

hydromorphic Denoting areas with waterlogged soils. *Gley soils form in such conditions.

hydromorphology A subfield of *hydrology that deals with the structure and evolution of Earth's water resources, and the origins and dynamic morphology of those water resources.

hydrosere A successional sequence of plants originating in water.

hydrosphere All the water on, or close to, the surface of the Earth. Some 97% of this water is in the Earth's seas and

oceans; of the rest, about 75% is in *ice-caps and sheets, about 25% in surface *drainage and *groundwater, and about 0.03% in the atmosphere.

hydrothermal vent A fissure on the sea floor, usually found in volcanically active areas of the ocean floor. Water that has been heated up to 400°C by underlying magma flows from these vents. This water usually contains dissolved minerals that precipitate out of the vent on contact with the colder ocean water, creating a chimney-like structure.

hygroscopic nuclei *Condensation nuclei which are hygroscopic, i.e. which tend to attract and condense ambient water vapour.

hyperconcentrated flow A much disputed term, this is a high-discharge flow of water; an intermediate condition between debris flow and normal streamflow. Try Xia et al. (2018) *Prog. Phys. Geog.* 3.

hypermobility Highly mobile travelling; the maximization of physical movement; a rapid switching of economic links among manufacturers, importers, and retailers.

hyperreality A state supposedly more real—more sensation-packed—than reality, which can seem flat and dull. A **hyperreal procedure** would not use observations taken from real things; rather, it would use photographic representations to construct a world; representations which begin to blur with material reality especially in highly commodified societies that depend heavily on advertising and image manipulation. These images, simulations, and signifiers are hyperreality. D. DeLyser, ed. (2010), p. 233 is excellent on this.

hyporheic zone A dynamic mixing zone at the interface between groundwater and surface water. Hydrologic exchange between streams and hyporheic zones is important in keeping surface water in close contact with chemically reactive mineral coatings and microbial colonies in the subsurface. For an exhaustive treatment of the topic, see Harvey and Wagner in J. Jones and P. Mulholland, eds (2000).

hypsography The study of landforms and processes in physical geography.

hypsometry The measurement of land height.

hysteresis The impact of its history on a physical system. The future development of the system depends on its history, for its progress is determined by the direction in which the reaction is occurring—that is, the value of one variable is dependent on whether the other variable is

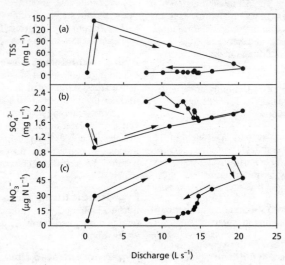

Hysteresis

increasing or decreasing. For example, when plotting the relationships between solute concentrations and river discharge during storm events, very *different* solute concentrations are measured at the *same* discharge on the rising and falling limbs of the storm hydrograph.

hythergraph A plot of monthly rainfall against monthly temperature over a year. *See* CLIMOGRAPH.

ice When the water content of snow falls below about 8% ice grain growth occurs by vapour flux of the air in the pores. The resulting growth of ice grains is very slow. In general, growth in wet snow speeds up with increasing water content.

ice, internal deformation The internal deformation of ice is achieved by *creep and large-scale faulting and folding. See Arenson et al. (2002) *Permafr. Periglac. Procs* 13, 117.

ice age A time when there are ice sheets on the continents, within which there may be *interglacial periods of milder climate. Ice ages last for some tens of millions of years, separated by intervals of *c*.150 million years. See Raymo and Huybers (2008) *Nature* 451, 284–5, DOI:10.1038/nature06589, on glaciation and Earth's orbit. A. Goudie (2000) is strong on stratigraphic information from ocean and ice cores, and the causes of climate change.

ice-albedo feedback *See* ALBEDO.

ice-cap A flattened, dome-shaped mass of ice, similar to an *ice sheet, but under 50000 km^2 in area.

ice contact feature A landform formed in contact with a glacier. For **ice contact terrace**, see *kame terrace. See Hickson (2000) *Geomorph*. 32, 3–4 on **ice-contact volcanism**.

ice fall An area of *extending flow marked by crevasses, where the gradient of a glacier steepens. The rapid flow at this point causes the ice to thin. See P. Knight (1999).

ice floe A flat section of floating ice which is not attached to ground ice. See Meylan et al. (1996) *J. Geophys. Res.* 101.

ice flow A glacier flows because the ice within it deforms in response to gravity. The major controls on ice flow are temperature and *shear stress, which is dependent on the ice thickness and the surface slope of the glacier. This can be summarized as:

$$\tau = \rho g(s - z)sin\ \alpha$$

where τ is the shear stress at a point within the glacier, ρ is the density of ice, g is the acceleration due to gravity, α is the surface slope of the glacier, s is the surface elevation, and z is the elevation of a point within the glacier.

The basal shear stress is given by:

$$\tau = \rho g h\ sin\ \alpha$$

where τ is shear stress, ρ is the density of ice, g is the acceleration due to gravity, h is the thickness of the glacier, and α is the surface slope of the glacier. See M. Bennet and N. F. Glasser (2009).

ice front The floating ice cliff at the seaward end of an *ice shelf. An ice front responds rapidly to temperature rise: in the Jakobshavn glacier in Greenland a floating **ice tongue** collapsed and caused the calving ice front to retreat several kilometres (http://www.geog.cam.ac.uk/research/projects/greenlandicesheet/).

ice lens An underground ice segment, often with convex upper and lower surfaces. *See* PALSA.

ice-marginal channel A channel cut by meltwater, running parallel to the ice margin, and found most often in a glacier's *ablation area, where rates of surface ablation and meltwater production are highest. Parallel sets of ice-marginal channels may form when the ice margin retreats. See Wadham et al. (2001) *Hydr. Procs* 15, 12.

ice mound In a *periglacial landscape, a swelling in the ground due to the expansion of an underground *ice lens. *Solifluction displaces the material at the top of the mound, so that when the ice lens melts a depression forms, often surrounded by a rampart. *See* PINGO.

ice patch, A small stretch of ice in the mountains that may have been in place for thousands of years; a residual, static ice mass with melting mechanisms which imply adjustments and small movements. See González-Trueba et al. (2011) *Geografiska Series A* 93, 57.

ice polygon (ice wedge polygon) A three- to six-sided polygon of *ice wedges, formed by *ice segregation and the drying and shrinking of sediments, and becoming more regular with age. See Matsuoka (1999) *Polar Geosci.* 12.

ice segregation The creation of separate bodies of ground ice in *periglacial conditions. Segregation develops well in silts (which seem to have the optimal pore size), depends on a moderate freezing rate, and often occurs under stones, which have high thermal conductivity. See Hall (2002)

S. Afr. J. Sci. 98 and Murton et al. (2006) *Science* 314.

ice sheet A continuous mass of glacier ice with an area over 50000 km^2; Typically an ice sheet is 1–3 km thick. Payne and Sugden (2006), *ESPL* 15, 7 find that both the vertical amplitude and the spatial distribution of bedrock basins and ridges are important in determining the pattern, rate, and extent of ice-sheet growth. An **ice core** is a column of ice extracted by drilling into an ice sheet.

ice shelf A sheet of ice extending over the sea from a land base, fed by snow falling on it, or from glaciers on the land.

ice stream Within a glacier or ice sheet, a stream of ice moving faster than, and not always in the same direction as, most of the ice. Ice streams respond to the weight of the ice, and to gravity. They are slowed by side drag from the slower-moving ice that flanks them, and by back pressure from the ice sheet itself (Stokes et al. (2007) *Earth Sci. Revs* 81, 3–4).

ice wedge In a *periglacial landscape, a near-vertical ice sheet, tens to hundreds of metres long, tapering into the ground to depths of 12 m. Individual wedges usually meet another wedge orthogonally. See Kanevskiy et al. (2017) *Geom.* 297. **Ice wedge casts** are sediment-filled, fossil ice wedges; see Harris et al. (2005) *Geomorph.* 71, 3–4 for the ice-wedge casting processes.

Ice-wedge networks form as ice fills the tension fractures resulting from cooling. They form in stable, aggrading or sloped surfaces, with winter temperatures from −15 °C to −35 °C. Modelled polygonal networks self-organize through interactions between fractures, stress, and re-fracture (Plug and Werner (2008) *Permaf. & Periglac. Procs* DOI: 10.1002/).

iconography The recognition, description, and interpretation of a set of symbols/images that people believe in, from the national flag to the culture taught in schools. For geographers, images of the landscape reveal symbolic meaning. However, landscape meanings change over time and between different groups; they are always open to debate and are often political (like the 'white cliffs of Dover'). See Leib et al. (2000) *GeoJournal* 52. Ponder on this: 'if Geography is prose, maps are iconography' (Lennart Meri 2013).

identity The individuality of a being or entity. National identity may be based on myths, and is actively constructed. Identity is strongly connected with place and space; individuals and groups tend to identify with the country, region, town, or village they live in. See Kaplan and Herb (2011) *National Identities* 13, 4 on the way geography affects national identities, while Arthur and Williams (2019) *PLoS1* 14, 4 use social media (Twitter) to quantify regional identity.

In physical geography, identity is established using 'methods of ontology based on meaningful collections of attributes' (Raper and Livingstone (2001) *TIBG* 26).

identity, geographies of The study of identities and space: 'we are in the midst of a redefinition of space. In the very moment that national and ethnic boundaries are breaking down we encounter paradoxical reinvestments in homeland, territorial integrity, localism, regionalism, and race- and ethnocentrism' (P. Yaeger 1996).

Identity politics is politics based around cultural identity, such as *disability *ethnicity, *gender, *race, or *sexuality.

ideology At its simplest, any ideology could be defined as a general view that informs people's actions.

idiographic Of, or relating to, the study of anything unique. The idiographic approach has been the cornerstone of *regional geography. In biogeography, **idiographic analyses** are detailed analyses of single species range; see Fattorini (2007) *Divers. & Distrib.* 13, 6.

igneous rock A rock formed from molten *magma, deep beneath the Earth's surface, subsequently coming to the surface as an *extrusion, or remaining below ground as an *intrusion. The plutonic mode of igneous rock formation is characterized by 'intrusive emplacement under compressional tectonic stress. The nature of the rock also depends on the rate at which it cooled; intrusions of magma cool slowly, allowing enough time for large crystals to form, while extrusions cool quickly, leaving little time for crystal growth.

illuviation The *downward translocation from the A *horizon and subsequent precipitation in the B horizon of clay-sized particles in a soil. The **illuvial horizon** is the B horizon, in which there is redeposition or entrapment of matter brought down from above.

image A picture built up by an individual from their social and physical *milieux, experienced from birth. Geography is a visual discipline with a past reliant on visual aids such as maps, globes, models, slides, and photographic illustrations. See A. Buttimer et al., eds (1999) on texts and images as mediators in the production of spaces.

imagined community This term was first used (B. Anderson 2006) to describe a nation, for even in the smallest nation, people will never know, meet, or hear about all their fellow-countrymen, but most people will have in their minds an image of their nation as a community. Thus, A. Latham et al. (2009) describe an imagined community as 'a group of people, united in the sense of their community, this unity being aided by newspapers, magazines, poster hoardings, cinema, radio, and television'.

imagined geographies (imaginative geographies) The ways that other places, people, and landscapes are represented by external observers, and the way these representations express both the perceptions, desires, fantasies, fears, and projections of those who create them, and the relations of power between subjects and objects of these imaginings. Thus, with independence, Singapore's ruling party sought to transform 'a group of people with divergent orientations . . . into a populace of loyal citizens' (L. Kong and B. Yeoh 2003). See Caroline Desbiens (2017).

imagined places Imagined places are not fantasy or fairytale; they have a connection with a place that exists geographically, for the word 'imagined' is used not to mean 'false' or 'made-up', but 'perceived', and refers to the perception of space created through certain images, texts, or discourse; images are not the reality, but are our only access to reality. Imagined places are constructed by their representations in art, film, literature, maps, and media; see Gorman-Murray (2007, *MC Magazine* 9, 3) on Sydney's King Street, imagined as a gay/lesbian place by the gay/lesbian media.

imbricated Of deposits, laid down in overlapping sheets (like roof tiles), as on steep slopes in *periglacial environments.

IMF *See* INTERNATIONAL MONETARY FUND.

immigration The movement into another country of someone as a permanent resident. Governments may try to restrict immigration by using quotas, discounting family ties, or being reluctant to accept asylum seekers or refugees. Skenderovic (2007) *Patterns Prej.* 41, 2 identifies the definition and exclusion of the Other as characteristic of radical-right politics. See Liu (2001) *Soc. & Cult. Geo.* 1, 2 on space, place, and scale, and the construction of racial-ethnic and immigrant identities.

imperialism An economic, cultural, and territorial relationship based on domination and subordination. Imperialism dominates the external trade of a subordinate nation, taking raw materials from the 'colony' and in return selling it finished goods, thus depressing manufacturing industry in the 'colony'; see Zanias (2005) *J. Dev. Econ.* 78, 1. It can exist without the creation of formal colonies. The **imperialist gaze** saw landscapes of subsistence farming as empty, despite the presence of indigenous people. Colonized spaces were 'stripped of preceding significations and then re-territorialized according to the convenience of the . . . imperial administration' D. Harvey (1989 and 2003).

import penetration The extent to which the country is dependent upon its imports, to reduce which governments may resort to *protectionism.

import substitution A strategy that encourages domestic manufacturing in order to reduce imports of manufactures, save foreign exchange, provide jobs, and lessen dependency. See Irwin (2020) *PIIE* on the rise and fall of import substitution, and Yu Zhou (2008) *World Devt* 36, 11, 2353 on export orientation and import substitution in China.

inceptisol *See* US SOIL CLASSIFICATION.

incidence matrix A matrix that shows the relationship between two classes of objects. Horner-Devine et al. (2007) *Ecology* 88, 6, 1345, who provide crucial help for the mathematically challenged, use presence-absence matrix in which each row represents a species or taxon, and each column represents a sampling site. Within each incidence matrix the presence of a species, taxon, or operational taxonomic unit (OTU) at a specific site is denoted by a 1, and its absence is denoted by a 0. Williamson (1978) *J. Ecol.* 66, 911 is recommended.

incised meander A *meander formed when a *rejuvenated river cuts deeper into the original meander; see O. Catuneanu (2006). An **intrenched meander** is an incised meander with a symmetrical cross-valley profile, an **ingrown meander** one with an asymmetrical profile (Rich (1914) *J. Geol.* 22).

inclusion 1. The action or state of incorporation, or of being added into a group or structure. The principle that binds together the present UK government's (and indeed Europe's) social policy is the reduction in the number of people and places experiencing '*social exclusion' from mainstream societal activities and the promotion of 'social inclusion'. Mitchell (2011) *AAAG* 101, 2 shows the way Marseille's city politicians have worked directly with both the representatives of major ethnic and religious groups and with the major capitalist players in the city, including elite entrepreneurs and members of the local mafia, to promote inclusion, based on ethnic ties, trust, and reciprocity, in marked contrast to the city of Paris.

2. A solid, liquid, or gaseous foreign body enclosed in a mineral or rock. **Melt inclusions** are small parcels of melt(s)—most are less than 0.0001016 cm across—that are entrapped by crystals growing in the magmas and eventually forming igneous rocks.

( ) SEE WEB LINKS

• *Crossing Paths: A Guide to The Pillars for Human Security and the 2030 Global Goals.*

indeterminacy The state of being vague and poorly defined. Indeterminism is a key idea in *chaos, catastrophe, and complexity theories. S. Lash, B. Szerszynski, and B. Wynne (1996), writing on environmental issues, note that 'a recognition of the intrinsic indeterminacy of the future, and of all our roles in shaping that future, imposes on us all the duty to take responsibility for the future we are making'.

index of centrality Many centrality indices are based on shortest paths, measuring, for example, the average distance from other vertices (Brandes and Pich (2006) *Int. J. Bifur. & Chaos* 17, 7). Simmons and Jones (2003) *Prog. Plan.* 60, 1 use it as a measure of the external market served by a city, and Channell and Lomolino (2000) *Nature* 403 use it to distinguish between 'central species' and 'peripheral species'. See Beauchamp (1965) *Behav. Sci.* 10 for an 'improved index of centrality'.

index of circulation *See* ROSSBY WAVES.

index of segregation A measurement of the degree of residential segregation between two sub-groups inside a larger population. See Johnston et al. (2005) *Urb. Studs* 42, 7.

index of vitality (I_v) An index to indicate the growth potential of a population:

$$I_v = \frac{(\text{fertility rate} \times \% \text{ aged } 20-40)}{(\text{crude death rate} \times \text{old} - \text{age index})}$$

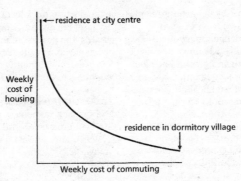

Indifference curve

indifference curve In economics, a graph of the various levels of *utility achieved at different prices through buying two commodities or services, for example, buying an expensive city-centre home, or paying for commuting from a cheaper, rural dwelling. At a given price, various combinations of the two would yield the same total utility, hence the term 'indifference'. The concept is used in the monocentric city model; see Bertaud and Brückner (2005) *Reg. Sci. & Urb. Econ.* 35, 2.

indegeneity Being characteristic of, or originating in, a particular region or country. Indigeneity is deeply historical, institutionalized, and shaped by powerful ontologies; see Radcliffe (2017) *PHG* 41, 2.

The UN Declaration of the Rights of Indigenous Peoples identifies and defines **indigenous rights**, whether individual or collective, embracing rights of ownership of identity, culture, ceremony, health, education, and employment. It emphasizes the rights of indigenous peoples to maintain and strengthen their own institutions, cultures, and traditions, and to pursue their development in keeping with their own needs and aspirations. It prohibits discrimination against indigenous peoples, and promotes their full and effective participation in all matters that concern them, and their right to remain distinct, and to pursue their own visions of economic and social development. UNESCO, in preparation for the International Decade of Indigenous Languages (2022–32), has issued an open call for nominations to indigenous peoples to be part of the Global Task Force entitled 'Making a Decade of Action for Indigenous Languages'.

Indigenous peoples are communities that live within, or are attached to, geographically distinct ancestral or traditional habitats or territories, identifying themselves as a distinct cultural group. Their cultural and social identities and their economic and political institutions are separate from the mainstream society or culture, and their rights are based on their historical ties to their territory.

Geographers focus on processes by which indigenous identities are produced, expressed, and diversified over time and space, and the continuing power relations between indigenous populations and 'colonial' powers. **Indigenous knowledge** describes those distinctive ideas, skills, and information held by indigenous peoples, usually contrasted with Western scientific knowledge. Briggs (2013) *Prog. Devt Studies* 13, 3, 231 notes that the promise of using indigenous knowledge in development strategies has not been fulfilled. Read his *Final Thoughts*, p. 39–40; well worth it.

indivisibility The minimum scale at which any technology can be used; almost all manufactured capital goods are indivisible. The indivisibility of a tractor makes it uneconomic for many small farmers in the developing world (V. Balasubramanian et al. 1998). It is crucial in finding optimum factory size and in economies of scale (Elliethy (2006) *J. Int. Dev.* 4, 4). Edwards and Starr (1987) *Amer. Econ. Rev.* 77, 1 hold that **indivisible labour specialization** brings scale economies.

induced demand In *transport geography, the additional traffic resulting from additions to road capacity (Levinson (2008) *J. Econ. Geog.* 8, 1). Goodwin and Noland (2003) *Appl. Econ.* 35 find that building new roads really does create extra traffic.

induration The hardening or cementing process that consolidates sediments to form sedimentary rocks. Sedimentary rocks usually become more and more indurated over time. See Hermann (2008) *J. Coastal Res.* 24, 6, 1357 on induration and sand dunes.

industrial cluster *See* CLUSTER.

industrial district *See* MARSHALLIAN INDUSTRIAL DISTRICT.

industrial diversification The deployment of investment and labour over a wider range of activities. Wei et al. (2007) *Econ. Geog.* 83, 4 show that diversification in China is led by local governments, and entrepreneurs

embedded in thick local networks. R. Boschma and G. Capone (2014) analyse the process of industrial diversification in the *EU-27 and European Neighbourhood Policy countries in the period 1995–2010.

industrial geography The study of the spatial arrangement of industry; the study of production and production systems. The central question of industrial geography is determining why industrial activities grow/decline in particular places.

industrial inertia The survival of an industry in an area after its original locational factors are gone. It is not unusual in industry, as sunk costs (costs that have already been incurred and cannot be recovered) are high, and, over time, enterprises tend to create *external economies (Rota and Vanolo (2007) *EURA Conf. Paper*).

industrialization The development of manufacturing industry from a predominantly agrarian society. Characteristic features of industrialization include the application of scientific methods to solving problems; mechanization and a factory system; the division of labour; the increased geographical/social mobility of the labour force; and capital deepening. These are also features of *capitalism, and capitalism is not the same thing as industrialization.

Industrialization is widely seen as the most important social and economic change of the last 200 years, with the power to modify or destroy any pre-existing social arrangement. Industrialization encourages urbanization; for example, Chen (2002) *China Econ. Rev.* 13, 4 finds that those Chinese provinces and municipalities with higher per capita GDP are more urbanized. As industrialization continues, there is a sectoral shift in employment. This change from manufacturing to commerce and services to a considerable extent means a change from more to less stable jobs.

Industrialization is often seen as a solution to poverty; Yamagata (2006) *Cambodian Econ. Rev.* finds that the Cambodian garment industry offers scope for the poor to increase their wages substantially. But Blumer (1969/70) *Studs Compar. Int. Dev.* 5, 3 holds that early industrialization brings 'a series of new demands on life' (such as a search for superior economic status) that lead to 'disorganization'.

industrial location policy The neoclassical view of the role of the state in economic activities is inherently suspicious of government interference in private sector decision making, and its justification for industrial location policy relies primarily on an increase in mobility, of both capital and labour. Behaviouralists are more particularly concerned with the effects of industrial incentives on capital mobility, and specifically whether or not industrial location incentives change the location preferences of entrepreneurs in favour of designated regions.

industrial location theory The general critical factors of industrial location are transportation, labour, raw materials, markets, industrial sites, utilities, government attitude, tax structure, climate, and community. Currently, the central issue of state industrial policy is to generate growth, innovation, and employment. Badri (2007) *J. Bus. & Public Affairs* 1, 2 reviews industrial location theory. **Industrial organization**—the structure of an industrial unit—is a component of industrial location theory; see

McCann and Sheppard (2003) *Reg. Studs* 37, 6–7.

industrial revolution Although there is some discussion about its timing, the industrial revolution is generally accepted as occurring in Britain in the late 18th and early 19th centuries. K. Pomeranz (2000) explains the industrial revolution though 'Europe's armed exploitation of other resource-rich regions, particularly in the Americas'.

industrial specialization The domination of a limited range of industries in a region. Decreasing transport costs will lead to an increase in specialization (Aiginger and Rossi-Hansberg (2006) *Empirica* 33). Svaleryd and Vlachos (2002) *J. Int. Econ.* 57 find that the financial sector has greater impact on industrial specialization among OECD countries than do differences in human and physical capital.

industry A sector of the economy in which firms create a group of related products and/or services. The extraction of natural resources, including agriculture, is **primary industry**; manufacturing is **secondary industry**; **tertiary industry** comprises services, including transportation and communication; **quaternary industry** deals with a range of producer services from banking to retailing to real estate; and **quinary industry** with consumer services such as education, health care, and government information. With economic development, industry moves from primary industry to progressively less material-oriented activities such as textiles and clothing, until, over time, tertiary industries dominate; the sequence is illustrated in the figure.

Phelps and Ozawa (2003) *PHG* 27, 5 propose four industrial phases.

inequality

	Protoindustrial	Industrial	Late industrial	Post-industrial
Sectorial basis	Agriculture-manufacturing	Manufacturing	Manufacturing	Services-manufacturing
Division of labour	Intra-firm, intra- and inter-sectorial	Inter-firm and inter-sectorial	Intra- and inter-firm, intra-sectorial	Inter- and inter-firm, inter- and intra-sectorial
Source of accumulation	Exports → domestic	Domestic → exports	Domestic → exports	Exports → domestic

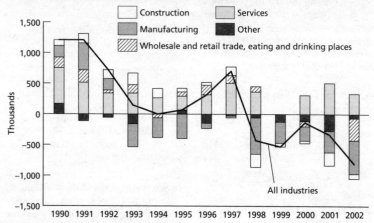

Industry. Industrial change in Japan 1990–2002
Source: Ström, 2005, *Soc. Sci. Japan J* 8, 2

inequality The fact of **global inequality** is not disputed; the world is routinely divided into developed/developing, North/South, more/less/least economically developed, high/medium/low development, and so on. The richest 2% of adults in the world own more than half of global household wealth. There is no consensus about the causes of global inequality, not least because of the difficulties in measuring inequality: if we scale income by some index of health, there is more inequality in the world than if we consider income.

Many writers blame globalization for global inequality: Ash (2004, *TIBG* 29, 2) argues that 'TNCs and international banks are deeply embedded in the production of inequality, through their influence over governments, their tax evasion schemes, their poor environmental record, their labour market practices, their locational mobility, their market power, their dumping practices.' Cornia and Court (2001, *Policy Brief 4* World Inst. Dev. Econ. Res.) claim that the liberalization of domestic banking associated with globalization has caused rises in income inequality.

There is also debate over the impact of *structural adjustment measures on inequality. In 2004, *SAPRIN claimed that structural adjustment polices are the largest single cause of inequality; at the other end of the spectrum, Ferreira (2004) *World Bank Policy Working Paper* 1641 asserts that, under a programme of structural adjustment, the population of Tanzania in poverty fell from 65% in 1983 to 51% in 1991.

infant mortality The number of deaths in the first year of life per 1000 children born. Poor sanitation is a major factor. Gyimah (2006) *Ethnicity & Health* 11, 2 finds that mortality is more of a reflection of socio-economic inequalities than ethnic-specific socio-cultural practices, but notes that wider birth spacing and longer breastfeeding also reduce infant mortality.

infiltration Water entering soil. Tillage increases soil porosity, and thus infiltration. Reduced tillage and stubble retention practices can greatly reduce surface sealing and improve infiltration rates. Soil type is significant—sands take in water faster than loams and clays. Infiltration may not occur if the speed of the water is too great.

Infiltration capacity is the maximum rate at which water can infiltrate the soil, and a key determinant of the volume of surface run-off, which occurs when rainfall exceeds a soil's infiltration capacity. **Relative infiltration capacity** is the ratio of mean precipitation intensity to saturated hydraulic conductivity.

informal sector (informal economy) '[E]nterprises owned by households that includes informal own-account enterprises, which may employ contributing family workers and employees on an occasional basis, and enterprises of informal employers, which employ one or more employees on a continuous basis. The enterprise of informal employers must fulfil one or both of the following criteria: size of unit below a specified level of employment,

and non-registration of the enterprise or its employees.' (ILO Report of the Fifteenth International Conference of Labour Statisticians, Geneva, 1993.)

The informal sector plays an important and controversial role. It provides jobs and reduces unemployment and underemployment, but in many cases the jobs are low-paid and the job security is poor. It bolsters entrepreneurial activity, but at the detriment of state regulations compliance, particularly regarding tax and labour regulations. It helps alleviate poverty, but in many cases informal sector jobs are low-paid and the job security is poor.

(((⊕))) SEE WEB LINKS
• The informal sector.

information city A city whose residents work mostly in education, consulting, banking, services, computer programming, computer manufacturing, and medical equipment. Those in the *creative industries could be considered as elements of both in some cases. Is Bangalore an information city? Liang (2005) *Wld Info. City Conf.* says yes, and writes of 'the other information city': 'the proliferation of non-legal media practices ranging from pirated VCDs, DVDs, and MP3s to grey market mobile phones and pirate book sellers on the streets.'

information society A society built on information storage, retrieval, and transmission, *time–space compression, *post-Fordism, *flexible accumulation, and the advance of finance capital, which is characterized by networking, *globalization, and the flexibility, individuality, and instability of work. 'Information technologies, both as a process and product, contain embedded social knowledge, and thus represent the construction of new social norms and institutions' (Eischen (2001) U. Calif.,

Santa Cruz, *W. Paper Series* 1009). 'The information society can be usefully characterized as a universe at the intersection of three distinct but interdependent spaces: the geographical space, the social space, and the informational space' (Ekbia and Shurman, *Information Society*, special issue).

information system An integrated assemblage of elements that collect, store, and process data and deliver information, knowledge, and digital products.

infragravity wave (IGW) A sea wave that is low in amplitude (frequencies of 0.05 to 0.003 Hz), has long periods (> 30 seconds), and can form standing waves that periodically oscillate the still water level of a surf zone. See Beetham and Kench (2011) *ESPL* 36, 14, 1872 for several explanations for the generation of IGWs.

infrastructure The framework of communication networks, administration, and power supply. **Infrastructural geographies** examine the material and organizational structures of social life in diverse settings, including the role of the state and other mediating institutions.

Infrastructure itself is an object of study, illuminating the materiality of space, the networks through which it is organized, and the interconnections between objects and bodies. It may also highlight the connections and dependencies that circulate in the production of distinct socio-economic phenomena. In addition, infrastructure is a means through which social and spatial wellbeing may be understood; it's a tool for governing social life in uneven and unequal ways.

ingrown meander *See* INCISED MEANDER.

inner city An area at or near the city centre with dilapidated housing, derelict land, and declining industry (also called the **twilight zone**, or **zone of downward transition**). Langlois and Kitchen (2001) *Urb. Studs* 38, 1 use a general deprivation index to locate deprivation in central Montreal (but note that it isn't confined to the inner city). The decline of US inner cities reflects *deindustrialization, combined with the suburbanization of a small number of middle-class African Americans.

innovation The introduction and adoption of something new. An **innovation wave** is the diffusion of an innovation from its point of origin. Beginning at the innovation epicentre, it is adopted by a very few, close to the focus. Through time, the peak of the wave moves away from the source as more distant locations adopt the innovation. Consolidation follows across the whole of the region concerned; finally, as the innovation saturates the region, diffusion slows down until the stage of maximum acceptance.

An innovation wave may occur when larger cities take up the innovation, which then filters down the urban hierarchy. The largest cities have benefited from many innovation cycles, and, in consequence, have developed a broader diversity of activities, and attained higher levels of social and organizational complexity—thus making them more likely to adopt further innovations.

input–output analysis A view of the economy that stresses the interdependence of different economic sectors—the output of one sector is often the input of another. See McBain et al. (2019) *J. Indus. Ecology* 23, 2.

InSAR (Interferometric Synthetic Aperture Radar) A remote sensing technique which uses radar satellite images. The US Geological Service has an excellent site on InSAR.

inselberg A steep-sided hill composed mainly of hard rock, which rises sharply above a plain in tropical and subtropical areas; see Römer (2007) *Geomorph.* 86, 3–4. *See also* BORNHARDT.

insolation The *solar radiation hitting the Earth's surface. Global variations in insolation are critical in the *general circulation of the atmosphere. 'Climatic changes result from variables in planetary orbits which modulate solar energy emission and change seasonal and latitudinal distribution of heat received by the Earth. Small insolation changes are multiplied by the *albedo effect of the winter snow fields of the Northern Hemisphere, by ocean-atmosphere feedbacks, and, probably, by the stratospheric ozone layer' (Kukla (1972) *Boreas* 1, 1). Insolation *forcing functions include: the eccentricity of the Earth's orbit, the angle from the vernal equinox to *perihelion, obliquity, and time.

instability 1. The condition of a slope just before the initiation of mass movement. See (2008) *Geomorph.* 94, 3–4, special issue.

2. The condition of a parcel of air which has positive *buoyancy, and thus a tendency to rise through the *atmosphere. It is the temperature of the parcel relative to the *ambient air which is critical, since this affects densities. A parcel remains unstable if it cools more slowly than the surrounding, stationary air.

Absolute instability occurs when an air parcel, displaced vertically, is hastened in the direction of the displacement; such a move will end when the temperature of the parcel is at one with its surroundings. The **potential instability** of a parcel of air

can be calculated using a *tephigram. **Conditional instability** occurs when a parcel of air would become unstable if lifted by some other force, such as a mountain, the moistening of lower-level air, or when a lower, stable parcel of air is overlain by an unstable layer. **Convective instability** is a tendency of an air parcel towards instability when it has been lifted bodily until completely saturated, which often occurs via the inevitable stirrings in the atmosphere. The atmosphere is said to be unstable when the fall of temperature with height of the environmental air is more rapid than that experienced by a rising air parcel, so that the parcel continues to be less dense than the ambient air. See E. Fedorovich et al. (2004).

institutional thickness The totality of the conditions that are crucial to economic transformation. It is associated with strong local institutions, high levels of interaction between local organizations, coalition, and a mutual awareness of being involved in a common enterprise. Institutional thickness includes trade associations, voluntary agencies, sectoral coalitions, concrete institutions, and local elites—their effects on local policy, and their consensus institutions: common agreements, shared views and interpretations, and unwritten laws. However, it is not clear how, exactly, institutional thickness promotes local economic development and prosperity. Further, how do we explain the fact that some very successful regions do not appear to be underpinned by a highly concentrated and coordinated matrix of institutions? Does institutional thickness stimulate growth, or vice-versa? See Zukauskaite (2017) *Economic Geog.* 93, 4.

integrated assessment (IA) An interdisciplinary process which combines, explains, and communicates knowledge from a range of disciplines in order wholly to weigh up an entire chain of causes and effects. Van Amstel (2005) *Env. Sci.* 2, 2–3 uses an integrated assessment model to evaluate six methane emission reduction strategies on future air temperature. For a second example, see Mendonça et al. in M. van Asselt et al., eds (2001).

integrated land-use and transportation models Models which attempt to represent human behaviour in a realistic way, measuring and representing the decision-making that leads to the organization of human activities in time and space. These range from daily household decisions—what to do, where and when to do it, and by which mode of transport—to long-term decisions relating to land use and transport networks. See Waldeck et al. (2020) *J. Transport and Land Use*, 13, 1.

integrated water management The coordinated development and management of water, land, and related resources, in order to maximize economic and social welfare, and ensure equity and sustainability based on the perception of water as an integral part of the ecosystem, a natural resource and a social and economic good.

The general objective is the environmentally sound, equitable, and sustainable utilization and development of water resources. This may include supply management, demand management (water conservation, transfer of water to uses with higher economic returns, etc.), water quality management, recycling and reuse of water, economics, conflict resolution, public involvement, public health, environmental and ecological aspects, socio-cultural aspects, water storage (including long-term storage or water banking), conjunctive use of surface water and groundwater, water pollution control, flexibility, regional approaches, weather modification, sustainability, and so on.

In a study of the *Loess Plateau, China, Chen et al. (2007) *PPG* 31, 4 list the factors to be considered: 'basic farmland construction, plantation of cash trees, firewood and conservation wood, roads, water banks, agricultural structure modifications, and local people's needs (such as daily fuel, daily drinking water, education, medical care and other factors regarding improving living conditions).' Remedial measures include the use of *check dams; afforestation and the planting of shrubs and grasses on slopes (*see* REHABILITATION ECOLOGY); and land closure (fallowing), in order for ecosystems to rehabilitate naturally, without further human intervention. Natural recovery by land closure is strongly recommended on the steep slopes.

In natural rehabilitation, the degraded soil may be improved and the biodiversity may be increased steadily over time. However, revegetation usually takes a long time, especially from land in an extremely degraded state; several decades to hundreds of years may be needed in the Chinese Loess Plateau. R. Quentin Grafton (2011) is the stand-out text.

integration 1. Social integration The process whereby a minority group, particularly an ethnic minority, adapts to the host society and where it is accorded equal rights with the rest of the community. Lemanski (2006) *GeoForum* 37, 3 reports on social integration in post-apartheid urban South Africa: 'different races are not only living peacefully in shared physical spaces but also actively mixing in social, economic and, to a lesser extent, political and cultural spaces.'

Lemanski notes a social continuum of integration (greeting in the street, visiting homes, inter-marriage) and a continuum of **spaces of integration**, for example physical space (shared neighbourhood), economic space (common employment type), social space (cross-race friendship), political space (common involvement in civic organizations), and cultural space (shared sense of belonging).

2. Economic integration can be the breaking down of trade barriers between nations. 'Economic integration has a direct effect on internationalisation by reducing transaction costs . . . Being in a currency union has been shown to have a significant effect on bilateral trade' (Rose (1999) *CEPR Discussion Paper* 329, London). **Horizontal integration** occurs across different sectors; **vertical integration** occurs within a hierarchy; see Farrington (2007) *J. Transp. Geog.* 15, 5.

intellectual property A creative work, or a device or invention that can be patented or copyrighted. Intellectual property rights laws can be understood as particular, culturally defined systems for codifying knowledge, employed to discipline objects, phenomena, and social relations (Parry (2002) *Antipode* 34, 4).

intensity The size of an earthquake, based on the impact on people, buildings, and the ground surface. Isoseismal lines represent equal earthquake intensity. The Modified Mercalli intensity scale is the world standard; the Medvedev–Sponheuer–Kárnik (MSK) is popular in Europe. Although differing in detail, the MM and MSK scales are much alike. As they are descriptive, and based on recorded damage, **intensity values** can be calculated and isoseismal lines drawn for present day and historic earthquakes (Chester (2001) *PPG* 25, 3).

intensive agriculture Agriculture with a high level of capital and labour inputs, and high yields. 'Intensive cropping systems, where large areas of land are cultivated permanently (or with relatively short fallow periods) through mechanization, the use of labour, irrigation water, and its dissolved nutrients, knowledge, and other inputs, develops concurrently with the evolution of socially and culturally complex human societies. Phillips (1998) *Appl. Geog.* 18, 3 provides a gloomy list of the effects of agricultural intensification in Italy: declining frequency of crop rotations; increased use of chemical fertilizers and pesticides; more up- and downslope tillage; and enlarged fields. Mechanization has led to a decline in draught cattle numbers, causing loss of agricultural employment, and a fall in soil fertility.

interaction The action between things, each upon the others. An **interaction model** describes the reactions of two or more processes or systems as they affect each other; see Eradus et al. (2002) *J. Transp. Geog.* 10, 2.

interception The retaining of raindrops by plant leaves, stems, and branches. Rainfall frequency is more significant than rainfall rate and duration in determining interception. Guevara-Escobar et al. (2007) *J. Hydrol.* 333, 2–4 evaluate interception around the canopy of a single tree.

interdependence In a heartening review of the term, Smith et al. (2007) *Geog. Compass* 1, 3 declare that interdependence conveys a sense of relying on and being responsible for others. 'Today we all gather here to declare our interdependence. Today we hold this truth to be self-evident: we are all in this together' (Will Smith, Live8, 2005).

interdisciplinarity The combination of two or more academic disciplines into one investigation; using skills and knowledge that cross traditional subject boundaries. See the special issue of *Area* (2009) 41, 4.

interface The zone of interaction between two systems or processes. Estuaries are interfaces between *fluvial and marine systems.

interflow Water that moves downslope through soil pores (as opposed to *throughflow, which is soil water moving downslope along impermeable soil horizons—it's difficult to separate the two processes in the field). Interflow is of major importance for the generation of run-off and groundwater recharge.

interglacial A long, distinct period of warmer conditions between *glacials when the Earth's glaciers have shrunk to a smaller area—the present, Holocene, period is an interglacial. Soon (2007) *Phys. Geog.* 28, 2 argues that the persistence of insolation forcing at key seasons and geographical locations, taken with closely related thermal, hydrological, and cryospheric changes, are enough to explain transitions in paleoclimates.

Intergovernmental Panel on Climate Change (IPPC) The leading international body for the assessment of climate change, established by the United Nations Environment Programme (UNEP) and the World Meteorological Organization (WMO) in 1988 to provide the world with a clear scientific view on the current state of knowledge in climate change and its potential environmental and socio-economic impacts. The IPCC is a scientific body under the auspices of the United Nations (UN). It reviews and assesses the most recent scientific, technical, and socio-economic information produced worldwide relevant to the understanding of climate change. It does not conduct any research nor does it monitor climate-related data or parameters.

intermediate beach A beach form somewhere between the high-energy *dissipative beach and the low-energy *reflective beach, and containing elements of both these two basic types of beach. For example, the upper part of an intermediate beach can be steep, and show characteristics of a reflective beach. Intermediate beaches tend to form near shore bars, which will dissipate some of the wave energy.

intermediate technology Mid-level technology, often the most suitable level for economic development. Chinese aid projects in Africa, for example, consist largely of intermediate technology: Chinese workers are paid at local rates, train their local equivalents, and aim to leave projects in a state where they can be locally managed, operated, and maintained so that even the spare parts could be locally produced. The *Intermediate Technology Group* is now called *Practical Action*.

internal colonialism The practice of exploitation by a nation-state within its own boundaries. In many states, indigenous communities suffer from poverty and isolation. *Dependency theorists argue that this situation arises when the state treats its indigenous people in the same way as the European powers treated their colonies.

internal deformation A distortion within a material, such as glacier ice: individual ice crystals react to stress by elongating, or by melting and recrystallizing, and ice may also shear along separate shear planes; see Hambrey et al. (2005) *J. Geophys. Res.*

110, F01006 and Mair et al. (2001) *ESPL* 23.

internal migration The temporary or permanent relocation of some of its population inside a *nation-state. Globally, China has the largest volume of internal migration; in 2003, the so-called floating population reached 140 million, most of them rural labourers moving from the countryside to cities and coastal areas. The rapid growth of rural–urban migration in China results from the privatization of agriculture, and the transition to a market economy (Ping and Pieke (2003) *Inst. Sociol., Chinese Acad. Soc. Scis*).

international division of labour The allocation of various parts of the production process to different places in the world. The world economy is organized through the horizontal and vertical linkages of an international division of labour. Whilst some countries have tended to operate within their own boundaries, the great bulk of the skilled labour force and the most complex labour processes are located in the advanced economies (the USA, western Europe, and Japan), and other areas have been mainly transformed into sources of a relatively cheap and disciplined, though less skilful, labour force (mainly East and South-East Asia). See Hardy (2007) *Eur-Asia Studs* 59, 5 on the new international division of labour and Poland.

International divisions of labour can involve transfers of services. Thus, Filipina workers take on the household tasks of middle-class women abroad, while they themselves may hire poorer workers to do their housework in the Philippines (Boyle (2002) *PHG* 26, 4). *Service offshoring* is an extension of outsourcing as firms apply an international division of labour to service tasks (Bryson (2007) *Geografiska B* 89).

To many Marxist scholars, the rise of East Asia as a global factory can often be seen as an empirical proof of the working of spatial fixes.

international financial system (global financial system) The collective name for the various official and legal arrangements that govern international financial flows in the form of loan investment, payments for goods and services, interest and profit remittances. The main elements are the surveillance and monitoring of economic and financial stability, and provision of multilateral finance to countries with balance of payments difficulties. With the decline of *Bretton Woods, much of the power over the international financial system, and especially power over how credit is created, bought, and sold, has transferred back to money capitalists. With the growth of electronic telecommunications, money capitalists can operate on a global scale, at speeds which nation-states find very difficult to match. In this new international financial system, the balance has shifted from a predominantly state-based financial structure with some transnational links, to a predominantly global system with some residual local differences in markets, institutions, and regulations.

International Fund for Agricultural Development (IFAD) A UN agency for ending rural poverty in developing countries.

(()) SEE WEB LINKS
• The IFAD website.

International Geosphere–Biosphere Programme (IGBP) A programme focused on acquiring basic scientific knowledge about the interactive processes of biology and chemistry of the Earth, including human modifications, as they relate to global change.

(⊕) SEE WEB LINKS
• Website of IGBP.

International Labour Organization (ILO)

An agency of the United Nations whose aims are: 'to promote and realize standards and fundamental principles and rights at work; to create greater opportunities for women and men to decent employment and income; to enhance the coverage and effectiveness of social protection for all; and to strengthen tripartism and social dialogue' (ILO).

(⊕) SEE WEB LINKS
• Website of the ILO.

International Monetary Fund

'The work of the IMF is of three main types. Surveillance involves the monitoring of economic and financial developments, and the provision of policy advice, aimed especially at crisis-prevention. The IMF also lends to countries with balance of payments difficulties, to provide temporary financing and to support policies aimed at correcting the underlying problems; loans to low-income countries are also aimed especially at poverty reduction. Third, the IMF provides countries with technical assistance and training in its areas of expertise. Supporting all three of these activities is IMF work in economic research and statistics. In recent years, as part of its efforts to strengthen the international financial system, and to enhance its effectiveness at preventing and resolving crises, the IMF has applied both its surveillance and technical assistance work to the development of standards and codes of good practice in its areas of responsibility, and to the strengthening of financial sectors. The IMF also plays an important role in the fight against money-laundering and terrorism' (IMF website).

The IMF doesn't get a good press from geographers, largely through its role in promoting *structural adjustment policies: as a result of austerity programmes sponsored by the World Bank and the IMF, 'the 1980s and 1990s were the "lost decades" for development in much of Africa' (Carmody (2008) *Geog. Compass* 2, 1). Watts et al. (2005) *PHG* 29, 1 describe the IMF as one of the ungoverned bodies that regulate global capitalism.

(⊕) SEE WEB LINKS
• The IMF website.

International Organization for Migration (IOM)

The leading inter-governmental organization in the field of migration working closely with governmental, inter-governmental, and non-governmental partners, the IOM works 'to help ensure the orderly and humane management of migration, to promote international cooperation on migration issues, to assist in the search for practical solutions to migration problems, and to provide humanitarian assistance to migrants in need, including refugees and internally displaced people. The IOM Constitution recognizes the link between migration and economic, social, and cultural development, as well as to the right of freedom of movement' (IOM Constitution). See Pécoud (2018) *J. Ethnic & Migration Studs* 44, 10.

(⊕) SEE WEB LINKS
• Website of IOM.

international trade theory

The theory that countries will produce and export goods which are relatively abundant, and import goods that require resources which are in relative short supply. A country or region which makes something relatively better than the competitors has a comparative advantage, and that comparative advantage dictates who trades what, and with whom. Andresen (2010) *Geog. Compass* 94–105 explains this very clearly.

internet A global network of connected computers. Rates of uptake are determined by costs, wealth, literacy, and telephone penetration rates. Zook (2002) *Jour. Econ. Geog.* 2, 2 observes that 'although both the internet and venture capital have been viewed as independent of geography, the development of this industry continues to highlight the continuing relevance of regions and place-based relations'. Thus, in the United States, website production is dominated by New York and San Francisco. The internet reduces communication costs, transportation costs, and search costs. The impact of reducing these costs varies across locations because it depends on preferences, the availability of substitutes, and the availability of complements. Thus, the diffusion of the internet benefits some locations more than others. See E. Leamer and M. Storper (2014) on the economic geography of the internet age.

See O'Kelly (2002) *Prof. Geogr.* 54, 2 on measuring access to commercial internet sites.

interpellation Interpellation is a process by which we encounter the values of a culture or ideology and internalize them. Attitudes towards, for example gender, class, race, religion, and patterns of consumption are presented and ideologies are communicated. This is how one class allows the other class to exert influence over them. Workers end up seeing themselves in the ways that the managers describe them. See McDowell et al. (2007) *Econ. Geog.* 83, 1. Interpellation is a factor in cultural geography which may affect decision-making. See also Davies and Brooks (2019) *Soc. & Cult. Geog.*

intersectionality The interconnections and interdependence of race with other categories. The originator of this term, Crenshaw (1991) *Stanford Law Rev.* 43, 6, considers the various ways in which race and gender interact to shape the multiple dimensions of Black women's employment experiences.

Wills (2008) *Antipode* 40, 1 believes that geography has real strength as a discipline for the study of intersectionality: 'to understand the ways in which class intersects with other social cleavages, with very different e/affects.'

interstadial A warmer phase within a *glacial which is too short and insufficiently distinct to be classed as an *interglacial.

intertextuality The way that any texts acquire meaning through the way they are described or evoked by other texts; every text is related to other texts and these relations generate the meanings of the text. Aitken in R. Kitchin and N. Thrift (2009) is very good on this concept.

inter-tropical convergence zone (ITCZ) That part of the tropics where the opposing north-east and south-east *trade winds converge. It is not a continuous belt, more like a necklace with groups of clouds as the 'beads'; in places there may be two or more 'strings'. The zone is narrower over the oceans, and broader over the continents, where other wind systems may be involved; in West Africa the ITCZ is the convergence of the Guinea *monsoon and the *Harmattan.

The ITCZ moves north and south; moving more over land, and arriving in the summer in each hemisphere. Its position is affected by the apparent movement of the overhead sun, the relative strengths of the trade winds, and the changing locations of maximum sea-surface temperatures (R. D. Thompson 1998). This means that the movements of the ITCZ are highly unpredictable. In May 2005, the African portion of the ITCZ was 1° south of the average; a period of drought near the Sahel. In July 2003, it was around 0.9° north of the average; heavy rains fell in Ethiopia, with flooding in Khartoum. The movements of the ITCZ also affect bush/forest fires in South-East Asia (P. Kershaw et al. 2001). Over the oceans, the ITCZ is broad, and often loses its identity.

intervening opportunities theory

Stouffer (1940) *Am. Soc. Rev.* 5, 6 argues that the volume of migration over a given distance is directly proportional to the opportunities at the migrants' destination, and inversely proportional to the number of opportunities between the place of departure and destination. Thus, prospective migrants might locate somewhere along the route instead of going to the originally planned destination. Many Jewish people, seeking to escape aggression in nineteenth-century Europe, planned to travel via Britain to America, but travelled only as far as London, where they found tempting economic opportunities. See Alix et al. (1999) *J. Transp. Geog.* 7, 3 on Montreal as an intervening transhipment opportunity between Quebec and Europe.

intrazonal soil A soil affected more by local factors than by climate, unlike a *zonal soil. For example, waterlogging creates *gley soils, and a limestone bedrock will produce a *rendzina.

intrenched meander See INCISED MEANDER.

intrusion A mass of igneous rock which has forced its way, as *magma, through pre-existing rocks, and then solidified below the surface of the ground; hence **intrusive rock**.

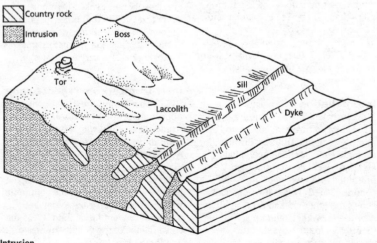

Intrusion

Igneous intrusions are classified according to their shape and size: *dykes are **discordant**, since they cut across the bedding or fabric of pre-existing rocks, *sills are **concordant**, since they intrude between layers. *Plutons are the largest intrusions, typically spheroidal, and ranging from *stocks to *batholiths.

invasion The establishment of species in ecosystems they are not native to; see, for example, Russell-Smith et al. (2004) *J. Biogeog.* 31, 8 on the invasion of rain forest into eucalyptus woodland in Australia. Campbell et al. (2002) *J. Biogeog.* 29, 4 outline a methodology which might be able to be used to predict invasions by alien plant species. 'Humans cause invasions, humans perceive invasions, and humans must decide whether, when, where and how to manage invasions' (Richardson et al. (2008) *PHG* 32, 2).

invasion and succession 1. A model of change used in *urban ecology to represent the effects of immigration on the social structure of an urban area. Invasion and succession involve a chain reaction, with each preceding immigrant wave moving outwards and being succeeded by more recent, poorer immigrants. This model saw immigrant enclaves as transitional stages on the road to eventual acceptance and integration in the larger (American) society (Clark in R. Waldinger and M. Bozorgmehr, eds 1996).
2. In plant ecology, the introduction and subsequent spread of a species.

invasive species A species is invasive if it: is not native to the biogeographical region in question; has adapted to its new location; is becoming more profuse/more widely distributed; is an irritant; and results from anthropogenic activity.

inversion The increase of air temperatures with height. (This is the reverse of the more common situation in which air cools with height.) Inversions occur: when strong, nocturnal, *terrestrial radiation cools the Earth's surface and therefore chills the air which is in contact with the ground; when cold air flows into valley floors, displacing warmer air; where a stream of warm air crosses the cool air over a cold *ocean current; where warm air rises over a cold front; and when air from the upper *troposphere, subsiding in a warm *anticyclone, is compressed and *adiabatically warmed. The boundary between the top of the cold air and the beginning of the inversion is an **inversion lid**. *See also* TRADE WIND INVERSION.

invisibles Imported and/or exported items that are not goods. Invisibles include government grants to overseas countries and subscriptions to international organizations, payments for shipping services, travel, royalties, commissions for banking and other services, transfers to or from overseas residents, interest, profits, and dividends received by or from overseas residents.

involution 1. The refolding of two *nappes differing in age so that parts of the younger nappe lie below older rocks.
2. The convolution of layers of ground under *periglacial conditions. To form thermokarst involutions, ice-rich permafrost must thaw, drainage conditions must be poor, and sediments must vary in texture or composition. In addition, the sediments should be susceptible to fluidization, liquefaction, or hydroplastic deformation. The repetition of differential heave and soft-loam settlement promotes decimetre-scale involutions in near-saturated soils subject to deep seasonal frost penetration.

ionosphere A layer of the
*atmosphere containing ions and free
electrons. The ionosphere is warmed as
it absorbs solar radiation, and it will
reflect radio waves. Three bands are
recognized: E, F, and F2, at about 110,
160, and 300 km above the Earth. The
ionosphere is in a state of constant
motion, and is affected by tidal forces
and by the Earth's magnetic field.

irradiance The rate of flow of radiant
energy through unit area perpendicular
to a solar beam. **Total solar irradiance**
is the dominant driver of global climate
(Mendoza (2005) *Advances Space Res.*
35, 5). See also Mishchenko et al. (2007)
Bull. Am. Met. Soc. 88, 5.

irrigation The supply of water to the
land by means of channels, streams, and
sprinklers in order to permit the growth
of crops. Without irrigation arable
farming is not possible where annual
rainfall is 250 mm or less, and irrigation
is advisable in areas of up to 500 mm
annual rainfall. To some extent,
irrigation can free farmers from the
vagaries of rainfall and, to that end, may
be used in areas of seemingly sufficient
rainfall because irrigation can supply the
right amount of water at the right time.

Writing on Ecuador, Cremers et al.
(2005) *Nat. Resources Forum*, 29, 1 judge
that bottom-up processes of awareness,
capacity-building grassroots' claims,
collective action, and mobilization are
critical in establishing water rights,
together with governmental political will
and an enabling political environment.
In a Malaysian study, Johnson (2000)
Geog. J. 166, 3 shows that tertiary
intervention has increased the capacity
of the farmers to unofficially control the
distribution and supply of the water
resource, resulting in a significant over-
supply of water; inefficient water use;
and a reduction in yields without a
reduction in incomes. Lam (1994), cited

by Kurian and Dietz (2004) *Nat.
Resources Forum* 28, 1, finds a negative
correlation between land-holding
inequality and irrigation management
performance. See Reinfelds et al. (2006)
Geogr. Res. 44, 4 on sustainable diversion
limits in river systems.

isentropic (isentropic surface) In
meteorology a surface of constant
potential temperature. Isentropic
surfaces slope very gently upwards
towards the cold air in the presence of a
horizontal temperature gradient, and
winds flow along these surfaces. See
Ambaum (1997) *J. Atmos. Scis* 54, 4 on
the isentropic formation of the
tropopause. In **isentropic analysis**, the
forecaster looks at the atmosphere in
three dimensions, considering constant
potential temperature surfaces. Air
parcels move up and down these
surfaces; therefore, the forecaster can
see where the moisture is located and
how much moisture is available. Try the
UW Isentropic Analysis and Modelling
Group home page.

island arc An arcuate island chain,
mostly of volcanic origin, formed when
*oceanic crust plunges into the *mantle
where it undergoes *subduction. The
*magma thus formed creates a chain of
submarine volcanoes—eventual islands;
see Jicha et al. (2006) *Geology* 34, 8 on
magma flow in the Aleutian Island arc.
See also Takahashi et al. (2007) *Geology*
35, 3.

island biogeography The
relationship between area and species
number on islands, as an equilibrium
between immigration and extinction.
See R. H. MacArthur and E. O. Wilson
(1967) and R. Whittaker (1998).

Island biogeography is also relevant to
continents, when plant and animal
communities are effectively reduced to
islands in a sea of cultivation or

urbanization; Meadows (2001) *PPG* 25, 1 writes on *habitat islands*—distinctive communities in a sea of other types of community.

isodemographic map A cartogram which represents nations, states, counties, and so on, in proportion to the size of their population. See A. Cliff and P. Haggett (1988).

isogloss A line on a map which represents the geographical boundary of regional linguistic variants. By extension, the term also refers to the dialect features themselves. For example, in the USA north of the Mason–Dixon line, the word greasy is pronounced with an 's'; south of the line, it is sounded with a 'z' (L. Campbell 2004). Isoglosses are usually highly simplified representations and do not depict an abrupt transition. Sometimes, a **bundle of isoglosses** may occur, where a number of isoglosses lie close enough together to indicate a true *dialect boundary; see W. Wolfram and N. Schilling-Estes (1998).

isoline/isopleths Any line on a map joining places where equal values are recorded: an **isohyet** connects points of equal rainfall; an **isophene** connects places with the same timing of similar biological events; an **isotherm** joins places of equal temperature; and an

isotim is a line drawn about a source of raw materials or a market where transport costs are equal.

isostasy The continental crust of the Earth has a visible part above the surface and a lower, invisible one. The balance between these two is isostasy. **Erosional isostasy** Isostatic readjustment in response to unloading (denudation). It is a 'plausible mechanism' (Maddy et al. (2000) *Geomorph.* 33, 3–4) for differential crustal movement, which may result from erosional unloading and depositional loading. However, the lack of evidence for significant surface denudation may suggest that erosional isostasy may be no more than a positive feedback response to uplift initiated by other influences (Maddy (1997) *J. Quat. Sci.* 12). **Glacio-isostasy** results from the decreasing weight of glacier ice on glacier melting, with unloading leading to uplift (Lambeck (1995) *J. Geol. Soc. London* 152). **Hydro-isostasy** is similar to glacio-isostasy; here, the loading/unloading of the continental shelves due to changes in sea level may have been significant in driving continental shifts.

isosteric Of that part of the atmosphere having uniform density.

isotropic Having the same physical properties in all directions.

jet stream A twisty, narrow ribbon of air that circles the globe at a height of around 11 km, sometimes at 400 km/h. Jet streams form where there is a major variation of air temperatures within a short geographical distance. The westerly jet streams in the Northern Hemisphere lie where significant breaks occur in the tropopause.

The **polar-front jet stream** is a *frontal wind, blowing west to east just below the *tropopause, parallel to the surface fronts and moving with them, and picking up the air rising from the fronts. It is strongest at the 200–300 mb level, swinging between latitudes 40 and 60 °N. Its speed and location vary from day to day with the Rossby waves. The polar-front jet marks the polar front, effecting *convergence and *divergence in the upper air. Divergence is more frequent in the downstream, poleward sector of the jet core, and is associated with the development of extratropical cyclones. At the 'jet entrance', the pressure gradient steepens, and the wind becomes super-*geostrophic, leading to high-level convergence. The polar-front jets often 'steer' the movement of major low-level air masses.

A strong polar-front jet is associated with rapidly moving *depressions, but the jet stream and its accompanying depressions in the surface can be diverted by *anticyclones. The jet then splits into the polar front jet and

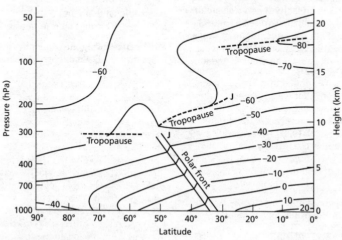

Jet stream. Identified by J on the chart

subtropical jet, which develop each year from March through to early Southern Hemispheric spring.

The **westerly subtropical jet** is at the poleward limit of the *Hadley cell, around 30°N and S. This is one of the most powerful wind systems on Earth, at times reaching speeds of 135 m s^{-1}, and it follows a more fixed pattern than the polar-front jet. It results from the poleward drift of air in the Hadley circulation and the *conservation of angular momentum. This jet is likely linked with the South Asian summer *monsoon.

The **easterly tropical jet** develops during the summer months at 15° N, and is strongest at the time of the summer monsoon. The **stratospheric subpolar jet stream** blows at a height of 30000m, being westerly in winter and easterly in summer.

Jevons Paradox This paradox, also known as the **rebound effect**, holds that as technology improves environmental efficiency, the gains will be eaten up by increased exploitation of the environment. For example, cars which are more fuel-efficient encourage more car travel. See Stapleton, Sorrell, and Schwanen (2016) *Energy Economics*.

joint A crack in a rock without any clear sign of movement either side of the joint.

jökulhlaup 'A "catastrophic" flood, often generated by an ice-dammed lake outburst'; for example, the volcanic eruption and concomitant catastrophic flooding from Grímsvötn, Iceland, in November 1996 (Tweed and Russell (1999) *PPG* 23, 1). See Alho (2003) *Geograf. Annal. A* 85, 3–4 on jökulhlaups and of land degradation.

Jurassic The middle period of *Mesozoic time stretching approximately from 190 to 136 million years BP.

(⊕) SEE WEB LINKS
• An overview of the Jurassic period.

just city Definitions vary, but most academic authors would agree that a just city is one where all its people have equitable and inclusive access to the opportunities and tools that allow them to be productive, to thrive, to excel, and to advance. Democracy, equity, diversity, growth, and sustainability are indispensable to the creation of just cities. But if the city is composed of diverse communities, conflict arises over whose definition of justice should prevail. P. Marcuse et al. (2009) provide an outstanding review of the different debates and theories that have emerged over time with respect to the 'just city'.

(⊕) SEE WEB LINKS
• MIT's 'Just Jerusalem'.

justice Geographical investigations of equity, fairness, and justice have centred on spatial equality, territorial justice, and minimum standards. However, what is just at one geographical scale for one set of workers might be unjust for other workers elsewhere; while aeronautical engineers working in Seattle might feel they are justly treated by Boeing, Mexican workers stitching seat-covers for Boeing aircraft may not enjoy equal pay and conditions.

American law is tethered to territory—that simply by moving an individual around in space, the rights that individual enjoys wax and wane; Osofsky (2008) *Villanova Law School* 53 explores '"wormholes" in the US legal system that transport people . . . into another timespace in which basic protections are absent'. He is referring to rendition flights. Of equal interest is Jeffrey's paper (2011) *TIBG* 36, 3, 344 on the Dayton Peace Accords over Bosnia and Herzegovina, where American legal processes remained remote from the locations where the crimes took place.

just-in-time system (JIT) A system of production which aims to deliver all the necessary inputs, such as raw materials, components, and labour, just in time for the appropriate stage of production. This is essentially a flexible system of manufacturing, linked with small production runs and flexible methods of production. It is largely credited with being a Japanese innovation. 'Dell [computers] has created a distinctive geography based upon a small number of geographically dispersed, strategically located assembly sites connected into close concentrations of suppliers needed for its just-in-time system' (G. Fields 2004).

Kamchatka current A cold *ocean current off the peninsula of the same name in Siberian Russia. See Cokelet et al. (1996) *J. Phys. Oceanogr.* 26, 7.

kame An isolated hill or mound of stratified sands and gravels which have been deposited by glacial meltwater. Lundqvist (2003) *Geografiska A* 85, 1 is recommended. **Kame terraces** are flat-topped, steep-sided ridges of similar *fluvio-glacial origin, running along the valley side. They are *ice contact features, formed between the side of a decaying glacier and a valley wall. **Moulin kames** form below *moulins (Carlson et al. (2005) *Geomorph.* 67, 3–4).

karre A collective name (pl. *karren*) for the shallow channels formed by solution on exposed limestone. **Kluftkarren** are enlarged joints; **rillenkarren** are very closely spaced small runnels; small radiating rillenkarren are the overflow of surface solution pans. **Rinnenkarren** are both longer and deeper, as much as 10 m in length and 0.5 m in depth. They may develop as a result of coalescence of small channels. Their walls are sharp, in contrast with the large, rounded hollows known as **rundkarren**. The types of karren can be explained as the results of the solution process under different hydro-dynamical behaviour (Ferrarese et al. (2003) *Geomorph.* 49, 1–2).

karst A landscape formed by the solution of soluble rocks like dolorite, limestone, and gypsum. Karst is a topography of sinking streams, sinkholes, and caves. **Labyrinth karst** are deep canyons of limestone formed by *carbonation. Initially the limestone shows the wide, deep tissues known as bogaz; these widen and deepen into long gorges known as karst streets, cross-cut with lines of erosion. The remnant of this carbonation is **tower karst**. See Zhu and Chen (2006) *Speleogen. & Evolution Karst Aquifers* 4, 1 on **cone karst**, tower karst, and tiankengs (giant dolines), and Harmon et al. (2006) *GSA Spec. Paper* 404 for perspectives on karst geomorphology, hydrology, and geochemistry. Karstic features do occur in non-carbonate rocks.

kata- From the Greek *kata-*, down-sinking, as in a *katabatic wind. The term is also used to describe the sinking of air in the warm sector of a *depression at a cold or a warm **kata-front**, bringing about a large-scale inversion of temperature at the fronts, which are fairly inactive. At a kata-warm front, cloud development is limited to cirrus and high stratus, and precipitation is restricted to light rain; at a kata-cold front strato-cumulus is common, and precipitation is similarly moderate.

katabatic wind A gravity-driven atmospheric current, which is forced by cooling air adjacent to a sloped surface. Descending, *adiabatically warmed katabatic winds are *föhn winds. Cold katabatic winds result from the slumping down of very cold, and hence dry, air. Coastal Antarctica is dominated by katabatic gales. The gentler katabatic flows of hill slopes encourage *frost hollows.

kettle hole Large masses of ice can become incorporated in glacial *till and may be preserved after the glacier has retreated. When one of these bodies of ice finally melts, it leaves a depression in the landscape; a kettle hole. **Kame and kettle topography** is hummocky terrain evolved by melt-out, comprising pitted or kettled outwash (*sandur).

Khamsin *See* LOCAL WINDS.

kinetic energy The energy of motion; in geomorphology, the energy used by wind, water, waves, and ice. Kinetic energy for channelled flow is defined as:

$$\frac{MV^2}{2}$$

where M is the mass of water, and V is the mean velocity.

kinship Family relationships: connection by blood, marriage, or adoption. Kinship networks can form an integral part of the individual's meaning and orientation in the world; see Megoran (2006) *Pol. Geog.* 25, 622.

knick point (nick point) A point at which there is a sudden break of slope in the *long profile of a river. In areas of uniform geology, the presence of a knick point may be evidence of *rejuvenation; the river is forming a new, lower profile cutting first from the mouth of the river and working upstream as *headward erosion takes place. See Crosby and Whipple (2006) *Geom.* 82. A **coastal knickpoint** occurs when a river or stream channel ends at a raised sea-cliff (Limbard and Bernard (2018) *Geom.* 306).

A **knickzone** is a steep reach caused by more resistant lithology, by an increase in shear stress downstream of a confluence, or by surface uplift (Bishop et al. (2005) *ESPL* 30, 6).

Knickzones form in response to base level fall, which is ultimately driven by *eustasy, *drainage capture, and *tectonics.

knowledge-based organizations Those organizations that successfully create new knowledge as the basis for new products and services, usually associated with large urban agglomerations. Geographical proximity is especially important for industries that rely on a synthetic or symbolic knowledge base, since the interpretations of the knowledge they deal with tend to differ between places. Industries drawing on an analytical knowledge base, which rely more on scientific knowledge that is codified, abstract, and universal, are less sensitive to geographical distance. See Martyn and Moodysson (2011) researchgate.net. Today, knowledge is considered among the most important resources, and an essential driving force of economic growth. The **knowledge-based economy (KBE)** generates information rather than goods and services. The production of information and of knowledge-intensive goods and services may be characteristics of post-industrial societies or a product of globalization.

Not everyone sees the rise of the KBE as beneficial. Von Osten in P. Spillmann, ed. (2004) points out that 'controlled access to knowledge goods and information . . . creates new global differences in power, new forms of resistance and subversive practices', and Chi-ang Lin (2006, *Eco. Econ.* 60, 1) is troubled by the persistent emphasis on knowledge and economic growth at the expense of poverty reduction and environmental conservation.

knowledge flows The mobility and migration of the highly skilled in knowledge-based economies is interwoven in complex ways with flows of knowledge. The internet has greatly contributed to the diffusion of codified knowledge and technology to China and the Asia Pacific Rim, but the distribution of knowledge is spatially uneven. Egger et al. (2007) *Ann. Reg. Sci.* 41 introduce factor mobility in the **knowledge-capital model** to explain why European nations are less specialized than US regions.

knowledge, geographies of These consider the unevenly distributed nature of cognition. See Vallance (2007) *Geog. Compass* 1, and Ibert (2010) *Regional Studs* 41, 1, on a geography of knowledge creation. See also Rutten (2017) *PHG* 41, 2 on the socio-spatial dynamics of knowledge production.

Kondratieff cycle An economic theory that states that Western capitalist economies are volatile as they expand and contract over the years. In contrast with what is referred to as the business cycle, the Kondratieff Wave (1935, *Rev.*

Econ. Stats 17) holds that these fluctuations are in fact part of much longer cycles known as 'super cycles' that last between 50 or 60 years or longer depending on factors such as technology, life expectancy, etc. and thus must be examined over their entirety to be best understood.

Each cycle lasts 50–60 years and goes through development and boom to recession. The first cycle was based on steam power, the second on railways, the third on electricity and the motor car, and the fourth on electronics and synthetic materials. I think there is now a fifth cycle, based on *knowledge. Alas, Smith (2003) *PHG* 27, 1 calls the Kondratieff cycle one of the 'big and spectacular accounts of the world', exaggerated in order to simplify, pretending that the partial is the whole.

Kuro Shio A warm *ocean current which, fed by the North Equatorial current, runs from the Philippines to Japan, thence feeding into the North Pacific current.

labour Manual or intellectual work, which is one of the *factors of production.

labour, geographies of The study of the spatial contexts in which workers live, and organize, and work.

Major themes are: the role of work and employment in contemporary capitalist society; workers' socio-geographical positionality within and beyond the workplace; and workplace structures and identity.

Gritsai (2005) *GaWC Res. Bull.* 162 and Winther (2001) *Urban Studs* 38 write on the evolution of a changing spatial division of labour.

labour geography Putting workers and their practices and interests at the heart of geographical analyses. Try Strauss (2020) *PHG* 44, 1.

labour intensive companies seek to find more profitable locations for their activities across the globe. This locational change has its greatest impact on 'intermediate' countries (such as Greece) which face competition from advanced economies on quality, and from developing countries on price, but the threat of capital flight has allowed corporations to play workers and nation-states against each other, undermining their bargaining power.

The **labour market** is the mechanism whereby labour is exchanged for material reward. All labour markets are territorially constituted, the results of the relationships between home, residential

setting and workplace, and between production and reproduction, and the geographical scales over which they operate have become increasingly intertwined and interactive. **Segmented labour market theory** sees the labour market as composed of self-contained sub-markets. Workers comprise different classes, genders, races, nationalities, and other groups that can become segmented. These segmented labour markets are socially and politically constructed, producing varied outcomes for workers.

Monastinotis (2007) *Area* 39, 3 presents a set of **labour market flexibility** indicators for the UK. The degree of labour market flexibility differs between high and low unemployment regions.

labour theory of value The Marxist contention that the value of a product reflects the amount of labour-time needed to make it. If the capitalist pays low wages that do not reflect the labour expended, he will obtain surplus capital. This may be seen as exploitation, which can lead to class conflict. Brown (2008) *Camb. J. Econ.* 32, 1, 1 25 is really useful.

labyrinth karst Deep canyons of limestone formed by *carbonation. The wide, deep fissures known as bogaz widen and deepen into long gorges (*karst streets*) with other, cross-cutting, lines of erosion. The remnant of this carbonation is *tower karst. In the late stages of evolution, labyrinth karst is replaced by limestone towers (Brook and Ford (1978) *Nature* 275).

laccolith An intrusion of igneous rock which spreads along bedding planes, forcing the overlying strata into a dome. Rocchi et al. (2002) *Geology* 30, 11 delineate the two-stage growth of laccoliths at Elba Island, Italy.

lacustrine Of lakes, especially in connection with sedimentary deposition. **Lacustrine plains** result from the in-filling of a lake. Soil parent materials are usually fine grained, well sorted, and often *varved. Ground surfaces are level to gently inclined and slightly concave. Landform elements include lakes, playas, some ox-bows, and some lagoons. For **lacustrine terraces**, see Korotkii et al. (2007) *Russ. J. Pacific Geol.* 1, 4.

lagoon A bay totally or partially enclosed by a spit or reef running across the entrance, known in the Baltic as a *haff*; see Miotk-Szpiganowicz et al. (2007) *2nd MELA Conf. Abstr.*

lag time (lagged time) The interval between an event and the time when its effects are apparent; for example, in the storm *hydrograph.

lahar A rapidly flowing, high-concentration, poorly sorted sediment-laden mixture of rock debris and water (other than normal streamflow) from a volcano. Lahars most commonly occur when a crater lake or an ice-dammed lake suddenly overflows; perhaps because of an eruption, the collapse of a dam, heavy rain, snow melt, or the mixing of a *nuée ardente with lake water. See (28 Sept. 2007) *GeoNet* on the lahar from Mt Ruapehu, New Zealand. Berti and Simoni (2007) *Geomorph.* 90, 1–2 propose a method for delineating lahar-hazard zones in volcanic valleys.

laissez-faire economics The view that a *market economy will perform most efficiently if it is free from government intervention, and is subject only to market forces. This view takes no account of environmental degradation

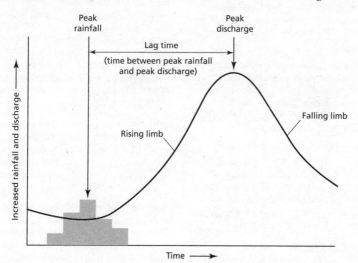

Lag time

or social justice (see Antle (1993) *Amer. J. Agric. Econ.* 75, 3). 'The free-market is not . . . a natural state of affairs which comes about when political interference with market exchange has been removed. In any long and broad historical perspective the free market is a rare, short-lived aberration. Regulated markets are the norm, arising spontaneously in the life of every society. The free market is a construction of state power' (J. Gray, 1998). Gray shows convincingly that laissez-faire capitalism has a very ambivalent relationship to democracy as it demands serious social control; Vojnovic (2003) *Geograf. Annal. B* 85, 1, in a study of Houston, Texas—'the archetype laissez-faire city'—finds that, despite the local laissez-faire rhetoric, government intervention in Houston's growth has been vital.

laminar flow A non-*turbulent flow where the movement of each part of the fluid (gaseous, liquid, or plastic) has the same velocity, with no mixing between adjacent 'layers' of the fluid. It may be seen at low velocities in a smooth, straight river channel (see Smith et al. (2007) *PPG* 31, 4).

land breeze A wind blowing from land to sea (an offshore wind) which develops in coastal districts towards nightfall. Pressure is higher above the land than above the sea, as the land cools more rapidly in the evening, and air therefore moves seawards along the *pressure gradient.

land capability The potential of land for agriculture and forestry depending on its physical and environmental qualities. The main factor investigated is soil type, but climate, gradient, and aspect are also considered. Present land use is not taken into account.

(⊕) SEE WEB LINKS
• A model of changes in land capability for Tasmania.

land classification Land classification is essential for geographers, planners, and, increasingly, for environmental scientists. There is no single correct way to classify land, and all classifications are subjective. Thus, the quality of the classification depends on the skill of the interpreter (Lofvenhaft et al. (2002) *Landsc. & Urb. Plan.* 58, 2–4). See the Food and Agriculture Organization of the United Nations Land Cover Classification System.

(⊕) SEE WEB LINKS
• The FAO Land Cover Classification System.

land consolidation A type of *land reform which aims to give each farmer one relatively large plot of land rather than scattered, small parcels of land. Issues in land consolidation include: the consent of the peasants; an assessment of the value of current farmland holdings; consolidation as the promoter of rural development; the assignment of new farmland for each household; the application of modern technologies; the control of corruption; an appeal system; the reimbursement of expenses by local or national government; and, controversially, population control. Kirit and Nagarajan (2004) *Occ. Paper, Nat. Bank Agric. & Rural Dev.* 31 record huge benefits from land consolidation in the Indian state of Tamil Nadu.

land cover The observed biophysical cover of the Earth's surface. Strictly speaking, it describes vegetation and man-made features, and omits bare rock and water, although, in practice, these elements are often included.

land degradation A noted decrease in, or any adverse influence on, the *carrying capacity of land, revealed by: decreased soil fertility; increased soil erosion; removal of vegetation cover; and general negative changes in land use which compromise the supply and quality of ground and surface water, biodiversity, the carbon-oxygen cycle, and general air quality.

In Australia, land clearing and the use of European farming systems have been major causes of severe land degradation, including *salinization and erosion. Douglas (2006) *Geog. Res.* 44, 2 links globalization in South-East Asia with land degradation, but finds that local changes may have a greater impact. Davis (2006) *Geog. J.* 172, 2 finds that land degradation in dryland Morocco is commonly blamed on overgrazing, despite existing evidence that ploughing marginal lands and over-irrigation are the primary drivers: 'all too often the outcomes of neoliberal reforms on the environment . . . include increased pollution of air, earth and water, and land degradation in the form of deforestation, soil exhaustion, salinization and erosion.'

Mambo and Archer (2007) *Area* 39, 3 use remote sensing to detect and map susceptibility to land degradation, and Kiunsi and Meadows (2006) *Land Deg. & Dev.* 17 present a land degradation model. See also A. Conacher (2001).

land economics The study of land use and of the factors which influence and shape it. Land values are central to this study as land use and land values are interlinked; see the journal *Land Economics*.

Land Information System A system for the capture, storage, manipulation, analysis, and display of land-use data. See Karikari et al. (2003) *Progress in Devt. Studs* 3, and try D. C. Lee on geospatial data sharing in Saudi Arabia.

landless workers' movement (landless movement) In Brazil, a social movement aiming to fight for access to the land for poor workers in general,

land reform, and to counter social issues hindering access to land, such as racism, sexism, media monopolies, and unequal income distribution. Ondetti (2006) *Latin American Pols & Soc.* 48, 2, 61 is very helpful.

land-locked state A nation with no access to the sea. Globally, there are twenty-six land-locked states ranging in size from the tiny Vatican City to Mongolia. Geography has put land-locked countries in a position of disadvantage. To be land-locked is more detrimental for a country's development now than at any time in the past; see A. Chowdhury and S. Erdenebileg (2006).

land management The way land is used is driven by the interplay of economic, social, and environmental factors. Land management is about finding the right balance of these, often competing, factors that allows sustainable land use. Whenever land is put to use, it is subject to stresses that can be quite high, as in the case of intensive agriculture. These stresses may not only affect the biodiversity of the land, such as through the loss of hedgerows, but can also have widespread social impacts—for example, by detracting from the land's natural beauty. Research on multifunctional land uses looks at land from a holistic perspective to understand how different patterns of use cause different stresses, what the consequences of these stresses are, and how they can be managed or corrected.

(⊕) SEE WEB LINKS
• EC Environment Research website.

land reform A sweeping change in land tenure. It usually involves the breaking-up of large estates and the widespread redistribution of the land into smallholdings, but may also be *land consolidation. In the Republic of

South Africa, the need for land reform arose from the racially discriminatory laws and practices which were in place for the largest part of the twentieth century. See Kloppers and Pienaar (2014) *AJOL* 17, 2.

land rent theory Any theory that explains rental value of parcels of land, for land rents are not exclusively determined by the intrinsic qualities of the land itself. *See* BID-RENT THEORY; VON THÜNEN MODEL.

land rights The legal entitlement to land. Indigenous and community lands are typically held under informal customary arrangements, which can leave the land vulnerable to outside commercial interests. Communities may therefore attempt to register their land rights with their own government. However, this process is slow and complicated, while commercial enterprises can gain title to land relatively quickly, and find shortcuts around regulations. See 'The Scramble for Land Rights' (2018) *World Resources Institute*.

landscape All the visible features of an area of land, the appearance of an area, or the gathering of objects which produce that appearance: 'the expression of interaction between humans and their environment' (C. Sauer 1925). See Minca (2007) *PHG* 31, 2; Revil (2007) *TIBG* 32, 2; and Henderson in C. Wilson and P. Groth, eds (2003) on landscape as social space.

A **landscape of governance** may be seen as the spatialization of governance, resulting from the interplay between natural-spatial conditions of place, public/private actors, and policy responses. It may not be the territorially defined spaces of the post-war nation states, so that governance increasingly has to rely on formal and informal

supra- and sub-national institutions of the nation states if it is to attain legitimacy. See Hashim (OECD 2002) on Malaysia, and Robinson and Shaw (2001) *Reg. Studs* 35, 5 on north-east England. With **landscapes of care**, 'landscape' can imply an overview, a climate, or indeed a geography; see Jonsson (2020) *Int. J. for Equity in Health* 19, 171 on landscapes of care and despair for Swedish youth.

landscape connectivity The way in which the compartments of a landscape fit together in a catchment. **Structural connectivity** refers to the physical relationship between landscape elements, while **functional connectivity** describes the degree to which landscapes actually facilitate or impede the movement of organisms and processes. The more connected a landscape, the greater the movement among its existing resources. The degree of landscape connectivity influences the probability of colonization, the risk of extinction, gene flow, local adaptations, and the potential for organisms to relocate in response to climate change. Go to the Institute for European Environmental Policy (2007) for guidance on the maintenance of landscape connectivity features for wild flora and fauna.

landscape conservation The planning and management of the scenic and wildlife resources in geographical and environmental systems. See Pinto-Correia (2000) *Landsc. & Urb. Plan.* 50, 1–3 on the possible integration of different interests and policies for nature conservation.

landscape ecology In ecology, a landscape is a composite land area comprising a cluster of interacting ecosystems that is repeated in similar form throughout. Landscape ecology specifically addresses the importance of the spatial element in ecological processes. The central themes are: detecting a pattern, and expressing it in quantitative terms; identifying and describing the agents of pattern formation; understanding the ecological implications of the pattern; spatially and temporally characterizing the changes in pattern and process; and managing landscapes to achieve human objectives. For ten differing definitions of landscape ecology, see the United States Regional Association for Landscape Ecology site.

landscape evaluation An attempt to assess the landscape in objective terms. Sometimes a consensus of views on the landscape is sought so that particular landscapes may be chosen as being outstandingly beautiful. Landscape description studies try to identify important items such as topography or buildings. Some kind of ranking method may be attempted to compare one landscape with another. See Hessburg et al. (2014) *USDA Forest Service*, on landscape evaluation and restoration planning.

landscape evolution The processes through which the landscape emerges. Earth's landscapes reflect the interaction of climate, tectonics, and denudational processes operating over a wide range of spatial and temporal scales; these processes can be considered catastrophic or continuous, depending on the timescale of observation or interest.

Perron and Fagherazzi (2012) *ESPL* 37, 1, 52 suggest a sequence of landscape evolution. Firstly, variability in initial conditions can give rise to steady-state landscapes in which the characteristics of individual landforms vary about the mean, even in a homogeneous system with constant forcing. Secondly, landforms can have a range of mean equilibrium dimensions, either because

different initial conditions evolve to different 'attractor' states, or because landscapes evolving from different initial conditions towards a single state converge too slowly to reach equilibrium under natural conditions. Thirdly, a landscape that experiences a perturbation or a change in process may begin to evolve towards a different equilibrium state, potentially leading to *hysteresis or rapid changes in topography.

A **landscape evolution model** is a mathematical theory describing how the actions of various geomorphic processes drive and are driven by the evolution of topography over time. Tucker and Hancock (2010) *ESPL* 35, 1, 28 provide a useful summary. Dymond and Rose (2011) *Geomorph.* 132, 1–2, 29 offer an alternative phenomenological approach to modelling landscape evolution, and compare their model to the highly erodible Waipaoa catchment in New Zealand. There is good agreement between the predicted and observed topography.

landscape matrix The most extensive and connected landscape type, playing the dominant role in the functioning of *landscape ecology. The characteristics of the matrix are the density of the patches (porosity), boundary shape, networks, and heterogeneity. The type of landscape matrix surrounding a *patch can mitigate the negative effect of habitat isolation for a given species, according to its degree of matrix habitat use (Barbaro et al. (2007) *J. Biogeog.* 34, 4). V. Ingegnoli (2002) describes three matrix types: continuous, with a single, dominant element type; discontinuous, with a few co-dominant element types; and web-shaped, with connected corridors of prevailing functions.

landscape morphology In *geomorphology, the form and spatial structure of the *landscape. Landscape morphology has a direct influence on water movements, soil, physical and chemical properties, and on the productivity of the vegetation cover. In *human geography it is the physical matter of the landscape; its shaping and reshaping, in which social structures and cultural worlds are enfolded.

landscape pattern The elements that are present in a landscape, in their proportions/relative amounts, and the way these elements are arranged. At every scale, real landscapes contain complex spatial patterns that vary over time; landscape pattern analysis quantifies these patterns and their dynamics. Spatial point patterns use the geographic locations of entities; surface patterns represent quantitative measurements that vary continuously across the landscape—there are no explicit boundaries; categorical/ thematic/choropleth map patterns represent data as a mosaic of discrete patches; and linear network patterns are self-explanatory. The go-to text is M. G. Turner and R. H. Gardner (2015).

*Landscape ecology seeks to associate ecological processes with landscape pattern.

landscape preference It is argued that most cultures have a preferred landscape: for example, Eleftheriadis et al. (1990) *Env. Manage.* 14, 4 find that, out of six European groups, Greeks preferred sea landscapes most, and Italians, Austrians and Yugoslavs (*sic*) least. Herzog et al. (2000) *En. & Behavior* 32, 3 observe that young children display higher landscape preference, since they see landscapes as 'playscapes', while teenagers are more interested in social concerns.

landscape sensitivity A measure of the resilience, or robustness, of a landscape to withstand specified change arising from development types or land management practices, without undue negative effects on the landscape, its visual aspects, and its value; the potential for landscape dysfunction as a result of human impacts; the likelihood of change. See Christine Tudor (2019). The Landscape East landscape-sensitivity analysis and recommendations set out a straightforward method by which decision-makers can assess the sensitivity of regional landscape character types to different change scenarios.

landscape studies The examination of the structure and organization of landscapes; of interrelationships between place-making, personal and social memory; of landscape as the location of cultural and ecological patterns, processes and histories; of the social life of landscapes—their reception by different societies; of 'how and why places come to be understood and experienced as thresholds of time-space in different societies, including how places constitute geographies of belonging through and beyond urban and national space' (Till (1999) *Ecumene* 6).

Landscapes may be viewed not simply as 'scenes' into which humans are inserted, but rather as the products of human activity, shaped through and shaping cultures. An important development in the study of the cultural landscape is the incorporation of social theory: to consider landscapes as part *of a process* of cultural politics, rather than as the outcome of that process. 'Within landscapes are particular sites— monuments or markers—which facilitate and direct the process of "collective memory" and through which social groups situate their identities in time and place' (Inwood et al. (2008)

Soc. & Cult. Geog. 9, 4). See *Site/Lines*, the journal and online forum of the Foundation for Landscape Studies.

(((⊕))) SEE WEB LINKS
• *Site/Lines*: the social life of landscapes.

Landschaft A concept of landscape which attempted to classify landscapes, usually distinguishing between the natural and the cultural landscape. See A. Baker (2003) *Geography and History: Bridging the Divide*.

landslide A form of *mass movement where the displaced material retains its form as it moves. Landslides may be divided into slides, falls, flows, topples, and lateral spreads. Probably the world's largest landslide occurred in south-west Iran in 1937, when a segment of the Kabir Kuh ridge, about 15 km long, 5 km wide, and 300 m thick, slid off the mountain, with enough momentum to travel 20 km.

Landslides are prompted by an increase in *pore pressure through snow melt, through precipitation, and through spring action. These all reduce the friction which binds the mass to the slope.

(((⊕))) SEE WEB LINKS
• Worldwide database of empirical rainfall thresholds for the possible occurrence of landslides.

land surface process models
These describe the energy, water, carbon, and nutrient fluxes on a local to regional scale using a set of environmental land surface parameters and variables. In principle many of these inputs can be derived through remote sensing. See Zhao and Li (2015) *Advances in Meteorology*.

land tenure The nature of access to land use. Common forms of land tenure are owner-occupied farms, and tenancies which basically involve

payment (in the form of labour, cash, or *share-cropping) to the landlord from the tenant. A *plantation is owned by an institution and uses paid labour. Collectives may own land together and work together, sharing any profits.

land use zoning The segregation of land use into different areas for each type of use: agricultural, industrial, recreational, and residential. Planners have sought to pursue *sustainable development using more mixed land use zoning in order to reduce the demand for travel, and greater coordination between transport and land use planning, including support for public transport. See Greed (2006) *Tijdschrift* 97, 3 on the problems for women in cities divided by traditional land use zoning, and Degg and Chester (2005) *Geog. J.* 171, 2 on land use zoning in the mitigation of seismic/volcanic hazards.

La Niña A state of unusually cool sea surface temperatures in the western Pacific, which are associated with lowered precipitation in the southern USA and western South America, but an increase in the frequency of *tropical cyclones in the Atlantic, and heavier rainfall in Indonesia and Australia. La Niña can be seen as the other extreme from *El Niño.

'There is strong statistical and modelled evidence that persistent, La Niña-like, cooler sea surface temperatures . . . produce multiyear droughts not only in the Great Plains and the Southwest, but also in the Mediterranean region of Europe, the Pampas region of South America, the steppes of Central Asia, and the outback of Western Australia' (Goodrich (2007) *Geog. Compass* 1).

lapili *See* PYROCLAST.

lapse rate In meteorology, the rate at which stationary or moving air changes temperature with a change in height. *See* ADIABAT; ENVIRONMENTAL LAPSE RATE.

latent heat The quantity of heat absorbed or released when a substance changes its physical state at constant temperature. The release of **latent heat of condensation** in the rising air of a *hurricane is the chief force fuelling that meteorological phenomenon. See Grossman and Rodenhuis (1975) *Monthly Weather Rev.* 103, 6. The **latent heat flux** is the flux of heat from the Earth's surface to the atmosphere that is associated with evaporation or transpiration of water at the surface and the subsequent condensation of water vapour in the troposphere. It is an important component of Earth's surface energy budget.

lateral accretion The build-up of sediments, as in a recurved spit (Kumar and Sanders (1974) *Sedimentol.* 21, 4), or a meander (Brooks (2003) *Geomorph.* 54, 3–4).

lateral erosion Usually of rivers; erosion of the banks rather than the bed. In a stream or river, it results in undercutting of the banks or terrace formation.

lateral moraine *See* MORAINE.

laterite Thick, red, and greatly weathered and altered strata of tropical ground. *Horizons are unclear and the nutrient status of the soil is low. Laterite is soft but gets brick-hard when exposed to the air. **Lateritic soils** (*latosols) are tropical/equatorial soils characterized by a deep weathered layer from which silica has been *leached, a lack of *humus, and an accumulation or layer of aluminium and iron sesquioxides. **Laterization** is the formation of lateritic soils, taking place in warm, wet climates. High temperatures and heavy rain cause strong weathering of rocks and minerals, and water moving through the soil causes *eluviation and *leaching. Almost all of the by-products of weathering are translocated out of the soil profile by leaching and eluviation, except for compounds of aluminium and iron; the latter give laterite its red colouring.

latifundium A large farm or an estate (pl. *latifundia*), particularly in Latin America. The estate is farmed with the use of labourers, who sometimes lease very small holdings from the landowner. See Martinelli (2012) *EHES Working Paper* 20 on the links between latifundia and inequality and inefficiency in Italy.

latitude Parallels of latitude are imaginary circles drawn round the Earth parallel to the equator. The parallels are numbered according to the angle formed between a line from the line of latitude to the centre of the Earth and a line from the centre of the Earth to the equator. Those regions lying within the Arctic and Antarctic circles, having values of 66.5° to 90°, are termed **high latitudes**. **Low latitudes** lie between 23.5° north and south of the equator, i.e. within the tropics. **Mid-latitudes**, also known as temperate latitudes, lie between the two.

latosol A major soil type of the humid tropics with a shallow A *horizon but a thick B horizon of clay, sand, and iron and aluminium sesquioxides which, respectively, give it a red or yellow colour. Much of the silica has been leached, and latosols tend to be of low fertility.

Laurasia One of the two original continents which broke from the supercontinent, Pangaea, by *continental drift. As a supercontinent, the other original continent, Gondwanaland, lasted much longer than Laurasia, and was hotter.

lava *Magma which has flowed over the Earth's surface. The *viscosity of lava depends on its silica content, pressure, and temperature. Temperature is the most important factor. **Basic lavas** have a low silica content and flow freely; **acid lavas** are more viscous. *See* EXTRUSION.

law A theory or hypothesis which has been confirmed by empirical evidence. The **geography of law** considers law and its location, looking at the way spaces are constructed on the terrestrial and marine surface of the earth by means of the law. Braverman et al. (2013) *Buffalo Legal Studies Research Paper Series 2013—032* is excellent.

law of retail trade gravitation *See* REILLY'S LAW.

law of the sea A framework, agreed to by the majority of maritime nations, for administering the seas. It recognizes seven administrative zones: internal waters, the *territorial sea, the *contiguous zone, the *continental shelf, exclusive fishing zones of up to 200 miles from a nation's coastline, exclusive economic zones of the same extent as the fishing zones, and the high seas. *See* UNCLOS.

layer An image of one geographic dataset in a digital map environment. Roads, national parks, political boundaries, and rivers might be considered different layers on a road map, for example.

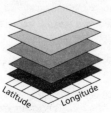

Census Tracts
Roads
Bus Routes
Shopping Centres
Industrial Sites

Latitude Longitude

layer

leaching The movement of water down the *soil profile. This results in the movement of cations, sesquioxides, *clay colloids, and *humus to the lower soil *horizons. Specific types of leaching include: lixiviation (the removal of the soluble salts containing metallic cations); the removal of *chelates; *lessivage; and, in tropical soils, *desilication.

lean production A method of production focused on getting the right things to the right place at the right time in the right quantity in order to achieve perfect work flow, while minimizing waste, being flexible, and able to change. Rutherford (2004) *PHG* 28, 4, 425 is good on this.

least developed country (LDC) To identify least developed status in a nation, UNCTAD—the United Nations Conference on Trade and Development—uses: low income (below $1018 in 2021); weak human assets, as measured through a composite Human Assets Index; and economic vulnerability, as measured through a composite Economic Vulnerability Index. Go to UNCTAD's Least Developed Countries Report.

lee The side (lee side) sheltered from the wind. Somewhat confusingly, a **lee shore** is the shore towards which the wind is blowing.

lee trough As a stable air column rises to cross a ridge, it shrinks in the vertical, therefore *diverging in the horizontal. This gives it a negative relative *vorticity. Accordingly, in the Northern Hemisphere, the flow is deflected anticyclonically: to the right. Since low pressure always lies to the left of an airstream in the Northern Hemisphere, the pressure is lower on the lee of the mountain. See Weisman (1990) *Monthly Weather Rev.* 118, 4.

lee wave A wave motion in a current of air as it descends below an upper layer of stable air, after its forced rise over a mountain barrier, which sets up vertical oscillations (also called a **standing, rotor, hill**, or **mountain wave**). The wavelength is *c*.5–15 km with an amplitude of *c*.500 m. Lee waves are often indicated by higher wind speeds and *lenticular clouds.

legal geography The study of the connections between location, and the law and legal system: where the law happens, and how social and cultural systems affect it. All judgments occur in both a time and place, so they are directly or indirectly influenced by what is occurring at that time, and in that place in society. In the United States, for example, state-by-state variations in laws on abortion and firearms are key concerns for legal geographers. See Bennet and Layard (2015) *Geog. Compass* 9, 7.

leisure, geography of The study of the demand, supply, planning, destination management, and impacts of tourism and recreation. Geographies of leisure study the relationships between leisure and other social practices and behaviours related to human movement; the spaces produced by leisure activities (gardens, heritage sites, parks, theme parks, and so on) and the meanings given to these spaces, together with the spatial patterns of people's behaviour in their free time. See Özger and Klaesson (2015) *Leisure Studies* 36, 2, on the new economic geography of leisure services, and Hubbard et al. (2008) *PHG* 32, 3 on the urban geography of 'adult entertainment' (yes, really).

lenticular cloud A lens-shaped cloud formed above a peak as air rises over mountain barriers and condensation occurs. *See* LEE WAVE.

lenticularis *See* CLOUD CLASSIFICATION.

lesbian geographies The ways in which lesbians organize themselves in both urban and rural landscapes. The spatialization of sexual lives is always gendered, and lesbians have certain cities, neighbourhoods, and small towns in which they are more likely to live; see Amin Ghaziani (2015) *Contexts*. Ferreira (2014) *Gender, Place and Culture* considers how lesbians and bisexual women negotiate same-sex displays of affection in public spaces. You may need to contact Ferreira to get this. See *Soc. & Cult. Geog.* (2007) 8, 1 for a collection of useful papers on this topic.

less economically developed country (LEDC) A country with low levels of economic development. Indicators include high birth, death, and infant mortality rates (typically over 20; over 30; and over 50 per thousand, respectively); more than 50% of the workforce in agriculture; and with low levels of nutrition, secondary schooling, literacy, electricity consumption per head, and GDP per capita—generally below $US1000 per capita. The word 'economically' was included in this term in order to redress perceptions of the people of financially poorer nations as uncivilized, irrational, ignorant, and unskilled.

lessivage The translocation of *clay colloids in a soil, with no change in their chemical composition. Characteristic features of lessivage are often observed in the soils without hydrological barriers hampering, or preventing, the vertical migration of soil water and mass transfer processes (Zaidel'man (2007) *Eur. J. Soil Sci.* 40, 2).

levée A raised bank of alluvium flanking a river. The bank is built up when the river dumps much of its *load during flooding. New Orleans was founded on a natural levée; elevations as high as 19 feet above sea level provided the driest land for settlement. French colonists created their own levées, and ever since permanent settlement began, residents have been expanding the levées. The US$14 billion network of levées and floodwalls that was built to protect greater New Orleans after Hurricane Katrina was a seemingly invincible bulwark against flooding. But the system will stop providing adequate protection in as little as four years because of rising sea levels and shrinking levées (*Scientific American* (2019)).

lexical diffusion In geolinguistics, phonological changes proceeding through the lexicon, word by word. See Britain (1997) *Paper Presented at NWAV-26*.

LGM last glacial maximum.

life course geography The way ideas of the life course are represented spatially. Key tenets of the life course approach are: socio-historical and geographical locations; the timing of lives; continuity or change; social ties to others; human agency and personal control; and how the past shapes the future. Over the course of a life, changing attachments to place have profound effects on how we conduct ourselves, and on how we constitute the spaces in and through which our lives are experienced. One technique is the use of life maps to enable people 'to be the tellers of their own stories' (Worth (2011) *Area* 43, I 4, 405). This method reinforces the *participatory focus of the research. See Pearce et al. (2018) *TIBG* 43, 4. An alternative or closely matching term is **life transition**, and the December 2011 issue of *Area* devotes much of its space to this latter term. Try Eliason et al. (2015) *Soc. Psychol. Q.* 78, 3.

life expectancy The average number of years which an individual can expect to live in a given society, normally derived from a national *life table. Life expectancy is usually given from birth but may apply at any age, and because, in all societies, mortality rates tend to be rather high in the first year of life, life expectancy at birth is usually significantly lower than at 1 year old. Women consistently have a longer life expectancy than men, especially in *more economically developed countries where the risks of childbirth are less than those in *less developed countries.

The lower life expectancies for less economically developed countries generally reflect high *infant mortality rates, but by the age of 70, the years of life remaining to an individual are, globally, very similar. Thus, the strong correlation between *GDP per capita and life expectancy becomes weaker as the age of an individual increases.

life space The limited time and space which an individual has in which to pursue a necessarily limited range of opportunities. Life space is the interaction of the individual with her or his behaviour setting. For example, the allocation of differing amounts of life space to different places will vary by the social class of a city's inhabitants.

life table A summary of the likelihood of living from one age to any other. In a life table, a hypothetical *cohort of 100 000 births is set up and then the loss by deaths is shown for each year of life. Averages of losses are calculated for a given year, and from this the actual diminution of the cohort is shown.

life world (lifeworld) The taken-for-granted dynamic of everyday experience that largely happens automatically, without conscious attention or deliberate plan. See Lorimer (2019) *TIBG* 44, 2.

lift force The upward force produced when fluid rises over a particle. In watercourses, the particle moves up from the bed into the flow when the lift force exceeds the gravitational force provided by the mass of the particle. See Cornelis (2004) *Geomorph.* 59, 1–4.

light industry The manufacture of relatively small articles (toasters as opposed to girders), using small amounts of raw materials. In consequence, the *material index is low, and such industries are more *footloose than heavy industries.

lightning An emission of electricity from cloud to cloud, cloud to ground, or ground to cloud, accompanied by a flash of light. It results from variations of electrical charge on droplets within clouds and on the Earth's surface. This variation may be caused by the break-up of raindrops, the splintering of ice crystals, or differences between splintered ice crystals and soft hail. As a *cumulus cloud develops, the frozen upper layer becomes positively charged, and most of the cloud base negatively charged, with positively charged patches. The negative charges are attracted to the earth, which has a positive charge. When the electrical field strength gets to about 1 MVm^{-1}, the electrical insulation of the air breaks down. The result is a *leading stroke*, or *stepped leader*, from cloud to ground, which creates a conductive path between them. The return stroke, from earth to cloud, follows the same pathway (only millimetres across), with a charge up to 10000 amps. The intense heat of the stroke engenders light, and a violent expansion of the air, making waves, heard as *thunder. Not all the negative charge may be released; there may be several return strokes, each prefaced by a downward dart leader which reactivates the channel. Where the path between ground and cloud is clearly visible, **forked lightning** is seen. The illumination of other clouds by a concealed fork is **sheet lightning**.

Gołkowski (2011) *Acta Geophysica* 59, 1, 183 investigates variations in the global atmospheric electric circuit.

Ball lightning has been described as a sphere of glowing light meandering through the lower air. Little is known about it, but there are a few fascinating 19th-century observations accessible from the American Meteorological Society website.

(()) SEE WEB LINKS
• American Meteorological Society.

limestone pavement A more or less horizontal, bare limestone surface, cut into by grikes (fissures) running at right angles to each other, leaving clints (upstanding areas of limestone) between them.

liminal In between, as in liminal geopolitics: between state and non-state, or official and unofficial diplomacy; see McConnell (2017) *TIBG* 42, 1, who outlines the advantages of liminality. J. Anderson (2012) uses this term to describe adolescents, who are in between childhood and adulthood.

limnion The lower layers in a body of water that are marked by low temperatures and insufficient light for photosynthesis. Levels of dissolved oxygen are low.

limnology The study of all the biological, chemical, meteorological, and physical aspects of freshwater ponds and lakes. See P. O'Sullivan and C. Reynolds, eds (2004, 2005) and Lau and Lane (2001) *PPG* 25, 2 on shallow lake ecosystems.

line In a *GIS, a one-dimensional feature or a feature that has length. Many line features also have width in real life, but as with points, GIS software assumes that they don't (M. N. DeMers 2009).

linear wave theory Linear wave theory was first presented by Airy (1845) to provide methods to calculate wave characteristics such as height, period, and speed using a sinusoidal model of the waveform. Based on the ratio of water depth, h, to deep-water wavelength L_o, three different wave regions can be identified: deep water ($h/L_o > 0.5$), intermediate water ($0.5 > h/L_o > 0.05$), and shallow water ($h/L_o < 0.05$). This is exceptionally well explained by Masselink in J. Holden, ed. (2012). Linear wave theory relies on the assumption of closed circular particle orbits, but this assumption breaks down in the transition zone approaching shore, where waveform changes occur.

line-haul costs The costs of transporting goods over a route, but not loading or unloading. The selection of any shipment route depends on the total sum of its network access, line haul, interlining, terminal transfer, and network exit costs. Sinha and Thykandi (2019) *MIT* tell you, clearly, more than you might want to know. The 200-year-long collapse in transport time has now been replaced by a plateau in general **line-haul speeds**, but time/space continues to collapse locally, for example through new fixed links or high speed trains (Knowles (2006) *J. Transp. Geog.* 14, 6).

linguistic distance The degree of contrast between two *dialects, measured by the percentage of items differing from a fixed set. See Gooskens (2005) *DiG* 13 on the geographical and demographic determinants of linguistic distance, and Scapoli et al. (2005)

J. Theor. Biol. 237, 1 on the link between linguistic similarity and linguistic distance.

link The route or line joining two *nodes.

linkages Flows of inputs and outputs to and from a manufacturing plant in association with other plants. Movements of matter are **material linkages** as opposed to **machinery** and **service linkages** (such as information, advice, and maintenance). Individual plants are also tied together by **forward linkages**—supplying customers, and **backward linkages**—with their suppliers. **Horizontal linkages** occur between plants which are engaged in similar stages of a manufacturing process.

literary geography The combination of the study of literature and a geographical perspective; how and why the geographical setting is important in the plot or mood of a text or texts, and how the setting itself advances the plot, as in E. M. Forster's *A Passage to India*; the study of textual meaning as it differs across space. See Sharp (2000) *Area* 32 on fictive geographies, and the journal *Literary Geographies*.

lithification Processes by which sediments are converted into hard rock. These include the expulsion of air, or the suffusion into the rock of *cementing agents in solution, like quartz.

lithology The physical character of a rock or rock formation; its composition, structure, texture, and hardness. A **lithogenous** sediment consists of particles derived from the physical and chemical breakdown of rocks and minerals.

lithosphere The Earth's *crust, and that upper layer of the *mantle which lies

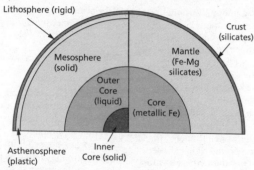

The Earth's constituent layers
Weber State University

above the *asthenosphere. The strength of the continental lithosphere is likely to be contained within the seismogenic layer; variations in the thickness of this strong layer determine the heights of the mountain ranges it can support. Constantin et al. (1996) *Geology* 24, 8 find that magmatic intrusion is a fundamental process in the construction of the oceanic lithosphere.

litter The layer of dead and dying vegetation found on the surface of the soil (some soil classifications assign the litter layer to the A *horizon, shown as A00). Cellulose, hemicellulose, and lignin make up 20–30, 30–40, and 15–40% of the total litter mass, respectively; the exact values are governed by plant type and age, onsite conditions, and yearly environmental variations (Yadav and Malenson (2007) *PPG* 31, 2).

Little Climatic Optimum The time, *c.* AD 750–1200, when Europe and North America were much warmer. See Nunn (2000) *NZ Geogr.* 56, 1 on the transition between the Little Climatic Optimum and the *Little Ice Age.

Little Ice Age The phase between AD 1550 and 1850 when temperatures were lower than they are at present, but not

globally. See Owens et al. (2017) *J. Space Weather and Space Climate* 7. **Little Ice Age glacierization** occurred over about 650 years and can be defined most precisely in the European Alps (*c.* AD 1300–1950) when extended glaciers were larger than before or since. **Little Ice Age climate** is defined as over a shorter time interval of about 330 years (*c.* AD 1570–1900) when Northern Hemisphere summer temperatures (land areas north of 20° N) fell significantly below the AD 1961–1990 mean (Matthews et al. (2005) *Geografiska A* 87, 1).

littoral drift The amount of sediment transported by *longshore currents.

lively commodities Commodities whose capitalist value is derived from them being alive and/or promising future life. Importantly, lively commodities do have a life of their own, which is not always recognized.

lixiviation *See* LEACHING.

load The matter transported by a river or stream. **Solution load** is dissolved in the water. **Suspension load** refers to undissolved particles which are held in the stream. On the river bed, the material of the *bed load jumps by *saltation, or rolls along the bed. The deposits forming a channel bed are known as **bed-material load**. Conventional theory is that meanders develop to an equilibrium form which is related to discharge and sediment load (Hooke (2007) *Geomorph.* 91, 3–4).

loam An easily worked, fertile soil, composed of *clay, *silt, and *sand, roughly in a ratio of 20:40:40. A **clay loam** has a clay content of 25–40%, a **silt loam** has more than 70% silt, and a **sand loam** has between 50 and 70% sand. Loams heat up rapidly, drain neither too slowly nor too easily, and are well aerated.

local For a term so frequently employed in geography, 'local' is difficult to define. If we consider 'local' as relating or restricted to a particular area, or one's *neighbourhood, the terms 'particular area' and/or neighbourhood need to be defined, so the problem is compounded. In terms of food production, Goodman (2003) *J. Rural Studs* 19 shows how the 'local' is associated with ecology, differentiation, and quality, with the latter being tied to relational ideas of tradition, trust and 'place'.

Most geographers acknowledge a continuum between local and global; thus 'whatever lure the local may hold for landscape geographers it must become one of understanding how particular places, particular landscapes fit into a larger, scalarly complex, mosaic' (Mitchell (2001) *PHG* 25, 2). And rather than positing the global and the local as a dichotomy, Ç. Keyder and his collaborators (1999) advocate a dialectic in which the global and the local are mutually constitutive.

local climate The climate of a small area such as a moorland or city—a mesoclimate—falling between a *microclimate and a *macroclimate. At this scale, such variables as local winds, *albedo, relief, slope, and *aspect are of considerable significance. *See* URBAN CLIMATE.

local economic development policy Enterprises promoted by the public sector, or by public–private partnerships, to attract new enterprises and investment. Local economic development is conditional in its geography on land use plans. Polèse and Shearmur (2006) *Papers Reg. Sci.* 85, 1 argue against using local economic development strategies as a means of arresting population and employment decline. See Bartik (2005) *Growth & Change* 36, 2 on local economic development incentives.

local food systems (LFS; **local food networks**) Local food systems include community gardens, community-supported agriculture, farmers' markets, food cooperatives, and community gardens, all of which may be described as direct agricultural markets, or **local food networks**. Local food systems will reduce food miles, and should make for fresher food, and support both localized production practices and local crops or livestock.

Anderson and Cook in J. M. Harris, ed. (2000) argue that local food systems can supplement and complement larger-scale food systems, for the realities of a local food system necessitate an integration of local and non-local, and conventional and sustainable in local food systems. Feagan (2007) *PHG* 31, 1 argues that the local in LFS will have to be contingent on the place—its social, ecological, and political circumstances—going on to warn against xenophobia, 'place purity', and anti-democratic orientations in local food systems.

localism A range of political philosophies that prioritize the local over regionalism and centralized government. Localism generally supports the local production and consumption of goods, local control of government, and the promotion of local history, local culture, and local identity. Regional geographers, humanistic geographers, and spatial scientists view localities as relatively natural phenomena. Marxist and political-economic geographers view localities as social phenomena produced by uneven capitalist development. Post-structuralist geographers view localities as characteristically open, plural, and dynamic; see Clarke (2013) *Policy Studs* 34, 5–6, which is strongly recommended. **New localism** refers to the concentration of power in urban centres which drive innovation and growth; see Collins (2016) *TIBG* 41, 2 on localism and civic pride.

localization economies Advantages arising from the localization together of a number of firms in the same type of industry. Localization economies are benefits which derive from being located close to other firms in the same industry, while urbanization economies are associated with closeness to overall economic activity and include knowledge externalities and a skilled labour market.

local knowledge A familiarity with and understanding of our day-to-day life; the tacit and explicit knowledge possessed and used by people who share the same culture. In order to learn from the local knowledge pool, firms need to be able to absorb the relevant knowledge and thus need to be cognitively close to the local environment; knowledge can only be transmitted when receiving firms are able to absorb it. See Ley (2003) *PHG* 27, 5, 537 for a social history of local knowledge.

local winds Local winds blow over a much smaller area than global winds

and have a much shorter time span. **Hot winds** originate in vast *anticyclones over hot deserts and include the Santa Ana (California), the Brickfielder (south-east Australia), the Sirocco (Mediterranean), the Haboob (Sudan), the Khamsin (Egypt), and the Harmattan (West Africa).

Cold winds originate over mountains or other snow-covered areas and include the Mistral, funnelled down the Rhone Valley, and the *Bora. Some local winds, such as the Southerly Burster of Australia, are associated with cold fronts. Other local winds include *land breezes and sea breezes. *See also* MOUNTAIN WIND; FÖHN.

location Absolute location is expressed with reference to an arbitrary grid system as it appears on a map. **Relative location** is concerned with a feature as it relates to other features.

Transnational companies can move their production plants, and their call centres, with ease from place to place—from continent to continent indeed—at the slightest whiff of lower costs. 'Rather than erasing the importance of material location … current trends in the world economy have in some senses reinforced its significance because, as companies can move about more easily, they can respond to even the smallest differences between places' (Massey (2006) BBC/OU Open2.net).

locational advantage Although it is true that transport costs have become less significant, and that new computer and telecommunication technologies have reduced the importance of location for production, locational advantage is still important in industrial location; for example, the overwhelming majority of logistics providers possess a geographic speciality (and even the worldwide integrated firms either have branches in, or have acquired firms with, strategic

geographic locational advantage.
See Murphy (2012) *Transportation Res. A* 46, 1, 91.

location-allocation model A mathematical model used to establish optimal locations. The model takes account of the location and demand of the customers, the capacity of the facilities, and operational and transport costs. See Location-Allocation in ArcGIS (Site Selection).

location coefficient Also known as the location quotient, this expresses the relationship between an area's share of a particular industry and the national share. Thus, the locational coefficient for a given region equals:

$$\frac{\% \text{ employed in a field in a given region}}{\% \text{ employed nationally in that field}}$$

See Guimarães et al. (2007) *J. Reg. Sci.* 47, 4.

location-specific Locational factors bestow *competitive advantages if they offer resources that are not easily transferred to other locations. Such resources are either tangible (like natural resources or labour) or intangible (like expertise, or specialized services).

Location-specific resources influence the locations of transnational corporations (TNCs), especially if they complement the expertise of the TNC; the more easily the TNC can influence the location-specific resources, the more likely it is to invest in the professional development of its human resources. The TNC is then less likely to substitute one location for another one. Geographic distance; transport infrastructure; specific human capital; management know-how; historic ties; a multicultural environment; and a high quality of life are major factors in location specificity.

location theory A group of theories which seek to explain the siting of economic activities. Various factors which affect location are considered such as localized materials and *amenity, but most weight is placed on transport costs. The go-to text is P. Dicken and P. E. Lloyd (1990).

Loch Lomond Stadial A stadial characterized by small ice caps and *cirque glaciers in the Scottish Highlands. The Loch Lomond Stadial was a period of glacial readvance during the overall shrinking of the British–Irish ice sheet. By its end, all the ice had gone.

((⊕)) SEE WEB LINKS

• The Loch Lomond Stadial, described at AntarcticGlaciers.org.

locked zone Along a tectonic *rift, an area where plates are still attached to each other. The Cascadia subduction zone stretches from California to British Columbia. All along this zone, the subducting plates are forced beneath the North American Plate, but around 30 km down the plates have become locked. Below this zone the plates are more pliable, allowing them to move more readily past each other. This freer movement deeper down causes strain to accumulate along the locked zone.

Once that strain is great enough to overcome the friction that keeps the plates locked, the fault will rupture, causing earthquakes.

lodgement The release and consolidation of debris from a glacier. Lodgement of particles occurs when the frictional drag between a particle and the glacier bed exceeds the shear stress resulting from the moving ice. The latter aligns fragments of this debris, known as **lodgement** *till, in the same direction as the flow of the glacier. Ruszczyńska-Szenajch (2001) *Quat. Sci. Revs* 20, 4 distinguishes between **hard lodgement**

till for deposits released from a glacier sole mainly due to friction, and **soft lodgement till** for sediments released from the base of a glacier due to the melting of ice in a more water-saturated subglacial environment.

loess (löss) Any unconsolidated, non-stratified soil composed primarily of silt-sized particles. The origin of loess is in dispute. Some writers believe the deposit to be wind-borne; others note the occurrence of the soil in *periglacial environments. Others stress the importance of glacial grinding in the production of silt-sized particles, or the importance of *salt weathering. The loess sequences of north-central China preserve the longest and most detailed record of *Quaternary climate change found on land. See Haberlah (2007) *Area* 39, 2 for a clear and concise summary on loess. Loess is very fertile but very difficult to conserve; soil erosion in most of the Loess Plateau is 5000–10000 Mg/km^2 per year, and 20000 Mg/km^2 per year in some places.

logit model A model used to represent choice between two mutually exclusive options; for example, a commuter may decide to drive to work or to use public transport. The **multinomial logit model** is mathematically simple (Andrew and Meen (2006) *Papers Reg. Sci.* 85, 3), and is widely used, but imposes the restriction that the distribution of the random error terms is independent and identical over the alternatives, causing the cross-elasticities between all pairs of alternatives to be identical, and this can produce biased estimates. The **nested logit model** allows the error terms of pairs or groups of alternatives to be correlated, but the remaining restrictions on the equality of cross-elasticities may be unrealistic (Hunt (2000) *Papers Reg. Sci.* 40, 1). Logit models which allow different cross-elasticity between pairs of alternatives include: the **paired combinatorial logit** (Koppelman and Wen (2000) *Transp. Res.* B4); the **cross-nested logit** (Vovsha (1997) *Ann. Transpt. Res. Board, Washington, D.C.*), and the **product differentiation model** (Filippini (1999) *Int. J. Econ. Bus.* 2). The major weakness of logit models is the implication that the choice between any two groups of alternatives depends solely on the characteristics of the alternatives being compared, and not upon the characteristics of any other alternatives in the choice set.

longitude The position of a point on the globe in terms of its *meridian east or west of the prime meridian, expressed in degrees. These degrees may be subdivided into minutes and seconds, although decimal parts of the degree are increasingly used.

longitudinal data Information on one or more areas over periods of time; for example, on changes in human reproduction.

long profile A section of the longitudinal course of a river from head to mouth, showing only vertical changes. The theoretically smooth curve shown by such a profile may be interrupted by breaks of slope which can result from bands of resistant rock or from *rejuvenation (Larue (2008) *Geomorph.* 93, 3–4).

longshore drift The movement of sand and shingle along the coast. Waves usually surge onto a beach at an oblique angle and their *swash takes sediment up and along the beach. The *backwash usually drains back down the beach at an angle more nearly perpendicular to the coast, taking sediment with it. Thus there is a zigzag movement of sediment along the coast. **Longshore currents**, initiated by waves, also move beach material along the coast. The term **littoral drift** is synonymous.

long wave *See* EARTHQUAKE.

lopolith A large *intrusion which sags downwards in the centre, forming a saucer-shaped mass. Wilson (1956) *GSA Bull.* 67, 3 would define lopoliths as funnel-shaped in cross-section with the most basic rocks at the bottom. See O'Driscoll et al. (2006) *Geology* 34, 3 on the Great Eucrite intrusion of Ardnamurchan.

Lorenz curve A cumulative frequency curve showing the distribution of a variable such as population against an independent variable such as income or area settled. If the distribution of the dependent variable is equal, the plot will show as a straight, 45° line. Unequal distributions will yield a curve. The gap between this curve and the 45° line is the inequality gap. Such a gap exists everywhere, although the degree of inequality varies.

See GINI COEFFICIENT.

Lösch model A model of central places developed by A. Lösch (1954) which is less narrow than that of Christaller's *central place theory, in that it treats the range, threshold, and hexagonal hinterland of each function separately. The resulting pattern of central places is much more complex, and yields a continuous, rather than a stepped, distribution of population sizes. See Esselbichler in A. Leyshon et al., eds (2011) pp. 33–4.

löss *See* LOESS.

low A region of low atmospheric pressure. In Britain, the term low is generally applied to pressures of below 1000 millibars.

low-order goods and services Goods and services with a low *range

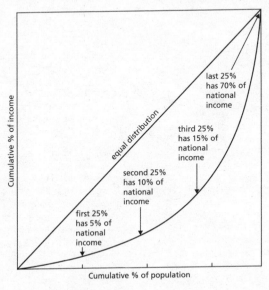

Lorenz curve

and a low *threshold population like daily newspapers, and hairdressing. The goods are often convenience goods.

Luminescence Dating (Optically Stimulated Luminescence Dating; OSL) All sediments include low levels of uranium, thorium, and potassium which produce ionizing radiation, which accumulates over time, and is absorbed and stored within the sediment. If this store is expelled from the sediment, it will generate luminescence, the amount produced being proportional to the accumulated dose. If the annual radiation dose is known, the age of the sediment can be calculated:

$$\text{Age} = \frac{\text{total accumulated radiation dose (Gy)}}{\text{annual radiation dose}}$$

Sediments rich in quartz and feldspar, which cannot be dated by conventional radiocarbon methods, can be absolutely dated ($\pm \sim 10\%$) within a range of 100 000 to 200 000 years. Aeolian sediments are ideal for luminescence dating, but samples from other sedimentary environments have been successfully dated. Li et al. (2012) *Holocene* 22, 397 describe the methodology well.

lunette *See* SAND DUNE.

lynchet In earlier, possibly prehistoric, times, small belts of uncultivated land were sometimes left between ploughed areas. These now can be seen in the landscape as part of a system of terraces. See Klingelhofer (1991) *Trans. Sm. Philos. Soc.* 81, 3.

lysimeter A device for the direct estimation of evapotranspiration, composed of a vegetated block of soil 0.5–1 m^3 to which the amount of water added is known, and from which the amount lost as run-off or percolation may be measured. Richards (2004) *PPG* 28, 1 explains further.

macro- Large-scale. Thus, a **macroclimate** is the general climate of a region extending across several hundred kilometres; and **macrometeorology** is the study of large-scale meteorological phenomena—from *monsoons to the *general circulation of the atmosphere.

macroecology The ecology of large scales; the study of ecology over broad spatial and long temporal scales. Macroecology has moved from valuing small scale to the large scale, but Blackburn and Gaston (2002) *Glob. Change & Biogeog.* 11, 3 argue that there is no general sense in which one scale of study is better than any other. Present trends in macroecology are predominantly statistical in approach, looking for rules and regularities. See Diniz-Filho et al. (2013) *Frontiers of Biogeography* 5, 3.

macrotidal Having a large tidal range; *Microtidal* is a range less than 2 m, *mesotidal* 2–4 m, and *macrotidal* over 4 m. Areas with a tidal range over 10 m are sometimes distinguished as *hypertidal*. **Macrotidal estuaries** have their dynamics dominated by tidal forcing, as compared to the forcing imposed by the river. Macrotidal sandy beaches are thought to represent optimal conditions for coastal dune development (Ruz and Meur-Ferec (2004) *Geomorph.* 60, 1–2).

magma The molten rock found below the Earth's crust which can give rise to *igneous rocks. Magma occurs in *subduction zones, at continental rift zones, *hot spots and *mid-oceanic ridges, and may rise as *mantle plumes. Magma composition is an important control on the geomorphology of lava flows and volcanoes. See Morgan and Ghen (1993) *Nature* 364 on magma supply. **Magmatic stoping** is the detaching and engulfing of pieces of the country rock, but its role in the location of plutons is disputed: see Žák et al. (2006) *Int J. Earth Sci.* 95 versus Glazner and Bartley (2006) *GSA Bull.* 118.

magnetic pole reversal The Earth's magnetic field resembles a bar magnet located at the Earth's centre, with its axis emerging at the magnetic poles. The north and south magnetic poles have repeatedly changed places, at irregular intervals, while the axis has stayed in place. When *igneous rocks form, they take up the prevailing pattern of the Earth's magnetic poles. As a result, areas of the ocean bed where *sea-floor spreading has taken place are characterized by **magnetic stripes**— parallel bands of igneous rock with differing magnetic polarity. See Nicolosi et al. (2006) *Geol.* 34 on the Marsili Basin, Italy.

magnitude Of an *earthquake, an expression of the total energy released.

malapportionment The existence of electoral districts with wildly unequal populations. Unintentional malapportionment happens when population growth in electoral districts varies widely. Redistricting is then needed to even out voter numbers; see Johnson et al. (2006) *Pol. Geog.* 25, 5.

Malthusianism In 1798, Thomas Robert Malthus (1766–1834) published his *Essay on Population* in which he put forward the theory that the power of a population to increase is greater than that of the Earth to provide food. He asserted that population would grow geometrically, while food supply would grow arithmetically. When population outstrips resources, **Malthusian checks** to population occur: misery, vice, and moral restraint.

Malthus' predictions were not borne out in 18th-century Britain, perhaps because of increases in food output, and emigration to the colonies. Malthusian theory has been deployed to underpin the theory of European historical superiority by arguing that Europeans, uniquely, have generally (and rationally) avoided the Malthusian disasters of overpopulation while non-Europeans (irrationally) have not done so and therefore not developed as Europe has.

Neo-Malthusian scholars argue that overpopulation will cause resource depletion and hunger which in turn would lead to political instability threatening Western interests and world peace. However, it has been argued that the neo-Malthusian promotion of family planning as the solution to hunger, conflict, and poverty has contributed to destructive population-control approaches, that are targeted most often at poor, racialized women; see *Uneven Earth* (2020).

mammatus Breast-shaped lobes of cloud, hanging from the undersurface of a *cumulo-nimbus anvil, signifying negative *buoyancy. **Mammilated** means smooth and rounded.

mandate A statement of empowerment, for example, being able to vote: 'representing humanity ultimately requires legitimation through some sort of people's mandate' (Dicken (2004) *TIBG* 29, 1). The term also describes a territory, once part of the German or Ottoman Empires, governed by a member of the League of Nations, 1919–39.

mangrove swamp A number of types of low trees and shrubs, growing on mud flats in tropical coastal areas where the tidal range is slight. Mangroves are significant agents of *progradation along tropical coasts, and are especially well developed in South and East Asia.

Manning equation Also known as the **Mark–Manning formula**, the **Gauckler–Manning formula**, or the **Gauckler–Manning–Strickler formula**, this is an empirical formula estimating open channel flow, or free-surface flow driven by gravity:

$$V = \frac{k}{n} R_h^{2/3} \, S^{1/2}$$

where V is the cross-sectional average velocity (L/T; ft/s, m/s); k is a conversion factor of (Length$^{1/3}$/Time), 1 m$^{1/3}$/s for SI, or 1.4859 ft$^{1/3}$/s U.S. customary units, if required. (Note: (1 m)$^{1/3}$/s = (3.2808399 ft)$^{1/3}$/s = 1.4859 ft$^{1/3}$/s); n is the **Manning coefficient**, and is unitless; R_h is the hydraulic radius (L; ft, m); and S is the slope of the water surface or the linear *hydraulic head loss (L/L) ($S = h_f/L$).

manor The smallest area of land held in the Middle Ages by a feudal lord, with its own court for minor offences. It usually consisted of a village, the lord's holding (*demesne*), and *open fields farmed on the three-field system.

mantle The mantle, with a density of up to 3.3 g cm^{-3}, and a thickness of some

2800 km, lies between the *crust and the Earth's *core. The upper layer, immediately below the *Moho discontinuity, is rigid, forming the lower *lithosphere; the lower layer is the *asthenosphere. See Husson (2006) *Geology* 34, 9 on upper mantle flow.

manufacturing industry Secondary *industry.

map A cartographic representation of selected spatial information; the cartographer decides what to include and what to leave out. 'Maps are active; they actively construct knowledge, they exercise power, and they can be a powerful means of promoting change' (J. Crampton 2010). We tend to restrict the 'map' term to paper maps, but spatial information can be shown on a computer screen, through braille, or a spoken description, and these may also be described as maps. **Map generalization** is the decreasing of the detail on a map when reducing its scale, for the map scale determines the size and number of graphic objects that can be put on a map. *See also* MENTAL MAP.

map projection Both a method of mapping a large area and the result of doing so. The Earth is a sphere; a map is flat, so that it is impossible to produce a map which combines true shape, true bearing, and true distance. 'The usefulness of a particular **map projection** and the justification behind its transformation from a three-dimensional spherical surface to a two-dimensional plane surface is all too often judged from the aesthetic or suspicious eye of a naïve map reader rather than by the science behind its creation' (M. Monmonier 2004). See also Monmonier (2005) *PHG* 29, 2.

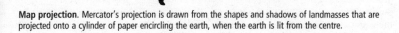

Map projection. Mercator's projection is drawn from the shapes and shadows of landmasses that are projected onto a cylinder of paper encircling the earth, when the earth is lit from the centre.

Mercator's projection exaggerates the size of the northern continents; it has been criticized as overemphasizing the importance of Europe and North America, although such was not Mercator's intention. No projection is perfect: for example, Mollweide's and Peters' are **equal area projections** (correct in area), but distort shapes. **Azimuthal projections** show true direction; **gnomic projections** show the shortest straight-line distance between two points; **orthographic projections** convey the effect of a globe. **Interrupted projections** show the Earth as a series of segments joined only along the equator. Details of the projection used are given below each map in a good atlas.

marginalization The process by which individuals and groups are prevented from fully participating in society. Marginalized populations can experience barriers to accessing meaningful employment, adequate housing, education, recreation, clean water, health services, and other social determinants of health. Both community and individual health are deeply affected by marginalization. The Ontario **marginalization index** is based on four criteria: residential instability, material deprivation, dependency, and ethnic concentration. **Geographies of marginalization** consider the way space is fashioned to privilege some groups and marginalize others; see Déry et al. (2012) *Hrvatski Geografski Glasnik* 74, 1, 5.

((⊕)) SEE WEB LINKS

• Ontario marginalization index.

marginal land Land which is difficult to cultivate, and which yields little profit.

margin of cultivation The distance from a market where the revenue from a product exactly equals total cost. If production costs are the same whatever the distance from the market, transport costs determine the margin of cultivation. See Imai (1985) *Annals Tohoku Geog. Assoc.* 37, 4, 279.

mariculture This is marine aquaculture; the cultivation of marine organisms in the open ocean, or in an enclosed section of the ocean, or in seawater-filled ponds or tanks. See Oyinlola et al. (2018) *PLOS One* 13, 1 for a global estimation of areas with suitable environmental conditions for mariculture species. Suryanata and Umemoto (2003) *Env. & Plan. A* 35, 2, 199 point out the problems associated with marine aquaculture in Hawaii.

maritime Of climates near the sea coast. Such climates have less extremes of temperature, both diurnally and seasonally, than their continental counterparts.

market An arena in which buyers and sellers exchange goods and services, usually for money; it doesn't have to be a physical location. **Market area analysis** A market area (also called trading area, service area, or catchment area) is the location where real, or possible, customers come from. Market area analysis is the investigation of how a firm's market area is established. The market area may be identified through empirical observations such as point-of-sale surveys (customer spotting), or by using mathematical market area models; see Löffler (1998) 45, 4, or Wieland (2016): *Multiplicative Competitive Interaction (MCI) Model* for a combination of both approaches.

Geographies of markets focus on the spatial nature of market forces, considering markets as arrangements of people, things, and socio-technical devices; the social and spatial underpinnings of everyday market activities; and diversity and difference in market contexts.

market cycle A series of periodic—usually one-day—markets such that a trader moves from one location to the next in a weekly cycle.

market economy An economy, characteristic of *capitalism, in which the major parts of production, distribution, and exchange are carried out by private individuals or companies rather than by the government, whose intervention is minimal. Market economies are characterized by specialized production, the freedom to exchange commodities between individuals, and the use of the *market mechanism to determine prices. Competition is the driving force, and supply and demand dictate how goods and services are manufactured or produced. This allows businesses and individuals to seek out goods and services of the highest possible quality for the lowest possible price. The disadvantage of a market economy is that it values these transactions more highly than the welfare of the individuals in society.

market mechanism The interaction of supply, demand, and prices, strongly influenced by the prevailing political climate. Thus, Page (2005) *TIBG* 30, 3 compares two approaches to water supply: 'that water is scarce and, like all scarce goods, it is best allocated through a market mechanism … to be sold at the full price of its production, including environmental externalities, [or] … that water, though scarce, is in its essence not commercial and that such a process of commodification … is little more than theft of a common good.'

market potential The intensity of possible contacts with markets. The so-called Harris market potential uses bilateral distances to weight surrounding locations' GDP:

$$MP_{\text{Harris}, \, i} = \sum_j \frac{\text{GDP}_j}{\text{dist}_{ij}}$$

A more sophisticated version of Harris' market potential uses road travel times for lorries as weights, instead of great circle distances:

$$MP_{\text{Traveltime}, \, i} = \sum_j \frac{\text{GDP}_j}{\text{traveltime}_{ij}}$$

See Holl (2011) *J. Econ. Geog.* DOI 0.1093/jeg/lbr030 on the productivity impacts of market potential and transport infrastructure investment in the manufacturing sector in Spain.

Markov chain analysis A probability process that can be used to analyse a chain of events through time, such as the housing rental market. One of the earliest geographers to use this probabilistic process is Clark (1965) *AAAG* 55, 2, whose instructions in this method reward careful study.

marl Clay, usually alluvial, rich in soft calcium carbonate. See Cerdà (1999) *Soil Sci. Soc. Am. J.* 63 on comparative erosion rates in marl, clay, limestone, and sandstone.

Marshallian industrial district A spatial concentration of specialized small and medium-sized firms, supported by a variety of regional institutions for promoting coordination, learning, and innovation. Dense interactions among a large number of competing and cooperating firms create an external economy favouring technological innovation and learning (Wang (2007) *J. Dev. Stud* 43, 6).

Marxist geography The use of Marxist theory and methodology to explain the political economy of the world—and change it. It highlights the dialectical processes (different contributions) of and between the natural environment, spatial relationships, and economic, social, cultural and political *formations*. At all scales, the capitalist mode of production assumes a centre–periphery pattern, leading to uneven development which, it is argued, is integral to capitalism. Rent, profit (and loss), urbanization (and counter-urbanization), devaluation (and speculation), gentrification (and revalorization), imperialism (and revolution), and enclosure (and globalization) are seen to be forceful, spatial-temporal, capitalist projects, which exploit the common person. See Castree and Gregory (2007) *AAAG* 97, 1.

Cambridge (2005) *PHG* 29 thinks that the Marxian critique of capitalism has been absorbed and normalized in geography, such that 'we are all political economists now'.

masculinity A category of identity, with its own internal relations of alliance, subordination and dominance. Geographies of masculinity are concerned with: work on the body and wellbeing, body size, male care-giving, men's experiences in diverse faith communities, and men and alternative spiritualities. It is argued that the development of geographical work on men and masculinities is important for helping to contest patriarchal structures and knowledge production. B. van Haven and K. Hörschelmann, eds (2004) is very well reviewed.

mashup Also known as **map hacking**, this is a website or web application that combines content from more than one source; a tool overwhelmingly suited to geographers. Google Maps Mania is free. You might find Ting (2009) *Access to Knowledge* 1, 2, helpful.

mass balance The relationship between the masses of the components of a landform, particularly in glaciers,

where zero mass balance describes a situation where accumulation balances ablation, a positive mass balance describes an increase in glacier mass, and a negative mass balance describes a reduction in glacier mass. See M. M. Bennett and N. Glasser (2009), p. 43, for an excellent summary of mass balance and global sea-level rise. Alexander et al. (2011) *Geograf.* 93, 1, 41 describe a simple steady-state mass-balance model for the Franz Josef glacier, New Zealand.

mass budget The balance between the accumulation and ablation of ice in a glacier. Its components are accumulation, ablation, and iceberg calving. A glacier that is not in balance with the ambient climate loses/gains mass and thereby raises/lowers global sea level (IAMU, 2006–7). The flow of solid ice governs the dynamical response of the glacier to a change in its mass budget. Machguth et al. (2007) *Geophys. Res. Abs.* 9, 06249 compare the output of three distributed mass-balance models of differing complexity.

mass conservation, principle of For a river, sediment continuity. Rivers adjust their width, depth, and slope to keep a balance between the water and sediment supplied from upstream and exported at the outlet; see Dietrich et al. (1999) *Hydrol. Procs* 13.

mass movement (mass wasting) The movement downslope of rock fragments and soil under the influence of gravity. The material concerned is not incorporated into water or ice, and moves of its own accord, but slides are often triggered by increase in water pressure on rocks and soil. **Granular mass movements** entail transport by large (>0.06 mm) solid grains, mixed with less-dense intergranular liquid or gas, and include rock avalanches, debris flows, pyroclastic flows. A widely used classification of mass movement uses the combination of types of movement (*falls, *topples, *slumps, *slides, and *flows) with the nature of the material (bedrock, *debris, and fine soil). Many cases of mass movement include more than one type of movement.

mass strength The strength of a rock's resistance to erosion. Mass strength will vary according to the innate strength of the rock, the jointing/ bedding of the rock, and its state of weathering. See Selby (1980) *Geomorph.* 24 on rock mass strength classification.

mass transfer In geomorphology, the movement of materials, especially of sediment. Sedimentary mass transfers move eroded sediments from their source area to an area of temporal storage or long-term deposition in sinks. Rates of sediment transfer are not only conditioned by competence of geomorphic processes but also by the availability of sediment for transport. Studies on mass transfers and sedimentary source-to-sink fluxes generally refer to the development of sedimentary budgets. See Beylich (2011) *Zeitschr. Geom.* 55, 2, 145 for a very clear account.

materiality (materialization) In the early 2000s, many scholars of human geography looked to 'rematerialize' the discipline, that is to say, to ground their studies in material culture, or to be concerned with socially significant differences of gender, class, race, sexuality, or (dis)ability; to ground them in 'reality' (Jackson (2000) *Soc. & Cult. Geog.* 1, 9).

mathematical geography The study of the interactions of geography and mathematics, such as the Earth's size and shape, time zones, and motion. See Withers (2006) *PHG* 30, 6.

matrix Array.
   See LANDSCAPE MATRIX.

Maunder minimum Between 1645 and 1715, a period of reduced solar activity (sunspots) that coincided with the coldest part of the *Little Ice Age. This may be a causal connection; variability in solar activity may be tied to terrestrial climate impacts in the form of warmer winters in some places and colder winters in others. See Diodato and Bellocchi (2012) *Holocene* 22, 589.

maximum sustainable yield The maximum yield of a renewable resource consistent with maintaining its stock. Arlinghaus et al. (2002) *Fish & Fisheries* 3, 4 review the concept.

meander A winding curve in the course of a river. A *sinuosity of above 1.5 is regarded as distinguishing a meandering channel from a straight one. The dimensions of a meander are related to the square root of water discharge, Q:

$$\lambda = k_1 Q^{0.5}$$
$$A_m = k_2 Q^{0.5}$$
$$w_c = k_3 Q^{0.5}$$

where λ is meander wavelength, A_m is meander amplitude, w_c is the channel width, and k_1, k_2, and k_3 are coefficients whose value varies with location. Bank resistance controls meander wavelength, and is positively correlated with meander wavelength. Meander bends with a low radius of curvature have deeper pools and more lateral migration; the ensuing bank erosion reduces pool depth via higher sediment inputs. Meander bends with smaller pools have less lateral migration, but use energy vertically, thereby deepening pools and fostering pool-riffle sequences.

   Eaton et al. (2006) *ESPL* 31 argue that slight initial variations in the shear stress distribution result in local net scour along the banks, increasing local

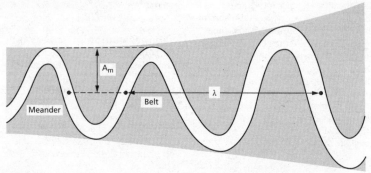

Meander

transport capacity. In a positive feedback, this leads to further local net scour, which is mostly vertical until the channel banks begin to fail, at which point bank erosion triggers negative feedback that increases channel sinuosity. Once one initial bend develops, others inevitably follow. See Millar (2005) *Geomorph.* 64 on equations for the meandering–braiding transition, and Pittaluga and Seminara (2011) *ESPL* 36, 1, 20 on nonlinearity and unsteadiness in river meandering. The **meander belt** is the total width across which the river meanders.

A **meander scroll** is a *point bar.

mean information field (MIF) In *diffusion, the field in which contacts

can occur. It generally takes the form of a square grid of 25 cells, with each cell being assigned a probability of being contacted. The possibility of contact is very high in the central cells from which the diffusion takes place, becoming markedly less so with distance from the centre, that is, there is a *distance decay effect. See Bivand (2008) *J. Reg. Sci.* 48, 1.

means of production The elements needed to produce goods and services: land, labour, and *capital. Any production process depends upon a particular material configuration of the means of production. Marx believed *capitalism to be characterized by a split between the capitalist owners of the

0.0096	0.0140	0.0168	0.0140	0.0096
0.0140	0.0301	0.0547	0.0301	0.0140
0.0168	0.0547	0.4431	0.0547	0.0168
0.0140	0.0301	0.0547	0.0301	0.0140
0.0096	0.0140	0.0168	0.0140	0.0096

Mean information field

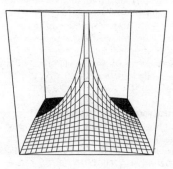

means of production, and the proletarians, with only their labour to sell.

mechanical erosion Erosion by physical means, with no chemical change. The main agents of physical erosion are wind, water, and ice. **Mechanical weathering** is the splitting up *in situ* of rock, with no chemical change.

media geographies The term 'media' includes animation, comic books, film, music, illustration, art, radio, newspapers, performances, television, video games, and virtual constructions. Media geographers study the relationship between geography and communication; the ways geographies are communicated through social media; media in spaces, spaces in media, media in places, and places in media. See Adams (2017) *Prog. Hum. Geog.* 41, 3.

medial moraine *See* MORAINE.

medicine, geography of (medical geography) A field of geography that considers the distribution of specific diseases and human characteristics in relation to the geographical and topographical features of regions, countries, or the world. In its initial incarnation, medical geography was an enquiry into the environmental correlations of diseases. Medical geographers also became interested in the spatial distributions of those medical facilities, from hospitals to clinics to GP surgeries, that society 'invents' to treat diseases. Key issues include geographical variations in incidences of disease, and the complex intersections between health, illness, place, and space. Medical geographers map/represent disease incidence (new cases) or prevalence spatially, to identify social or environmental factors associated with

disease and locate areas where risk is particularly high. See Sui (2007) *Geog. Compass* 1, 3 on GIS and medical geography, and Smyth (2008) *PHG* 32, 1 on the geographies of health inequalities.

Mediterranean climate A climate of hot, dry summers and warm, wet winters, characteristic not only of Mediterranean lands, but also in California, central Chile, and the extreme south of Africa. In summer, the climate is dominated by *subtropical anticyclones, and *trade winds prevail. In winter, mid-latitude depressions bring rain. *Local winds are of great significance.

Mediterranean soils Formed in *Mediterranean climates, these soils owe their red colour to the *leaching of clays in the wet winters, which releases iron. Leaching is rare in summer so that a carbonate *horizon often builds up.

megalith Any free-standing large stone sited by humans, mostly dating from *c.*3000 to 2000 BC. See Krzemińska et al. (2018) De Gruyter Open Access, on the significance of megalithic monuments in the process of place identity creation, and in tourism development. C. Holtorf (2001) points out that all interpretations of megaliths, whether as tombs or 'mnemonic markers', can support the idea that megaliths were meant for a future time.

megalopolis A many-centred, multi-city, urban area of more than 10 million inhabitants, characterized by complex networks of economic specialization. The *Metropolitan Institute Census Report Series* 05:01 (2005) is an excellent resource on this.

melting pot A description of the fusion of a number of ethnic groups, cultures, and religions to produce new cultural and social forms. Under mild conditions of interconnectedness, a society proceeds towards homogenization, with waves of foreign immigration introducing unconnected groups into the population, providing short-term heterogeneity. As links form between these new groups and the rest of the population, the homogenization process typically reasserts.

melt-out till Subglacial melt-out *till reflects the amount of sediment in transport at the time of melting; supra-glacial melt-out till is located at the margins of a glacier, and includes lateral and terminal moraines.

meltwater Glacial meltwater is produced by the melting of glacier ice at the surface, or by pressure and *geothermal heat at the base, forming *subglacial streams. Some surface meltwater may percolate through the ice to emerge at the base. Both subglacial and proglacial **meltwater erosion** are extremely effective agents of landscape change. Water often flows under pressure within and below decaying ice, pressure that enables meltwater to flow upslope, and carry large quantities of debris, which promotes *abrasion.

Glacial meltwater landforms form subglacially; at or near the ice margin; and in proglacial settings. **Ice margin meltwater** follows the edge of the glacier when drainage routes have been cut off by ice. **Tunnel valleys** occur below the ice and can cut steep-sided, flat-floored valleys; see Cutler et al. (2002) *Quat. Int.* 90, 1. **Spillways** are channels cut by streams overflowing from *proglacial lakes; see Thieler et al. (2007) *Pal. Pal. & Pal.* 246, 1. **Coulees** are canyons which result from the sudden and violent release of water from ice-dammed lakes when the barriers which impound them are breached. See Shaw (2002) *Quat. Int.* 90, 1 on the

meltwater hypothesis for subglacial bedforms, and Hannah and Gurnell (2001) *J. Hydrol.* 246, 1 and Pohl et al. (2005) *Atmos-Ocean* 43, 3 for models of meltwater action.

memory We construct meaning about our lives and worlds through the places we create, inhabit, or visit, such that places become mosaics of memory, metaphor, matter, and experience, which create and mediate social spaces. Look for O. Jones and J. Garde-Hansen (2012).

Tolia-Kelly (2004, *TIBG* 29, 3) writes on **re-memory**: 'a conceptualization of encounters with memories, stimulated through scents, sounds and textures in the everyday ... souvenirs ... [and] signifiers of "other" narrations of the past, not directly experienced'.

mental health, geographies of
Essentially, how and why the social and physical environments matter for mental health and psychological well-being in human populations; the geographic variability of the occurrence of psychological problems, and of access to therapies. See Lowe and DeVerteuil *Health & Place* 28 on austerity Britain and mental health. Parr (2000) has a useful paper in *Health & Place*, which is the most important journal on this topic.

mental map A map of the environment within an individual's mind that reflects the knowledge and prejudices of that individual. A mental map indicates the way an individual acquires, classifies, stores, retrieves, and decodes information about locations. Castellar, Maria, and Juliasz (2018) *Procs ICA*, 1 analyse the role of mental maps in the development of concepts of city and landscape. Ojanen and Vuohula (2007) *J. Common Mkt Studs* 45 describe Finland as lacking a prominent presence on the mental map of other EU governments.

Mercalli scale A measurement of the intensity of an *earthquake.

Modified Mercalli scale of earthquake intensity:

I. Felt by very few, except under special circumstances.

II. Felt by a few persons at rest, especially on the upper floors of buildings.

III. Felt noticeably indoors, although not always recognized as an earthquake. Vibration like passing lorry.

IV. Felt by many indoors during daytime, but by few outdoors. Some awakened at night. Vibration like lorry striking building.

V. Felt by nearly everyone; many awakened. Some breakages, disturbances of trees, telegraph poles.

VI. Felt by all; many run outside. Some heavy furniture moved.

VII. Everyone runs outside. No damage in well-built buildings; moderate damage in ordinary structures; considerable damage in poorly constructed buildings.

VIII. Considerable damage except in specially constructed buildings. Disturbs people driving cars.

IX. Damage even in specially designed structures. Buildings shifted from foundations; ground cracked; underground pipes broken.

X. Ground badly cracked. Railway lines bent. Landslides considerable.

XI. Few brick-built structures remain standing, if any. Bridges destroyed. Broad fissures in the ground.

XII. Total damage. Waves observed on ground surface. Objects thrown upward into the air.

Adapted from US Geological Survey.

mercantilism The view, current in early modern Europe, that one nation's gain depends on another nation's loss; a trading nation prospers only if it encourages the export, but discourages the import, of manufactures. See Ballinger (2011) *SSRN* on mercantilism and the rise of the West.

meridian An imaginary circle along the world's surface from geographic pole to geographic pole; a line of longitude. Meridians are described by the angle they form west or east of the **prime meridian**, which has a value of 0° and runs through Greenwich, England.

meridional circulation In meteorology, air flowing longitudinally, across the parallels of latitude. **Meridional flow** occurs in *atmospheric cells, resulting in part from longitudinal temperature variations. See Song et al. (2008) *J. Climate* 21, 19. The **meridional temperature gradient** is the change in temperature along a *meridian.

mesa A steep-sided plateau or upland, formed by the erosion of nearly horizontal *strata. A mesa resembles a *butte, but is very much wider; see Brookes (2001) *Geomorph.* 39, 1–4.

mesoclimate *See* LOCAL CLIMATE.

mesometeorology The study of middle-scale meteorological phenomena; between small features, like *cumulus clouds, and large features, like anticyclones. **Meso-scale** describes systems, or patterns of systems between small and *synoptic scale *c.*10–100 km across.

Mesozoic The middle *era of Earth's history, from around 225 to 190 million years BP.

(⊕) SEE WEB LINKS

• An overview of the Mesozoic era.

metadata This describes data, specifically the nature of a dataset: who wrote it and when; how the data were created and why; the computer network where the data were created; and the standards used.

metageography An individual's geographical perspective of world geography, springing from their cultural upbringing. Metageography is the spatial framework on which we hang our knowledge; a framework we take for granted. We speak, for example, of Europe, but rarely ponder what that means. Do we define Europe by membership of the European Union? Or by entries to the Eurovision Song Contest? A **metageographic community** is a network of like-minded individuals, institutions, and organizations where physical nearness is not essential for the existence of collective identity. It is metageographic because it exists only as the result of interaction over distance; it is not bound by the proximity of its members in localized physical space (see Nashleanas (2011) *AAAG* 101). Metageographical may also refer to concepts that lack specific spatial coordinates, and contain incidental conceptual 'baggage'.

metagovernance The organization by the state of the context and ground rules for governance; the process in which the discussion, formulation, and application of norms, principles, and values for governance takes place. See Morrison (2016) *International Planning Studs* 21, 3.

metamorphic aureole An area of rock altered in composition, structure, or texture by contact with an igneous *intrusion. See Richards and Collins (2002) *J. Metamorph. Geol.* 20, 1. **Metamorphic rocks** have been changed from their original form by heat or by pressure beneath Earth's surface; for example, limestone to marble, shale to slate, slate to schist. **Dislocation metamorphism** (a widespread but non-universally recognized concept—see Sobolev et al. (2002) *Bull. Voronzeh State Univ.* 2: *Geology*) occurs through friction along fault planes or *thrust planes. **Regional metamorphism** (also called **dynamic metamorphism**) occurs during an *orogeny; see Patison et al. (2001) *Australian J. Earth Scis* 48, 2.

metaphor Geographers are interested in the metaphors people use to explain their perception of a place or a process. For example, Hulme (2012) *TIBG* 37, 3, 346 looks at some of the ways in which writers, geographers, and artists have imagined the island of Cuba. He quotes A. Gaztambide-Geigel (1996) in imagining Cuba as a key; 'if the American continent lies behind the door of its eastern coastline, then Cuba—the largest island in the Caribbean—looks as if it's being inserted into the lock'. This metaphor partly explains the US thinking behind the Cuban Missile Crisis of 1962.

metapopulation In ecology, a place-specific assemblage of species that are connected.

metasomatism The change in *country rock brought about by the invasion of fluid; see Storey et al. (2004) *Geology* 32, 2. Granites are formed at great depths by invading granulizing fluids; see Langer (1966) *Annals NY Acad. Scis* 136, 1.

meteoric water Groundwater originating from *precipitation.

meteorology The study and science of all aspects of the atmosphere— weather and climate—aiming to understand the physical and chemical nature of the atmosphere, its dynamical behaviour, and its complex interactions with the surface. It includes both short- and long-term weather forecasting, and the determination of past and future climatic change. Meteorology now includes the study of the atmospheres of other planets.

metrology The scientific study of measurement, especially the standardization and definition of the units of measurement in science. See Pike (2001) *Prof. Geogr.* 53, 2.

metropolis A very large urban settlement, usually with accompanying suburbs. No precise parameters of size or population density have been established. The growth of metropolises during the 20th century reflected a constant increase in the mobility and interconnections among places, regions, cities, and countries around the world.

Soja, in G. Bridge and S. Watson (2012) believes the metropolis era to be at an end.

metropolization The process through which institutionally, functionally, and spatially fragmented urbanized regions become integrated as coherent metropolitan systems. See Cardoso and Meijers (2021) *Urban Geog.* 42, 1.

micelle A sub-microscopic aggregate of molecules within a *clay colloid.

microclimates The climates of those parts of the lower *atmosphere directly affected by the features of the Earth's surface. The height of this part of the lower atmosphere varies according to the size of the influencing feature, but could be four times the height of the feature (Barry (1970) *TIBG* 49). Local factors commonly override the influence of larger scale macroclimatic factors (Vitt et al. (1994) *Arct. & Alpine Res.* 26). See Lindberg (2006) *6th Int. Conf. Urb. Climate* on modelling the urban microclimate.

micro-credit Lending small amounts of money to very poor households at commercial rates, rather than at the 'usurious rates of loan sharks' (S. Buckingham-Hatfield 2000). The classic model is the Grameen Bank, initiated in Bangladesh by Mohammed Yunis. Similar micro-credit schemes have been set up throughout the developing world, and have been enormously effective, not only in alleviating *poverty and improving child nutrition, but also in increasing the voluntary use of contraception. N. Burra et al., eds (2005) assess the impact of micro-credit. Poverty trends would probably not be reversed, 'but booming micro-credit would at least speak for a world where justice would have a greater place' (Santiso (2005) *Int. J. Soc. Sci.* 57, 185).

(⊕) SEE WEB LINKS
• Grameen Bank website.

microsimulation A computer analysis of individual behaviour, be it, for example, people or vehicles: the creation of large-scale population microdatasets and the analysis of policy impacts at the microlevel. Microsimulation methods aim to examine changes in the life of individuals within households, and to analyse the impact of government policy changes for each individual and each household. See Ballas et al. (2005) *Population, Space and Place* 11.

mid-latitude depression An area of low atmospheric pressure occurring between 30° and 60°. This low is some 1500–3000 km in diameter and is associated with the removal of air at height and the meeting of cold and warm *air masses in the lower atmosphere. At the fronts between the air masses, a horizontal wave of warm air is enclosed on either side by cold air. The approach of the warm front is indicated by high cirrus cloud. The cloud then thickens and lowers, and rain falls. As the warm sector passes over, skies clear and the temperature rises. The cold front is marked by heavier rain and a fall in temperature. The warm front advances at 20 to 30 miles per hour, whilst the cold front can move forward more quickly at 40 to 50 miles per hour. In consequence, the cold front eventually pinches out the warm air, lifting it bodily from the ground to form an *occlusion. See R. Barry and R. Chorley (2003).

mid-oceanic ridge *See* OCEANIC RIDGE.

migration The movement of people from one place to another. The terms **in-migration** and **out-migration** are used for **internal migration**; **voluntary migration** refers to unforced movements, **compulsory migration** describes the expulsion of minorities from their country of birth by governments, or by warring factions. In the case of *commuting, migration is a daily act, but, because there is no change of residence, a purist would not call commuting a migration, preferring the term *mobility.

In **innovative migration** people move to achieve something new. **Economic migration** is the movement of people from a poor to a richer area. The gains for economic immigrants do not cause economic problems in the receiving nations: farmers in destinations from New Zealand to New Mexico benefit from the hard work of immigrant workers. Technological institutions from CERN (the European Organization for Nuclear Research) in Geneva to Silicon Valley in California innovate thanks to ingenious immigrants. Native-born workers also gain on average because they benefit from the complementary skills that immigrants bring, or because they are consumers of the products and services immigrants provide. Study after study proves that increased labour mobility leads to large gains for the immigrants and positive overall gains for the destination country.

However, the compelling evidence on the economic gains and social benefits of economic migration is not reflected in the attitudes to economic migration in the receiving countries. In political opinion polls in host countries, economic in-migration is among peoples' worst fears. People worry about what migrants and refugees would do to jobs and wages, welfare programmes, crime, schools, and their national identity.

Hoey (2005) *J. Contemp. Ethnog.* 34, 5 writes on **non-economic migration**.

Some political geographers, such as J. Hyndman (2005) refer to **migration wars**—conflicts, usually by nation-states—over who to let in and who to keep out. As globalization proceeds, countries have invested heavily in border enforcement, increasing patrols along borders, building fences, and using infrared technology to keep people from crossing borders illegally. Mitra and Murayama, *Inst. Developing Econs Res. Paper #137*, provide a wealth of detail on the inter-state migration for India.

migration–development nexus
The connections between *migration and *development, particularly arising from outmigration from the Global South. See Boese and Moran (February 2021) *Frontiers Sociol.* on the regional migration–development nexus in Australia.

migration systems theory This suggests that migration flows acquire a measure of stability and structure over space and time, allowing for the identification of stable international migration systems, which are

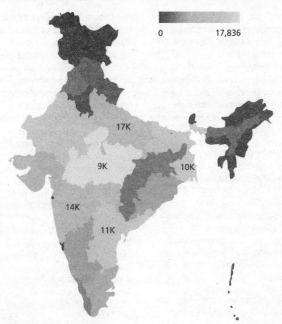

0   17,836

17K

9K   10K

14K

11K

India—the number of migrants by state

characterized by relatively intense exchanges of goods, capital, and people between certain countries and less intense exchanges between others. An international migration system generally includes a *core receiving region, which may be a country or group of countries, and a set of specific sending countries linked to it by unusually large flows of immigrants. See de Haas (2021) *Comparative Migration Studies* 9, 8.

Milankovitch cycles There are three interacting, astronomical cycles in the Earth's orbit around the sun: in the shape of the elliptical orbit (*c.*95 000 years); in the axis of rotation (*c.*42 000 years); and in the date of perihelion (the time of year when the Earth is closest to the sun, about 21 000 years) (M. Milankovitch 1930). There is also

a 100 ka cycle; see Berger (1978) *J. Atmos. Scis* 35.

The medium-scale cycles are poorly known: Campbell et al. (1998) *Geology* 26 recognize up to eight cycles, but cannot explain their origins. Ware (1995) *Fisheries Oceanog.* 4 demonstrates four short-term cycles: 2–3 years, 5–7 years (El Niño–Southern Oscillation), 20–25 years (bidecadal oscillation), and a poorly resolved, very low-frequency oscillation with a 50–75-year frequency. See Harris (2005) *PPG* 29, 2 for an excellent summary. Imbrie and Imbrie (1980) *Science* 207 show that Milankovitch cycles satisfactorily explain the major fluctuations in ocean temperature.

milieu The sphere in which individuals live and which they are affected by, including tangible objects and people, social and cultural phenomena, and the *images which influence human behaviour. This is a huge theme in Canadian geography; see *Canad. Geog./ Géog. canad.* (2000) 44, 1; (2002) 46, 1 and 46, 4; (2004) 48, 3; and (2006) 50, 4.

In economic geography, the local milieu is important for knowledge transfer. See Gareau (2008) *Antipode* 40, 1 on the *neoliberal milieu.

military geography The spatial impacts and expressions of military power; a political geography focused on the spatiality of armed conflict; the political economies and sociocultural geographies of militarism, particularly in nonconflict situations. Go to the final report on geography and engagement with the military, *AAG Geography and Military Study Committee*: issues, status, and findings.

military–industrial complex In a nation, those industries that provide *materiel* for the military, together with the military establishment itself. Try Barnes (2008) *Geoforum* 31, 9, 3; it is very wordy.

mimesis The production of as accurate a reflection of an entity/the world as possible. In other words, 'realism'. Some writers (starting with Aristotle), have stressed that mimesis produces something that is separate from the being it represents; where one thing symbolically stands for another. See Malmberg and Maskell (2002) *Env. & Plan. A* 34 on spatial clusters, spin-offs, and mimesis.

mineralization The breakdown of organic residues in soils; an essential process in humus formation. See Andriesh (2006) *18th World Congr. Soil Sci.*

mineralized zone An enriched zone of mineral deposits around an *igneous intrusion.

minimum efficient scale of production (mes. of production) The smallest possible size of a factory compatible with profitable production, below which *economies of scale don't apply. Mes. can be defined in terms of production, or of employment. The turnover of a single product is ultimately bounded by the minimum efficient scale and consumer demand for a specific product. A firm may lower its mes. by buying in components, or through technological change.

misfit stream A stream that appears to be too small to have eroded the valley in which it flows; a stream whose volume is greatly reduced or whose meanders show a pronounced shrinkage in radius. It is a common result of drainage changes effected by capture, glaciers, or climatic variations (Beaty (1990) *Canad. Geogr./Géog. canad.* 34, 2).

mist A suspension of water droplets which restricts visibility to between 1 and 2 km.

mixed farming The farming of crops and livestock under unified management.

mixed methods research Combining qualitative and quantitative methods in a single study. De Lisle (2011) *Caribbean Curriculum* 18, 87 explains further.

mobility Mobility means movement, and it is a key concept in geography, along with space and place, for mobility/ movement creates places and social relationships. Geographers of mobility study: the movements of peoples; the movements of objects; imaginative travel (as on television); virtual travel (internet); and communicative travel (phone, Zoom, etc.). Of interest, too, are how mobilities are powered, and how they can be tracked and traced. See

Prout et al. (2018) *TIBG* 43, 4 on the politics of mobility for indigenous Australians. The **mobilities paradigm/ mobilities turn** emphasizes that all places are tied into networks of connections that stretch beyond each such place, and mean that nowhere can be an 'island'. It challenges the imagery of 'terrains' as spatially fixed geographical containers for social processes, and questions classifications such as local, regional, and global. Look for Simon Cook on this.

modal split The varying proportions of different transport modes which may be used at any one time. The choices of modes may be determined by the costs, destinations, capacities, and frequencies of the modes together with the nature of the goods carried and their destinations. In commuter behaviour, the key variable determining modal split and trip length is the distance the residence was located from the *central business district. Location, urban form, levels of car ownership, and the mixture of different social classes in the ward of residence are important determinants of modal split. See Titheridge and Hall (2006) *J. Transp. Geog.* 14, 1.

model A representation of some phenomenon of the real world made in order to facilitate an understanding of its workings. A model is a simplified and generalized version of real events, from which the incidental detail, or 'noise', has been removed; an 'abstraction of an object, system or process which helps us understand reality.

In the 1960s, the quantitative revolution in geography saw an emphasis on models: basic mathematical equations and models, such as gravity models, deterministic models, such as von Thünens and Weber's location models, and stochastic models that use probability. The 'cultural' turn—the counter-positivist response from human geography—stressed the weaknesses of models.

In physical geography, the basic requirement of a model is that it includes the important factors controlling a system, yet is restricted in complexity. One solution is to use a lumped model, which is constructed using average or 'lumped' representation of the relevant phenomena. Thus a lumped model of a catchment area would assume that catchment is homogeneous, by using representative soil properties and precipitation inputs. All this is reasonably clearly set out in Maurer et al. (2010) *J. Amer. Water Res. Ass.* 46, 5, 1024. Or try V. P. Singh and M. Fiorentino, eds (1996), section 10.2.1.

mode of production The way in which society organizes production. K. Marx and F. Engels (1974) claimed that under the **capitalist mode of production**, bourgeois democratic society is naturalized, so that the way people act and feel is universalized as 'human nature', in order to preserve the status quo. Geographers recognize that each mode of production creates its own geography: for example, capitalism seems to be inextricably wedded to *uneven development (see, among many, Walker (1978) *Rev. Radical Pol. Econs* 10, 3).

moder A humus which is less acid than *mor but more acid than *mull. See Fons and Klinka (1998) *Geoderma* 86, 1 on mormoders and leptomoders.

modernity The knowledge, power, and social practices which developed in 17th-century Europe, along with beliefs in rationality and progress. Larsen (2006) *Geografiska B* 88, 3 sees Western modernity as characterized by the binary oppositions of self-other, nature–culture, and future–past, while Taylor (2007) *TIBG* 32, 2 sees the subordination of cities to territorial states as the most important geographical attribute of modernity. Murphy (2006) *AAAG* 96, 1 argues that a core tenet of modernity is an abrupt break with the past, but D. Harvey (2003) argues against the 'modernity as break' thesis.

modernization theory All modernization theories are based on the assumption that less-developed areas are merely at an earlier stage along a single development path already experienced by the most-developed capitalist economies; Rostow's model is an example, and the *flying geese theory is an East Asian variety of modernization theory.

modifiable areal units Areas studied in geography, such as states, counties, census areas, or *enumeration districts, which may be arbitrary units unconnected with any existing spatial patterns developed—the boundaries of many African states, for example, cut across cultural and ethnic regions. In many cases the investigator has to use the units for which data are available rather than the units which are more suited to the investigation. Guo and Bhat (2004) *Transportation Res. Record* 1898 contend that the fundamental issue is the inconsistency between the analyst's definition of areal units, and the decision-maker's perception of residential neighbourhoods, and propose a way out of the problem.

Moho discontinuity (Mohorovičić discontinuity) The boundary in the Earth's interior between the *crust and the upper *mantle, marking a change in rock density from 3.3 (crust) to 4.7 (mantle). The oceanic Moho separates partially serpentinized ultramafic rocks,

above, and fresh tectonized ultramafic rocks below; see Clague and Straley (1977) *Geology* 5, 3

The Moho occurs at about 35 km below the continents (but can be depressed to nearly 50 km depth, and at around 10 km beneath the oceans.

Moh's scale A relative scale of hardness, based on the ability of a harder mineral to scratch a soft one. The ten minerals in the scale are: talc, gypsum, calcite, fluorite, apatite, feldspar, quartz, topaz, carborundum, and diamond. Use Moh's scale by testing an unknown mineral against one of these standard minerals. Whichever one scratches the other is harder, and if both scratch each other they are both of the same hardness.

moist coniferous forest Forest found only on the west coast of Canada and the USA where the climate is cool and moist throughout the year. Very large evergreens such as redwoods and Douglas fir grow, and undergrowth is sparse.

moisture budget (moisture balance) The balance of water, as represented broadly by the equation:

$$\text{balance} = \text{precipitation} - (\text{runoff} \\ + \text{ potential evapotranspiration} \\ + \text{ the change in soil} - \text{moisture})$$

Moisture balance values of less than 1.0 indicate an annual excess of PET over P, a moisture deficit. Areas where values fall below 0.5 are regarded as drylands.

moisture index A measure of the water balance of an area in terms of gains from precipitation (P) and losses from *potential evapotranspiration (PE). The moisture index (MI) is calculated thus:

$$MI = \frac{100(P - PE)}{PE}$$

See Smakhtin and Hughes (2007) *Env. Model. & Software* 22, 6 on drought indices.

mollisols In *US soil classification, soils with a rich humus content, developed under grassland. *Rendzinas and *chestnut soils fall into this category. For a clear and useful account, see Kravchenko et al. (2011) *Chin. Geogra. Sci.* 21, 3, 257.

monadnock An isolated peak, the relict of long-term subaerial denudation, now standing above the *peneplain. The term comes from Mount Monadnock, New England; see Fowler-Billings (1949) *GSA Bull.* 60, 8.

monetary policy, geography of The aim of monetary policy is to control the supply of money, often targeting a rate of interest for the purpose of promoting economic growth and stability. The official goals usually include maintaining relatively stable prices and low unemployment, but monetary policy and aims vary spatially, and this spatial variation is the preserve of geographers. See Chandra Mekha (2011) *Int. Conf. on Humanities, Geog. & Econs.* on Indian monetary policy.

money, geography of The interactions between money and *place and *space, ranging from the global to the regional and local, such as variations in access to *financial services. Of considerable interest is the continued predominance, despite *time–space compression, of certain financial centres, such as Tokyo, New York, Frankfurt, London, and Amsterdam.

It may be argued that the globalization of money has undermined the monetary power of sovereign nation-states: this has been weakened by the growing frequency and volume of the decision-makers in specialized investment houses and banks, readily shifting hot money around the globe, constructing geographies through their assessments of internal conditions in emerging markets and their power to engage with them.

There is surprisingly little on fraud in the geographical literature. *ID Analytics* examine how US identity fraud rates vary by geography, finding that the highest rates of identity fraud occur in New York state, followed by California, Nevada, Arizona, and Illinois, while the least risky states tend to be in the upper Midwest and upper New England.

monoculture A farming system given over exclusively to a single product. Writing on globalization and agrarian change in the Pacific Islands, Murray (2001) *J. Rural Studs* 17 lists some disadvantages of monoculture: 'increasing pollution, soil degradation and ground water depletion; further concentrating economic power, property ownership and social inequalities; and contributing to urbanization as displaced small growers migrate to towns and cities. Moreover, the Tongan economy has been left more vulnerable to global economic fluctuations.' However, monocultures are also linked to high yields, while their implied opposite—a world of mixed, non-chemical agriculture—might not so readily feed future populations.

monogenetic Of a volcano, formed by a single, short-lived eruptive event.

monopoly The provision of a good or service by a single supplier, who then has the power to set prices, since there is no competition. Look for M. P. Feldman (2020) *Kenan Institute* on how tech monopolies are exacerbating income inequality across America; see also Holmes and Stevens (2004) *J. Econ. Geog.* 4 on monopoly, geographic concentration, and business size.

monsoon Colloquially, a sudden wet season within the tropics, but, more explicitly, a seasonal shift of air flows, cloud, and precipitation systems. Monsoons have been described in West and East Africa, northern Australia, Chile, Spain, and Texas, but the largest is the south-west monsoon. This Asian monsoon is the atmospheric response, complicated by the presence of water vapour, to the shift of the overhead sun, and therefore zone of maximum heating, from the Tropic of Capricorn in late December to the Tropic of Cancer in late June. Associated with this response are major changes in *jet stream movements and a meridional shift of the rain-bringing *inter-tropical convergence zone.

In winter, pressure is high over central Asia; winds blow outwards, and some depressions, guided by upper-air westerlies, move from west to east, bringing rain. During the spring, a *thermal low develops over northern India. Rains, related to a trough in the upper air, enter Burma in April/May, and India in late May/early June. In early summer the direction of the upper air changes, and the tropical easterly jet stream is semi-permanent about 15° N for the rest of the summer. With this change, the monsoon 'bursts', giving heavy rain across the southern half of the subcontinent.

By late June there is a continuous southerly flow of warm, moist air into the monsoon trough lying across northern India. This flow is a continuation of the south-east *trades, altered to south-westerlies by the *Coriolis force as they move north, across the equator, bringing huge quantities of water vapour, and reaching the west coast of India as the south-west monsoon. Here, the rainfall is *orographically enhanced. *Monsoon depressions are formed in association with these air flows, steered by the now easterly jet. Subsiding upper air prevents rainfall in the Thal and Thar deserts of the north-west of the subcontinent. By autumn, the easterly jet stream is replaced by a narrow band of the westerly subtropical jet stream, which follows the southern Himalayas. The south-west monsoon begins to retreat in September. Thereafter, the north-east trade winds dominate, and *hurricanes are common in the Bay of Bengal. See O'Hare (1997) *Geo* 82, 3 and 4 for an exegesis on the Indian monsoon.

monsoon depression A summer low pressure system affecting South Asia, 1000–2500 km across, with a *cyclone circulation about 8 km. Lasting two to five days, it comes roughly twice monthly, bringing 100–200 mm of rain per day. See Daggupaty and Sikka (1977) *J. Atmos. Scis* 34, 5.

monsoon forest A tropical forest found in *monsoonal regions. Trees are semi-deciduous or evergreen, in order to withstand drought. Monsoon forest is more open, and has more undergrowth, than *tropical rain forest. Kuznetsov and Kuznetsova (2013) *Biology Bull.* 40, 2, 187 provide a comprehensive and detailed survey of tropical monsoon forest in Vietnam.

Monte Carlo model A simulation model using computational algorithms, and relying on repeated random sampling to obtain numerical results; it lets the user see all the possible outcomes of any decisions made and assess the impact of risk. This can be done in Excel.

mor A forest humus characterized by an organic matter on the soil surface in matted horizons, reflecting mycogenous decomposers, and with an abrupt boundary between the organic horizon and the underlying mineral soil. See Olsson (2005) *Glob. Change* 11, 10.

moraine The rocks, boulders, and debris that are carried and deposited by a glacier or ice sheet. Moraine is often partly stratified, since some may have been formed under water. **Ablation moraine** is common on retreating glaciers, and coarse, because meltwater has washed out finer particles. As the glacier shrinks, lateral moraines are deposited at the side of the glacier. Where two lateral moraines combine, a central, **medial moraine** may be formed. **Ground moraine**, also called a *till sheet, is a blanket covering the ground. Other moraines have been moulded by ice parallel to the direction of ice movement. These include **fluted moraines** which are long ridges, possibly formed in the shelter of an obstruction. *Drumlins are streamlined moraines. **Dump moraines** are ridges formed approximately transverse to flow from material delivered to the margin of a glacier by ice flow. They mark the stationary position of an ice margin. The size of dump moraines depends on the rate of ice flow and the debris content of the ice. A **hummocky moraine** is a strongly undulating surface of ground moraine, with a relative relief of up to 10 m, and showing steep slopes, deep, enclosed depressions, and meltwater channels, which have formed from the meltout of supraglacial or englacial material. Boone and Eyles (2001) *Geomorph.* 38, 1 suggest that the growth and decay of ponds in repeated cycles of subglacial till failure and landform development can produce hummocks, but Lukas (2006) *PPG* 30, 6 notes that the term hummocky moraine can be problematic.

End moraines/terminal moraines are till ridges, usually less than 60 m high, marking the end of a glacier. In plan, they are crescentic, corresponding with the lobes of the glacier; a well-developed end moraine shows that the ice front was there for some time. End moraines often constitute alluvial fans, formed at the ice margin by the redeposition of supraglacial material.

Recessional moraines mark stages of stillstand during the retreat of the ice; a transverse moraine is a recessional if the up-glacier surface shows streamlining. **Rogen moraines** are fields of transverse moraines, 10–30 m in height, up to 1 km long, 100–300 m apart, and often linked by cross-ribs. Their origin is uncertain. **De Geer moraines** form where a glacier meets its *proglacial lake, and consist of till, layered sand, and lake deposits (Evans et al. (2002) *J. Quat. Sci.* 17, 3). **Push moraines** occur when a glacier is retreating in the melt period but advancing in the cold season. Glaciohydraulic supercooling may aid the development of moraines created by melt-out from basal ice.

moral geography Human impacts on the Earth system have profound moral consequence, and moral geographies look at global variations in moral beliefs and the way these interact within spaces and places; geographical scale plays a crucial role in the construction and maintenance of moral frameworks. Moral questions for geographers include: does distance diminish responsibility? Should we interfere with the lives of those we do not know? Is there a distinction between private and public space? Which values and morals, if any, are absolute, and which cultural, communal, or personal? And are universal rights consistent with respect for difference? Human impacts on the Earth system have profound moral consequences. The uneven generation and distribution of harms and the acceleration of human forces now altering how the Earth system functions also trouble moral accounts of belonging. See Schmidt (2019) *TIBG* 44, 4.

morbidity Ill health. **Morbidity rates** are of two types: the *prevalence rate* (the numbers suffering from a specific condition at any one time), and the *incidence rate* (the numbers suffering from a particular condition within a given time period, usually a year).

more economically developed country (MEDC) A country with: low birth, death, and infant mortality rates (characteristically around 10; around 10; and under 12 per thousand, respectively); less than 10% of the workforce in agriculture; high levels of nutrition, secondary schooling, literacy, electricity consumption per head; and average incomes above $15 000 per person, per year. 'Economically' was included in this term some 30 years ago, to redress the perceptions of financially richer nations as more civilized, and more rational.

more-than-human world (more-than-human geographies) The non-human world, and its geographies; a contemporary geographical iteration (repetition and reworking) of the way the social, the cultural, and the political are developed in relation to a range of non-human forces and potentials. The more-than-human approach emphasizes the ways in which humanity totally depends on non-human things, which are often beyond human control or understanding: 'physical things are more than just inert matter that humans act upon. Rather, objects can influence, channel, and have impact upon our lives'; see Ray and Thomas (2018). See also Srinivasan (2019) *TIBG* 44, 2.

morphodynamics In geomorphology, the study of the interdependence of processes and morphology; the interaction between them. This might be the study of the evolution of landscapes in response to the erosion and deposition of sediment, at scales from the instantaneous to the long term.

morphogenesis The process or processes that initiate the development of a landform; hence **morphogenetic factors**. See Kompani-Zare et al. (2011) *Catena* 86, 3, 150.

morphogenetic region Any region whose geomorphology and geomorphological processes are recognizably distinct. For an example, see Abtahi et al. (2013) *Euro. J. Experiment. Biol.* 3, 4.

morphological impact assessment system (MImAS) A tool for assessing the impact of modifications to a river on river morphology and associated ecological quality. The inputs to the tool are:

- length and type of engineering modification
- length of the river or reach to be assessed
- channel type.

Each type of modification has a particular impact rating for each channel type, which is multiplied by the length of channel affected, and scaled to the reach assessment length:

$$\text{impact} = \left(M \times (R/L) \times 100 \right)$$

where M is the modification impact rating, L is the length of the modification, and R is the length of the reach being assessed. See Gilvear and Jefferies in J. Holden (2012) p. 360.

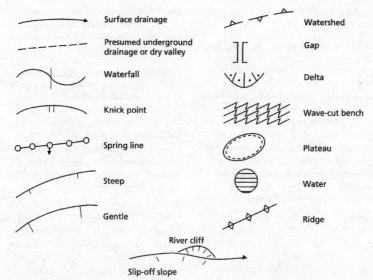

Surface drainage		Watershed	
Presumed underground drainage or dry valley		Gap	
Waterfall		Delta	
Knick point		Wave-cut bench	
Spring line		Plateau	
Steep		Water	
Gentle		Ridge	
	River cliff		
	Slip-off slope		

Morphological mapping

morphological mapping Mapping landform segments and facets using breaks of slope and inflections. 'Landforms are discrete units that can readily be defined and verified at different scales by proven techniques, are therefore acceptable integrated classifiers of the landscape, and can be used to divide it into discrete segments' (Bocco et al. (2001) *Geomorph.* 39, 3–4).

morphology 1. The study of form; the term is now used as a synonym for the form itself, as in the morphology of the landscape or of the city.

2. In geolinguistics, the phonological shapes of words that adapt to special grammatical functions. See U. Ammon (2005).

morphometric analysis The quantitative depiction and analysis of a landform.

morphometry The process of measuring the external shape and

dimensions of landforms (for example, Federici and Spagnolo (2004) *Geografiska A* 86, 3 on cirques), or living organisms (for example, Wilfert et al. (2006) *J. Biogeog.* 33, 11 on termites).

morphotectonics Also known as tectonic geomorphology, this is the study of *tectonic impacts on landforms. These include faulting, folding, or crustal subsidence. See Thapa, Ravindra, and Sood (2008) *Int. Arch. Photogrammetry* 37 for a helpful abstract and paper.

mother cloud A cloud from which another cloud develops; for example strato-cumulus cumulo-genitus (from cumulus). See Curic et al. (2002) *EGS XXVII Abstract* #3084.

moulin (glacier mill) A rounded, often vertical hole within stagnating glacier ice. Meltwater, heavily charged with debris, swirls into the hole. Some of this debris settles out at the base of the moulin, forming a **moulin kame** after the ice retreats. Moulins are ephemeral features.

mountain building The creation of uplands by movements of the Earth's crust. This is often, but not necessarily, associated with an *orogeny. Possible causes of the uplift of mountains are: a decrease of the density of the crust causing it to rise, possibly above a *hot spot; thickening of the crust at *collision margins; subduction below continents, causing them to rise; overthrusting of sedimentary rocks during collisions; and compression at *converging or at *conservative plate margins.

mountain ecology A change in altitude up a mountain equals a latitudinal shift away from the equator, and vegetation types change accordingly. The sequence in the San Francisco mountains is hot desert cactus at 200 m; oak scrub at 800 m; coniferous forest at 2000 m; and alpine tundra at 4000 m.

mountain meteorology Upland areas are cooler than lowlands of the same latitude, since temperatures fall by 1 °C/150 m. The heating of a valley floor may cause *anabatic winds, and colder and heavier air at the peaks may spill down as a *katabatic wind. When warm, moist air rises over mountain barriers, *orographic rain falls; when this air has passed over the barrier, it descends and is adiabatically warmed, which may bring *föhn winds. The warmed air can now hold the remaining moisture, so a 'rain shadow' of drier air develops in the lee of mountains. Lee depressions form downdraught of mountains, and major barriers such as the Himalayas affect the upper atmosphere. See C. D. Whiteman (2000).

mountain wind A wind formed when dense, cold air flows downslope from mountain peaks, or from a glacier. See KATABATIC WIND; BORA.

mud flat An accumulation of mud in very sheltered waters. Mud is carried into *estuaries and sheltered bays and settles at low water.

mud flow The *flow of liquefied clay. Debris of all sizes, including large blocks of rock, may be transported by mud flows. Mud flows in small drainage basins are usually triggered when the rainfall has reached a value equal to 10–20% of the local mean annual rainfall (MAR). Mud flows in larger basins are triggered when rainfall is equivalent to 15–30% of the local MAR.

mudslide A mass movement of *debris lubricated by melting ice and snow, or by heavy rain. Extreme *precipitation, together with anthropological features, such as felling and deforestation, and periodic bush fires are the major factors initiating mudslides. See also JÖKULHLAUP; LAHAR.

mull In soil science, a mild humus produced by the decomposition of grass or deciduous-forest litter. Earthworms and other soil fauna mix this humus thoroughly with the mineral content of the soil. See Ampe and Langohr (2003) Catena 54, 3.

multicell storm A severe local storm comprising a succession of convective cells. The first cell is formed by a powerful updraught of moist air. As the ensuing downdraught stifles this updraught, a new updraught develops to the right of the first. This sequence continues until the surface air cools. See Fovell and Tan (2000) QJ Royal Met. Soc. 126, 562.

multiculturalism Broadly speaking, the principle that different cultural or ethnic groups within a society have a right to remain distinct rather than assimilating to 'mainstream' norms; this is 'the diverse policies and ways of thinking about societies characterised by cultural plurality' (Nash (2005) *Antipode* 37, 2). 'The political accommodation by the state and/or a dominant group of all minority cultures, defined first and foremost by reference to race or ethnicity; and more controversially, by reference to nationality, aboriginality, or religion' (Modood in I. McLean and A. McMillan, eds (2003)).

Liberal multiculturalism focuses on cultural diversity, celebrating ethnic variety, and teaching tolerance. Liberal multiculturalism encourages private, individual choices of identity.

Critical multiculturalism is concerned with 'majorities' as much as 'minorities', and considers the institutions and practices forming the whole society. It sees inequalities of power, and racism, as central, and emphasizes recognition and rights. Try Nyland (2006) *J. Progressive Human Services* 17, 2.

multi-model inference (MMI) A method of simultaneously comparing the evidence for multiple hypotheses, each represented as a model. Millington et al. (2011) *Geog. Compass*, DOI:10.1111/j.1749-8198.2011.00433.x review two complementary MMI techniques—model selection and model averaging—and highlight examples of their use by biogeographers.

multinational corporation *See* TRANSNATIONAL CORPORATION.

multinational state A state which contains one or more ethnic groups, as identified by religion, language, or colour. For example, citizens of Tashkent, Uzbekistan maintain the pre-Russian Revolutionary cultural traditions of Muslim central Asia within the aesthetic of a modernized, multinational, Russianized society, based on Soviet-era institutions.

multiple nuclei model A representation of urban structure based on the idea that the functional areas of cities develop around various points rather than just one in the central business district. Some of these nuclei are pre-existing settlements, others arise from urbanization and *external economies. Distinctive land use zones develop because some activities repel each other; not all land users can afford the high costs of the most desirable locations, and some require easy access.

multiplier The economic consequence of an action, intended or otherwise. Once a location becomes established as a production centre, economies of agglomeration (which increase with increasing regional size) tend to give it permanent cost advantages over other locations. These scale and cost advantages are reinforced by the relatively higher wages paid to workers in the scale-intensive industries, which are thought to act, by means of a Keynesian income multiplier, as a stimulus to local markets, resulting in additional scale economies, which, in turn, lead to further growth of regional exports (Leichenko (2000) *Econ. Geog.* 76, 4).

In a study of China's expanding cities, Huang et al. (2021) *Appl. Sci.* 11, find that an inverted U-shape relationship is detected between employment multiplier and city size, namely the larger the city, the greater the employment multiplier, but when the city size exceeds a certain value, the employment multiplier begins to decline.

See Peck (2002) *J. Econ. Geog.* 2 on the multiplier effects of major factory closures: reduced demand, spending power, savings, and investment.

multi-scale landscape morphometry Digital images of

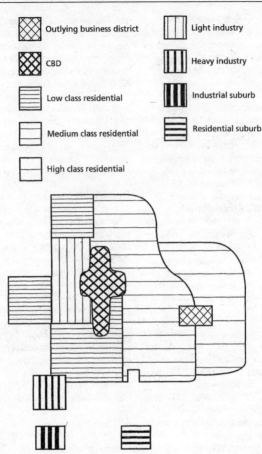

Outlying business district

CBD

Low class residential

Medium class residential

High class residential

Light industry

Heavy industry

Industrial suburb

Residential suburb

Multiple nuclei

landscape features may be captured in a sequence of successively coarsened resolutions. Hierarchy theory (see O'Neill et al. in M. G. Turner and R. H. Gardner 1991) suggests that patterns observed at any given resolution will, to some extent, weaken the patterns observed at finer resolutions and will themselves be constrained by coarser resolution patterns. Establishing the range of scales over which landscapes show hierarchically nested spatial patterns makes it possible to identify the scales at which fine-scale processes affect global scale patterns and vice versa. The selection of appropriate scales of measurement is fundamental to this process, and is generally guided by principles of hierarchy theory, since hierarchy incorporates relational links between nested levels. Tay et al. (2005) *Geosci. & Rem. Sens. Letts, IEEE* 2, 4 provide a simple framework to generate multi-scale digital elevation models and extract topologically significant multi-scale geophysical networks.

mythical geography Not 'unreal' geography, but the geography of myths: where they are recounted, and, where possible, the geographical locations of mythological events. Myths are the carriers of belief and value systems, and symbolic places that celebrate history and invest locations with mythical meaning provide a sense of identity in place and time; they fuse history and geography in terms of myth and memory. See Essebo (2019) *PHG* 43, 3.

naive geography The usage of the knowledge that people have about their surroundings, including *GIS and cartography, to improve their understanding of spatial information, and to help in decision-making (here, naive signals instinctive, not stupid, geography). It is the set of informal ideas of geographical relationships based on the human mind (see M. A. Nelson (2003)). For a comprehensive review of naive geography, see Egenhofer et al. in A. Frank and W. Kuhn, eds (1995).

nappe Part of a broken recumbent fold which has been moved forward over the rock formations beneath and in front of it, finally covering them; see van Hinsbergen et al. (2005) *Geology* 33, 4 on nappe stacking in the Aegean region.

narratives of place *See* PLACE.

nation The ethnic nation has ethnicity as the basis of membership, with culture as a key cohesive element. The cultural interpretation regards 'nation' as something organic, evolving on the basis of an aesthetic, primordial community passed down from the past, and hence the modern nation may be related to historical ethnicity. Culture and language are crucial cohesive elements of this community. In contrast, the territorial or civic nation has territory as the basis of membership and citizenship as the cohesive force. Although many *multinational states exist, E. Gellner (1983) argues that a mismatch between national and political units engenders

nationalist movements; see Sekulic (1997) *Nats. & National.* 3, 2.

nationalism This refers to the territorial expression of identity: a sense of belonging to a group or community associated with a particular territory. Often national identities do not correspond with the territory of the nation-state; many French-Canadians, for example, view Quebec as a distinct nation and believe it should become its own nation-state. Most states in the world are multinational states, meaning that they contain a variety of nationalities.

National Park An area less exploited by humans, containing sites of particular scenic or scientific interest, and protected by a national authority. 'Geography, and our identification with it, give us a sense of place. Geography also affects our national identity, and for many National Parks is the fundamental reason for their establishment as parks' (United States National Parks Authority).

There may be conflicts between the conservation of the natural environment and public access; 'where this happens, priority must be given to the conservation of natural beauty' (Lord Sandford, 1974). Imrie and Edwards (2007) *Geog. Compass* 2 record protests by locals over the Maya Biosphere Reserve, Guatemala. See Milian (2007) *Cybergeo* on the politics of 'sites naturels' in the Pyrenees.

Attempting to resolve post-colonial tensions, the management of many national parks in the developing world is shared between indigenous communities and the non-indigenous; see Waitt et al. (2007) *TIBG* 32, 2 on the Uluru National Park.

nation-state The term 'nation-state' is a mix of two linked, though different concepts, 'nation' and 'state'. Nations are groups of people who believe themselves to be linked together in some way, based on a shared history, language, religion, other cultural practices, or links to a particular place. States are legal and political entities, with power over the people living inside their borders. In this way, states are associated with territorial sovereignty. The concept of a nation-state fuses together the nation—the community— and the state, yet the nation-state rarely occurs in practice. There are more nations than states; several presumed nation-states are actually empires; there are diverse and significant political units in addition to states, including the EU; and political units overlap, and rarely politically 'saturate' the territory where they are located. Nonetheless, the nation-state is an important crossroads between the global and the local; it is the critical site for geopolitical study, and Rodrik (2013) *Econ. Geog.* 89, 1, 1 argues that the nation-state is the main determinant of the global distribution of income, the primary locus of market-supporting institutions, and the chief repository of personal attachments and affiliations. **Nation-building** has as its aim the unification of its people, via state action, to promote stability and viability. See Schatz (2003) *Kellogg Institute Working Paper #303* on nation-building in Kazakhstan, which, at final independence comprised 39% ethnic Kazakhs.

natural area 'A geographical area characterized both by a physical individuality and by the cultural characteristics of the people who live in it' (Zorbaugh in J. Lin and C. Mele, 2005). Most cities contain numbers of natural areas, which are delimited by informal boundaries, such as canals, parks, railways, rivers, and main roads.

natural increase The surplus or deficit (negative natural increase) of births over deaths.

natural region A region unified by its physical attributes, especially its latitude, relief and structure, and location. The 'natural' region is a highly contested concept. Even so, in Canada, federal and provincial park and protected area system plans adopt natural region representation approaches (Lemieux and Scott (2005) *Canad. Geog./Géog. canad.* 49, 4).

natural resource Any property of the physical environment, such as minerals, or natural vegetation, which can be exploited by humans. Natural resources may be classified as *renewable and *non-renewable. The journal *Natural Resources Forum* covers technical issues of extraction, development, and efficient exploitation of resources, and international, multidisciplinary issues related to sustainable development and management of natural resources, particularly in developing countries and countries in transition.

natural resource evaluation Assessing the economic value of the natural environment. Key considerations include: an assessment of the importance of each resource to human life in a particular context, how far one resource can be replaced by another, whether the resource is a stock or a flow, and whether it is exhaustible or renewable. For the analytic tools used in natural resource evaluation, see S. Schueller et al. (2006).

natural resource management (NRM) The management of natural resources to promote development that is economically viable, socially beneficial, and ecologically *sustainable. For the methodologies used in natural resource management, see Viergever et al. (2006) *Geomatics World* 4 and G. Herath and T. Prato (2006).

natural vegetation In theory, the grouping of plants which has developed in an area without human interference; since most landscapes have been changed by humans, it is argued that there are relatively few areas of truly natural vegetation left; many biogeographers use, instead, the concept of **potential natural vegetation**—the vegetation that would exist today if humans were removed from the scene and if the plant succession after their removal were telescoped into a single moment. The time compression eliminates the effects of future climatic fluctuations, while the effects of humans' earlier activities are permitted to stand. The potential natural vegetation reveals the biological potential of a site.

nature Commonly biophysical processes and the morphology of the landscape unchanged by man; space, the air. Nature is not a relatively fixed forcing factor, rather, it co-evolves with society. 'In a world with multiple realities of "nature" that are constructed by diverse groups with differing motivations and access to power, decision-making in environmental management can become a contest over whose knowledge is "right"' (Pedynowski (2003) *PHG* 27, 6).
See Sheppard (2011) *Dialogues Hum. Geog.* 1 46: it's clear and thought provoking.

nearest neighbour analysis The study of a spatial phenomenon in order to discern any regularity in spacing by comparing the actual pattern of that phenomenon with a theoretical random pattern. For the methodology, see Mak (2003) *Geog.* 516, 4. Alternatively, most *GIS software can do it.

nearshore In a coastline, the area comprising the swash, surf, and breaker zone; the area in which waves are forced to break owing to the shallowing of water closer to the shoreline.

nearshoring The relocation of work by a transnational corporation to a neighbouring country that has lower costs, such as labour, taxes, energy, or land. This proximity allows for the sharing of common cultural norms, and having similar time zones. Within Europe, there are major nearshoring centres in Bulgaria, the Czech Republic, Hungary, Poland, Slovakia, and Romania. Meyer (2006) mpra.ub.uni-muenchen.de/5785/, is strong on relevant data for eastern and central Europe.

nebkha *See* SAND DUNE.

nehrung A synonym for *barrier beach.

neighbourhood A district forming a community within a town or city, where inhabitants recognize many of each other by sight. See C. Moughtin (2005). Different neighbourhoods feel different not just because of their residents' socioeconomic status, demographics, or built environments, but also because of the actual personality of their residents: extraversion, neuroticism, openness to experience, agreeableness, and conscientiousness. See Jokela et al. (2014) *Proc. Natl Acad. Sci. USA* 112. Schnur (2005) *Tijdschrift* 96, 5 shows that local social capital can make the crucial difference in neighbourhood development.

neighbourhood effect The power of the environment to structure social contacts. Johnston et al. (2004) *Polit. Geog.* 23, 4 find that neighbourhood effects are among the strongest influences on voting patterns. There is also experimental evidence of long-term neighbourhood effects on adult income attainment, health, and other important outcomes; moves to a neighbourhood that is less poor during childhood are associated with higher adult earnings, for example. See Chetty et al. (2016) *JAMA* 15, 16.

neoclassical economics
Neoclassical economics focuses on the behaviour of individual consumers and firms. Consumers and firms interact in markets, which set the prices of goods and services. Taking into account their limited incomes and the prices set in the market, consumers buy the goods and services that maximize their well-being. Taking into account the prices set in the market, firms make decisions, including location decisions that maximize their profits. Firms compete with one another, and this competition ensures that they will seek out and take advantage of profitable opportunities.

Neoclassical economics suggests that surplus labour and *capital will move to areas where labour and capital are in short supply, so that uneven development will be eradicated (Taylor (2007) *Globalizations* 4, 4). Other important ideas include *cluster development and trickle-down economics—but see Basu and Mallick (2007) *Camb. J. Econ.* 32, 3.

neo-colonialism The control of the economic and political systems of one state by a more powerful state, usually the control of a developing country by a developed one. It is marked by the export of *capital from the least economically developed countries

(LEDCs) on the periphery to the controlling more economically developed countries (MEDCs) at the core, a reliance by the LEDCs on imported manufactures from the MEDCs, and adverse *terms of trade for the periphery or satellite. 'Neo-colonialism is the regressive impact of unregulated forms of aid, trade and foreign direct investment; and the collaboration of African leaders with foreign leaders to ensure that the interests of both are met with little concern for the development, sustainability and poverty reduction and wellbeing in African countries. The relationship is asymmetrical or at the cost of African states and their people, who are dependent rather than inter-dependent and do not profit through development or sustainability. They face destruction of their culture, religion and education through continued advancement of foreign culture, religion and language to supplant the African and growing radicalisation of the population' (Segell (2019) *Insight on Africa*). The means of control are usually economic, including trade agreements, investment, and the operations of *transnational corporations, which are often seen as the primary instruments of neo-colonialism.

neo-conservatism A particular version of conservatism that holds that security is best attained by using state power to spread freedom and democracy, if necessary by force and without international cooperation; a view strengthened in the USA after 9/11. Elden (2007) *Int. Politics* 44, 1 writes that the war on Iraq violated territorial sovereignty, even though neo-conservatives see territorial sovereignty as an absolute. See also Yamazaki (2002) *Geopolitics* 7, 1 on neo-conservatism in Japan.

Neogene The current geological period of the *Cenozoic era, consisting of the Miocene, Pliocene, *Pleistocene, and *Holocene epochs. The Neogene began 23 million years ago (E. Martin and R. Hyde 2008).

neogeography New forms of geographical knowledge, the proliferation of Web-based geographic information technologies, and the parallel opening up of mapping to non-expert cartographers. See Leszczynski (2014) *AAAG*, 104. These neogeographical data range from place tags on virtual globes to GPS data that can be mapped and combined with other data to create large, dynamic, open datasets. Connors et al. (2012) *AAAG* 102, 6, 1267 have a long and detailed survey, with many case studies.

neoglacial A time of increased glacial activity during the *Holocene. Lamoureux and Cockburn (2005) *Holocene* 15 suggest that increased moisture, perhaps driven by shifts in oceanographic conditions, were important in North American neoglacials.

neoliberalism It seems generally to be agreed that this term refers to the liberalizing of global markets associated with the reduction of state power: state interventions in the economy are minimized; privatization, finance, and market processes are emphasized; capital controls and trade restrictions are eased; free markets, free trade, and free enterprise are the buzzwords. Beyond that, definitions become more partial; D. Harvey (2005), for example, speaks of 'the doctrine that market exchange is an ethic in itself, capable of acting as a guide for all human action, and the reduction of the obligations of the state to provide for the welfare of its citizens ... unfettered individual rights'. Mainstream scholars depict

neoliberalism as an innovation of capitalism, while *political economists view the rise of neoliberalism as a response to a capitalist crisis of accumulation. Additionally, Birch and Mykhnenko (2009) *J. Econ. Geog.* 9, 3, 355 argue that whereas **neoliberal ideology**—based on abstract economic concepts such as free market efficiency—represents a global discourse that developed at diverse sites around the world, neoliberalism as a state-led project produced national varieties of neoliberalism, in which deregulation, privatization, and trade liberalization were pursued for different political reasons, in different ways, and to different extents. Neoliberalism has been hybridized and mutated as it travels around the world; see Springer (2010) *Geog. Compass* 4, 8.

Case studies of the impacts of neoliberalism abound; see, for example: Budds (2004) *Sing. J. Trop. Geog.* 25 on Chile; McCarthy (2006) *AAAG* 96, 1 on British Columbia; Ryan and Herod (2006) *Antipode* 38, 3 on Australia and Aotearoa; Gökarkisel and Mitchell (2005) *Global Networks* 5, 2 on Turkey; Fisher (2006) *Soc. Justice* 33 on race, neoliberalism, and 'welfare reform' in Britain; and Emery (2006) *Soc. Justice* 33, 3 on contesting neoliberalism in South Africa. See Liverman and Vilas (2006) *Ann. Rev. Env. & Resources* 31 on neoliberalism and the environment.

Attention has also been paid to **governance/governmentalities of neoliberalism**: see Larner and Butler (2005) *Studs Polit. Econ.* 75 on New Zealand; McCarthy (2004) *Geoforum* 35, 3. **Spaces of neoliberalism** are addressed by N. Brenner and N. Theodore, eds (2003), R. Basu (2004), W. Larner and W. Walters (2004); and **scales of neoliberalism** by Kohl and Warner (2004) *Int. J. Urb. & Reg. Research* 28, 4.

neotectonics The study of the causes and effects of the movement of the Earth's crust in the Neogene, i.e. the late *Cenozoic era. See Székely et al. (2002) EGU *Stephan Mueller Spec. Pub. Sers.* 3 on neotectonic movements and their geomorphic response.

nesting The way in which one network fits into a larger one. For example, Inkpen et al. (2007) *Area* 39, 4, see scale as a **nested hierarchy**. Species communities often exhibit **nested assemblages**, the species found in species-poor sites representing subsets of richer ones (Schouten et al. (2007) *J. Biogeog.* 34, 11).

net primary productivity In ecology, the amount of energy which primary producers can pass on to the second *trophic level; see Zhao et al. (2012) *PLoS ONE* 7, 11 DOI:10.1371/journal.pone.0048131.

net radiation The balance of incoming solar radiation and outgoing terrestrial radiation (also called **net radiative balance**).

(()) SEE WEB LINKS
- Michael Pidwirny on net radiation and the planetary energy balance.

network A system of interconnecting routes which allows movement from one centre to the others. Most networks are made up of nodes (*vertices*), which are the junctions and terminals, and links (*edges*), which are the routes or services which connect them. **Situated networks** are simply networks in which the nodes are placed in space at their actual geographic locations.

   Network connectivity is the extent to which movement is possible between points on a network—cities' housing firms with high service values will record large measures of network connectivity.

See McLaughlin (20 September 2018) Center for Data Innovation.

network society A society whose social structure is made up of networks powered by communication technologies (Castells (2004), p. 3). According to Castells, the network society has emerged through: the restructuring of industrial economies to accommodate an open-market approach; the freedom-oriented cultural movements of the late 1960s and early 1970s, including the civil rights movement, the feminist movement, and the environmental movement; and the revolution in information and communication technologies. Hampton and Wellman (2001) *Amer. Behavioral Scientist* 45, 476 have an interesting paper on Netville, Ontario: see especially pp. 491–2.

neural network An artificial neural network (ANN) is a way of processing information that is inspired by the way biological nervous systems such as the brain process information. It is made up of a large number of highly interconnected processing elements (neurones) working together to solve specific problems, such as pattern recognition or data classification. Like people, ANNs learn by example, adjusting the connections that exist between the neurones. 'Neural networks, with their remarkable ability to derive meaning from complicated or imprecise data, can be used to extract patterns and detect trends that are too complex to be noticed by either humans or other computer techniques.' This quote is from Christos Stergiou and Dimitrios Siganos, 'What is a Neural Network', Imperial College.

(()) SEE WEB LINKS
- Imperial College computing website.

neural turn The description of a trend by some geographers to consider the intersections of geography and neuroscience; cognition, mental processes; the way geographical concepts have been approached within neuroscience. See Pykett (2018) *TIBG* 43, 2.

névé An alternative term for *firn.

new economic geography An economic geography emphasizing relational, contextual, and social economic behaviour, and, particularly, the importance of tacit knowledge. It is explicitly concerned with understanding uneven spatial economic development, and particularly the spatial agglomeration of economic activity; 'a body of research which fundamentally attempts to explain the formation of economic agglomeration (or concentration) in geographical space' (Fujita and Krugman (2004) *Papers Reg. Sci.* 83, 139). See also Garretsen and Martin (2010) *Spatial Economic Analysis* 10, 2.

new economy Most new-economy approaches argue that computers, and in particular networked PCs, have changed things in a fundamental way. The new economy is typified by new styles of production, novel forms of consumption and organization in everyday life, new horizons of planning and logistics of mobility, and new forms of materiality built around smaller enterprises distributed in a much more dispersed geography. It relies on the old economy for physical capital and a portion of demand, and the location of business activity. It also relies to a growing extent on telephone-based networks for production and delivery, but also has leading-edge layers that require face-to-face contact. The key players in the new economy are service activities.

new industrial spaces Agglomerated production systems that coordinate inter-firm transactions and entrepreneurial activity. See Hansen and Winther (2007) *Danish J. Geog.* 107, 2 on Copenhagen.

new international division of labour (NIDL) A global division of labour associated with the growth of *transnational corporations and the *deindustrialization of the advanced economies. The most common pattern is for research and development in *more economically developed countries, and cheap, less skilled labour in *less economically developed countries. The impacts of the NIDL have been uneven: between nations, where some benefit more than others, and within nations, in locally specific ways. See Marin (2006) *J. Eur. Econ. Ass.* 4, 2–3 on outsourcing.

newly industrializing country (NIC) A country that is still developing economically, through industrialization and urbanization, more strongly than most other developing countries. Most have seen rapid industrialization since the 1970s. See Sum in H. Beynon et al., eds (2001) on East Asian newly industrializing countries. See also N. Forbes and D. Wield (2002).

new political economy The economic study of politics, with a macroeconomic focus; a focus on the whole economy. Due to its 'ample nature', different economists have different understandings of what 'macroeconomic' means, so their definitions may clash; see Almeida (2018) *CHOPE Working Paper* no. 2018-16. See Bradshaw (2008) *New Political Economy* 13, 2.

new regionalism A process-based, diverse, collaborative, and open mode of governance, not territorially fixed. Instead of stressing the importance of local versus central, it emphasizes both the causal power of the endemic uncertainty of the system, and under institutionalization. It seems to argue that a great deal of economic and political change might best be thought of as endogenous to local units (a product of local realities), rather than simply different reactions to national priorities.

E. LeSage and L. Stefanick (2004) provide a useful summary table:

New regionalism	Old regionalism
*Governance	Government
Process	Structure
Open	Closed
Collaboration	Coordination
Trust	Accountability
Empowerment	Power

new town A newly created town, planned to relieve overcrowding and congestion in a major conurbation. In the UK, new town housing was arranged in *neighbourhood units of around 5 000 people with their own facilities such as shops, schools, and medical centres, and housing densities were low—about five houses per hectare. All British new towns were planned to give a balance of social groups. See Nakhaei et al. (2015) *Cumhuriet Science Journal* 36, 3 for an analysis of the development of new towns in response to metropolitan needs, and Ward (2002) on planning the 20th-century city.

new urbanism A way of fostering a community by basing its plan on an old European village, with homes and businesses clustered together. Residents can walk to shops, businesses, theatres, schools, parks, and other important services. It is a movement dedicated to improving the quality of life in cities by changing urban form. This movement, according to Talen (1999) *Urb. Studs* 36, 8, 1361, holds to 'an unswerving belief in

the ability of the built environment to create a "sense of community"'. Read her review of this belief.

niche A set of ecological conditions which provides a species with the energy and habitat which enable it to reproduce and colonize. A niche is usually identified by the needs of the organism. See Chave et al. (2002) *Am. Nat.* 159 on niche differentiation and the maintenance of species diversity. Etienne and Olff (2004) *Am. Nat.* 163 develop a simple model uniting a neutral community model with niche-based theory.

nimbo-, nimbus Referring to clouds bringing rain, as in nimbo-stratus clouds.

nitrogen cycle The cycling of nitrogen and its compounds through the *ecosystem. Most plants obtain the nitrogen they need as inorganic nitrate from the soil solution. Animals receive the required nitrogen they need for metabolism, growth, and reproduction by the consumption of living or dead organic matter containing nitrogen in molecules.

(⊕) SEE WEB LINKS

• Michael Pidwirny on the nitrogen cycle.

nitrogen fixation The alteration of atmospheric, molecular nitrogen to nitrogen compounds. The fixation mechanisms are: biological micro-organisms (such as those in the root nodules of leguminous plants), *lightning and other natural ionizing processes, and industrial processes.

nivation The effects of snow on a landscape. The main processes are backwall failure, sliding and flow, niveo-aeolian sediment transport, supra- and ennival-sediment flows, niveo-fluvial erosion, development of pronival stone pavements, accumulation of alluvial fans and basins, and pronival solifluction. For the factors influencing the effects of nivation, see Raczkowska (1995) *Geograf. Annal. A* 77, 4. Nivation may produce the shallow pits known as **nivation hollows**. In time, these hollows may trap more snow and may deepen further with more nivation so that *cirques or *thermocirques are formed; see Kariya (2005) *Catena* 62, 1.

node The accessibility of a node is given by the travel cost between a node and another node, or a weighted average of travel cost to a number of nodes in the network. Transport nodes show a distinct hierarchy, and the importance of a transport node appears to correlate strongly with its effects on the surroundings. Factors that determine the importance of a transport node are: the situation of the transport node with respect to major metropolitan regions; the number and type of the converging transport modes; the number of destinations that can be reached from the transport node, and the frequency of the transport moves. The more important a transport node, the greater the potential regional economic development. See de Langen and Visser (2005) *J. Transp. Geog.* 13, 2.

nomadism A form of social organization where people move from place to place. Nomadism incorporates the advantage of mobility; traditionally nomadic groups were able to exploit natural resources at dispersed locations. 'In the discourse of modernization and social change, nomadism's place is usurped by agriculture' (Kreutzman (2003) *Geog. J.* 169, 3). As international boundaries are increasingly well defined, with border guards, nomadism declines: N. M. Shahrani (1979) uses the term **closed frontier nomadism**, with sedentarization and confined migration cycles.

Governments try to immobilize nomads for the purposes of taxation, as well as to improve their nomads' health and literacy; although the mobility of capital and labour in the age of empire produced a spatial order where nomadism may be seen as a cultural problem, some local authorities keep some families mobile and excluded. B. Jordan and F. Düvell (2003) argue that **global economic nomadism** requires a redefinition of citizenship beyond national borders which involves shared duties for those who have access, and rights for those who remain on the outside.

nominal On a *GIS map, a geographic feature that has a name only.

nomothetic A scholarly approach that stresses the value of general principles that apply everywhere in space and time.

nonconformity A series of sedimentary strata overlying an igneous or metamorphic rock.

non-ecumene The uninhabited or very sparsely populated regions of the world. It is not easy to draw boundaries between the *ecumene and the non-ecumene as regions of dense occupation merge into sparsely populated regions. If there is a boundary, it is not static.

non-linear (nonlinear) Describing a system in which the output is not uniformly related to the input (not a straight-line graph). All the component parts of the Earth system interact on each other, which is an expression of non-linearity.
  See Phillips (2003) *PPG* 27, 1 on non-linearity in geomorphic systems.

non-place An indistinct location, with nothing to distinguish it: a shopping mall, airport, supermarket; anywhere and nowhere.

non-renewable resource A finite mass of material which cannot be restored after use, such as natural gas. (S. Barr and P. Chaplin (2004) argue that cultural heritage is a non-renewable resource.) Each non-renewable resource has a theoretical depletion time—a ratio of its stock and rate of use. Non-renewable resources may be sustained by *recycling.

non-representational theory Representational describes the ways that words, images, or other media capture/represent the characteristics of things. However, M. Shapiro (1988) argues that representations do not 'imitate' reality but are the practices—visual, aural, oral, quantitative, or qualitative—through which things take on meaning and value. This means that academics have to move beyond mere representation. M. Doel (1999) provides a helpful parallel. The task of a painter, he says, is not to paint an object, or even to represent it—but to be that object's effects. This does not mean that descriptions are no longer admissible. Close descriptions can still be offered while attending to the situated, embodied sense-making work being undertaken by the peoples involved, the work that makes those encounters what they are.

Non-representational theorists argue that 'representation' generates an unwavering, deadened picture of the world. They emphasize knowing stuff through connection and participation; the spotlight is on the process, rather than the outcome—'it ain't what you get but the way that you got it'.

This theory chimes with the concept of *hybrid geographies. Lorimer (2005) *PHG* 29, 1 offers the term 'more-than-representational'. Try Colls (2012) *TIBG* 37, 3, 430 on feminism and non-representational geography.

normative In general, relating to an ideal standard (normal), judging behaviour and actions as 'right' or 'wrong', good or bad, permissible or impermissible, desirable or undesirable. More specifically, describing any academic argument that criticizes current arrangements and calls for the creation of a better future.

North Following the terminology of the Brandt report (1979) a portmanteau term used to describe the advanced economies/more economically developed countries.

North Atlantic Deep Water (NADW) A water mass largely formed in the Labrador Sea and the Greenland Sea by the sinking of highly saline, dense water. It is part of the 'conveyor belt' model of the *thermohaline circulation. It has been suggested that global warming may cause sufficient freshwater to be released from the melting of ice in the northern latitudes to slow or even to stop the formation of the NADW by reducing salinity to the level where it is insufficiently dense to sink. If this occurred it would stop the Gulf Stream flowing north, and thus cause the oceanic conveyor belt, which is transferring heat from the equator, to cease. This would have a dramatic cooling effect on the climate of north-west Europe (Krom in J. Holden (2012), p. 61).

North Atlantic Drift Also known as the Gulf Stream, this is a warm surface ocean current originating in the Gulf of Mexico, which widens to several hundred kilometres, and slows to less than 2 km/h, and then divides into several sub-currents, one of which is the North Atlantic Drift. The intensified westerlies in the northern North Atlantic and strengthened northerlies along the western part of the Nordic Seas force the North Atlantic Drift to follow a more easterly path, giving an asymmetric sea surface temperature response in the northern North Atlantic, and thereby maintaining the properties of the Atlantic Water entering the Nordic Seas. See Rasmussen and Thomsen (2004) *Pal. Pal. & Pal.* 210, 1 on the role of the North Atlantic Drift in millennial glacial climate fluctuations.

North Atlantic Oscillation (NAO) A large-scale alternation of atmospheric mass between the Icelandic low- and the Azores high-pressure area. When the pressure difference is large, with a deep Icelandic low and a strong Azores high, the NAO is said to be high or positive, and said to be negative when the pressure difference is less (or occasionally negative) as a result of persistent blocking in the Iceland–Scandinavia area. Strong positive phases of the NAO tend to be associated with above-normal temperatures in the eastern USA and across northern Europe, and below-normal temperatures in Greenland and across south-eastern Europe and the Middle East. Opposite patterns are observed during negative phases (Perry (2000) *PPG* 24, 2—an excellent source).

The name **Arctic Oscillation** was introduced to highlight the fact that the pressure anomalies associated with the NAO span most of the Arctic. But 'oscillation' is something of a misnomer, as the northern and *southern annular modes oscillate regularly in time. For this reason, in recent years the NAO/AO nomenclature has been increasingly supplanted in the dynamical literature by the phrase **Northern Annular Mode (NAM)**. Annular denotes the longitudinal scale of the pattern, and annular mode suggests the NAM reflects dynamical processes that transcend a particular hemisphere or, for that matter, planet.

North Pacific Current An eastward current, originating east of the Emperor Seamounts (170° W) and forming the northern part of the North Pacific subtropical gyre. It maintains the Arctic Polar Front with the Pacific Subarctic current. See Dinniman and Rienecker (1999) *J. Phys. Oceanog.* 29, 4.

nuclear family The small family unit of parents and children. It may be that the institutionalization of the nuclear family household emerges from the processes of capitalism, as well as existing patriarchal processes. But D. Lal (1998) claims that 'capitalism is not inevitably connected with the nuclear family or even individualism, although its genesis depended on both'.

nuclear power A form of energy which uses nuclear reactions to produce steam to turn generators. Naturally occurring uranium is concentrated, enriched, and converted to uranium dioxide—the fuel used in the reactor. This undergoes nuclear fission, which produces large amounts of heat. Some of the highly radioactive spent fuel may be reprocessed while the bulk must be disposed of. Both are costly and hazardous undertakings. The main advantage of nuclear power is the low CO_2 emissions: 'nuclear power is a low-carbon form of electricity generation that can make a significant contribution to tackling climate change' (Meeting the Energy Challenge: A White Paper on Nuclear Power URN:08/525).

A major disadvantage is the risk of contamination. There are also very high construction and decommissioning costs. Furthermore, nuclear power stations have a short lifespan. Parkins and Haluza-DeLay (2011) are excellent on the social and ethical considerations of nuclear power development.

nucleated settlement A settlement clustered around a central point, such as a village green or church. 'The problem is to determine where dispersed settlement finishes and nucleated settlement begins' (B. Roberts, 1996). Nucleation is fostered by defence considerations, localized water supply, the incidence of flooding, or rich soils so that farmers can easily get to their smaller, productive fields while continuing to live in the village. 'Nucleated settlement does not in itself necessarily induce a sense of community . . . nucleated settlement in itself is only a part of the process of constructing and maintaining a sense of common interest' (C. Dalglish 2003).

nuclei Minuscule solid particles (sing. *nucleus*) suspended in the *atmosphere. Three types of atmospheric nuclei are distinguished: **Aitken nuclei**, of radii less than 0.1 μm (Heggs et al. (1991) *J. Geophys. Res.* 96, 18), **large nuclei**, radii 0.2–1 μm, and **giant nuclei**, with radii greater than 1 μm. See Clarke (1992) *J. Atmos. Chem.* 14.

These nuclei may be scraps of dust, from volcanic eruptions or dust storms, or salt crystals, or given off when bubbles burst at the surface of the sea. Atmospheric nuclei can scatter sunlight enough to lower temperatures, if enough are present, and play an important role in *cloud formation; see Kulmala et al. (2001) *Tellus B* 53, 4.

nudge theory argues that subtle behavioural cues, social norms, default positions, and even architectures may be used to empower citizens to make healthier decisions. This is from Jones, Pykett, and Whitehead (2014) *Env. & Planning C*, 32, but it's not an easy read. Nudge theory is of interest to those geographers who study the link between human decision-making and its spatial outcomes.

nuée ardente 'A glowing cloud of volcanic ash, *pumice, and larger *pyroclasts which moves rapidly downslope, typically starting as large blocks, and then fragmenting to a mixture of dust through to house-sized particles' (Mursik et al. (2005) *Rep. Prog. Phys.* 68).

numerical modelling The use of numbers and equations to describe physical and human phenomena. 'Numerical models can be useful for explaining poorly understood phenomena or for reliable quantitative predictions. When modelling a multi-scale system, a "top-down" approach—basing models on emergent variables and interactions, rather than explicitly on the much faster and smaller-scale processes that give rise to them—facilitates both goals' (Murray (2007) *Geomorph.* 90, 3–4).

nunatak A mountain peak which projects above an ice sheet, generally angular and jagged due to *freeze–thaw, contrasting with the rounded contours of the glaciated landscape below. Confirming a former nunatak is generally difficult. See Kleman and Stroeven (1997) *Geomorph.* 19, 1–2.

nuptiality The frequency of marriage within a population, usually expressed as a marriage rate. R. Woods (2002) suggests that the role of nuptiality in influencing fertility during the late 19th century was geographically variable and could not be attributed to some overarching explanatory factor.

nutrient cycle The uptake, use, release, and storage of nutrients by plants and their environments; the transformation of inorganic compounds into organic forms due to biological uptake, and then back to an inorganic state. The nutrient cycle is the main functional process maintaining the stability and production of an ecosystem, and land cover is the source and sink of the material and energy supporting the biosphere.

Schematic nutrient cycle

Nutrient cycling processes

A. Sources (inputs)
   4. Atmospheric inputs (e.g. dust, aerosol, biological fixation, precipitation)
   5. Weathering from parent material (primary minerals from rocks)
   6. Deposition of minerals through the movement of animals (e.g. movement of ungulates in Serengeti), soil (down slope), and seepage (drainage waters)

B. Storage capacities (pools)
   1. Living biomass
   2. Detritus (e.g. litterfall, dead/decaying roots, fallen logs and branches)
   3. Soil

C. Internal ecosystem cycling
   10. Litter inputs (above- and below-ground)
   11. Decomposition and mobilization processes
   12. Uptake, utilization, and redistribution processes by higher plants (e.g. trees primary sink for nutrients – $1/3$ of the total annual requirement of nutrients acquired from internal redistribution. Excluding 70% of N removed from the leaves before litterfall)

D. Losses (outputs)
   7. Volatilization to the atmosphere
   8. Harvest transport, and erosion
   9. Leaching below the rooting zone

Nutrient cycle

oasis A watered spot in an arid area, occurring when the *water-table reaches the surface.

objectivity A state of mind free of prejudice.

occlusion A stage in a *mid-latitude depression where the cold front to the rear catches up with the leading warm front, lifts the wedge of warm air off the ground, and meets the cold air ahead of the warm front. If this overtaking air is colder than the cold air ahead of it, it will undercut it, forming a **cold occlusion**. If it is warmer than the cold air which is ahead of it, it will ride over it.

occupancy rate The number of people dwelling in a house per habitable room (except for kitchens and bathrooms). In the UK, a rate of minus 1 implies that there is one room too few for the number of people living in the household. In 2011, England and Wales had 2 million households with an occupancy rate of minus 1 or lower.

occupational mobility The ease with which a worker can change jobs, job profile, or job content.

ocean–atmosphere interaction (ocean–atmosphere coupling) Oceans and the atmosphere constantly interact with each other; surface winds drive the *ocean currents, moving warm water polewards and cold water equatorwards, and evaporation from warm oceans removes latent heat from the atmosphere. This latent heat is released when the vapour condenses with height.

Ocean–atmosphere oscillations are interactions which switch suddenly from one phase to another; the best known being the *El Niño–Southern Oscillation.

ocean current A permanent or semi-permanent horizontal movement of unusually cold or warm surface water of the oceans, to a depth of about 100 m. Ocean currents are generated by the thrust of the wind, the spin of the Earth, and the Moon's gravity, and are also affected by variations in water

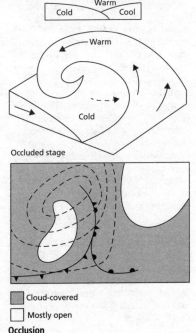

Cloud-covered

Mostly open

Occlusion

temperature and density, and by the *Coriolis force. These currents are an important factor in the redistribution of heat between the tropics and the polar regions; *see* THERMOHALINE CIRCULATION. The Southern Ocean, for example, is a crucial cog in the global heat engine. An increase in the sea-surface temperatures of 1–2 °C in the North Atlantic current may intensify tropical cyclones.

Cold currents can modify temperatures up to 100 km inland. As tropical air streams move over these currents, advection *fog forms over the sea, stripping the air streams of most of their moisture; onshore winds are therefore dry. Conversely, **warm currents** originate in tropical waters, bringing unusually warm conditions to coastal areas in higher latitudes. The *North Atlantic Drift, for example, creates a difference of around 4 °C between the winters of coastal Canada and coastal western Europe of the same latitude.

oceanic core complexes (OCC)

Massifs in which lower-crustal and upper-mantle rocks are exposed at the sea floor, formed at mid-ocean ridges through slip on detachment faults rooted below the spreading axis. The formation of these large, shallow sea-floor features appears to be an episodic outcome of plate rifting and accretion at slow spreading ridges. An outstanding example of an oceanic core complex has been mapped at the eastern intersection of the Mid-Atlantic Ridge and the Atlantis transform fault; see Planert et al. (2010) *Geophys. J. Int.* 181, 1, 113.

oceanic crust

That portion of the outer, rigid part of the Earth which underlies the oceans. It is made mostly of a basalt layer, rich in iron and magnesium, average density around 2.9, and 5–6 km thick, but see Rodger et al. (2006) *Geology* 34, 2.

oceanic ridge

An underwater mountain range developed where *magma rises up through a cracking and widening ridge in the oceanic crust. Some magma cools below the crust, some forces into fractures, and much flows out to form new crust, which is then pushed away from the ridge. The disparity in spreading rates between the Northern and Southern Hemispheres is correlated with the rate of latitudinal circumferential extension.

As new crust is created each side of the ridge, it takes up the prevailing *magnetic polarity of the Earth, which reverses from time to time. As a result, symmetrical bands of crust, with alternating polarity, develop on either side of the ridge. These magnetic patterns are used to calculate the rate of the *sea-floor spreading. The term **mid-oceanic ridge** properly refers to the ridge at the centre of the Atlantic Ocean, which comes to the surface at points such as Tristan da Cunha and Ascension Island. Other ridges, such as the Pacific–Antarctic ridge, are not truly at the centre of the ocean.

oceanography

The study of the oceans. This covers the shape, depth, and distribution of oceans, their composition, life forms, ecology, and water currents, and their legal status.

ocean trench

A long, narrow, but very deep depression in the ocean floor where two adjacent plates converge, and one descends beneath the other. For this reason ocean trenches are also known as convergent plate boundaries. An alpinotype orogeny is also termed an **ocean trench orogeny**.

offshore

1. In meteorology, moving seawards from the land, as in an offshore wind.

2. In geomorphology, the zone to the seaward side of the breakers.

offshore bar

See BAR.

offshore financial centre (OF)
Better known as 'tax havens', offshore financial centres—such as the Cayman Islands or Panama—are jurisdictions with a disproportionate level of financial activity for non-residents. This financial activity is usually dominated by the provision of intermediary services for larger neighbouring countries. By providing competition for domestic banks, offshore banks also lower the costs for anyone who had been using the domestic financial system. For example, simply by being close to Andorra and Monaco, France has a more competitive banking system, which provides more credit at lower interest rates. The indirect competitive benefits of offshore financial centres more than offset their users' costs.

offshoring For businesses, transferring portions of their work to outside, foreign suppliers rather than completing at home. Frequently, work is offshored in order to reduce labour expenses. See Jensen and Pedersen (2011) *J. Management Studs* 48, 2 on the economic geography of offshoring. *See also* NEARSHORING.

ogives Bands of dark and light ice that stretch across glaciers, in the shape of a pointed arch. Multiple shear zones are formed, through which basal ice is uplifted to the glacier surface to produce the dark, foliated ogive bands (Goodsell et al. (2002) *J. Glaciol.* 48, 161).

okta A measure of cloud cover, from 1 okta (scant) to 8 oktas (complete). Oktas are shown on *synoptic charts by a circle which is progressively shaded in as cloud cover increases.

oligotrophic Poor in nutrients, usually of lakes, soils, and peat bogs.

ombrotrophic Pertaining to vegetation or soils that gain nutrients and water from precipitation only.

onion weathering *See* EXFOLIATION.

ontogeny The origin and development of a phenomenon.

ontology The study of being. Ontology is made up of theories about knowledge. Each discipline has its own ontology; for example economists tend to think about labour in terms of wages and human capital; geographers typically see labour location as a result of worker attachments to complex social institutions and networks. See Boschma and Martin (2007) *J. Econ. Geog.* 7, 5 on ontology for evolutionary economic geography.

((⊕)) SEE WEB LINKS
• UK Ordnance Survey ontologies.

ooze A deposit either of algae, or fine inorganic sediments, developed on the sea floor, or at the bottom of lakes. See H. Armstrong and M. Brasier (2005). **Calcareous oozes** are made up of the plankton in the surface waters of the oceans.

opencast mining A system of mining which does not use shafts or tunnels and is, hence, cheaper. The layer above the mineral seam is removed and the exposed deposit is extracted using earth-moving machinery. The overburden may then be replaced. See Ghose and Majee (2001) *J. Env. Manage.* 63, 2, 193 on air pollution caused by opencast mining and its abatement measures in India.

open data Data that are freely available to everyone without mechanisms of control, such as copyright or patents.

open field A distribution of farmland associated with *feudalism in Europe. Each manor had two or three large open fields and each farmer was awarded a number of strips within each field, so that no one had all the good or all the poor land. 'From late in the seventeenth century both the open fields and land held in common . . . began to be enclosed' (Pounds (2005) *J. Hist. Geog.* 31, 3).

open source software Computer software that is distributed along with the code that is used to create the software, under a special software licence that allows users to use, change, and/or improve the software's source code, and to redistribute the software either before or after it has been modified. If any modifications are made to the source code, other developers must be told what has been changed, and how. Open source software is freely available, but not necessarily *free* in financial terms.

open system A system that is not separated from its environment, but exchanges material or energy with it. Soil may be seen as an open system (see Trudgill and Briggs, 1977, PPG). See Nordstrom (1994) *PPG* 18 on an open system model of beach change, and Douglas (1981) *PPG* 5 on the city as 'an integrated open system of living things interacting with their physical environment'.

ophiolite suite A sequence of rocks consisting of deep-sea marine sediments overlying a range of igneous phenomena, including basalts and *dykes. Ophiolite sequences are indicators of sea-floor spreading, as tectonic movements pull and thin the crust, allowing magma to rise up and create new crust. When the magma is cooled by the sea water, it forms a variety of igneous rocks and structures. See Saha et al. (2010) *J. Earth Sys. Sci.* 119, 3, 365 on a completely preserved ophiolite suite from South Andaman, India.

opportunist species (opportunistic species) Species which survive by the rapid colonization or recolonization of a *habitat: for example, the roach in Ireland (Caffrey and McLoone, *Environ 2003*). Such species have very good powers of dispersal in order to seek new habitats. Life is short, fecundity and reproduction are high, and these species are able to withstand difficult conditions. The term **fugitive species** is also used.

opportunity cost The second choice that you give up when choosing the best; for example, in choosing a better-paid job further away, commuting time is the opportunity cost. Baldwin and Okubo (2006) *J. Econ. Geog.* 6, 3 show that regional policies tend to attract the least productive firms since these have the lowest opportunity cost of leaving an agglomerated region. See Willis et al. (1998) *Transp. Res. D* 3, 3 on land prices and **social opportunity cost**.

optimization model A model used to find the best possible choice out of a set of alternatives. A simple example might be finding the most efficient transport pattern to carry commodities from the point of supply to the markets, given the volumes of production and demand, and transport costs. See Ward et al. (2003) *Annals Reg. Sci.* 37 on evaluating urban change.

optimizer A decision-maker who seeks to maximize profits; in choosing an industrial location, the profit maximizer is assumed to know all the relevant factors, including the cost of assembling materials and distributing products, the price of labour, and *agglomeration and *external economies. Among many who find optimizer a static and simplistic concept are Feola and Binder, *Tropentag 2007. Compare with* SATISFICING.

optimum location The best location for a firm in order to maximize profits; 'in many circumstances, the optimum location of the firm and the optimum production relationships can be shown to be interdependent. See Bradshaw (2001) *Australian Geogr.* 32, 2.

oral history The collection of living people's first-hand knowledge of persons, events, and places. See Riley and Harvey (2007) *Soc. & Cult. Geog.* 8, 3, 345 on oral history and the practice of geography.

ordinal Any phenomenon that can be compared by rank.

ordnance datum (OD) In the UK, mean sea level at Newlyn, Cornwall, from which all other spot heights on Ordnance Survey maps—and hence all *contours—are established.

Ordovician A period of *Palaeozoic time stretching approximately from 500 to 430 million years BP.

( ) SEE WEB LINKS
• An overview of the Ordovician period.

organic acid Acid compounds of carbon, such as acetic acid, which are produced when plant or animal tissues decompose. Organic weathering plays a significant role in the release of nutrients to plants. See Jones et al. (2003) *Plant & Soil* 248, 1–2.

organic weathering The breakdown of rocks by plant or animal action or by chemicals formed from plants and animals, such as *humic acids. Pope et al. (2002) *Geomorph.* 47, 211, while not central to the topic, should be useful.

Organization for Economic Cooperation and Development (OECD) A group of nations, comprising most of western Europe, together with Australia, Canada, Chile, Colombia, Costa Rica, Iceland, Israel, Japan, Mexico, New Zealand, South Korea, and the USA, and formed to develop

strategies which will boost the economic and social welfare of the member states.

( ) SEE WEB LINKS
• OECD homepage. The country-by-country reports are a treasure trove of data.

Organization of African Unity (OAU) An association of independent African states constituted in 1963, and designed to encourage African unity, to discourage *neo-colonialism, and to promote development.

Organization of American States (OAS) An association of Latin American states with the USA, constituted in 1948. It is based on democracy, human rights, security, and development.

( ) SEE WEB LINKS
• The OAS website.

orientalism A description of the West's depiction of the cultures of the Eastern Hemisphere, orientalism became a political term through the work of Edward Said (1978), for whom orientalism means the European academic and popular discourse about the Orient, expressed through colonial bureaucracies and styles, doctrines, imagery, scholarship, and vocabulary. Said argued that 'the Orient has helped to define Europe (or the West)'; to Western eyes, orientalism is what the West is not; so by representing the East as an exotic, bizarre, and inferior attachment, the Orientalists made colonial conquest a natural and logical extension of the rise of the West. Samuel Huntington (1993) *Foreign Affairs* 72, 3, for example, divides huge sections of the Earth into 'civilization groups', with Western civilizations at the top of the hierarchy.

H. Bhabha (1995) dismisses a clear separation between colonizers and colonized, recognizing 'the ambivalence, *hybridity, and mimicry found in colonial representations of Orientalised places and, at the same time, to the same phenomena in the self-representations of colonized peoples'.

orogeny A movement of the earth which involves the folding of sediments, faulting, and metamorphism. A **cordilleran orogeny** begins with sedimentation at a passive continental margin, ranging from coarse sand and silt near the shore, to limestone reefs in tropical seas. Fine-grained clastic sediments accumulate on the deeper continental slopes, forming shales and greywackes. The deposition of these *geosynclinal deposits is followed by *subduction and compression. Folds, thrust faults, and a volcanic arc form. Lateral growth continues through igneous activity, and metamorphism, uplift, and deformation result from continuing plate convergence. See Şengör (1999) *GSA Spec. Paper* 338.

Initially, a **continental collision-type orogeny** is similar to the cordilleran type. However, when two continental plates collide, both are too thick and too buoyant for subduction. Consequently, the plates are welded together to produce a large mountain chain. See the excellent Andersen et al. (2002) *Abstr. & Procs Norw. Geol. Soc.* 22, 12.

orographic precipitation (orographic rainfall) Also known as **relief rainfall**, this forms when moisture-laden air masses are forced to rise over high ground. The air is cooled, the water vapour condenses, and precipitation occurs. See Gray and Seed (2000) *Meteorol. Applics* 7 on the characterization of orographic rainfall. The step-duration **orographic intensification coefficient method** may be used to estimate the effects of topography on storm rainfall; see Lin (1989) *Atmospheric Deposition, IAHS Publ.* 179.

orthogonal At right angles. Orthogonals plotted through the crests of waves in plan illustrate the process of *wave refraction.

osmosis The passage of a weaker solution to a stronger solution through a semi-permeable membrane. In soils, the more dilute soil moisture passes by osmotic pressure into plant roots. *Salinization causes **reverse osmosis** in soil, where higher concentrations of salt in the soil water draw water from the plant roots, causing crops to wither, and perhaps die. See Kotzer (2005) *Desalinisation* 185, 1–3.

otherness Difference. Otherness occurs as identities are constructed through exclusion or denigration. Kuus (2004) *PHG* 28 finds 'a broadly orientalist discourse that assumes essential difference between Europe and Eastern Europe', but J. Wedel (2001) observes that the assumption of otherness is not only inscribed on east-central Europe by the West but is also appropriated by east-central Europeans themselves. Generally, cultural otherness, or the othering of certain groups within a society as inferior and non-existent, perpetuates social and economic inequalities. Thus, the *favelados (shantytown dwellers of Brazil)* are unable to obtain 'personhood' because they are 'othered' as inferior degenerates within Brazilian society; see J. Perlman (2011).

outer space The physical universe beyond the Earth's atmosphere. MacDonald (2007) *PHG* 31, 5 aims to establish outer space as a mainstream concern of critical geography. 'More than half a century after humans first cast their instruments into orbit, contemporary human geography has been slow to explore the myriad connections that tie social life on Earth to the celestial realm. My starting point is a return to an early-modern geographical imagination that acknowledges the reciprocity between heaven and earth.' Of course, although satellites circulate in outer space, their origins and impacts are very much on the ground.

outlet glacier A fast-moving section of glacier ice bordered by rock, differing from an *ice stream, which is bordered by ice. For an example, see Glasser and Hambrey (2002) *Sedimentol*. 49, 1. An ice stream is a fast-flowing 'river' of ice within more slowly moving ice sheet walls.

outsourcing For businesses, the transference of portions of their work to other companies. See Slobozhan (August 2017) *freshcode*, on the geography of outsourcing software development.

outwash (outwash sands) and gravels Sorted deposits which have been dropped by *meltwater streams issuing from an ice front. The material deposited is coarse near the ice front, becoming progressively finer with distance from it. Outwash fabric tends to be well *bedded, and *current bedding is common; *see* SANDUR. The finest elements of the outwash may be deposited in *proglacial lakes to form *varves. A series of streams may produce several alluvial fans which coalesce to form an **outwash plain** (valley train).

overdeepening A phenomenon found in *cirques and in the steps of *glacial troughs, where the middle section(s) of the feature are lower than the mouth. One suggestion is that the overdeepened section is the zone of maximum ice thickness, another that overdeepening coincides with less resistant rocks. Evans (1999) *Annals Glaciol*. 28 ascribes overdeepening to rotational flow, while Hooke (1991) *GSA Bull*. 103, 8 describes a positive feedback process, as overdeepening causes crevassing at the surface, resulting in erosional forces that accentuate that overdeepening.

overland flow The flow of water over a hillslope surface. When water infiltrates a soil and all the pore spaces are full, the soil is saturated and the water table is at the surface. Overland flow will then occur. This is **saturation-excess overland flow**, which can occur at much lower rainfall intensities than those required to generate **infiltration-excess overland flow** (also known as *Hortonian overland flow). Saturation-excess overland flow can occur at the foot of a hillslope even when it is not raining. Overland flow may be laminar, turbulent, and transitional, or consist of patches of any of these flow states. Horton (1945) *GSA Bull*. 56 suggests that overland flow is likely to be laminar near watersheds, but with distance downslope overland flow becomes more turbulent; he explains that these changes result from the microtopography of the soil surface: run-off gathers into depressions, which increase in size downslope. As the depressions deepen they capture more flow through the cross-grading of the hillslope surface. Surface run-off deepens as a result.

The Darcy–Weisbach, Chézy, and Manning equations are used to predict overland flow velocity (Nunnally (1985) *Env. Manage*. 9, 5). Smith et al. (2007) *PPG* 31, 4 review the problems of measuring overland flow, and evaluate some solutions.

overlay At its simplest, a stack of transparent map layers, revealing where things co-occur. See Ahlqvist in R. Kitchin and N. Thrift (2009) for overlays in *GIS.

overpopulation Too great a population for a given area to support. 'What might be termed overpopulation in one context might not in another because of differences in the standard of living aspired to by different populations.

Perhaps the foremost diagnoses of overpopulation come from P. Ehrlich (with A. Ehrlich 1990). Neo-Malthusians support his argument; see Lempert (1987) *E. Afr. Econ. Rev.* 3, 1. Marxists, however, view overpopulation as the result of the maldistribution and underdevelopment of resources (Marx, *Capital*, vol. iii; see also Y. S. Brenner 1969). In the developed world, some would suggest that pollution and the desecration of the countryside are indicators of overpopulation.

overspill 1. The population which is dispersed from large cities to relieve congestion and overcrowding, and, possibly, unemployment. It occurs with redevelopment in the city where new building is at much lower densities so that some people—the overspill—cannot be housed in the city.

2. The condition of a fluvio-glacial lake when water levels rise above the barrier impounding it; see Kaiser et al. (2007) *Earth Surf. Procs Ldfms* 32, 10.

overthrust A nearly horizontal fold subjected to such stress that the strata override underlying rocks; see Price (1998) *GSA Bull.* 100, 12 and Hubert and Ruby (1978) *GSA Bull.* 70.

over-urbanization Population growth in an urban area which outstrips its job market and the capacity of its infrastructure; 'urbanization without industrialization' (M. Davis 2006). In 1988, 100000 migrants from mainland China flooded into Hainan. The provincial government was swamped by more than 180000 applications for only 30000 jobs, with the result that those unable to find work turned to street hawking. Meanwhile, in-migrating peasant workers congregated in sprawling urban 'village' settlements that lacked basic amenities. The infrastructure of urban centres in Hainan sagged under these immense pressures (Go and Wall (2007) *Sing. J. Trop. Geog.* 28, 2).

ox-bow lake A horseshoe-shaped lake once part of, and now lying alongside, a meandering river with a narrow 'neck' between meander loops. When the river breaks through this narrow stretch of land, the old meander becomes a temporary lake; ox-bow lakes quickly fill up and become hollows in the landscape.

oxidation The absorption by a mineral of one or more oxygen ions. Oxidation is a major type of chemical weathering, particularly in rocks containing iron; see Price and Bevel (2003) *Chem. Geol.* 202, 3. Oxidation occurs in soils when minerals take up some of the oxygen dissolved in the soil moisture; see Kayak et al. (2007) *J. Env. Qual.* 36.

oxisol A soil of the *US soil classification. *See* FERRALLITIZATION.

ozone A form of oxygen, and an atmospheric trace gas, made by natural photochemical reactions associated with solar ultraviolet radiation. Ozone has three atoms of oxygen combined in one molecule, rather than two atoms, as in free oxygen. The proportion of ozone in the atmosphere is very small, but it is of vital importance in absorbing solar ultraviolet radiation. The **ozone layer** (**ozonosphere**), is an ozone-rich band of the atmosphere, at 10–20 km above the Earth, but is at its most concentrated between 20 and 25 km. See Schiermeier (2007) *Nature* 449. When the ozone layer thins (the 'hole' over Antarctica is an example) increased solar ultraviolet radiation reaches the surface of the Earth, with consequent damage to human health. See European Space Agency (2004), 5 October.

pacific coast A coastline where the trend of ridges and valleys is parallel to the coast. If the coastal lowlands are inundated by the sea, a coast of interconnected straits parallel to the shore may result.

Pacific Decadal Oscillation (PDO) A long-term fluctuation in sea surface temperatures in the extratropical north Pacific, waxing and waning every 20–30 years. The 'cool'/'negative' phase is characterized by a wedge of unusually low ocean temperatures in the eastern Pacific, with a warmer 'horseshoe' of unusually high ocean temperatures in the north, west, and southern Pacific. In the 'warm'/'positive' phase, the western Pacific cools and the east warms. In western North America, positive phases of the PDO are associated with decreased winter precipitation, snowpack, and streamflow in the northwest, and higher precipitation in the southwest. Conditions reverse during negative PDO phases. Schneider and Cornuelle (2005) *J. Climate* 18, 21 argue that the north Pacific sea surface temperature anomaly and the PDO are 'a response to changes of the north Pacific atmosphere resulting from its intrinsic variability, remote forcing by *ENSO and other processes, and ocean wave processes associated with the ENSO and the adjustment of the north Pacific Ocean by *Rossby waves'. See also Goodrich and Walker (2011) *Phys. Geog.* 32, 4, 295.

pack ice Large blocks of ice on the surface of the ocean, formed when an ice field is broken up by strong waves. Since 2002, the East Greenland pack ice is much less, or even non-existent (Brooks (2005) *Geol. Today* 21, 6).

pahoehoe A type of lava flow which spreads in sheets, associated with highly fluid, basic lava, such as that ejected from *Hawaiian volcanoes. The surface is a glassy layer which has been dragged into ropy folds by the movement of the hot lava below it. Whether the lava is aa or pahoehoe depends mainly on viscosity and strain rate, which in turn depend on crystallinity, dissolved gas content, temperature, bubble content, slope, eruption rate, and lava composition. See Self et al. (2005) at *OpenGeoscience 1* 5, 283.

((⊕)) SEE WEB LINKS
- Website of OpenGeoscience, British Geological Survey.

palaeoclimate The climate of a particular period in the geological past, before historical records or instrumental observations. Palaeoclimatology uses evidence from glaciers and ice sheets, geology, sediments, and tree rings; even the micromorphology of leaves to determine Earth's past climates.

palaeoecology The reconstruction of past environments from fossils evidence. See Seppä and Bennett (2003) *PPG* 27, 4 on progress in palaeoecology, and Rull (2010) *The Open Ecology Journal*, 3 on the nature of palaeoecology.

palaeoenvironmentology The study of prehistoric and historic environments. Traditionally, the most widely applied technique in Quaternary studies has been that of *palynology. Seidenkrantz (2013) *Quaternary Sci. Rev.* 79 uses micropalaeontological proxy data, together with geochemical and sedimentological data, to infer palaeoenvironmental conditions in the Labrador Sea.

palaeohydrology The study of past hydrology. Evidence for hydrological changes during previous periods of Earth history comes from the alteration, deposition, and erosion of rocks from these periods. Palaeohydrology includes changes in flora and fauna assemblages that have been strongly influenced by changes in hydrology. See Willard et al. (2011) *The Holocene* 21, 2, 305 on a pollen-based palaeohydrologic reconstruction from North Carolina.

palaeolimnology The reconstruction of the paleoenvironments of inland waters and the changes experienced in these waters due to events such as climatic change, internal ontogenic processes, or human impacts, for example, *eutrophication, or *acidification. Dong et al. (2012) *The Holocene* 22, 107 provide a useful example.

palaeomagnetism *See* GEOMAGNETISM.

palaeoseismology The study of ancient earthquakes through the observation and analysis of those rocks and sediments which have undergone continuous sediment creation for the last few thousand years; for example, lakes, river beds, swamps, and shorelines. Volume 326 of *Geomorphology* is wholly dedicated to the topic.

palaeosol A soil exhibiting features which reflect past conditions and processes. The presence of a palaeosol may have *palaeoclimatic significance. See Spinola et al. (2017) *Catena* 150.

Palaeozoic The era stretching approximately from 570 to 225 million years BP.

(🌐) SEE WEB LINKS
• An overview of the Palaeozoic era.

palimpsest A palimpsest is a parchment from a scroll or book from which the text has been scraped or washed off, and which can be used again. With the passing of time, the faint remains of the former writing would reappear enough so that scholars could discern the text, and decipher its meaning. Geographers really like the analogy of the palimpsest; Graham (2010) *Tijdschrift Econ.* 101, 4, 422 writes 'All places are palimpsests. Among other things, places are layers of brick, steel, concrete, memory, history, and legend. The countless layers of any place come together in specific times and spaces and have bearing on the cultural, economic, and political characteristics, interpretations, and meanings of place'. In geomorphology, **palimpsest landscapes** are composed of a mosaic of active and relict (inactive) landforms of different ages.

(🌐) SEE WEB LINKS
• 'Development of Palimpsest Landscapes', *Vignettes: Key Concepts in Geomorphology*.

Palmer Drought Severity Index (PDSI) A drought index that uses temperature and precipitation values from the climate division dataset of the National Climatic Data Center along with other components of the water balance equation to measure the departure of soil moisture supply from normal. The output of the PDSI consists of positive (wet) or negative (dry) values centred on 0 (normal) with values above +4.0 or below −4.0 generally considered extreme. Monthly maps of drought conditions in the contiguous US as measured by the Palmer Drought Severity Index, Palmer Hydrological Drought Index, Palmer Modified Drought Index, and Palmer Z-Index (Palmer, 1965) are provided for January 1900–present; go to the NOAA site. Yan et al. (2013) *Mathematical Problems in Engineering* present a modified PDSI.

SEE WEB LINKS
• NOAA Historical Palmer Drought Indices.

palsa There is no generally agreed definition for this term, but it appears to be chiefly used to refer to mounds of peat with a frozen core, found in wetlands in the subarctic northern hemisphere.

palynology The study of pollen grains as an aid to the reconstruction of past plant environments. Kiage and Liu (2006) *PPG* 30, 5 combine evidence from palynology, lake sediments, and associated records, to provide a more accurate and complete assessment of the paleoenvironmental changes. One drawback is that most of the pollens found come from wind-pollinated species, and animal-dispersed pollen is under-represented.

pan A large, shallow, flat-floored depression found in arid and semi-arid regions. Pans may be flooded seasonally or permanently. See Viles and Goudie (2007) *Geomorph.* 85, 1–2.

panarchy A model expressing the complex organizations and interactions between people and nature. While hierarchies describe top-down systems, panarchies occur in nested adaptive cycles.

pandemic An epidemic over a large area; throughout an entire country, continent, or the whole world. COVID-19 is the most recent example, but see Chandra et al. (2013) *Int. J. Health Geographics* 12, 9 on the influenza pandemic of 1918–19, where over 15 million died in less than a year.

Pangaea A supercontinent comprising all of Earth's landmass, before it was split by *continental drift. See W. M. Marsh and M. M. Kaufman (2012), p. 246 on the movements of India during the break-up of Pangaea.

parabolic dune *See* SAND DUNE.

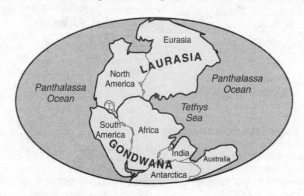

Pangaea

paradigm The prevailing pattern of thought in a discipline or part of a discipline. Perhaps the most powerful Western paradigm has been the 'scientific method'. T. Kuhn (1972) argued that the evolution of a new paradigm marks a new stage in thinking. This new paradigm persists until too many anomalies occur, when it is replaced by a new paradigm that is able to explain the anomalies. In geography, for example, the regional geography paradigm was superseded by the *quantitative revolution, and then the *cultural turn.

R. Inkpen (2005), pp. 18–21 is strongly recommended for paradigms in physical geography; Lomolino (2000) *Glob. Ecol. & Biogeog.* 9, 1 offers a new paradigm for island biogeography.

paraglacial processes The non-glacial Earth-surface processes, sediment accumulations, landforms, landsystems, and landscapes that are directly conditioned by glaciation and deglaciation. This distinguishes it from the term 'periglacial' which is defined as 'cold, non-glacial' and is applied to environments in which frost-related processes and/or permafrost are either dominant or characteristic. See Iturrizaga (2008) *Geomorph.* 95, 1–2 on paraglacial debris landscapes. For the difference between proglacial, periglacial, and paraglacial, see Slaymaker (2009) *Geological Society, London, Special Publications* 320.

parallel drainage *See* DRAINAGE PATTERNS.

parallel slope retreat The evolution of a hillslope when the slope angle remains constant. It requires some mechanism to remove weathered material from the foot of the slope in order to maintain the existing topography. See Ollier (1995) *Prog. Phys. Geog.* 19, 3.

parenting To be or act as a mother or father to someone. It is Aitken in R. Kitchin and N. Thrift (2009), 'an emotional and day-to-day practice that is 'embedded' in the parents' social and spatial settings; ways of parenting vary from place to place. Luzia (2010) *Soc. & Cult. Geog.* 11, 4, 359 writes on the 'actual places and spaces of parenting'.

parent material The rock or deposit from, and on, which a soil has been formed. The nature of the parent rock will largely determine the nature of the *regolith, and hence the soil texture—basalt tends to produce clay soils; granite sandy soils. See also Mahaney (1993) *PPG* 19, 3.

Pareto optimality A situation where it is impossible to improve the economic lot of some without making others worse off. Sen (1970) *J. Polit. Econ.* 78 shows that, within a system of social choice, it is impossible to have both a commitment to minimal liberty and Pareto optimality. *See* ZERO-SUM GAME.

park Originally an enclosed area used for hunting, in the 18th century the term applied to the grounds of a country house. It now refers to open land used for recreation in a town or city. Within the UK, parks were a feature of 19th-century *urbanization, in order to relieve the 'artificiality' of urban living. However, Brainard et al. (2006) *CSERGE,* ECM-2006–05 argue that parks have usually been sited with little regard for the geography of where different social groups live: within Birmingham, UK, 'even within the most deprived areas, whites have better access to park areas than non-whites'.

participatory GIS The gathering of data using traditional methods such as interviews, questions, and focus groups, all using some form of paper maps to allow participants to record spatial detail, thereby giving greater privilege and legitimacy to local or indigenous spatial knowledge. See Radil and Anderson (2019) *Prog. Hum. Geog.* 43, 2 for a stern review of the process.

participatory research The co-construction of research between researchers and people affected by the issues under study, such as patients and/or decision-makers who apply the research findings (e.g. health policymakers). See Brasher (2020) *J. Cultural Geog.* 37, 3. **Participatory Rural Appraisal (PRA)** uses the same approaches; see Yusuf (2007) *J. Agric. Res. & Dev.* 6, 37.

particulates (particulate matter) Atmospheric particles—dust, pollen, soot, salt, sulphur, etc. See Zhang et al. (1994) *J. Atmos. Scis* 33, 7 on the impact of dust on particulate nitrate and ozone formation.

Partition When the British left the Indian subcontinent in 1947, they partitioned (split up) the 'British Raj', creating the separate countries of India and Pakistan to accommodate religious differences between Muslims and Hindus. Pakistan has a majority Muslim population and India is primarily Hindu. See M. Hasan (1994) for a collection of essays, extracts, memoirs, and a short story to create an outline of the events preceding and surrounding India's Partition.

(🌐) SEE WEB LINKS

• Partition in postcolonial studies.

passive margin In plate tectonics, a margin between *continental and *oceanic crust contained entirely within a single *plate. Tectonic activity here is minimal. See Leroy et al. (2007) *Geophys.*

J. Int. on the uplift and evolution of passive margins.

pastoralism Raising of livestock; pastoral farmers grow crops to feed their livestock, and pastoral nomads raise their livestock on natural pasture.

An approach from complex system theory stating that the history of a system partly determines the future state of the system In landslides, path dependency means that the history of landslides at a certain location affects the susceptibility of future landslides at or near that location.

patch An area of vegetation, between 0 and 10 hectares, that has a uniform structure and composition, and differs from the surrounding vegetation. Watson (2002) *J. Biogeog.* 29, 5–6 distinguishes between fragments—remnants of a previously widespread habitat—and islands, which have always been restricted and isolated in their spatial extent.

path In *behavioural urban geography, and within a city, a channel people move along.

path-dependence When any current turn of events depends on the paths of previous consequences, rather than simply on current conditions. This term seems, to some extent, to resemble *hysteresis, or *contingencies. Greener (2002) *Manage. Decisions* 40, 6 explains path dependence as the entry into a system from which it can't escape without the involvement of an outside force/shock. Samia et al. (2017) *Geom.* 292 consider path dependence in landsliding. See MacKinnon on path dependence and economic geography (*Geog. Compass* 2008).

patriarchy A gender system of the social control of women by men that has pervaded all aspects of human existence, including politics, industry, the military, education, philosophy, art, literature, and civilization itself (K. Millett 1970). This includes the 'private' spheres of love, sexuality, marriage, and children—the family is politicized as the foundation of patriarchal power; see Bennett (2006) *Gender, Place & Cult.* 13, 2.

Patriarchy and masculinity are not eternal but (like every other concept) constructed and open to change. Erman (2001) *Int. J. Urb. & Reg. Res.* 25, 1 shows how an urban community reproduces traditional patriarchal authority over in-migrants. Adams and Ghose (2003) *PHG* 27, 4 observe that Bollywood idols reflect not only patriarchy but subethnic and colonial power relations. 'It is a sad fact that one of the few profoundly non-racial institutions in South Africa is patriarchy' (A. Sachs, no date). Hern (2006) *Pol. Geog.* 28, 6 defines **transpatriarchies** as forms of patriarchy extending across and between national societies. 'Remasculinization of public rhetoric, and privileging economic and military concerns have sharp impacts on women's livelihoods and political spaces' (Radcliffe (2006) *PHG* 30, 4). See Waitt (2006) *AAAG* 96, 4 on white gay patriarchy.

patterned ground Most commonly in *periglacial areas, the arrangement of stones into polygons, isolated circles, and concentrations of circles known as nets, steps, and stripes. Polygons and circles are more common on level surfaces, stripes generally form on slopes; the patterns are made of coarser stones, separated by much smaller stones (*fines*).

Together with convection, desiccation and the formation of needle ice within the *active layer produce patterned ground. Matsuoka (2003) *Geomorph.* 52, 1–2 attributes patterned ground to the upward injection of slow-freezing, waterlogged silts at particular points,

while Kessler and Werner (2003) *Science* 299, 5605 note *self-organization in patterned ground.

peace, geographies of Also termed **pacific geopolitics**, this is a way of thinking geographically about world politics in order to promote peaceful and mutually enriching human coexistence. Peace is a contested spatial process and political discourse; see Courtheyn (2018) *PHG* 42, 5.

peak land-value intersection (PLVI) The point in a *CBD, often, but not always, at a road intersection, where land values are at a maximum.

peasant A farmer whose activities are dominated by the family group. The family provides all the labour and the produce is for the family as a whole; occasional surpluses are sold in the open market. Landholdings are small, sometimes owned by the family, but often leased; see Kull (2006) *AAAG* 96, 3 on the perceived rights of peasants versus state control. 'Peasants are not a declining class but rather are undergoing a variety of transformations. In some settings they will indeed disappear, in others experience stresses that lead to their semi-proletarianization and, in yet others, peasants were, and still are, being created by capitalism' (Johnson et al. (2005) *Antipode* 37, 5). Although peasants have been characterized as backward and resistant to change, peasant strategies can be highly rational in a society where there is little margin for error; see Faminow et al. (1999) *World Animal Rev.* 93. See Desmarais (2008) *Hum. Geog.* 1 on peasant resistance to neoliberalism.

peat A mass of dark brown or black, partly decomposed plant material, formed in waterlogged ground, where temperatures are low enough to slow down the decomposition of plant residues. It is characteristically found in cold climates, and as a relict feature in temperate zones. Peat may be used as fuel, and is widely used by gardeners; so much so that peat bogs are now under threat. See M. Evans and J. Warburton (2007). See also Evans et al. (2019) *Geo: Geography and Environment*; e00075 on peatland carbon stocks.

ped In a soil, an aggregate of silt, sand, and clay of characteristic shape, resulting from the formation of hydrogen and ionic bonds between soil particles. Peds may be further developed by plant roots, by *polysaccharide gums secreted by soil fauna, and by alternate freezing and thawing, or wetting and drying. Ped properties can be reflected in the structure of pore space (Pachepsky and Rawls (2003) *Eur. J. Soil Sci.* 54, 3). *See* SOIL STRUCTURE.

pedalfer Any soil high in aluminium (Al) and iron (Fe), and from which the bases such as calcium and magnesium carbonates have been *leached. Pedalfers generally occur in regions with an annual rainfall of more than 600 mm.

pediment A low, *concave element at the foot of a hillslope (also called a *concave*, or *waning*, slope). See Twidale (2019) *Géomorphologie* 20 1º on pediments and platforms: problems and solutions.

pedocal A soil high in calcium carbonate and magnesium carbonate.

pedogenesis The formation of soils. **Pedogenic processes** are soil-forming processes. See Foss (2006) *Soil Sci.* 171, 6, Suppl. 1 on milestones in pedogenesis. The chief pedogenic factors are time, *hydrology, *parent rock, climate, fauna and flora, and terrain, but terrain variables may be ineffective predictors of soil characteristics where a dense canopy hides a non-uniform erosional

environment (Williamson et al. (2006) *Soil Sci. Soc. Am. J.* 7). **Pedology** is the science of soils: their characteristics, development, and distribution. Try Vasu et al. (2016) *Soil Research* 55, 3.

pedostratigraphy The study of the *stratigraphy and spatial relationships of surface and buried soils; particularly useful in interpreting *paleosols. Zhao et al. (2020) *Catena* 187 could be useful.

pedon The smallest three-dimensional roughly hexagonal volume of soil that can be recognized as a distinct type. Pedons range from 1 to 10 square metres in area, depending on the variability in the horizons.

pelagic Of marine life, belonging to the upper layers of the sea.

peneplain An erosional plain, near to base level. 'Despite more than a century of effort, no convincing example of a contemporary peneplain has been identified' (Phillips (2002) *Geomorph.* 45, 3–4).

perception The manner by which we make sense of the world; a process shaped by culture, acting as a filter, which moulds our 'take' on life. One such filter is the level of understanding; where once disasters were interpreted as 'Acts of God', today they are more often perceived as the result of human irresponsibility or malevolence.

Of particular interest is **hazard**, or **risk**, **perception**; what science deems to be an acceptable level of risk may not match what risks society finds acceptable. 'When the differences between social and scientific notions of risk become acute, then the outcome is a *social amplification of risk* by the public' (Herrick (2005) *Area* 37, 3). However, Crozier et al. (2006) *Area* 38, 2 find that people in high-risk earthquake zones tend to be fatalist, or at least resigned to the consequences. Perception also shapes economic decision-making; see Farley (2007) *AAAG* 97, 4 on the perception of 'good' forests, and Warren (2007) *PHG* 31, 4 on the clash between perception and ecology. *See* ENVIRONMENTAL PERCEPTION.

perched water-table A partly saturated, isolated, confined *aquifer underlain by an impermeable rock, with the main *water-table below the two.

percolation The filtering of water downwards through the *bedding planes, *joints, and pores of a permeable rock.

percolines An underground network of water seepage zones: old root channels, soil cracks, and animal burrows enlarged by *interflow. **Percoline flow** is laterally concentrated diffuse interflow. See A. Lerman and M. Meybeck (1988).

perennial stream A watercourse which flows throughout the year.

perfect competition Under conditions of perfect competition, there are many suppliers; a perfectly elastic supply of the *factors of production; no collusion between suppliers; and buyers and sellers are fully aware of the prices being charged throughout the market. This is an unlikely state of affairs; imperfect competition is much more common. Try Mossy (2003) *Chaos, Solutions & Fractals* 18, 3: technical, but interesting.

perforation kame *See* KAME.

performance The acting out of our identities and roles, through street theatre, community theatre, and legislative theatre. Eyerman (2004) *Thesis Eleven*, 79, 25 argues that performance implies that action is also symbolic, and public as well as social.

But social actions and practices are also performances that are open to interpretation; people seek to create particular spatialized identities and spaces, which might change with different performances. See Gerlach (2015) *TIBG* 40, 2 on the performance of politics.

performativity Using speech and gestures to act out/create a situation. J. Butler (1993), who created this concept, describes it as the 'reiterative power of discourse to produce the phenomena that it regulates and constrains'. In other words, that repeated discussion can bring about the desired effect: what we say and do about something shapes what it is. A good example is given by Waitt (2008) *Soc. & Cult. Geog.* 9, 1, who illustrates the way that aggressive attitudes towards the ocean by Australian shortboard surf-riders creates a kind of masculinity. In a **performative perspective**, all subjects are still in process, and to study a subject is to analyse the way it is produced (Kuus (2007) *TIBG* 32, 1)—meaning, I think, to look at the ways a subject creates itself.

peridotite A coarse-grained igneous rock, made mostly of olivine and pyroxene, and thought to be the main constituent of the *mantle.

periglacial Referring to the processes and landforms of any area with a *tundra climate, or where frost processes are active and *permafrost occurs in some form. See the entire issue of *Geom.* (2017) 293b. **Periglacial climates** are arid, with temperatures below 0 °C for at least six months, and summers warm enough to allow surface melting to a depth of around 1 m.

   Periglacial processes include *abrasion, *freeze–thaw, *nivation, and *solifluction, and are responsible for the formation of new deposits, the alteration of existing unconsolidated deposits, and the modification of existing landforms by *mass movement. You could not do better than Rowley et al. (2015) *Developments in Earth Surface Processes*, chapter 13.

periodicity This refers to recurrences at regular intervals. The major external cycles affecting the Earth are lunar (with daily, monthly, and 18.6 year cycles), planetary cycles, with 20 000- and 400 000-year recurrences (*see* MILANKOVITCH CYCLES), and solar cycles; Lean (2010) *WIREs Clim. Change*, 1: 111 is excellent on this.

periphery The edge, or margin. Murray and Challies (2006) *Asia Pacif. Viewpt* 47, 3 describe New Zealand as a **resource periphery** and Chile as **semi-peripheral**.

permafrost Areas of rock and soil where temperatures have been below freezing point for at least two years. Permafrost need contain no ice; a sub-zero temperature is the sole qualification. **Continuous permafrost** is present in all *periglacial areas apart from small, localized thawed zones, while **discontinuous permafrost** exists as small, scattered areas of permanently frozen ground. A **frost table** marks the upper limit of permafrost, which is overlain by the *active layer. Ice develops mainly in upper levels in **epigenetic permafrost**; in **syngenic permafrost** it is regularly distributed throughout the whole thickness of the permafrost; go to the US Army Permafrost tunnel for both types.

　Permafrost aggradation (growth) decreases the thickness of the active layer, is responsible for the formation of *pingos, and may be caused by the freezing of *taliks. **Permafrost degradation** (decline) plays a key role in the development of *thermokarst; see

Jorgensen et al. (2001) *Clim. Change* 48, 4. Permafrost is a very sensitive system; small mistakes in constructing buildings in this environment can have catastrophic effects; see Schneider et al. (2021) *The Cryosphere*, 15 on the consequences of permafrost degradation for Arctic infrastructure.

Permian The latest period of *Palaeozoic time, stretching approximately from 290 to 248 million years BP. The **Permian mass extinction** occurred between 290 and 252 million years ago, during which 95% of all marine life on Earth was killed, and 70% of all land families became extinct.

(⊕) SEE WEB LINKS
• An overview of the Permian period.

persistence (geomorphological persistence) The persistence of a landform varies with the magnitude and recurrence interval of the event which brought it about. Calver and Anderson (2004, *TIBG* 29, 1) suggest the terms **static persistence** where 'spatial positioning is constant, [and there is] virtually no change in the feature over time', and **dynamic persistence** where 'the type of feature remains in evidence . . . but individual representations are formed and/or degrade'. See also Wilcock and Iverson, eds (2003) *Geophys. Monog. Ser.* 135.

personal space The zone around individuals—usually that used when speaking with a normal conversational voice and for friendly interaction—which they reserve for themselves and their group. Associated with personal space are the concepts of privacy, autonomy, rights, and transgressive invasions. The extent of a personal space around an individual is reckoned to be 1–1.5 m for an Anglo-Saxon. When women are empowered, they are more able to create their own personal space

p

(George (2007) *Gender, Space & Cult.* 14, 6). Barker (2000) *Area* 32, 4 only hints at the 'globalization of personal space'.

p-forms and micro-channel networks Small-scale landforms of glacial erosion—usually less than 1 m in width and depth—on glaciated bedrock surfaces. Sichelwannen are sickle-shaped bedrock depressions, with an open end pointing in the direction of ice flow, and are the most common. See Shaw (1994) *Sed. Geol.* 91. The term 's-form' applies if a glacial meltwater origin can be established.

phase space 1. Very simply, a continuum representing two variables: A and B. The continuum ranges from 100% A, 0% B, to 0% A, 100% B. See, for example, Hooke (2003) *TIBG* 28, 2 on the phase space of meander movement and bend curvature. Melton (1958) *J. Geol.* 66 describes a mature drainage basin phase space with four variables.

2. Very much less simply, an abstract concept that captures all the possible spaces in which a spatio-temporal system might exist *in theoretical terms*; not just what happens but what might happen. 'The geography of phase space is flexible, but not totally arbitrary: the main possibilities are "already there", constrained by contextual realities' (I. Stewart and J. Cohen, 1997). This use of the concept is not totally distinct from 1; there are just many more factors involved. Martin (in Harrison et al. (2006) *Area* 38, 4) sees phase space as a middle road between relational space and more fixed notions of spatiality.

Hooke (2007) *Geomorph.* 91, 3–4 uses **phase space plots** to help uncover emergent behaviour in meandering rivers.

phenomenological analysis An investigation into things and experiences which illustrates their fundamental nature, in their own terms. For the geographer, this approach is best seen as abstract modelling, as in the *von Thunen and *Alonso models, which reduce people to unemotional digits.

phenomenon instance In *geomorphology, a **phenomenon instance** is an entity whose form and behaviour varies over time and space. Thus, for example, a coastal spit may only 'act as such' at certain stages of the tide cycle, at other times more resembling an embryonic *sand dune. *See* EMERGENCE.

phenotype The outward appearance, physical attributes, or behaviour of an organism, which develop through the interaction of nature and nurture. For **phenotypic variation**, see Robertson and Robertson (2008) *J. Biogeog.* 35, 5.

Phillips curve A negative exponential curve demonstrating the relationship between the percentage change in wages and the level of unemployment. The theory is that high wages cause high inflation, and the lower the rate of unemployment, the higher the rate of inflation. See Parkinson (February 2018) *Monthly Labor Review*, for an examination of the Phillips curve using city-level data.

photic zone Those upper levels of a water body that are penetrated by light.

photochemical smog Nitrogen dioxide (NO_2) is emitted from petrol engines. Ultraviolet light splits this into nitric oxide (NO) and monatomic oxygen (O). The hydrocarbons emitted from the burning of *fossil fuels react with some of the monatomic oxygen to form photochemical smog. See Hart et al. (2004) *13th Joint Conference on the Applications of Air Pollution Meteorology*. Photochemical smog is most common where the sunshine is strong and long-lasting and where car use is high; see Zhang and Oanh (2002) *Atmos. Env.* 36, 26 on photochemical smog in Bangkok. Photochemical smog damages plants (Arbaugh et al. (2003) *Env. Int.* 29, 2–6), and irritates eyes and lungs (Fukuoka (1997) *Int. J. Biometeorol.* 40, 1).

photogrammetry The use of aerial photographs and images from remote sensing to measure terrestrial features. See *J. Photogrammetry & Remote Sensing*. Photogrammetry employs high-speed imaging and the accurate methods of remote sensing in order to detect, measure, and record complex 2-D and 3-D motion fields; it is 'the art, science, and technology of obtaining reliable information from noncontact imaging and other sensor systems about the Earth and its environment, and other physical objects and processes through recording, measuring, analyzing and representation' (*International Society for Photogrammetry and Remote Sensing*).

(🌐) SEE WEB LINKS
- Website of International Society for Photogrammetry and Remote Sensing.

photography and geography Geography is a visual discipline, working with maps, globes, models, slides, and photographic illustrations. D. Cosgrove (2008) argues that the 'truth' of any image can be seen by looking through its deceptive surface to the society it's seen in.

phreatic Describing *groundwater below the *water-table. The **phreatic zone** is permanently saturated. A volcanic **phreatic eruption** has *meteoric water mixed with the lava, and may create steam, or a *geyser (Fontaine et al. (2003) *Earth & Planet. Sci. Letters* 210, 1–2). A **phreatophyte** is any desert plant with roots down to the phreatic

zone; see Ridolfi et al. (2007) *J. Theoret. Biol.* 248, 2.

phylogenetic Referring to evolutionary development; the **phylogenetic diversity** (**PD**) of a species is a measure of the biodiversity and taxonomic distinctness. It can be estimated by looking at the phylogenetic relationships among taxa. Try Chao et al. (2015) *Methods Ecol. & Evolution* 6, 4 for the methodology. PD can then be used to determine conservation priorities at various biogeographical scales (McGoogan et al. (2007) *J. Biogeog.* 34, 11).

phylum The second highest category (of seven) in the scientific system of classification for organisms (pl. *phyla*), consisting of one or several similar or closely related classes.

phylogeography The study of those principles and processes that control the geographical distributions of genealogical lineages, especially those within, and among, closely related species. See Chevolot (2006) *Molecular Ecol.* 15, 12 for an example.

physical geography The study of the processes shaping the land-surface of the Earth, with particular emphasis on the spatial variations that occur and the temporal changes necessary to understand the environments of the Earth. One purpose is to understand how the Earth's physical environment is the basis for, and is affected by, human activity. Physical geography was conventionally subdivided into *geomorphology*, *climatology*, *hydrology*, and *biogeography*, but is now more holistic in systems analysis of recent environmental and Quaternary change. It uses mathematical and statistical modelling and remote sensing, to develop research for environmental

p

management and environmental design, and benefits from collaborative links with many other disciplines such as biology (especially ecology), geology and engineering (see K. Gregory (2002)).

Between 1850 and 1950, the main ideas that had a strong influence on the discipline were uniformitarianism, evolution, exploration and survey, and conservation. In the 1960s, there was a new concern with the dynamic processes of earth systems. This new concern, which has evolved to the present, is founded on basic physical, chemical, and biological principles and employs statistical and mathematical analysis. It has become known as the 'process approach' to physical geography. More recently, physical geographers have turned to emerging ideas in the natural sciences about nonlinear dynamical systems and complexity to explore the relevance of these ideas for understanding physical-geographic phenomena (see Rhoads (2004) *AAAG* 94, 4). Advances in remote sensing, geographical information systems and information technology have enabled a more global approach; a second new development has been the advent of a more culturally-based approach throughout many branches of physical geography. Since 2000 there has developed an increasingly holistic trend, a greater awareness of a global approach and of environmental change problems (see Gregory (2001) *Fennia* 179, 1).

physical quality of life index (PQLI) An attempt to measure the quality of life or well-being of a country. The numerical value is derived from the adult literacy rate, infant mortality rate, and life expectancy at age 1, all equally weighted. It is one of a number of measures created because of dissatisfaction with the use of GNP as an indicator of development. See Zinovyev and Gorban (2010) arXiv. However,

Rodrik and Brodsky (1981) *World Development* 9, 7 suggest problems with this indicator.

physiological drought When soil water, although present, is unavailable, as when it is frozen, or when *evapotranspiration exceeds plants' uptake of water. See Sausalito and Bellinger (2007) *Silva Fennica* 41, 2 on morphological and physiological drought.

phyto- Of a plant; relating to plants; hence **phytoextraction**, the removal of unsafe material from soil or water by plants. See Ma et al. (2016) *Catena* 136.

phytogeography The study of the distribution of plants, or taxonomic groups of plants, to account for their distribution, with regard to their origin, dispersal, and evolution; their present, and past, spatial relationships. For an example, see Arruda (2013) *An. Acad. Bras. Ciênc.* 85, 2.

piedmont glacier A glacier formed from the merger of several *alpine glaciers as they emerge from the mountains; see Hall and Denton (2002) *Holocene* 12.

pingo 1. open system pingo, hydrostatic pingo A large *ice mound formed under *periglacial conditions, so called because it is formed from an unfrozen pocket confined by approaching *permafrost.

2. Closed system pingo An ice mound, as above, but formed as water, under *artesian pressure within or below permafrost, which causes it to buckle upwards. MacKay (op. cit.) argues that not all such pingoes form in a truly closed system. Gurney (1998) *PPG* 22, 3 proposes a third category: polygenetic (or 'mixed') pingoes.

Pingo

1. Open system pingo (East Greenland type)

a. Permafrost advances

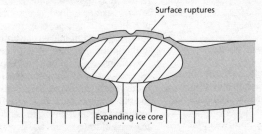

b. 'Free' water freezes and expands, lake floor mounds up, lake water drains away

Surface ruptures

Expanding ice core

2. Closed system pingo (Mackenzie type)

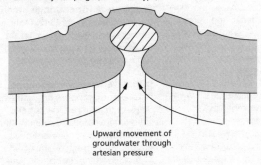

Upward movement of
groundwater through
artesian pressure

3. Collapsed pingo (open or closed system)

Rampart   Previous mound

Solifluction

Ice core
melted

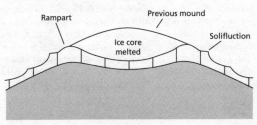

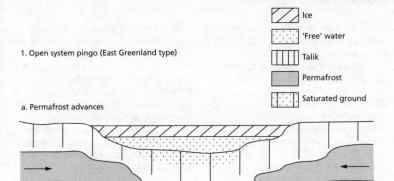

Ice

'Free' water

Talik

Permafrost

Saturated ground

pioneer community The earliest *sere developing on a raw site. A **pioneer species** is a plant species dominating a community in the first stage of succession.

pipe 1. A volcanic channel or conduit filled with solidified *magma. Sometimes the hard pipe rock is exposed after erosion. See Barnett and Lorig (2007) *J. Volcanol. Geotherm. Res.* 159, 1–3. For **pipe eruption**, *see* CENTRAL VENT ERUPTION.

2. In *hydrology, a subsurface channel, often near horizontal, through which water passes. Pipes can transfer water underground as a rapid route for subsurface storm flow. See Carey and Woo (2000) *J. Hydrol.* 233, 1–3.

pipeline *See* GLOBAL PIPELINE.

place Five conceptualizations of place may be distinguished: place as physical location or site, place as a cultural and/ or social location, place as context, place as constructed over time, and place as a social process. We make space through interaction at all levels, just as places, regions, nations, the local, and the global are made in this way. H. Raffles (2002) understands places as spatial moments that come into being and continue being made at the meeting points of history, representation, and material practice.

Try Larsen and Johnson (2012) *AAAG* 102, 3, 632 for a comprehensive summary of the major geographical work on 'place'. With the **relational reading of place**, the term 'relational' indicates a focus on the ways things are connected, with respect to each other. Thus, a relational reading of place is concerned with the social, political, and material processes by which people create their own settings through their spatial connections. Try Pierce at al. (2011) *Trans. Inst. Brit. Geogs* 2011, 36 if you're feeling strong.

Narratives of place ascribe meaning to places; they are the way we envisage or think about places socially and politically. This process can transform empty space into a meaningful construct. Over time, narrative representations of place have the potential to become deeply imbedded in us, and shape the way we understand, and interact with, the world. (You might recognize the pithy narrative 'where there's muck, there's brass', a 19th-century narrative claiming that dirty, smoky places were also places where someone was making money.)

place frames Places are 'made' by factors such as individual emotional attachment, language, election campaigns, economic opportunities, and political economy—the physical conditions of a location and the daily life experiences of its residents which unite them. These, among others, constitute the place frame. Such factors may be local and specific, linked to particular places and times—that is, by *contingencies. In this way, place frames mould **place formation** to which Phillips (2018) *Prog. Phys. Geog.* 6 adds further considerations, including individualism, multiple causality, and place as a contingent process.

place dependent Located and bound in time and space by, for example, fixed *capital, institutions, labour skills, and shared conventions. Products and commodities can take on the qualities of the places from which they come; think of the French *apellations controlées*.

place identity People's bonds with places. Peng et al. (2020) *Front. Psychol.* provide a comprehensive, but not very clear, summary of definitions for this term. Some authors argue that place identity is imposed, created, and manufactured, while *sense of place* remains personal, found and grounded in lived experience (Carter et al. (2007) *Soc. & Cult. Geog.* 8, 5).

placelessness E. Relph (1976) claimed that, with mass communication, and increasingly ubiquitous high technology, places become more and more similar, so that locations lose a distinctive 'sense of place'. Place means particularity and locality, while placelessness means blandness and similarity. In fact, novel cultural regions and specialized cities have been emerging even as a latter-day version of regionalism, and examples of neo-localism are taking root in many parts of the United States; see Wilbur Zelinsky (2011).

Gertler in J. Peck and H. Yeung demonstrates the limited ability of head offices to impose bland uniformity across their operations overseas.

place-making The set of social, political, and material processes by which people create and recreate the geographies they live in; the planning, design, management, and programming of public spaces, rooted in the community. Place-making is a networked process, created by the social and spatial relationships that link individuals together through their common location.

place marketing 'Branding' a location in order to 'sell' it, that is, to make it more attractive; a tool to promote the attractiveness of a locality in such a way that target groups (prospective users) know its distinctive features and are persuaded to go there; the set of tactics used to market a geographical location to improve its business. See Hospers (2011) *Tijdschrift* 102, 3, 369. Hospers (2010) *Place Branding and Public Diplomacy* 6, 280 is also useful on the limits of place marketing.

place names The study of the early forms of present place names may indicate the culture which gave the name together with the characteristics of the site. In the UK, for example, *ey* meaning a dry point and *ley* meaning a forest, wood, glade, or clearing appear in many place names such as Chelsea and Henley-in-Arden.

Place names label, define, and represent places and people; 'a place name sometimes fills up its territory with sense of place and homogenizes it. Fixing a boundary on a map reinforces a real differentiation and might segregate the dwellers' (Okamoto (2000) *PHG* 24, 328). In the Hawaiian Islands, place names change 'from being reflections of Hawaiian geographic discourse, to being encoded within Western approaches to knowledge, commodification of the environment, and control of territory' (Hermann (1999) *AAAG* 89, 1). Place names mark the spatiality of power relationships; see Usher (2003) *Canad. Geogr./Géog. canad.* 47, 4 on aboriginal land claims in Canada.

It is common for revolutionary regimes to create new state symbols to 'remove evidence of the deposed regime, and to establish an identity for the usurper'. During the Soviet period a special framework of laws and instructions was created to regulate a renaming policy. With the fall of the Soviet Union, renaming began in earnest; in Armenia, for example. By 1988 some 598 out of 980 place names in the Armenian SSR had been renamed.

(())) SEE WEB LINKS
• Arseny Saparov on the alteration of place names in Soviet Armenia.

places of memory Sites where symbolic imaginings of the past interweave with the present. Places of memory are both constructed and contested (Rose-Redwood (2008)

Soc. & Cult. Geog. 9, 4). K. Till in J. Agnew et al., eds (2003) explores places of memory, topographies and sites of social memory, and conflicts over national places of memory. See Forest and Johnson (2002) *AAAG* 92, 3 on Soviet memorials, and Alderman (2003) *Area* 35, 2 on street names.

place utility The significance a place has to a particular function or people. Adams and Adger (2013) *Environ. Res. Lett.* 8, 1 investigate the attachment people form to place, and the role of the environment in creating that attachment.

planation surface *See* EROSION SURFACE.

planetary boundary layer The lowest 500 m of the troposphere; the layer most strongly influenced by the land, or sea, beneath it.

planetary winds The major winds of the Earth such as the westerlies, *trades, etc., as opposed to local winds.

planeze One of a series of triangular facets facing outward from a conical volcanic peak. The planezes are separated by radiating streams which run down the flanks of the cone. Try Coltorti et al. (1999) *Geol. Soc. Special Publication* 162.

planform The outline of an object viewed from above.

plankton Minute organisms which drift with the currents in seas and lakes. Plankton *algae, various animal larvae, and some worms; the animals are **zooplankton** and the plants are **phytoplankton**. See W. Lampert and U. Sommer (1997) for a discussion of the 'top-down' versus 'bottom-up' plankton controversy.

planning As practised by local or national government, the direction of development. See S. Ward (2004) for a well-presented and easily read history of planning in the United Kingdom. S. Campbell and S. Fainstein, eds (2003) provide a valuable stimulant for constructive debate and reflection.

Planning blight is the adverse effect of a proposed development, such as a motorway, which could cause a drop in house prices. If the landowner cannot dispose of the property, or cannot make as much use of it as was previously possible, he or she may serve a purchasing notice on the planning department of the local authority. See R. Eddington (2006).

plant 1. In a system, the buildings, machinery, and land that inputs enter, and output exits.
   2. In industrial geography, an individual factory producing power or manufactured goods.

plantation A large agricultural system, generally a monoculture, employing labour on a large scale to produce tropical and subtropical crops. Early stage processing often takes place on site. Plantations may be seen as a spatial expression of *imperialism and *capitalism (B. Warren 1980). See Caney and Voelks (2003) *PHG* 27, 2 on landscape legacies of plantations in Brazil, and try Kenny-Lazar (2020) TIBG 45, 2.

plant community An assembly of different species of plants growing together in a particular habitat; the floral component of an *ecosystem. The concept of community can be applied to a range of scales from a small pond to the Amazon rain forest. Kent et al. (1997) *PPG* 21, 3 summarize developments in the conceptualization of the plant community. Burns (2007) *J. Biogeog.* 34, in a study of the assembly of an island plant community off the west coast of Vancouver Island, BC, finds that his results provide evidence for both the deterministic and individualistic viewpoints on community structure, and suggests that both chance and determinism contribute to the assembly of this island plant community.

plant recruitment The establishment of new seedlings and small plants.

plant succession The gradual evolution of a series of plants within a given area. This occurs in a roughly predictable order, while the habitat changes. **Primary succession** is 'ecosystem development on barren surfaces where severe disturbances have removed most biological activity' (L. Walker and R. de Moral 2003). **Secondary succession** is the replacement of a community after a disturbance; see Halpern et al. (1997) *Ecology* 70 for a summary of the variables influencing secondary succession. In **autogenic succession** the plants themselves are the genesis of change (Ellis and Coppins (2006) *J. Biogeog.* 33, 9); in **allogenic succession** the changes are driven by forces outside the ecosystem. The two are not mutually exclusive; see Francis (2006) *Area* 38, 4.

plastic flow Movement of material, especially rocks and ice, under intense pressure, when it flows like a very *viscous substance and does not revert to its original shape when pressure is removed. As it moves, it *shears. In ice, plastic flow is due to pressure at depth; a thickness of at least 22 m is needed for plastic flow *temperate glaciers. Plastically flowing ice will flow around and over an obstacle, which may cause deposition in the lee of the obstruction: this is **plastic moulding**; see R. LeB.

Hooke (2005). **Plastic deformation** is an irreversible change in the shape of material, without fracturing, resulting from compression or expansion. See Moeyersons et al. (2006) *Geomorph.* 76, 3–4 on plastic deformation in a *vertisol.

plate A rigid segment of the Earth's crust which can 'float' across the heavier, semi-molten rock below. The plates making up the continents—**continental plates**—are less dense but, at up to 35 km deep, are thicker than those making up the oceans—the **oceanic plates**—which are up to 5 km deep. A plate is a part of the *lithosphere which moves over the plastic *asthenosphere. The boundary of a plate may be a *constructive, *destructive, *conservative, or, more rarely, a *collision margin. The theory of **plate tectonics** submits that the Earth's crust is made up of six large plates: the African, American, Antarctic, Eurasian, Indian, and Pacific plates, and a number of small plates, the chief of which are the Arabian, Caribbean, Cocos, Nasca, Philippine, and Scotia plates.

The movement of plates causes global changes, such as *continental drift and a remodelling of ocean basins and the creation of major landforms: *oceanic ridges, *fold mountains, *island arcs, and *rift valleys, together with *earthquakes and *volcanoes, which occur at a destructive plate boundary where one plate plunges below another. See Cloos (1993) *GSA Bull.* 105, 6 on lithospheric buoyancy and subduction of oceanic plateaus, continental margins, island arcs, spreading ridges, and seamounts.

Plates are driven, at least partly, by forces from the convecting mantle. Oceanic driving forces are exerted when a lithospheric plate is being carried by a faster-moving *asthenosphere. *Ridge-push* is the gravitational sliding of a plate off a mid-ocean ridge, in response to the *ridge-push force*, caused by the

p

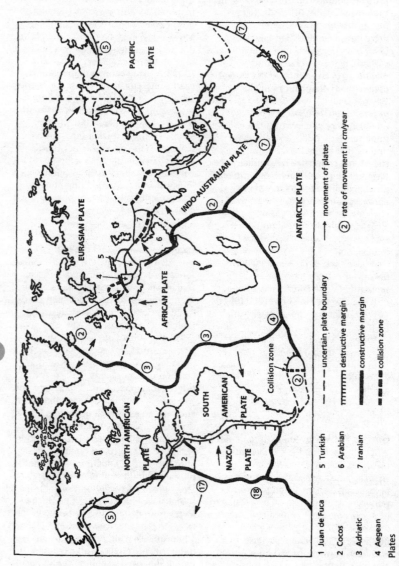

PACIFIC PLATE

PaCIFIC PLATE

EURASIAN PLATE

INDO-AUSTRALIAN PLATE

ANTARCTIC PLATE

AFRICAN PLATE

Collision zone

NORTH AMERICAN PLATE

SOUTH AMERICAN PLATE

NAZCA PLATE

Plates

1 Juan de Fuca   5 Turkish
2 Cocos   6 Arabian
3 Adriatic   7 Iranian
4 Aegean

→   movement of plates

②   rate of movement in cm/year

---   uncertain plate boundary

✲✲✲   destructive margin

▬▬   constructive margin

▬▬▬   collision zone

horizontal spreading of the near-surface asthenosphere at constructive margins. Anderson (2001) *Science* 293 holds that only upper mantle convection may be driven by plate tectonics; Bokelman (2002) *Geology* 30, 11 finds that, in the case of the North American plate, the mantle plays an important role in driving the plates. The sinking of a cold, dense slab of oceanic plate provides a *slab-pull force*. Conrad and Lithgow-Bertelloni (2002) *Science* 298 estimate the relative importance of 'pull' versus 'suction', finding that the pull from lower mantle slabs is not coupled to the surface plates. Eagles et al. (2009) *Tectonophysics* offer an animation of the Southern convergent margins.

plateau An extensive, relatively flat upland. Some plateaus are formed structurally, from resistant, horizontal rocks; some are *erosion surfaces; some made of **plateau lavas**; see Bondre et al. (2004) *Procs Ind. Acad. Scis, Earth Planet. Sci.* 113 on the Deccan Plateau.

playa A flat plain in an arid area found at the centre of an inland drainage basin, such as the Salinas Grandes, Jujuy Province, Argentina. Within such an area, lakes frequently form. Evaporation from the playa is high and alluvial flats of saline mud form. The term is also used to describe a lake within such a basin. The deepening of the basins by deflation is favoured by salt weathering processes. See Gutiérrez-Elorza et al. (2005) *Geomorph.* 72, 1–4.

Pleistocene An *epoch of the Quaternary period, also known as the Ice Age, from 1.8 million to ~10000 years ago.

(((⊕))) **SEE WEB LINKS**
• An overview of the Pleistocene epoch.

plinian eruption A highly explosive volcanic eruption with dense clouds of gas and *tephra propelled upwards for many kilometres. *Fissure eruptions are common, as are the destruction of the crater walls, and the fragmentation of solidified lavas. 'The magma mass discharge rate governs the eruption magnitude of plinian eruptions, and depends strongly on conduit geometry' (Varekamp (1993) *J. Volc. & Geothermal Res.* 54, 3–4). Arce et al. (2003) *Geol. Soc. Amer. Bull.* 115, 2, 230 describe the Plinian eruption of Nevado de Toluca volcano, Mexico.

plinthite A hard capping or crust at the surface of an unconsolidated soil. See D. Steila and T. Pound (1998) on the stages of plinthite development, and Eze et al. (2014) *Pedosphere* 24, 2.

plucking *See* QUARRYING.

plume An upwelling of molten rock through the *asthenosphere to the lower *lithosphere. The *hot spot thus formed shows up in volcanic activity at the surface. At the Yellowstone Hotspot, a rising mantle plume has a massive head 800 km in diameter; see US Geological Survey. As continents move over the hot spot, there forms a chain of volcanoes at the surface, since the plume is stationary with respect to the *mantle. See Constantini et al. (2007) *J. Volc.*

plunge pool A pool at the base of a waterfall, formed by *eddying, *hydraulic action, *cavitation, and *pothole erosion.

pluralism 1. A situation where no particular cultural, ethnic, ideological, or political group is dominant. When individuals affected by a given issue in the same way band together, and each group competes, negotiates, and compromises with other groups for influence, resources, such as money, or public support' (Eisenberg (1996) *Soc. Sci. Info.* 35, 2). **Engaged pluralism** is defined as a lively and respectful engagement across theoretical, methodological, and topical 'borders' in order to increase diversity and build mutual respect among scholars; see Rosenmann et al. (2020) *Prog. Hum. Geog.* 44, 3.

2. The cultural diversity of a plural society. Clifford (2006) *Area* 38, 4 speaks of 'pluralism, which has been much talked about but rarely practised', and Smith (2005) *Antipode* 37, 5 has a merry tilt at pluralism. R. Eversole and J. Martin (2005) observe the underlying assumption in Australia that a recognition of pluralism may be able to overcome potentially conflicting political interests in the regions, or actually sustain some communities that may be becoming economically and environmentally non-viable or socially dysfunctional. Mouffe (2000) (IHS) discusses **agonistic pluralism**—in which *the political* refers to the antagonism that is constitutive of society and *politics* refers to practices, discourses, and institutions that order and organize society but that are always open to the possibility of antagonisms because they are embedded in the political. Kenny (2004) discusses the **new pluralism**, finding that the preservation of difference tends to become a condition of any group's inclusion in the political community in a way that undermines democracy.

3. An academic stance that does not focus on one approach only; see Clare and Siemiatycki (2014) *Prof. Geogr.* 66, 1, 4.

pluralist geography Any geography that allows people with differing approaches to debate with one another without assuming that one must eventually 'win' by dominating the other, or that anything goes. Smith (2005) *Antipode* 37, 5 sees pluralism as the core politics of neo-critical geography.

pluton A solidified, underground, igneous mass, varying in size from *batholiths to *sills and *dykes. See McNulty et al. (2000) *GSA Bull.* 112, 1 on the Mount Givens pluton, Calif., and Stevenson et al. (2007) *J. Geol. Soc.* 164, 1 on the Eastern Mourne pluton, Ulster. A **plutonic rock** is an igneous rock which has cooled and crystallized slowly, at great depth.

pluvial A time of heavier precipitation than normal. See Pinson (2008) *Geoarchaeology* 23, 1 on evidence of more than one pluvial during the last million years. Pluvials have created current desert landforms, including wadi terraces.

podzol (podsol) A soil characteristic of the coniferous forests of Russia and Canada. **Podzolization** occurs when severe *leaching leaves the upper *horizon virtually depleted of all soil constituents except quartz grains. Clay minerals in the A horizon decompose by reaction with *humic acids and form soluble salts. The leached material from the A horizon is deposited in the B horizon as a humus-rich horizon band or as a hard layer of sesquioxides. See Sauer et al. (2007) *J. Plant Nutrit. & Soil Sci.* 170, 5 for a review of knowledge on the development, threats, and functions of podzols in 2007.

poetics The linguistic, literary, and theoretical resources used in *representation; hence **geographical poetics**—the varieties of language in which geography has been, continues to be, and ought to be, written. See Kong (2001) *PHG* 25, 2 for an example.

poikilotherm An organism which has its body heat regulated by the temperature of its surroundings. See Lindsey (1966) *Evolution* 20 on latitudinal variations in the body sizes of poikilotherm vertebrates.

point In a *GIS, a zero-dimensional feature. Because every object takes up space, no geographic feature can really be without length or width. Instead, points can be used to represent geographic features on maps; only its name and location are needed. See M. N. DeMers (2009).

point-bar deposit The accumulation of fluvial sediment at the *slip-off slope on the inside of a *meander. Van de Lageweg et al. (2014) *Geology* 42, 4, 319 explain the formation of point bars.

point pattern analysis For an examination of the method, potential, and limitations of two classes of point pattern analysis, see Bishop (2007) *Area* 39, 2.

polar Applying to those parts of the Earth close to the poles. The **polar area**, in particular the Arctic, is considered to be the forerunner of climate change (Schiermeier (2006) *Nature* 441, 146). **Polar air masses** (**P**) originate in the mid-latitudes (40°–60°) and are characteristically cold.

polar front The discontinuous, variable *front which forms over the North Atlantic and North Pacific, where polar maritime air meets tropical maritime air. The formation of *mid-latitude depressions at the polar front is connected with the development of troughs in the **polar front jet stream**, a band of high-velocity winds in the wider *Rossby waves. The polar front *jet stream moves southwards in winter and northwards in summer. During an *El Niño event, it shifts south, bringing cooler air, which might suppress tornado activity in the US Midwest. Mid-latitude aridity can result from the effect of mountains on the polar jet stream. See Dupont et al. (2005) *Geol.* 33, 6 on the sensitivity of the Namibian upwelling to latitudinal shifts in the Southern Ocean

polar front zone; Ingvaldsen (2005) *Geophys. Res. Letter* 32, LI6603 on the link between the polar front and the North Cape Current; and Hayes and Zenk (1977) *NOAA Technical Report ERL PML* 28 on the Antarctic Polar Front.

Polar easterlies blow east–west, from *polar highs to sub-polar lows. A **polar high** is a mass of cold, heavy air, centred at the poles, and produced by downward, vertical air currents from the *polar vortex, bringing high pressure at high latitudes. For **polar low**, *see* COLD LOW.

polar vortex A rotating polar high pressure system, formed as a distinct column of cold air in the middle to lower stratosphere, developed during the long polar night; wind speeds may reach 100 metres s^{-1}. Upper *troposphere air is sucked into the vortex, then sinks to cause *polar highs. See Chshyolkova et al. (2007) *Angeo* 25 on the Arctic Polar vortex. The **stratospheric polar vortex**, located 5–30 miles above the Earth's surface, only appears in winter; the **tropospheric polar vortex** is year-round and occupies the lowest layer of the atmosphere that touches the surface. Go to NORA Climate.gov; it's a great site.

polder Land reclaimed from the sea, lakes, or river deltas; the land is bounded by a dyke, drained, and is maintained by pumping.

political ecology The study of the relationships between nature and society; the way political, economic, and social factors affect, and are affected by, environmental issues. Some of these issues may be interactions between society and resources; for example, poverty may be brought about by poor management, which leads to environmental degradation, which, in its turn, deepens poverty. (It cannot be assumed that the poor have an intrinsic propensity to degrade environmental resources.) Examples of political ecology include the cultural *eutrophication of Lake Erie and the deepening poverty of Masai herders when blocked from traditional grazing.

Walker (2005) *PHG* 29, 73 noted that 'while political ecology has thrived, its coherence as a field of study and its central intellectual contributions remain the subject of sometimes contentious debate'; one of the recurrent, and unresolved, questions has been 'Where is the ecology in political ecology?' Go to *Progress in Human Geography* (2019) 43, 1 for two articles on political ecology. Do enter 'some grumpy thoughts on political ecology & biophysical science' into your search engine.

Global political ecology concerns itself with the globalization of environmental problems and responses, such as the role of the political economy of global capitalism in the political ecology of a series of environmental disasters. See Richard Peet et al. (2011).

political economy The study of the principles that govern the production and distribution of goods under capitalist systems, which includes the roles that individuals and political entities such as states play in the production and consumption of goods. In a useful article, Sheppard (2010) *J. Econ. Geog.* 11, 2, 319 explains that geographical political economists consider a wide-ranging set of theories of the capitalist space economy, connecting agency with socio-spatial structure, through extensive multi-method case study research, as well as theoretical analysis.

political geography The study of the geographical and spatial aspects of politics; the spatial expression of political ideals and the consequences of decision-making by a political entity. Of particular interest are the geographical factors which influence voting patterns; for example, Morrill et al. (2007) *Pol. Geog.* 26, 5 on the red (conservative/republican) and blue (liberal/democrat) divide in the United States, or Buhaug (2006) *Pol. Geog.* 25, 3 on local determinants of African civil wars. For clear reviews, see Sparke: (2004) *PHG* 28, 6; (2006) *PHG* 30, 3; (2008) *PHG* 32, 3; and Staeheli (2010) *PPG* 34, 1, 67.

Studies on quantitative methodologies and spatial analyses based on GIS include Buhaug and Lujala (2005) *Pol. Geog.* 24, 4, who find that the relation of a conflict affects its duration. For critical, feminist, and popular geopolitics, see Hyndman (2004) *Pol. Geog.* 23, 3; for peace and conflict studies. See Sene (2004) *African Geopol.* 15–16, on the federating role of soccer, and territoriality.

politics of place An emphasis on the importance of place and locality as arenas for political activities; far from being bounded, homogeneous, and stable entities, localities are actively shaped by political economic processes (capital, and localized modes of political regulation), and cultural processes. See Elmhirst (2001) *Sing. J. Trop. Geog.* 22, 3, a clear and helpful paper; see also D. Massey (2005) and (2004) *Geografiska B* 86 and Cumbers et al. (2008) *PHG* 32, 2.

polje In *karst terminology, a flat-floored, steep-sided, enclosed basin up to 65 km long and 10 km wide. Streams can be ephemeral or permanent; usually the water drains into *streamsinks. Most poljes are aligned with underlying structures such as folds, faults, and troughs. See Pisano et al. (2020) *Journal of Maps* 16, 2.

pollen analysis *See* PALYNOLOGY.

pollution The presence in, or introduction into, the environment of a substance that has harmful or poisonous effects. 'As far as the geographical distribution of pollution is concerned: pollution is concentrated in agglomerations, as in [major] Third World cities . . . environmentally harmful activities are located in peripheral regions, as is the case with nuclear power stations in industrialised countries . . . [and] in some cases . . . pollution is rather uniformly distributed over the country' (Rauscher and Bouman (1998) www.feem-web.it/worldcongress/abs2/rausch2.html). See Lange and Quaas (2007) *BEJ Econ. Analys. & Policy* 7, 1 on the effect of environmental pollution on agglomerations, and the special issue of *Climate* (2019) on environment pollution and climate change.

Diffuse pollution is pollution from widespread activities with no one discrete source; for example from pesticides or from urban run-off. Go to the OECD site (2017): on Diffuse Pollution, Degraded Waters.

Biodegradable pollutants are the waste products, and remains, of animals and plants, including natural oil and gas, together with certain chemicals. They are non-polluting if adequately dispersed.

Non-biodegradable pollutants such as lead, may be concentrated as they move up the food chain; this is **biological magnification**—see J. Nathanson (2002). Within western Europe, air pollution, associated with basic industries such as oil refining, chemicals, and iron and steel, as well as with motor transport, is probably the principal offender, followed by water and land pollution. Other forms include noise, and thermal pollution.

Present-day problems of pollution include *acid rain and the burning of *fossil fuels to produce excessive carbon dioxide. The **pollution haven** (or ecological dumping) hypothesis states that if free trade occurs between countries with different environmental standards, countries with lower standards will develop a comparative advantage in relatively 'dirty' industries. See Ranocchia and Lambertini (2021) *Env. & Resource Econs* 78.

pollution dome A mass of polluted air in and above a city or industrial complex which is prevented from rising by the presence of an *inversion above it. Insolation responds to pollution domes—Tomalty and Brazel (2000) *Procs Urb. Envs*, Conf. AMS, Davis, record inner-city values as 15% lower than outside values. Winds may elongate the dome into a pollution plume; see Nemecek et al. (1995) *Int. J. Rock Mech. &c.* 32, 7 on pollution plume containment. The term has been extended to describe **light pollution domes**.

polygon 1. In geomorphology a sand wedge polygon is a micro-relief landform, free of vegetation and identifiable by textural changes in the soil. Ice-wedge polygons are formed after a temperature drop so extreme that the ground contracts and cracks, forming a narrow gap. In winter, snow blowing across the tundra surface fills in the crack and freezes to form a thin vein of ice. The following spring this ice vein thaws, but it refreezes the following winter and the ground cracks again along the same ice vein. New snow blows into the crack and freezes, gradually 'wedging' open the ground.

2. In *GIS, a closed area bounded by a connected sequence of paired x and y coordinates, the first and last coordinate pairs being the same, and all other pairs being unique. More commonly known as an area.

polysaccharide gum The sticky by-product of the decomposition of roots by micro-organisms, which can bind soil minerals into aggregates, such as *peds.

polythermal glacier A thermally complex glacier with both warm and cold ice. Warm ice usually occurs where the ice is thickest, as a result of geothermal heating, while the snout and margins of the glacier are frozen to the bed. See Roberson et al. (2011) *Geograf. Series A* 93, 71 DOI: 10.1111/j.1468-0459.2011.00420 on the physical properties of a polythermal valley glacier in Norway.

pool and riffle The alternating sequence of deep pools and shallow riffles along the relatively straight course of a river. As the *thalweg meanders within the channel, pools typically form near the outside bank of bends, and riffle areas usually form between two bends, where the thalweg crosses over from one side of the channel to the other. Gravel- and cobble-bedded streams usually have regularly spaced pools and riffles. Coarser sediment particles are found in riffle areas, while smaller particles occur in pools.

The distance between the pools is 5–7 times the channel width. Look for Emma Bucknell (2015) on the formation and adjustment of a pool–riffle sequence in a gravel-bed flume.

population A group of individuals of the same species within a *community. Life tables provide a picture of survival and mortality in populations. Cohort life tables—whereby individuals are tracked from birth—are appropriate for plants and sessile organisms. Static life tables— which record age at death within a certain period of time—are appropriate for mobile and long-lived organisms. For the 'laws' of population ecology, see Turchin (2001) *Oikos* 94, 1.

population density The ratio of a population to a given unit of area.

Lagerlöf and Basher (2006) *MPRA Paper* 369 make the bold claim that population density correlates with per capita income in Canada and the United States. On firmer ground is Sato (2007) *J. Urban Econ.* 61, 2, who argues that migration from areas of lower to higher population density in Japan maintains regional variations in fertility. **Crude population density** is simply the number of people living per unit area and is not necessarily a development indicator.

population dynamics The study of the numbers of populations and the variations of these numbers in time and space. The main determinants are birth and death rates and migration; which, in turn, reflect levels of development, and socio-economic and cultural factors. See Millington et al. (1999) *App. Geog.* 19, 4 on population dynamics and socio-economic change; Bhandari and Grant (2007) *J. Rural & Comm. Dev.* 2 on population dynamics and land use; Ahn in H.-R. Kim and B. Song (2007) on population dynamics and urbanization in Korea. See also H. Macbeth and P. Collinson, eds (2002); try L. M. Hinter (2000) on the environmental implications of population dynamics. For geolinguists, see N. Komarova and M. Nowak (2002) on the population dynamics of grammar acquisition.

population ecology The study of how the population sizes of species living together in groups change over time and space. It includes population dynamics and distribution, evolutionary ecology, ecological genetics, theoretical models, conservation biology, agroecosystem studies, and bioresource management. See the *Journal of Population Ecology*.

population equation The future size of a population depends on a range of functions. Thus:

$$P_{t+1} = P_t + (B - D) + (I - E)$$

where P_t and P_{t+1} are the sizes of population in an area at two different points in time, t and $t + 1$ are those points, B is the birth rate, D is the death rate, I is the immigration, and E is the emigration. For an example of its use, see Roos (2005) *J. Econ. Geog.* 5, 616.

population geography The study of human populations; their composition, growth, distribution, and migratory movements, with an emphasis on the last two. It is concerned with the study of demographic processes which affect the environment, but differs from demography in that it is concerned with the spatial expression of such processes. *Population, Space and Place* is the journal of the UK Population Geography Research Group.

population growth The rate of increase of the world's population has become ever faster:

'Future population growth is highly dependent on the path that future fertility takes. In the medium variant, fertility is projected to decline from 2.6 children per woman today to slightly over 2 children per woman in 2050. If fertility were to remain about half a child above the levels projected in the medium variant, world population would reach 10.6 billion by 2050. A fertility path half a child below the medium would lead to a population of 7.6 billion by mid-century. That is, at the world level, continued population growth until 2050 is inevitable even if the decline of fertility accelerates (Population Division, Dept. Econ. & Soc. Affairs, UN Secretariat, *2004 Revision*). Ninety per cent of the differences across countries in total fertility rates are accounted for solely by differences in women's reported desired fertility. Other key determinants of population growth are: mortality rates (life expectancy), the initial age profile of the population (whether it is relatively old or relatively young to begin with), and migration. For world population trends, see the United Nations Population Fund. *See also* POPULATION PROJECTION.

year	population in billions
1850	1.2
1900	1.6
1950	2.55
2000	6.1
2050	9.7 (estimated)

population projection The prediction of future populations based on the present age–sex structure, and with the present rates of fertility, mortality, and migration. Wilson and Rees (2005) *Pop. Space Place* 11 review developments in population projection methodology.

population pyramid *See* AGE–SEX PYRAMID.

pore In geomorphology, a minute opening in a rock or soil, through which fluids may pass. Porous rocks allow water to pass through or be stored within them. **Pore pressure** is the pressure applied by water in the pores to soil and rock particles, reducing internal friction; when the rock or soil is saturated, pore water pressure can be so great that slope failure results; see Zen-Hua Zhou et al. 2018 *Geom.* 308. However, when the material dries out, only the smallest pores are water-filled, resulting in **negative pore-water pressure**. The water is held in these pores through strong suction, so that negative pore water pressure can increase the pressure, especially with clays. Negative pore water pressure in soil is referred to as soil moisture tension (or suction).

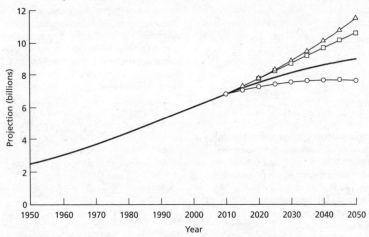

Population projection. Population of the world 1950–2050 by projection variants

porosity The ratio of the volume of pores to the volume of matter within a rock or soil, expressed as a percentage. Iverson et al. (2000) *Science* 20 conclude that relatively small variations in porosity can influence landslide behaviour profoundly. Koulouri and Giourga (2007) *Catena* 69, 3 use soil porosity as one variable in determining soil erodibility. Chemical weathering increases rock porosity under virtually all circumstances. Porosity is one of the best indicators of soil structural quality.

positionality Where you are coming from must reflect your own personal viewpoint. Thus, positionality is the recognition and declaration of one's own position. Gold (2002) *Ethics, Place & Env.* 5, 3 looks useful.

positive discrimination A policy designed to boost some deprived region or minority group and to redress, at least in part, inequality and/or *uneven development. Policies of positive discrimination have been criticized for treating the effects of inequality rather than tackling its causes. Other criticisms made are that not all of the minority or region needs help, and that many deprived people are outside the catchment area.

positivism A positivist method is a philosophy of science of simply sticking to what we can observe and measure. It uses repeatable research methods, so that the same tests can be performed again as a quality control.

possibilism The observation that, although the physical environment may promote certain actions or rule them out, humans also play their part in choosing their strategies; such choices are not pre-determined by that environment. For example, while the natural environment may impose constraints on human life, culture is determined, independently of nature, by human social conditions. See Marshall Sahlins (1976).

post-colonialism A sub-discipline of geography which awaits a consensus on its definition; some definitions and observations follow: 'The geographically

dispersed contestation of colonial power and knowledge' (A. Blunt and C. McEwan, eds 2000). 'Much of what passes for postcolonial theory in British geography reinforces new forms of colonial epistemologies and colonial hierarchies, while destabilizing their older forms' (Gilmartin and Berg (2007) *Area* 39, 1). 'Post-colonialism studies might best be regarded now as a term for a body of diverse and often contesting formulations of the cultural production of colonized people rather than as a discipline or methodology *per se*' (B. Ashcroft et al. 2002). 'Post-colonialism has two key meanings: temporal—the period after colonialism; and critical—the study of colonialism and its aftermath' (Gilmartin (2009) in C. Gallaher et al.). J. P. Sharp (2008) provides an excellent introduction for students of colonialism and post-colonialism.
Post-colonial urban theory is the study of the conditions of post-imperial places, particularly the examination of colonial inscriptions on cities and the reinscriptions that follow decolonization. In this sense, 'inscription' and 'reinscription' can also signify ascribing identity, heritage, and meaning. Short (2012) *AAAG* 102, 1, 129 tackles this, if only indirectly.

postdevelopment theory A political acknowledgement of the injustices and unequal power relationships of the practice of development, and a view that development cannot offer sufficient tools for transformation. Instead, it suggests resistance to *globalization provided by grassroots organizations and non-governmental organizations, alternative forms of livelihood, and local forms of knowledge and practice. See A. Ziai (2017) *Third World Quarterly* 38, 12. See also Sidaway (2007) *PHG* 31, 3.

post-Fordism A system of production and *accumulation characterized by flexibility of both labour and machinery, the vertical break-up of large corporations, small-batch production, better use of links between firms so that subcontracting is increasingly used, *just-in-time production, niche consumption, commodified culture, and the popularity of monetarist ideologies among governments. Post-Fordism is associated with *agglomeration, which simplifies interaction between associated economic activities. See Bryson and Henry (2000), in J. Sidaway, P. Daniels, M. Bradshaw, & D. Shaw (eds). Linda McDowell (2010) *TIBG* 16, 4, 400 is somewhat brisk on 'the new gender order of post-Fordism'. *See* FLEXIBLE ACCUMULATION; FLEXIBLE SPECIALIZATION.

post-humanism A concept of humanity in which the 'human' is merely one life form, whose worth and workings are not morally superior; thus, a rejection of anthropocentrism. There are several implications; not least is the view that humans themselves, with biotechnology, will change. A post-humanist geography might consider the role of the non-human—plants, animals, energies, and technologies—in any investigation; it might be characterized by a new ethical appreciation of our non-human 'cousins', or explore the construction of boundaries between human and non-human nature; see Williams et al. (2019) *TIBG* 44, 4. Braun (2004) *Geoforum* 35, 269 provides a clear discussion on post-humanism in geography.

post-industrial city A city exhibiting the characteristics of a *post-industrial society. Service industries dominate with a strongly developed *quaternary sector and *footloose industries abound, often on pleasant open space at the edge of the city. Post-industrial cities are also characterized by large areas of office blocks and buildings for local government administration. These cities often exhibit a marked inequality of income distribution because of the contrasts between those who are appropriately skilled—professionals, managers, administrators, and those in high-technology service industries—and the poorly paid service workers who look after their needs, together with the unemployed. The former can afford high house prices, and, in fact, contribute to them; the latter cannot. Phelps and Ozawa (2003) *PHG* 27, 5 present a taxonomy of forms of agglomeration in proto-industrial, industrial, and post-industrial urban contexts, commenting that contemporary, post-industrial agglomerations are larger than preceding ones, with an increasingly complex pattern of specialization within and between urban areas. O'Connor and Go (2013) in T. Flew (ed.) have an interesting paper on Manchester as a post-industrial city. *See also* ENTREPRENEUR; INFORMATION CITY.

post-industrial society A post-industrial society has five primary characteristics: the domination of service, rather than manufacturing, industry, the pre-eminence of the professional and technical classes, the central place of theoretical knowledge as a source of innovations, the dominating influence of technology, and levels of urbanization higher than anywhere else in the world. Mahon (2006) *TIBG* 31, 4 argues that childcare arrangements are central to post-industrial societies.

postmodernism A philosophical stance which claims that it is impossible to make grand statements about the structures of society or about historic causation, because knowledge is partial and situated, and no one interpretation is superior to another. Postmodernism in geography is expressed in the recognition that space, place, and scale are social constructs, not external givens. Of particular interest is the way that *time and space have been 'compressed' by modern transport systems, especially by jumbo jets; hence **postmodernity**, which can apply to the ever swifter movements of capital and information in postmodern society. This development has reduced the *friction of distance for most *advanced economies. D. Harvey (1989) sees *time–space compression as a fundamental social feature of postmodernism. Some geographers claim that postmodernism challenges the dominance of time and history in social theories; instead stressing the significance of geography and spatiality (see E. Soja 1989). The postmodern tradition also stresses, and indeed champions, difference.

Among geographers, interest in postmodernism is waning.

post-productivist transition A transformation in rural economies since the mid-1980s characterized by: the production of food within an increasingly competitive international market; the progressive withdrawal of state subsidies for agriculture; reductions in food output; growing concerns over food quality; the growing environmental regulation of agriculture; and the creation of a more sustainable agricultural system. In New Zealand, post-productivist agriculture means farming while also protecting indigenous vegetation and wildlife; in this it shows how the attitude of farmers towards native habitat and wildlife mirrors changes in wider New Zealand society (Mairi Jay Department of Geography, University of Waikato). Post-productivist elements of course vary from one part of the world to another.

In the UK the *Countryside Stewardship Scheme attempts to bring about a transformation from a *positivist to a more post-positivist approach to agriculture. But Amsted in L. Landmark and C. Sandstrom, eds (2013) takes issue with this concept. *See* THIRD FOOD REGIME.

post-socialist city Discussions of the post-socialist city can founder on the definition of post-socialism (see Andros (2001) *Anthropoids*). Lin (2004) *Eurasian Geog. & Econ.* 45, 1 argues that the interaction of the socialist legacy of industrialization 'with the new forces of marketization and globalization has given rise to a peculiar pattern of simultaneous industrialization and tertiarization, differing from the Western norm of linear progression'. Tsenkova (2008) *Urb. Geog.* 29, 78 is richly detailed; see also Hirt (2007) *Urb. Geog.* 28, 8, and try K. Stanilov (2007).

poststructuralism Poststructuralists believe that knowledge production is political, and focus on the way the formation of dominant and weaker groups is made normal, and justified through the use of language. Dominance and 'otherness', for example, are often based on the way people are defined/described in society and how they work with those identities/descriptions. Try Marcus Doel (1999) for a sound review of poststructural geographies. See also Jonathan Murdoch (2005).

potential evapotranspiration (PET) The maximum continual loss of water by evaporation and transpiration, at a given temperature, given a sufficient supply of water. PET often outstrips actual evapotranspiration.

pothole Loosely, a vertical cave system; more precisely, a more or less circular hole in the bedrock of a river. The hole enlarges because pebbles

inside it collide with the bedrock as the water swirls. Potholes carved into streambeds can be important components of channel incision. Springer et al. (2005) *Geomorph.* 82, 1–2 demonstrate that the pothole dimensions of radius and depth are strongly correlated.

poverty 'A pronounced deprivation in well-being.' A. Sen (1985, 1993) argues that what is important to well-being is the 'capability to function in society'. These capabilities include access to food, shelter, clothing, education, security, and good health: capabilities which the poor lack. The poor also lack a voice in society, are vulnerable to risk, and are powerless.

SEE WEB LINKS
- World Bank Poverty and Equity data.
- The Human Development Report.

power The power structure of a society is reflected in its social organization and in its economy: *capitalism is controlled by a minority who dominate the *factors of production, and in a *centrally planned economy it is the state that dominates. Different political systems tend to have their own spatial expression. See Boudreau (2007) *Env. & Plan. A* 39, 11 on making new political spaces.

Geographies of power study the spatial distribution of power over others across multiple geographical scales, from the body to the globe. Territories are continually subjected to resistance, contestation, renegotiation, *deterritorialization, and reterritorialization; see, for example, Mawdsley (2007) *Geog. Compass* 1, 3 on China and Africa as emerging challenges to the geographies of power.

Themes include state formation, where power is transmitted through social channels to territorial ones; see, for example, Jessop (2007) *Pol. Geog.* 26, 1 and the relationship of minority cultures and empowerment, as in Mavroudi (2008) *Pol. Geog.* 27, 1, on Palestine. See Kobrin (2001) in A. Rugman and T. Brewer (2003).

power law A mathematical relationship between two quantities. When the frequency of an event varies as a power of some attribute of that event (e.g. its size), the frequency is said to follow a power law. Thus, Hack's law (Hack, (1957), *US Geol. Surv. Prof. Pap.* 294-B:45) states that the length l of the main stream of a river basin scales like a power h of the basin's area a:

$$l \propto a^h$$

and Li et al. (2011) *Geomorph.* 130 (2011) 221 find a power law relationship between rainfall and the frequency of landslides in Zhejiang province, China.

prairie A large area, found outside the tropics, with grassland and occasional trees as natural vegetation, as in the prairies of North America, the South American pampas, the Russian *steppes, and the South African veld. A **prairie soil** is a soil of the wetter prairies, resembling *chernozem in its high humus content and its development under grassland. Prairie soils are *leached of calcium, and are slightly acid. See Barrett and Randall (1998) *Soil Sci.* 163, 6.

praxis Often designating *accepted* practice, at its most simple, praxis is practice as opposed to theory; Kitchin et al. (2013) *Dialogues Hum. Geog.* 3, 1, 56 argue that new social media produce new forms of public geography and digital praxis.

Pre-Boreal A warmer division of the *Flandrian. For the rapid climate change at the Younger Dryas–Pre-Boreal transition, see Starkel (2003) *Glob. & Planet. Change* 35, 1.

Precambrian The oldest *era in Earth's history dating from about 4600 million years BP. 'To speak of "the Precambrian" as a single unified time period is misleading, for it makes up roughly seven-eighths of the Earth's history' (UCMP).

((⊕)) SEE WEB LINKS
• The divisions of Precambrian time.

precession The wobbling of the Earth on its polar axis, which has a return period of 26 000 years. This change in movement and orientation alters the amount and location of solar radiation reaching the Earth.

precipitation In *meteorology, the deposition of moisture from the atmosphere onto the Earth's surface in the form of *rain, *hail, *frost, *fog, sleet, or *snow. Initially, cloud droplets grow around nuclei through condensation and diffusion. In warmer clouds the larger droplets then grow by collision and coalescence with the smaller ones. In colder clouds the *Bergeron–Findeisen mechanism is thought to operate, probably in conjunction with the growth of ice crystals: through *accretion, as supercooled water droplets freeze on impact with the ice; and aggregation, as smaller ice crystals stick to larger ones. Much precipitation begins in the form of ice crystals, but melts as it falls, to become rain. Variations in the intensity, amount, timing, duration, and frequency of precipitation, whether frozen or liquid, have important implications on the physical, chemical, and biological processes on Earth. See Zhang et al. (2007) *Nature* 448 on *anthropogenic precipitation; Holmer (2007) *Geograf. Annal.* 89, 4 on rainfall change; and Malby et al. (2007) *Hydrol. Scis J.* 52,

2 on long-term variations in
*orographic rainfall.

predator–prey relationships In
theory, there should be an equilibrium
between predators and prey. Thus, when
predators are scarce, the numbers of
prey should rise. Predators would
respond by reproducing more and,
possibly, by changing their hunting
habits. As the population of predators
rises, more prey is killed and their
numbers fall. Many of the predators then
die; thus numbers of predators and prey
oscillate between two extremes. See
Gilpin and Rosenzweig (1972) *Science*
177, 1358 for the classic model. Gilg et al.
(2003) *Science* 302, 866 should be useful.
A **keystone predator** helps define an
entire ecosystem; without its keystone
species, the ecosystem would be very
different; it might even cease to exist.
Gray wolves are keystone predators
in the Greater Yellowstone
Ecosystem, USA.

pre-industrial city A model of
the pre-capitalist city, proposed by
G. Sjöberg (1960). The city centre is
occupied by a small elite. The lower
classes occupy the concentric zone
surrounding the centre, and the outcasts
are consigned to the outer edges of the
city. Abbott (1974) *Amer. Soc. Rev.* 39, 4
finds that the distribution pattern of
socio-economic status in pre-
revolutionary Moscow was consistent
with Sjöberg's model.

preservation The protection of
human features in the landscape, as
opposed to *conservation which is
concerned with the protection of the
natural landscape. This distinction is not
always made; see, for example, Shafer
(2004) *Lands. & Urb. Plan.* 66, 3 on the
voluntary preservation of America's
natural heritage.

pressure gradient The rate of
change in *atmospheric pressure
between two areas, providing a force
which moves air from *high to *low in an
effort to even up the unequal mass
distribution of the air. On a global scale,
the most powerful pressure gradients are
in a *meridional direction, caused by
meridional disparities in *insolation. The
pressure gradient wind is the
movement of air in response to pressure
differences, blowing from high to low. It
is modified, however, by the action of
the *Coriolis force. The term is also used
in fluvial geomorphology; see Powell
(1998) *PPG* 22, 1.

pressure melting point In
glaciology, the temperature, often well
below 0 °C, at which ice under pressure
will melt. In basal ice, flow at the
pressure melting point is partly from
basal sliding and partly from shear
deformation. Warm glaciers have bases
at or above the pressure melting point of
their ice.

pressure release The expansion of a
rock formed under pressure when that
pressure is released (also known as
dilatation). A glacier may remove the
overburden, and the revealed rock
'bursts' open; see the classic Lewis
(1954) *J. Glaciol.* 2. Twidale (1973) *Rock
Mech. & Rock Eng.* 5, 3 attributes sheet
jointing to pressure release.

price mechanism The system where
the prices of commodities are
determined by demand and supply; the
regulation of economic activity by
supply. **Pricing policies** are the
arrangements whereby prices of
commodities to the consumer are
determined; see, for example, Arentze
and Timmermans (2007) *J. Transp. Geog.*
15, 1.

primary industry (primary activity)
Economic activity, such as fishing,

p

forestry, and mining and quarrying, concerned with the extraction of natural resources.

primate city The largest city within a nation which dominates the country not solely in size—being more than twice as large as the second city, as in London and Birmingham, UK—but also in terms of influence. Tammarau (2000) *Tijdschrift* 91, 1 explores this concept in centrally planned, developed, and developing economies.

principal components analysis (PCA) Probably the oldest and best known of the techniques of multivariate analysis, this procedure reduces the dimensionality of a data set in which there are a large number of interrelated variables, while retaining as much as possible of the variation present in the data set. This is achieved by transforming to a new set of variables, the principal components, which are uncorrelated, and which are ordered so that the first few retain most of the variation present in all of the original variables. Demšar et al. (2013) *AAAG* 103, 1, 106 provide an excellent overview of PCA in spatial settings.

private space Spaces, such as private properties, that are not open to all. Bondi (1998) *Urb. Geog.* 19, 2 shows how divisions between public space and private space operate at different scales and take different forms in different neighbourhoods. These forms illustrate how gender and class are interwoven in demarcations between, and connotations of, public and private spaces.

process model 1. A way of looking at the world based on the processes involved, rather than starting with the object. This means that a process model is not based on a research question.

Hollyer explains this approach very clearly at Geosoft.com.

2. A process-based software program model. Wergen et al. (2012) *Geomorph.* use a 3-D morphodynamic, process-based numerical model (Delft3D) to investigate the processes governing the evolution of major morphological features in alluvial estuaries as they evolve through the interaction between tidal movement, the available sediments, and the geometry of the tidal basin.

(⊕) SEE WEB LINKS
- Delft3D Open Source Community website will give all you need for fluvial, marine, and sedimentary processes.

process–response system A natural system in which energy and matter flow through a set of interacting things. The flows interact with the entities; there is a two-way feedback between process and form. Castedo et al. (2011) *Geomorph.* 177–78, 128 give an excellent example of a process–response coastal recession model of soft rock cliffs.

producer An organism which can fix energy from the sun and transform it by photosynthesis into food. **Producer services** are services such as accountancy, marketing, advertising, market testing, and commercial lending provided by one business to another. *See* OUTSOURCING.

production In ecology, the increase of body mass, as food is converted into new living material.

production, geographies of The study of global divisions of labour, production processes, and regional worlds of production. 'Changing geographies of production are a product of the inter-play of (inter alia) corporate, state, and trades union strategies, as companies pursue profitability, trades unions and workers seek new employment and/or protect existing jobs, and states attempt [to pursue] *accumulation in their territory' (Hudson (2002) *TIBG* 27, 262). Coe (2012) *PHG* 36, 3, 389 provides a really useful A–Z of the distinguishing features of **Global Production Network** analysis.

productivist Describing an agricultural system which is *intensive, expansionist, and based on the expansion of world trade in food, ever-increasing farm sizes, and the use of technology to increase output; factory farming and *agribusinesses are both examples. In Brittany, for example, the removal of hedges has decreased *infiltration and *groundwater storage. Annually, some 30000 tonnes of nitrates are leached into British rivers, and slurry from livestock has added to pollution.

productivity The output of an economic activity, in terms of the economic inputs. There is a central link between knowledge capital and productivity, while investing in highly dense regions increases congestion, thereby reducing productivity. A **productivity rating** is an estimate of an area's ability to support plant growth; see Yang et al. (2005) *Rev. Agric. Econ.* 27, 1 on conservation reserve enhancement.

proglacial Situated in front of a glacier. A **proglacial lake** is formed between the terminus of the ice and the higher ground which is often in the form of a terminal *moraine. Glacial Lake, Wisconsin, was a large proglacial lake. Glacial recession caused the ice dam to fail catastrophically and the lake drained rapidly (Clayton and Knox (2008) *Geomorph.* 93, 3).

progradation The accumulation of beach material when there is an excess in the supply of sediment; a feature of, for example, *delta and *mangrove coasts; Hori et al. *Geomorph.* 41, 2 find that the average progradation rate of the Yangtze delta rose from 50 to 80 km/year about 2000 years BP.

project In economic geography, a time-limited operation; a 'one-off' activity, such as film-making. See Kloosterman (2008) *J. Econ. Geog.* 8, 4 on **project-based organizations**.

projection *See* MAP PROJECTION.

protalus rampart A thin ridge, about a metre high and many metres long, composed of rock fragments, running along the foot of a mountain face and thought to be produced from rock fragments sliding over a snow surface. Also known as **pronival rampart**: Margold et al. (2011) *Geograf. Series A*, 93, 2, 137.

protection 1. The imposition of *quotas or *tariffs on foreign imports by a government: to become self-sufficient, to protect new industries, or as a bargaining tool. See Bai et al. (2003) *W. Davidson W. Paper* 565, and Young (2000) *Quart. J. Econ.* 115, 4 on China.

  2. A form of copyright; European legislation has provided legal protection to designated regional foods. See Parrot et al. (2002) *Eur. Urb. & Reg. Studs* 9, 3, 241.

protest, geographies of A protest registers an objection, either expressed individually or organized collectively. Geographies of protest are concerned with: mapping the distribution of protests; the detailed disposition of protestors and authorities; and the protest camps and occupations themselves. These days, internet-based protests are adding to and extending more traditional forms of objection.

p

proximity Nearness in space, time, or relationship. Oerlemans et al. (2001) *Tijdschrift* 92, 1 explain the **proximity effect** as benefiting from localized ties. Abramovsky and Simpson (2011) *J. Econ. Geog* 11, 6, 949 find evidence of co-location (within 10 km) of R&D facilities in pharmaceuticals with high research-rated chemistry departments, consistent with geographically localized knowledge spillovers and the importance of accessing academic knowledge for pharmaceutical firms.

proxy (proxies) Datum/data on areas that do not permit direct measurement; for example, in studies of past climates where tree-rings, borehole temperatures, and pollen abundance are used. Different proxies preserve the climate information in different ways and are sensitive to climate variables at different time scales.

psychic income The enjoyment, which cannot be measured in financial terms, that people derive from a location or particular place. See Reardon and McCorkle (2002) *Int. J. Retail & Distribution Manage.* 30, 4 on psychic income and shopping. Garnham (1996) *Festival Manage. & Event Tourism* 4 argues that major events improve self-image and raise morale, but Olds (1998) *Festival Manage. & Event Tourism* 11 suggests that the psychic income of a hosting community is not always positive.

psychogeography The study of specific effects of the geographical environment on the emotions and behaviour of individuals, consciously organized or not; the landscape of atmospheres, histories, actions, and characters that charge environments. See I. Sinclair (2003).

public administration, geographies of The study of the roles of space and place in contemporary public administration; the implications for public policy-making of reconfiguring spaces and places. Historically, the organization of government activity has been based on the location together of actor (public administrator), act (administration), and geographical setting (administrative unit).

Globalization brings about new spatial forms of organization and processes in a 'virtual world', and these new spatial forms of organization and processes represent new 'public spaces'; see Toonen (1998) *Public Admin.* 76, 2.

public geography Geography for, by, or with the non-academic public. This may include making geographical research more accessible, making geographical knowledge together with members of the public, and perhaps giving greater recognition to 'lay' geographical knowledge in its own right. But Fuller and Askins (2011) *PHG* 33, 4, 654 explain that definitions of this term can be more complicated than the one here.

public goods Goods freely available, either naturally, like air, or from the state, like education in most developed countries. Krause and Krause (2011) *Urb. Studs* 49, 11, 2399 argue that climate protection is a global public good. See also R. Baldwin (2005).

publics Small groups of people who follow one or more particular issue very closely. Different kinds of publics occupy different kinds of spaces: 'for example, a public space that would foster a liberal-economistic version of publicity might be one that facilitates state functioning and makes it easier for the state to safeguard and regulate individual rights and to promote economic development and growth' (Staeheli and Mitchell (2007) *PHG* 31, 6). 'Publics are deeply embedded in social and machinic complexes involving the mobilities of people, objects, and information' (Sheller (2004) *Env. & Plan. D* 22, 1). For Fuller and Askins (2007) *Antipode* 39, 4 publics, while multiple and in flux, are also created.

public sector That part of a national economy owned and controlled by the state, including nationalized industries, national and local government services, and public corporations. Mrinska (2007) *Northern Economic Agenda*, paper 2, concludes that public intervention is necessary to revive the economy of economically depressed regions, resulting in increased employment and spending, which should stimulate growth in the private sector.

public space The space that the public values: 'space to which it attributes symbolic significance and asserts claims. Citizens create meaningful public space by expressing their attitudes, asserting their claims, and using it for their own purposes. See Qian and Lu (2019) *TIBG* 44, 4 on square dancing in urban China. These purposes include getting drunk and urinating/vomiting; see Jayne, Holloway, and Valentine (2006) *PHG* 30, 4, 451. At a more elevated level, see Springer (2011) *Antipode* 43, 2, 525, and Ramadan (2013) *Eur. Urb. & Reg. Studs* 20: 145 on Tahrir Square, Cairo.

Public spaces, including parks, squares, and markets, are 'co-produced', and the key principles for their development include: leaving room for self-organization, diversifying activities to encourage diverse people to participate, and maintaining access and availability. Strategies intended to 'design out crime', such as cutting down bushes, installing vandal-proof street furniture, and closing public toilets, affect the attractiveness and damage the usefulness of public spaces. Veronis

(2006) *Env. & Plan. A* 38, 10 finds that the use of public space is central to the cultural politics of migrants in Toronto; Ortiz et al. (2004) *GeoJournal* 61, 3, report on women's use of public space in Barcelona; and Mackintosh (2007) *J. Hist. Sociol.* 20, 1–2 writes on the engagingly titled bourgeois geography of bicycling.

pull factor A positive factor exerted by a locality that people move to. These have included: new land grants for farmers (the Great Plains), assisted passages and other government inducements (for Australia, see P. Burroughs and A. J. Stockwell 1998), freedom of speech or religion (see N. Philbrick 2006), and material inducements (Hong Kong; see Lin (2002) *Asia Pac. Viewpt* 43, 1).

pumice A very light, fine-grained, and cellular rock produced when the froth on the surface of lava solidifies. See Calcaterra et al. (2007) *Geomorph.* 87, 3. A **pumice raft** comprises pumice fragments floating on the sea; one was spotted in 2006 (*Geol. News*, Nov. 2006).

pumped storage scheme Electricity cannot be stored, so when demand is low, at night, some can be used to pump water from a lower to a higher reservoir. At peak demand, the water is allowed to fall back to the lower level, passing through turbines which turn generators. See Bruch (2007) *UNEP Dams and Development*, 92–3 on the Palmiet Pumped Storage station, South Africa.

push factor In *migration, any adverse factor which causes out-migration. Examples include: famine, changes in land tenure (the Highland Clearances, 1790–1850, A. MacKenzie 1999), political persecution (Tamil separatists, Sri Lanka, Stokke (2000) *Growth & Change* 31, 29), and mechanization which made agricultural workers redundant and which made factory products cheaper than those of cottage industry (see Pisani and Yaskowitz 2002, *Soc. Sci. Quarterly* 83, 2 on rural depopulation in Portugal). Relatively few migrations are spurred by push factors alone. See Barnett and Adger (2007) *J. Polit. Geog.* 15, 4 on climate-change induced migration.

push moraine *See* MORAINE.

puy The French term for a volcanic neck, revealed by differential erosion. The type location is the Puy de Dôme. The Hohentweil, Hegau, Germany, is a further example.

pyramidal peak Synonym for horn or aiguille.

pyramid of numbers (ecological pyramid) A diagram of a *food chain which shows each *trophic level as a horizontal bar, drawn in proportion to the *biomass. There is a large fall from *producers and primary *consumers; thereafter the decreases are smaller. The animals on the higher levels are generally larger and rarer than animals lower down the pyramid.

pyroclast A fragment of solidified lava, ejected during explosive volcanic eruptions. Classification of pyroclasts is by size. Fragments less than 4 mm across are ash; compacted ash is *tuff*. Material between 4 and 32 mm is *lapili* and fragments larger than 32 mm are *blocks*. Collectively, these fragments are *tephra. Pyroclasts formed from lava produce volcanic bombs and volcanic breccia. **Pyroclastic flows** are also known as *nuées ardentes. They result from the bursting of gas bubbles within the magma, which fragments the lava. Eventually, a dense cloud of fragments is thrown out to form a mixture of hot gases, volcanic fragments, crystals, ash, pumice, and shards of glass. See Adas et al. (2006) *GSA Bull.* 118, 5.

qualitative Concerned with meaning, rather than with measurement. The emphasis is on subjective understanding, communication, and empathy, rather than on prediction and control, and it is a tenet that there is no separate, unique, 'real' world. All qualitative researchers are positioned subjects, and as such, the rigour of any researcher's work depends on the suitability of the methodology, the use of multiple methods, the inclusion of verbatim quotations, and on its credibility and transferability. Bradshaw (2001) *Area* 33, 2 outlines 'some choices which might help address possible difficulties in qualitatively researching the powerful'. For methods in qualitative research, see Davies and Dwyer (2007) *PHG* 31, 2; 2008, *PHG* 32, 3; and for a checklist for evaluating qualitative research, see Baxter and Eyles (1997) *TIBG* 22, 4. See Suchan and Brewer (2000) *Prof. Geogr.* 52, 1 on qualitative methods in cartography.

quality of life The well-being of individuals and societies; either perceived, or as identified by an indicator, such as *life expectancy at birth. Go to the UN Development Index.

quantification The numerical measurement and analysis of processes and features.

quantitative revolution In geography, the intellectual movement beginning in the 1950s that explicitly introduced to the discipline scientific forms of theorizing and techniques of empirical verification, transforming geography into an analysis-oriented scientific discipline. The quantitative revolution is generally considered to have emerged from a general dissatisfaction with regional geographic study, and a consequent shift in focus towards more systematic and specialized approaches. The success of this new geography may have been that it was able to collapse an initially disparate set of concerns into a single analytical framework (the regression or multiple regression equation). Quantification began, flourished, and was, in turn, criticized: 'much that characterised the quantitative revolution was about theorising and employing statistical procedures in the office to generalise about change everywhere' (Bedford (2005) *Asia Pac. Viewpt* 46, 2). Quantification was attacked for being unrealistic and bloodless, turning humans into automata, for being too deterministic, and for ignoring the importance of subjective experience. Was the quantitative revolution a conspiracy? No, says Johnston (2007) *TIBG* 32, 3.

quarrying The removal of already-loosened bed material and/or material resulting from bed failure. Since the tensile strength of ice is low, *glacial quarrying is not possible unless the rock is shattered.

There is now general agreement that quarrying is favoured beneath thin, fast-flowing ice. Drake and Shreve (1973)

Procs Royal Soc. London, Ser. A 332, propose a heat pump effect: water that is melted in high-pressure areas flows away and does not refreeze at the immediately adjacent low-pressure area, and this leaves cold patches that advect downglacier. Fluctuations in basal water pressure also play an important role in the formation of glacially quarried landforms; see Glasser and Bennet (2004) *PPG* 28, 1. Landforms of glacial quarrying include roches moutonnées, rock basins, and zones of *areal scouring.

Quaternary (Pleistogene) The most recent period of geological time, covering the last 2 million years, and including the *Pleistocene and the *Holocene.

quaternary industry (quaternary activity, quaternary sector) Also known as the knowledge economy, this is economic activity concerned with information: its acquisition, manipulation, and transmission. Into this category fall law, finance, education, research, and the media.

queer geographies A major concern of queer geographies has been the critical role of place and space in the production of sexual identities, practices, communities, subjectivities, and embodiments, and the scholarship of queer space includes its location, nature, and definition, its history, memories, events, and subcultures, and the relationship between space and social justice. This is examined by looking at spaces such as closets, communities, and cruising grounds; 'finding or creating spaces in which to know ourselves and become known to others' (Sember (2003) *Space and Culture* 6, 3). Through repeated acts of 'coming out', gay men and lesbians map where and how to be 'out' and when and where not to be. These environments are largely concealed. Correspondingly, there is a focus on the centrality of sexualities in constituting social space. Although gay men have often produced clearly visible territorial enclaves in inner-city areas, lesbian spaces have been comparatively invisible. Knopp (2007) *Prof. Geogr.* 59, 1 focuses on the potential of feminist-inspired and allied queer geographies to rethink a variety of ontologies. Hubbard (2000) *PHG* 24, 2 writes that 'everyday space' is experienced as aggressively heterosexual by lesbians and gay males; that heterosexuality has served to create and justify other forms of oppression and confinement in Western cities.

A. Tucker (2009), on queer geography in the Republic of South Africa, is recommended.

quickclay Clay whose structure collapses completely on remoulding, and whose shear strength is thereby reduced to almost zero (Eilertsen et al. (2008) *Geomorph.* 93, 3–44).

quickflow That part of a storm rainfall that moves quickly to a stream channel. While quickflow runs largely through cracks, *base flow flows slowly through a porous medium—or it may be, more simply, that quickflow is just faster than slowflow during basin-scale response. See Schwartz (2007) *J. Am. Water Resources Ass.* 43, 6.

quota A limit on the import or export of a particular product imposed by a government. Import quotas may be imposed as protectionism; export quotas may be imposed in countries that depend on the export of a particular raw material, as a means of stabilizing prices. Quotas are usually controlled by the issue of licences. *transnationals can relocate production from one country whose annual quota is exhausted to another whose quota has not been filled or that is not bound by quotas, in the process playing off different producers and governments against each other.

race A classification system used to categorize humans into large and distinct populations. Race is not to be confused with *ethnicity; the answer to the question 'does race exist?' is, biologically no, but socially yes, for race is a social construct which, on occasion, can be a matter of life and death. Winlow in R. Kitchin and N. Thrift (2009) is excellent on this. **Geographies of racism** range from examining geographies of segregation and racism, to exploring the cultural politics and social practice of racism, to everyday geographies of identity and experience. See C. Dwyer and C. Bressey, eds (2008).

radial drainage *See* DRAINAGE PATTERNS.

radiation Energy travelling in the form of electromagnetic waves: X-rays, ultraviolet, visible, infra-red, microwaves, or radio waves.

radiation fog *See* FOG.

radiative forcing The increase in the trapping of outgoing *terrestrial radiation by greenhouse gases. **Radiative cloud forcing** is the change of the radiative energy flux caused by the presence of liquid water and ice in the atmosphere.

radical geography An analysis of the processes by which inequalities—in race, class, gender, or age—are produced and maintained. Radical geographers argue that studying the visible geography of spatial relationships

is not enough; the power relationships, specifically the political and economic structures, are fundamental to understanding them. For example, any study of industrial location would be worthless, unless the operations of *transnational corporations, and tariff and trade agreements, are taken into account. Radical geography includes attempts to change the situation on the ground for everyday people in everyday places. See D. Fuller and R. Kitchin (2004) for an exploration of the role of the radical academic in society. **Radical cartography** is the practice of mapmaking to subvert conventional notions and thus actively promote social change. See A. Bhagat and L. Mogel (2008), who proclaim that all maps have an inherent politics that often lies hidden beneath an 'objective' surface.

radioglaciology The use of ice-penetrating radar to investigate the thickness, roughness, motion, and debris of a body of ice, and to measure and detect crevasses and sub-ice lakes. See Plewes and Hubbard (2001) *PPG* 25, 2 for a detailed account of the investigative processes.

radiometer A passive remote sensor, sensitive to *terrestrial radiation of one or more wavelengths of the visible and infra-red. Advanced Very High Resolution Radiometer currently flies aboard many US and European satellites, and its data may be analysed using the AVHRR Hydrological Analysis system, together with ILWIS.

SEE WEB LINKS
- AVHRR Hydrological Analysis system.

radiosonde A free-flying balloon carrying meteorological instruments. The balloon climbs to a height of 20–30 km above mean sea level, sending information from these sensors to ground stations, whereupon it bursts, returning the equipment to the ground.

rain A form of *precipitation consisting of water droplets ranging from 1 to 5 mm in diameter. The type of rain produced reflects the circumstances in which it formed. A mass of warm air rising at a warm front will develop layered clouds and produce steady rain. Air forced to rise quickly at cold fronts will bring heavier rain. These are both examples of frontal rain. *Convection rain occurs when warm, unstable air rises rapidly. Air forced to rise over mountains may form *orographic (relief) rain. *See also* BERGERON–FINDEISEN THEORY; COALESCENCE THEORY.

raindrop erosion The dislodging of soil particles by large drops of rain. See Zhao et al. (2012) *Trans. Chinese Soc. Agric. Engineering* 28, 19.

rainfall intensity The rate at which rain falls, usually measured in millimetres per hour. Intense rainfall is associated with **convectional rain**, notably in thunderstorms and tropical regions, where intensity may be over 100 mm per hour. (British rainfall intensity is normally of the order of 2 mm per hour.) The intensity of rainfall is normally inversely proportional to its duration.

rainfall run-off The overland and downslope flow of rainwater into channelled flow when the rock or soil is saturated. In **rainfall run-off**

modelling, hydrological models are categorized as *lumped*—which treat the catchment area as a single unit—or *distributed*—which represent the catchment as a system of interrelated subsystems, both horizontally and vertically. For **deterministic rainfall models**, see Ramos et al. (1995) *Water Resources Res.* 31, 6. For **fuzzy conception rainfall models**, see Ozelkan and Duckstein (2001) *J. Hydrology* 253, 1. For a very clear guide through a **parametric rainfall-run-off** simulation model see Dawdy, Lichty, and Bergmann (1972) *US Geol. Surv., Prof. Paper* 506-B. For the use of *artificial neural networks to relate meteorological variables to run-off, see Antar et al. (2006) *Hydrol. Procs* 20, 5. *See* RUN-ON.

rainfall threshold The amount of precipitation needed to trigger slope failures. See Gabet et al. (2004) *Geomorph.* 63, 131–43 on rainfall thresholds for landsliding in Nepal.

rain forest An area of luxuriant forest, found where rainfall exceeds 1000 mm yr^{-1}. Although rain forest does develop in temperate latitudes, most of the world's rain forest is tropical. Hill and Hill (2001) *PPG* 25, 3, 326 explain why tropical rain forests are so species-rich, and Puyravaud et al. (2003) *J. Biogeog.* 30, 7 applaud the use of vegetation thickets as restoration tools in rain forests.

rain shadow An area of relatively low rainfall to the lee of uplands. The incoming air, forced to rise over the high land, causes precipitation on the windward side. The descending air is subject to *adiabatic warming, which increases its capacity to hold much of the remaining water vapour, further reducing rain on the lee side. See Ralph et al. (2002) *Conf. Mount. Meteo.* on the impact of a prominent rain shadow in California.

rainsplash The impact of raindrops on the soil may break down soil *peds, loosen soil particles, and cause *turbulence in the *sheet wash of water flowing downslope. 'Rainsplash . . . is not a single process, but a combination of several discrete but interacting soil particle detaching and [soil particle] transport mechanisms' (Terry (1998) *Austr. J. Soil Res.* 36, 3). As overland flow becomes deeper, the effect of raindrop impact on the soil decays significantly (Mosley (1973) *Zeitschrift* 18). Furbish et al. (2007) *U. Vanderbilt* argue that soil grain transport by rainsplash is largely an *advection–dispersion process.

raised beach A former *beach, recognizable by beach deposits and marine shells, often accompanied by rock platforms and dead *cliffs, which now stands above sea level some metres inland. At sites of *isostatic rebound, several raised beaches may be seen at different levels; the raised beaches of Scotland range from 6 to 14 m OD. Raised beaches result from tectonic uplift; see van Vliet-Lanoe et al. (2000) *J. Geodynam.* 29, 1. Gallagher and Thorpe (1997) *Irish Geog.* 30, 2, 68 explain how to date a raised beach.

ramparts As *ice wedges grow, they push up the soil at the edges into ramparts; see M. Shahgedanova (2002). **Protalus ramparts** are coarse, angular, non-stratified thin ridges of rock fragments, about 1 m high, lying along the base of a mountain face. They are associated with persistent snow banks. Shakesby et al. (1999) *Geografiska A* 81, 1 show that they can be formed by snow-push.

range of a good or service The maximum distance/travel time an individual will travel to obtain a given good or service; rare speciality shops have a larger range than shops selling *convenience goods.

ranker An *intrazonal soil, not yet fully developed. This soil is shallow, with a fibrous O *horizon, and an A horizon directly on top of loose, non-calcareous rock.

ranking of towns Various methods have been used to devise a *hierarchy of towns, including comparing the number and type of functions, using multivariate analysis, measuring the interaction between a settlement and its field, ranking settlements by the size of their *spheres of influence; and using graph theory. Recent rankings have been based on quality of life or on urban competitiveness. The 2007 China Urban Competitiveness Study (Hong Kong) found, unsurprisingly, that Hong Kong was at the top of the Chinese hierarchy.

rank-size rule This 'rule' predicts that, if the settlements in a country are ranked by population size, the population of a settlement ranked n will be $1/n$th of the size of the largest settlement. When settlement size is plotted against rank, on normal graph paper, a concave curve results; plotted on logarithmic scales, a straight line emerges—this is the *rank-size pattern*.

Many developing countries show *primacy*: a sharp fall between the largest, *primate city and the other cities. The *binary pattern*, found mostly in federal countries such as Australia, shows a concave curve. The *stepped order* pattern shows a number of settlements at each level, with each place resembling others in size and function. Fonseca (1989, Institute of Mathematical Geography) explains why the rank-size rule 'works'.

Raoult's law This states that the presence of a solute will lower *saturation vapour pressure. *See* CONDENSATION NUCLEI.

rapids Areas of greatly disturbed water across a river, rapids have a continuous

r

and relatively gentle slope, rather than a sudden vertical drop.

raster data These data represent the landscape as a rectangular matrix of square cells, such as the pixels that make up a digital photograph. *See* VECTOR.

rational choice theory A theory arguing that individual self-interest is the fundamental human motive, and that individual actors pursue their goals efficiently (see Clarke in M. Tonry and N. Morris, eds 1983).

rationality The process of using reason or logic to solve a problem. Behavioural economists claim that humans as a species are rational, but without much proof that the maximization of *utility is the fundamental basis of human behaviour. In the UK, for example, strong assumptions are being made about the economic rationality and decision-making competence of individuals in areas of government policy such as health, pensions, and education (Strauss (2008) *J. Econ. Geog.* 8, 2).

Ravenstein's 'laws' of migration E. Ravenstein's laws of migration, 1889 (1995, *J. Royal Stat. Soc.* 48) introduced the notion that people move in order to better themselves economically. In this view, migration is considered as the individual's response to regional differentials in economic development. Ravenstein's fifth 'law' of migration suggests that women are more mobile than men, at least across short distances—a contention supported by Faggian et al. (2007) *J. Reg. Sci.* 47, 3.

reach (**river reach**) A short length of river channel. Unstable river reaches that display rapid planform change are described as active reaches; but there is a notable lack of consistency in its definition.

reaction time The time between any kind of change and the response it elicits in a system. In geomorphology it is the interval between a disturbance or change in controls and observable morphological change. In hydrology, the bigger a catchment, the longer its reaction time. Harrison (2003) *Am. Geophys. Union* C22A-01 extends this term to glaciers; but there the reaction time may depend on the glacier/climate history in a non-transparent way.

realism At its most simple, realism describes the belief that what is 'real' is what is immediately present to and open to the notice and understanding of the basic human senses, especially to sight and touch. However, one characteristic realist idea is that there is a difference between appearance and reality, and that it is the task of inquiry to find access beyond what *appears* to be the case to what is *really* the case. In considering realism in geography, Sack (1982) *TIBG* 7, 4, 504 argues for the ideas of natural necessity, and the existence in nature of underlying structures, potentials, or tendencies: 'the features that are essentially realist point to deeper structures in nature than are seen on the sensory surface'. Try Sayer in S. Aitken and G. Valentine (2006) p. 154, who is refreshingly honest.

recession The decline in river flow after a storm event has passed. On a *hydrograph, the **recession limb** records the fall in *discharge after the river has reached peak flow due to a storm event. The slope of the recession limb reflects the amount of water stored in the basin and the way it is held in the *catchment area. See Stravs et al. (2008) *Geophys. Res. Abstr.* 10, EGU2008-A-02978 on recession limbs and flow forecasts.

recessional moraine *See* MORAINE.

reclamation The process of creating usable land from waste, flooded, or derelict land, or a re-evaluation of some feature. For example, Wilson (2011) *AAAG* 101 presents a highly political analysis of the 'reclamation' of the American West and the location of Japanese 'concentration camps' during the Second World War.

recoding Placing an item of data into a different classification class. This is often done when two different data sets have different category boundaries.

recovery rate The time taken for a *diffusion wave, as in the diffusion of a concept, to re-form, having been blocked by a diffusion barrier. The recovery rate of a wave front is directly related to both the type and size of the barrier it encounters. See Cliff and Hagget (2006) *J. Geogr. Syst.* 8, 3.

recreation Activities which are undertaken as leisure. C. M. Hall and S. J. Page (1999) is well reviewed. *See also* *tourism. See* LEISURE, GEOGRAPHY OF.
   Recreation carrying capacity is the amount of recreation which a site can take without any deterioration of its qualities. See, for example, W. Theobold (2005).

rectangular drainage *See* DRAINAGE PATTERNS.

recurrence interval The length of time between events of a given magnitude. In hydrology, the return period has an inverse relationship with the probability that the event will be exceeded in any one year. See Keylock (2005) *Adv. Water Resources* 28, 8 on flood recurrence intervals. See also Srikanthan and McMahon (1985) *Hydrol. Scis* 30, 2 on calculating the recurrence interval of drought through the stochastic analysis of rainfall and streamflow data.

recycling The reuse of *renewable resources in an effort to maximize their value, reduce waste, and reduce environmental disturbance. A *cycle economy*, based on the economical and responsible treatment of limited resources, is a key concept of today's Japanese environmental policy (Kenichi (2002) *Mining & Materials Processing Inst. Japan* 118, 9).

REDD+ (removals of greenhouse gas emission by forests) In developing countries, a programme for lowering emissions by stopping deforestation and forest degradation with an international payments scheme, the slowing of habitat fragmentation, and the conservation of forest biodiversity and soil. See Lu et al. (2012) *Chin. Geogr. Sci.* 22, 4, 390.

redevelopment The demolition of old buildings and the creation of new buildings on the same site. Redevelopment can solve existing problems of congestion and poor design but, for residential areas in particular, it is seen to be wasteful of resources, destroying communities. Fang and Zhang (2003) *Asia Pac. Viewpt* 44, 2 detect the hidden agenda of local elites during urban redevelopment.

redistricting The process of re-establishing electoral district boundaries for electing members to a legislative body. Redistricting is conducted periodically for legislative bodies in countries that elect candidates from discrete geographic districts; for example, Canada, the UK, and the USA. See Leib, in B. Warf (2007).

redlining Limiting or charging more for things such as banking, insurance, health care, or even supermarkets, to people who live in what are seen as high-risk districts. In the USA, it began in the 1930s, when federal agencies encouraged lenders to rate neighbourhoods for their mortgage risk. Since the 1960s, it has been associated with racial discrimination, and neighbourhood decline. Historical evidence indicates that across Canada the first areas to be redlined were the less-desirable suburbs. The suburban origin of redlining in Canada perpetuated social class diversity in Canadian suburbs; see Harris (2003) *Canad. Geogr./Géogr. canad.* 47, 3. Eisenhauer (2001) *GeoJournal* 53, 2 describes the location of out-of-town superstores as 'supermarket redlining'.

red rain The washing out of fine dust particles over mid-latitudes. For Europe, the major sources of dust are the Sahara and the fringes of the Sahel. This dust is picked up by air streams which rise into the upper *troposphere, and carried north and west until it is washed out by precipitation. McCafferty (2008) *Int. J. Astrobiol.* 7 examines historical and mythical accounts of red rain.

reduction The loss of oxygen from a compound. For example, the sesquioxide ferric oxide can be reduced to the monoxide ferrous oxide by bacteria. *See* GLEY SOILS.

reductionism Explaining any phenomenon in terms of only one causal factor.

reflexivity By a researcher, the habit of reflecting upon her/his own identity or standpoint in relation to her/his scholarship; self-transformation and self-examination in the pursuit of knowledge, and between the researcher and the phenomena studied; 'reflexivity that aims, even if only ideally, at a full understanding of the researcher, the researched, and the research context' (Rose (1997) *PHG* 21, 3, 305).

reforestation Replanting a previously wooded area that has been

felled; see Vallauri et al. (2002) *Restoration Ecol.* 10, 1.

refugee 'A person who, owing to a well-founded fear of persecution for reasons of race, religion, nationality, membership of a particular social group or political opinion, is outside her or his country of nationality and who is unable or unwilling to return' (UN Protocol 1976). Ramadan (2013) *TIBG* 38, 1, 65 has an interesting paper on the refugee camp as a distinctive political space. *See also* ASYLUM MIGRATION.

refugium Refuge areas (pl. *refugia*). Haffer (1969) *Science* 165:131 argues that 'during several dry climatic periods of the Pleistocene and post-Pleistocene, the Amazonian forest was divided into a number of smaller forests which were isolated from each other by tracts of open, nonforest vegetation. The remaining forests served as "refuge areas" for numerous populations of forest animals, which deviated from one another during periods of geographic isolation. The isolated forests were again united during humid climatic periods when the intervening open country became once more forest-covered, permitting the refuge-area populations to extend their ranges (Baker *Amer. Scientist*, nd). This is the **refugia hypothesis** which T. T. Veblen, A. Orme, and K. Young, eds (2007), p. 125, cast doubt on.

reg A North African term for *desert pavement.

regelation Within a moving glacier, the melting of the basal ice at points of increased pressure and refreezing at points of decreased pressure.
Regelation slip occurs when a glacier moves over an obstacle. Pressures build up on the up-glacier side, causing localized pressure melting. The meltwater flows around the obstacle and

then refreezes on the down-glacier side where pressure falls.

regime The seasonal variation in river flow which tends to be repeated each year. A river is said to be **in regime** if the morphological characteristics of a given channel, such as its size, fluctuate around a mean condition over time.

regime theory A set of empirical equations relating channel shape to discharge, sediment load, and bank resistance. The theory proposes that dominant channel characteristics remain stable for a period of years and that any change in the hydrologic or sediment regime leads to a quantifiable channel response (such as erosion or deposition).

   Urban regime theory attempts to locate a phenomenon, such as a city, in its wider spatial, political, and economic context, arguing that growth and development results from groups of actors (people), working cooperatively together. See Mossberg and Stoker (2001) *Urb. Affairs Rev.* 36, 6, 810 on the development of urban regime theory.

region Any tract of the Earth's surface with either natural or man-made characteristics which mark it off as being different from the areas around it. Any given regional map is made up of, and reflects, an unevenly developing, often superimposing mosaic (see Martin and MacLeod (2004) *TIBG* 29, 4).

   Paasi (2001, *Eur. Urb. & Reg. Studs*) describes regions as collective institutional structures created by societies, their social practices and their discourse (verbalizations).

   Many geographers have attempted to distinguish regional boundaries; for example, Slaymaker (2007) *Sing. J. Trop. Geog.* 28, 1 discusses whether or not South-East Asia is a legitimate physical geographical region. However, topography is only one component of geographical identities. A place might be said to realize or to embody a regional character, but usually only in comparison with other regions. Jones (2006) *AAAG* 96, 2 finds international region building 'a messy, problematic, and highly contested activity for parcelling, regulating, and representing geopolitical space'.

regional climate model Any software system designed to model climates between global and local level. Relationships are derived from large-scale variables (predictors) and local/regional variables (predictands). For example, RegCM3 is a three-dimensional, hydrostatic, compressible, primitive equation, σ-coordinate regional climate model which is maintained at MIT.

regional development 1. The provision of aid and other assistance to regions which are less economically developed.
   2. The differential in economic outcomes. Ezcurra and Rapún (2006) *Eur. Urb. & Reg. Studs* 13, 4 explore the relationship between regional *inequality and economic development level in fourteen western European countries for the period 1980–2002, finding that, beyond a given level of per capita GDP, regional inequality can be seen to decrease. Liard-Muriente (2007) *Area* 39, 2 highlights the potential opportunities and pitfalls when designing incentive strategies.

regional geography The study in geography of *regions and of their distinctive qualities. A precondition of this study is the recognition of a *region, its naming, and the delimitation of its boundaries. One approach has been to identify 'natural' regions while another was to establish economic regions based on agriculture and/or industry. Often there was an intimation of a link between the two types of region. Once the keystone of geography, the status of regional geography has been in decline. 'However, the great number of feasible ways of dividing space into regions and time into periods opens the door for alternative narratives, including those which challenge conventional Eurocentric interpretations' (Wishart (2004) *PHG* 28, 3).

regional inequality A disparity between the standards of living applying within a nation. It is difficult to quantify the prosperity or poverty of a region, but there are two basic indicators: unemployment (which has been used in Britain as a symptom since the 1920s), and per capita income, which in England generally falls with movement to the north and west. Some would assert that economic development brings about regional inequality, but Fan and Sun (2008) *Euras. Geog. & Econ.* 49, 1 reveal that both interregional and intraregional inequalities in China have declined since 2004. *See also* UNEVEN DEVELOPMENT.

regional innovation system Very generally, a set of interacting private and public interests, formal institutions, and other organizations that promote the generation, use, and dissemination of knowledge. The effects of these can encourage firms within the region to develop specific forms of capital that is derived from social relations, norms, values and interaction within the community, in order to reinforce regional innovative capability and competitiveness. See Stejskal (2018) in Stejskal, Hajek, and Hudec (eds), on the Czech Republic, and Pino and Ortega (2018) *Cogent Business & Management* 5, 1.

regionalism Regions may be seen as 'sub-national or transnational areas differentiated from their larger host state(s) by regional claims to distinctive historical, cultural, political and/or economic experiences. Regionalism, then, is a set of claims or an identity made on the basis of the region. Giordano (2000) *Pol. Geog.* 19, 445 distinguishes between **institutional regionalism**, which relates more to the processes of 'regionalization' that have taken place within and between European states, and **autonomist regionalism**, which refers to the forms of minority, separatist, and ethnic regionalisms that have gained increasing exposure in recent decades. Also, the two categories are not mutually exclusive. Jones and MacLeod (2004) *TIBG* 29, 4 write on **spaces of regionalism**, which (re)assert 'national and regional claims to citizenship, new forms of political mobilization, and cultural expression, and the formation of new types of territorial government'; see J. Agnew (2002), for example.

regionalization The demarcation of regions. Although regionalization is a slippery concept that varies from writer to writer, regardless of definition, the intention of regionalization programmes is to improve access, quality, cost, and equity, ideally being organized around a well-defined localization, urban concentration, or activity complex, or some combination of these. realized with a regional spatial structure which is more dispersed. Wishart (2004) *PHG* 28, 3 thinks regionalization should be seen as part and parcel of the art of *representation. See Wang et al. (2008) *Nat. Hazards* 44, 169 on the regionalization of urban natural disasters in China.

regional multiplier In the framework of a simple regional multiplier model, the first round of a successful new enterprise creates additional regional income. In a second round, the additional expenditures of the firm in the local economy will create more income. See Keet in J. Adésínà et al. (2006) and A. Hirsch (2005), both on African topics, and Bathelt (2007) *Geog. Compass* 1, 6 on a knowledge-based multiplier model of *clusters.

regional science An interdisciplinary study which concentrates on the integrated analysis of economic and social phenomena in a regional setting. It seeks to understand regional change, to anticipate change, and to plan future regional development, and draws heavily on mathematical models. The relevant journal is *Regional Science and Urban Economics*. See also Plane (2016) *Annals Regional Sci.* 57.

regolith Derived predominantly from rocks, regolith consists of detritus and chemically altered residues of physically and/or chemically altered components of rocks, as well as many new minerals precipitated during the chemical alteration of parent rocks. The depth of the regolith varies with the intensity and duration of the weathering process; within the tropics it may be hundreds of metres deep. Soil is simply regolith with added organic material.

regression analysis A technique for identifying the relationship between a dependent variable y, and one or more independent variables x. While correlation 'makes no assumption as to which is the dependent and which the independent variable, it simply assesses the degree of association between the variables. Regression attempts to describe the dependence of a variable on one or more explanatory variables. It assumes that there is a one-way causal effect from the explanatory variable(s) to the response variable. *Principal components analysis is a regression technique.

regressive In geomorphology, describing a fall in sea level relative to the land.

regulation theory Capitalism is always prone to periods of crisis, but occasionally displays periods of long-run stability known as regimes of accumulation. These are formed by the coupling of accumulation (production) and some form of social regulation, such as institutions, habits, customs, and norms. This coupling occurs by chance, through class struggle. Two examples of regulation theory are the *Fordist post-war period of capital expansion (which broke down in the early 1970s), and the post-Fordist (or after-Fordist) capitalist period based on flexible specialization, flexible accumulation, and reflexive accumulation. For regulation theory 'at work', try Touzard and Labarthe (2016) on regulation theory and the transformation of agriculture. Not, however, an easy read.

Reilly's law The principle that the flow of trade to one of two neighbouring cities is in direct proportion to their populations and in inverse proportion to the square of the distances to those cities.

rejuvenation The renewed vigour of a once active process. The term is generally applied to streams and rivers which regain energy due to the uplift of land through *isostasy or by a fall in the base level. Rejuvenation may also apply to tectonic movements (Valdiya (1993) *Current Sci.* 64).

relational geography A geography which sees things—places, people, flows, events—as part of a web, or webs of links, rather than those things in their physical location. It is very difficult indeed to find any academic articles on this topic which actually mention any place or places.

r

relative humidity (U) The ratio of the actual vapour density (which indicates the amount of water vapour present in the air) to the theoretical maximum (saturation) vapour density at the same temperature, expressed as a percentage. This may be expressed as:

$$U = 100 \, e / e'_w$$

where e is actual vapour pressure and e'_w saturation vapour pressure with respect to water at the same temperature.

Saturated air has a relative humidity of 100%. Air with a relative humidity in excess of 100% is said to be **supersaturated**. Relative humidity varies both diurnally, with a dawn maximum and an afternoon minimum (Priante-Filho et al. (2004) *Glob. Change Biol.* 10, 5), and, less conspicuously, annually, both variations being in opposition to the pattern of temperatures.

relaxation time The time taken by a geomorphological system to adjust to a sustained change in the nature/intensity of an external process. For the Yellow River, Wu et al. (2008) *Geomorph.* 100, 3–4 find a mean relaxation time of about five to six years for channel adjustment.

relict landform A geomorphological feature which was formed under past processes and climatic regimes but still exists as an anomaly in the changed, present-day conditions. Only a few very young landforms are the result of currently operating geomorphic processes. See Stroeven et al. (2002) *Geomorph.* 44, 1–2 on tors, boulder fields, and weathering mantles in north-eastern Sweden. Relict surfaces contain information on past surface processes and long-term landscape evolution (Goodfellow et al. (2008) *Geomorph.* 93, 3–4).

relief rain *See* OROGRAPHIC PRECIPITATION.

religion, geographies of The study of the impact of place and space on religious belief and vice-versa. This includes: the mapping, distribution, and diffusion of religions; the study and analyses of the different sites of religious practice, and the ways in which religion shapes human response to them; and the significance of religions for different populations at different spatial scales. Recent geographies of religion examine the emergence of more variegated and complex religious landscapes in many countries as a result of migration; Kong (2010) *Prog. Hum. Geog.* 34, 6 is well worth reading.

relocalization Not a removal, but a strengthening preservation of local values and services; an alternative to *globalization. Resistance to the agro-food distanciation (e.g. reducing air miles) is at the core of relocalization within local food systems. See the special issue on food relocalization, *Cambr. J. Regions, Econ. & Soc.* 3, 2.

rematerialization The grounding of geographical analysis in the concrete world of actual physical matter. Very simply, rematerialization in academic geography means work on real places and real people.

re-memory *See* MEMORY.

rendzina An *azonal soil rich in humus and calcium carbonate, developed on limestone. The A *horizon is dark and calcareous, but usually thin; the B horizon absent, and the C horizon is chalk or limestone.

renewable resource A recurrent *resource which is not diminished when used but which will be restored, such as *wind energy. Renewable resources may

residential mobility

be consumed without endangering future consumption as long as use does not outstrip production of new resources.

rent gap The gap between the actual rent paid for a piece of land and the rent that could be collected if the land had a 'higher' use. The idea is central to, but does not entirely explain, *gentrification; see N. Smith (1996).

rent gradient The decline in rents with distance from the city centre, reflecting the cost of transport from the outlying districts to the centre.

replacement rate The fertility rate required to hold the population constant in the absence of immigration or changes in longevity.

representation The ways in which meanings are formed, conveyed, and shared among members of social groups. These representations can be defined as *culture, and cultural forms, notably language, and to some extent shape the reality they represent. Given that it is a working, creative, language-driven process, representation cannot be neutral or without power relations. Geographers are interested in the relations between representation and space and place, for places and the human landscape are neither blank nor neutral but laden with meaning and symbolism. *Landscapes may be representations, from which people choose, rearrange, and specify their meanings; they 'are responsive to the geographical and historical specificities of the context within which they are created and consumed' (Markwick (2001) *Geog.* 86, 1).

representational *See* NON-REPRESENTATIONAL THEORY.

reptation A method of sand transport whereby grains are set into a low motion

due to the high-velocity impact of a descending saltating grain. In this way, grains are set in motion at wind velocities lower than those required to move them by wind alone.

research and development (R and D) Research on new discoveries for industrial processes. Generally, R and D in *transnational corporations remains home-nation oriented, and in many cases locally embedded, but there are forces eroding this orientation. TNCs may innovate domestically, but the need for maximizing returns on investment means that R+D will spill over internationally. Home base has become insufficient to generate technological advantage.

reserves The proportion of resources, notably mineral resources, which can be extracted using the prevailing technology.

residential differentiation (residential segregation) The evolution of distinct neighbourhoods, recognizable by their characteristic socio-economic and/or ethnic identity. *See also* SEGREGATION; ETHNIC SEGREGATION.

residential mobility Moving house. A disequilibrium model is commonly used as a theoretical framework for understanding residential mobility. In this model, people decide when the current residence becomes suboptimal. As long as it is not suboptimal, people will stay where they are, since moving incurs adjustment costs and other losses. 'Optimality' relates to the housing unit's characteristics, its location, and the neighbourhood surroundings relative to the household's needs and preferences. Of course, all this is subject to cost and income constraints. Housing that may have been optimal can become suboptimal due to changes in household composition or circumstances, housing or neighbourhood quality, household income, or the cost of housing. See Coulton et al. (2012) *Cityscape*, on residential mobility and neighbourhood change.

resilience 1. In *geomorphology, the ability to recover towards a predisturbance state, which is directly related to dynamic stability. Dungeness Foreland—a large sand and gravel barrier located in the eastern English Channel—has, for 5000 years, demonstrated remarkable geomorphological resilience in accommodating changes in relative sealevel, storm magnitude and frequency, variations in sediment supply, as well as significant changes in back-barrier sedimentation. See Long et al. (2006) *Geomorph.* 82, 309.
   2. Of an economy, the ability to withstand an economic shock, such as the closure of a factory, or the loss of a key market, or more global events such as the global financial crisis of 2007. Resilient economies are able to bounce back quickly from the negative effects of the shock. See Martin (2012) *J. Econ. Geog.* 12, 1, 1.

resistance The ability of a system to avoid or minimize responses to externally imposed changes. Resistance has two main contributors: some indication or measure of strength, chemical or mechanical stability, or susceptibility to modification; and absorption. In hydrology, resistance is the friction force counterbalancing the downslope gravitational force. Resistance is a function of the surface area in contact with the flow, and so flow becomes more efficient when the ratio of the flow cross-sectional area A to the wetted perimeter P is high (defining the *hydraulic radius R). See Smith et al. (2007) *PPG* 31, 4 on applying flow resistance equations to overland flows.

resource An available supply of something that is valued because it can be used for a particular purpose, usually to satisfy particular human wants or desires. Giordano (2003) *AAAG* 93, 2 develops a typology classifying common resources into one of three categories—open access, fugitive, and migratory—based on spatial relationships between resources and resource users. *See also* NATURAL RESOURCE.

resource allocation The assessment of the value of a resource or of the effects of exploiting a resource. Spatial decision support systems (SDSS) are designed to make complex resource allocation problems more transparent and to support the design and evaluation of allocation plans. For a decision support algorithm for managing resource allocation problems involving competing interests, see K. Kyem et al. (2001) *Trans. GIS* 5, 2.

resource-frontier region A newly colonized region at the periphery of a country which is brought into production for the first time. Hacquebord and Avango (2009) *Arctic Anthrop.* 46, 1–2 use a *core–periphery model in order to understand the general trends in the history of natural resource exploitation in an Arctic Resource Frontier Region.

resource management The allocation and conservation of *natural resources. The main emphases are on: an understanding of the processes involved in the exploitation of resources; the analysis of the allocation of resources; and the development and evaluation of management strategies in resource allocation. Sustainable development and environmental protection are major goals. *See also* NATURAL RESOURCE MANAGEMENT.

resource nationalism The policy of a nation-state to have political and economic control over its natural resources, in order to increase or maintain national access to, and revenue from, its resources. Resource nationalism may also be described as the complete set of strategies that a host state uses to increase control over natural resource wealth at the expense of foreign participation and investment. See Arbatli (2018) *Energy Research & Soc. Sci.* 40, 101–8 on resource nationalist strategies in the oil industry. The HM Government Horizon Scanning Programme 2014: Resource Nationalism is full of useful information.

response time The speed of a reaction to a change in environment. See Dyke and Peltier (2000) *Geomorph.* 32, 1–4 on response times and sea-level changes. For land ice, the dynamic response time is proportional to the size of the ice mass. J. Bamber and A. Parsons (2004) have a table of estimated response times for various components of the climatic system. *See also* REACTION TIME; RELAXATION TIME.

restructuring A change in the economic make-up of a country. It may involve: reordering production to achieve *economies of scale; a switch of investment from one sector to another (*see* DEINDUSTRIALIZATION); a change in the spatial distribution of industry; or a change in the economic system; see D. Harvey (2005) on the restructuring of capitalism. Restructuring may be imposed by an authority, like the World Bank, to improve the ailing economy of a nation which is in debt to that authority. *See also* STRUCTURAL ADJUSTMENT.

retailing, geography of The study of spatial patterns of retail and consumer behaviour. This includes the analysis of retailing within a city; hierarchies of retail centres based on *central place theory, and the relationship between out of town malls and city shopping centres. Models are used to forecast retailing and consumer decisions, mostly at the intra-urban scale. These methodologies treat space as neutral, or independent.

In contrast, in the '**new economic geographies of retailing**', space and retail activity work together. In this way, retail capital can structure spaces, with shopping streets, markets, and malls, but capital is also influenced by socio-spatial processes. Key themes in economic geographies of retailing include: the reorganization of working corporate structures, retailer–supplier interactions; the social relations of production; the organization and technology of retail distribution; and the workings of retail capital.

In the developed world, shopping is now part of the expression, construction, and contestation of identity; consumption may define who we are. At the same time, much of current thinking is that consumption is intrinsically evil, and morally corrupting.

Perhaps the supreme incarnation of shopping as leisure is the out-of-town shopping centre/shopping mall. Shopping malls offer consumers a safe and climate-controlled alternative to the perceived dangers and unpredictability of city-centre shopping, transforming public spaces into sanitized, privatized, often very similar zones. Mall corridors, consumers and shops could be almost anywhere—Los Angeles, Singapore, Moscow, Rio, or Johannesburg, and M. Sorkin (1992) claims that 'the globalisation of retailing is creating the *ageographical city*, a city without place attached to it'.

re-territorialization The restructuring of a territory/place that has experienced *deterritorialization. See Bunnell et al. (2006) *Int. J. Urb. Reg. Res.*

30, 1 on the complex re-territorializations on the Indonesian island of Bintan.

return flow Water which has seeped through the soil as *interflow but which backs up the hillslope when it has reached a saturated layer. See Guebert and Gardner (2001) *Geomorph.* 39, 3–4.

return period The length of time between events of a given magnitude.

reurbanization The development of new homes, businesses, and community facilities within existing urban areas; a process in which cities stop losing population, stabilize, and regrow. It is, however, not clear if reurbanization represents a long-term process of urban living or a short-term stage. See Kabisch et al. (2019) *Pop. Space & Place* 25, 8.

revanchism Revenge. The urban geographer Neil Smith has argued that while post-1960s urban policies tended towards anti-poverty legislation and affirmative action, subsequent *neoliberal politics were characterized by a discourse of revenge against environmental activists, feminists, gays and lesbians, the working class, and recent immigrants: 'the public enemies of the bourgeois political elite and their supporters'. See Hennigan and Speer (2019) *Urban Studies* 56, 2.

revegetation Re-establishing and developing plant cover, either by artificial or natural means; see for example Owen et al. (2001) *J. Biogeog.* 28, 5, which reports that most revegetation following anthropogenic and/or environmental interference is through vegetative reproduction. One of the key components of any restoration programme is to adequately and efficiently identify potential sites for revegetation. As with any restoration activity, economic cost plays a major role (Lawson et al. (2007) *Geogr. Res.* 42, 4).

Reynolds number (R$_e$) Four factors combine to determine whether the flow of water within a channel is *turbulent or *laminar: the density, velocity, and viscosity of the water, and the hydraulic radius of the channel. Since the density of water is 1, the Reynolds number expresses this combination as:

$$R_e = \frac{VR}{\mu}$$

where V = velocity of the liquid, R = hydraulic radius, and μ = viscosity of the liquid. The Reynolds number is a dimensionless quality; in streams, the maximum number for laminar flow is between 500 and 600, depending on temperature, and at high Reynolds numbers, above 2000 to 25000, flow is turbulent. Smith et al. (2007) *PPG* 31, 4 note that the flow is laminar for Reynolds numbers below 2300 (in circular pipes), and above 4000 in turbulent. 'An ill-defined region between these two limits is known as the transitional zone. The critical Reynolds number transition to turbulent flow depends on the exact flow configuration and should be determined experimentally.'

rheology The study of flowing materials, such as mudslides. See Egar et al. (2018) *Geom.* 315.

rhythmic In *geomorphology, a series of alternating processes, such as *aggradation and degradation.

ria The seaward end of a river valley which has been flooded as a result of a rise in sea level. The name is from the type location in Galicia, Spain.

ribbon development A built-up area along a main road running outwards from the city; a location that combines cheaper land with high accessibility, and the chance of attracting trade from passing traffic. Look for the Council for the Preservation of Rural England (CPRE) on ribbon development.

Richter scale A scale of the *magnitude of earthquakes, ranging from 0 to (in theory) 10. Richter (1935) *Bull. Seism. Soc. Am.* 25 equated the magnitude of an earthquake to the base 10 logarithm of the ground motion in millimetres, measured on a certain type of seismograph, plus a correction factor related to the distance of the earthquake. The distance is calculated from the difference in arrival time for different types of waves that travel at different speeds. So, for a constant distance between an earthquake's hypocentre and the seismograph, the ground motion has to increase by a factor of 10 to cause an increase of 1 on the Richter scale. Professor Richter was trying to characterize the energy of a seismic event, not damage. Newer measurements, referred to as the 'seismic moment' and 'moment magnitude', have been developed to address some of the Richter scale's shortcomings (Johnson (2004) US Dept Energy). *See also* MERCALLI SCALE, for measurements of earthquake *intensity.

ridge and furrow A set of parallel ridges and depressions formed during the period of strip cultivation in the Middle Ages. As the land was ploughed always to the same pattern, the plough threw up earth to make ridges which often survive in the present landscape. Ridge and furrow landscapes are still clearly visible in the landscapes of midland England. See Whittington and Brett (1979) *J. Hist. Geog.* 5, 1.

rift valley A long strip of country let down between normal faults, or between a parallel series of step faults (also called a *graben*). There is a maximum rift valley relief, controlled by the competition between isostasy and lithospheric geometry. The biggest terrestrial rift valley system is the East African system, at 3000 km long. *Plate tectonic theory suggests that rift valleys are the result of large-scale doming above a mantle *plume, followed by fracturing along the crest of the dome as plates diverge. Many rifts, like the Rhine rift valley, have the Y-shaped pattern characteristic of a triple junction, indicating that they arise from plate separation; see Horner-Johnson et al. (2003) *Eos* (*Trans. Am. Geophys. Union*) 84, 46.

right to the city Henri Lefebvre created this concept in 1968, defining this as the rights of all urban dwellers, regardless of citizenship, ethnicity, ability, gender, and so forth, to participate in shaping the city. David Harvey sees the right to the city as a right to change ourselves by creating the city we want. It's a collective, not an individual, right because it requires the inhabitants to have power over the processes of urbanization. The intrinsic worth of cities to their inhabitants is not merely their instrumental or economic value; 'the right to the city is the right to make full use of the city and to live a richly urban life' (Painter (2005) ICRRDS, U. Durham). For an elaborate critique, see CASUM, What Participation? Whose Community? at http://casumm.files.wordpress.com/2008/02/community-particpation-law.pdf.

rills Small *channels, between 5 and 2000 mm in width, and very closely spaced. They develop well in areas with heavy rainfall, especially upon weaker rocks, such as volcanic ash. Sediment yields increase with slope gradients and flow rates (the former having more effect than the latter) and greater flow rates cause more rill erosion and soil loss under the same slope gradient. Rills may widen and deepen to form *gullies. A power relationship exists between width and total flow discharge in rills. See Torri et al. (2006) *Geomorph.* 76, 3–4.

rime *See* FROST.

ring dyke *See* DYKE.

riparian Relating to a river bank. The **riparian zone** is composed of the area of land adjacent to streams and rivers. This zone is often a narrow strip between the channel and hill slope in headwater areas but also occurs as a more extensive floodplain adjacent to larger rivers. A river and its surrounding riparian vegetation are two dynamic systems that interact through several hydrological, geomorphological, and ecological processes; see Perucca et al. (2007) *River Resources Res.* 43, 3 on the role played by vegetation on meandering river morphodynamics. Riparian vegetation dynamics are driven by allogenic (external) hydrogeomorphological factors, with autogenic (self-generating) influences affecting both plant dynamics and the river environment from the earliest stages of plant establishment, and becoming more important as landform stability is achieved; see Francis (2006) *Area* 38, 4.

rip current A strong current moving seawards in the near-shore zone. Rip currents are probably formed by onshore wave mass transport, as water piled up against the beach provides a hydraulic head for an offshore return flow. For a continuity equation to explain velocities in a rip channel, by which the discharge out of the rip channel is equal to incoming wave-induced transport over the transverse bar, see Brander and Short (2001) *J. Coast. Res.* 17, 2. See also MacMahan et al. (2005) *Marine Geol.* 218, 1–4.

rising limb The section of a *hydrograph from the start of increased discharge to the maximum flow.

risk The likelihood of possible outcomes as a result of a particular action or reaction. Risk differs from *uncertainty in that the likely outcomes of risks can be assessed as a series of different odds, while there is no calculation of probabilities in uncertainty. Risk is socially constructed because the interpretation of physical threats is not just a subjective process worked out by individuals but is also strongly affected by the mores, norms, values, and institutions held in common by social groups. What science deems to be an acceptable level of risk may not match the social perception of acceptability. When the differences between social and scientific notions of risk become acute, the public tends to amplify the risk. Watson and Stratford (2008) *Soc. & Cult. Geog.* 9, 4 recognize three socio-spatial orderings of risk (displacement, replacement, and reorientation). **Risk analysis** includes risk assessment, risk characterization, risk communication, risk management, and policy relating to risk, in the context of risks of concern to individuals, to public- and private-sector organizations, and to society at a local, regional, national, or global level. See the Society for Risk Analysis (SRA) site, and go to the Deloitte & Touche site 'Risk Assessment in Practice'; it's clear and comprehensive. *See also* UNCERTAINTY.

river 'Rivers are truly perceived not as "things in space" but as "processes through time"' (Goodwin (1999) *Stream Notes*).

river basin management Conserving, managing, and developing the water, land, and related resources within a river basin, with the aim of maximizing potential economic and social benefits derived from it. Where necessary, this might include the restoration of freshwater ecosystems. Restoration, rehabilitation, river basin management, and their derivatives are practised in at least 21 different countries because of the exploitation and subsequent deterioration of the riverine environment. However, different political decision-makers tend to generate costs, benefits, and risks that are unevenly distributed across social groups. Since the adoption of the EU Water Framework Directive (look it up!), all the All European Union nations use a river basin approach for water management. Also see Hirsch and Wyatt (2004) *Asia Pac. Viewpt* 45, 1 on the Se San River.

river capture *See* CAPTURE.

riverine Relating to, similar to, or formed by, a river. Hydrological connectivity (the exchange of matter, energy, and biota via water) plays a major role in sustaining **riverine landscape diversity**.

river pattern, discriminant A component of river pattern classification based on the *Darcy–Weisbach equation, reasonably clearly explained by Song and Bar (2015) *Catena* 135.

river restoration The recovery of ecological 'soundness' in a corrupted river, or river basin, by replacing lost, damaged, or degraded biological components, and restoring, as far as possible, the pre-existing hydrologic, geomorphic, and ecological processes. Restoration activities include: cutting a new, meandering course, creating stepped banks to provide diverse ecological niches, removing dams to restore fish movement, reconnecting flow by cutting linking channels, setting back or breaching *levees, and dredging.

McDonald et al. (2004) *TIBG* 29, 3 argue that the purpose of restoration has shifted from simple utilitarian needs for flood and erosion/sedimentation control towards the incorporation of ecological concerns.

( ) SEE WEB LINKS
- The UK River Restoration Centre has case studies.
- The European Centre for River Restoration has a useful bibliography.

River Styles framework A method of river maintenance based on four key principles: respect for river diversity; working with river dynamic and change; working with the linkages of biophysical processes; and using geomorphology as an integrative physical template for river management activities. Brierley et al. (2011) *Applied Geog.* 31, 1132 provide a case study.

( ) SEE WEB LINKS
- Website of River Styles® framework.

river terrace A bench-like feature running along a valley side, roughly parallel with the valley walls. Bridgland and Westaway (2007) *Geomorph.* argue that uplift is essential for the formation of river terraces. Bridgland (2002) *Quat. Sci. Revs* 19, 13 suggests a six-stage model:

1. the incision phase in which terrace generation occurs, occurring at the transition to interglacial conditions (discharge is high as a result of melting permafrost);

2. an aggradation phase, again at the glacial–interglacial transition, seen mainly in the lower reaches of valleys;

3. the interglacial phase, in which fine-grained sedimentation (rarely preserved) is predominant;

4. a further phase of incision at the interglacial–glacial transition;

5. the main aggradational phase, at the interglacial–glacial transition, as a result

r

of considerable sediment being liberated by the decline of vegetation;

6. a phase of glacial climate during which there is relatively little activity, much of the potential discharge being locked up in permafrost.

river training The attempt to 'flood-proof' a river/channel. This may include the straightening, narrowing, and shortening of a river course, removing river gravels, installing new flood banks and river-bank walling, and improving floodplain drainage. See Korpak (2007) *Geomorph.* 92, 3–4 on the geomorphological impacts of river training in mountainous channels.

road pricing A strategy to reduce *urban *congestion, first used in Singapore in 1978. An electronic road pricing (ERP) system can electronically monitor, and track, vehicles entering a restricted zone to control the flow of inbound traffic. The system is capable of automatically imposing a demand-sensitive congestion toll on every vehicle without requiring them to slow down or stop, when the congestion level in the restricted zone exceeds a preferred threshold level.

The ERP system was first tried out in Hong Kong from July 1983 to March 1985, but not implemented due to public rejection, arising from concerns over the privacy of movement. In Singapore, on the first day of its implementation (1 April 1998), the usual morning rush hour traffic along one of the heavily congested highways decreased by 17%. Lower-income groups find road pricing less acceptable than high-income groups, not surprisingly. A. Mahendra, ed. (2010) is useful on the impacts of road pricing on business.

rock avalanche The mass movement of a minimum volume of 500000 m³ of rock derived from the collapse of

competent masses of coarse-grained, hard bedrock. For rockslide-rock avalanches in mountain landscapes, and the landforms associated with them, backed up with useful pictures and really clear explanations, see Hewitt (2006) *PPG* 30, 3.

rock cycle The conversion of rocks to other forms; under pressure and heat, *igneous and *sedimentary rocks can become metamorphic rocks. All rock types can erode to form layers of sedimentary rocks and all rocks can be completely melted, cooling and hardening at or near the Earth's surface to form igneous rock.

rock fall The detachment and downward fall of rock from a cliff face to its base. Rock fall may be triggered by permafrost degradation, seismicity, and *glacial debuttressing, but it is its geological structure that makes the rock face more or less prone to rock fall. See Ravanel and Deline (2011) *Holocene* 21, 357.

rock flour Silt- and clay-sized particles of debris formed from grinding due to *abrasion within and at the base of a glacier, and carried by *meltwater streams.

rock glacier A very slowly moving river of angular rock debris, characteristic of high mountain permafrost and made up of debris and ice that creeps on hillslopes at a typical rate of a few decimetres per year. Ikeda and Matsuoika (2006) *Geomorph.* 73, 3–4 distinguish between **boulder rock glaciers**, with an active layer of matrix-free boulders derived from crystalline rocks, and massive limestone; and **pebbly rock glaciers**, made of matrix-supported debris derived from less resistant shales, and platy limestone.

rockwall weathering Frost weathering of resistant rock walls which occurs when seasonal freezing promotes ice segregation in the rock. See Matsuoka (2008) *Geom.* 99, and Domènech (2018) *Geomorph.* 306.

Rossby waves Long ridges and troughs in the westerly movements of the upper air, with a wavelength of around 2000 km, discovered by C. J. Rossby in 1939. Four to six waves girdle the Northern Hemisphere at any one time. Some are a response to relief barriers, like those east of the Rockies, and east of the Himalayas.

Rossby waves are thought to be a reaction to the unequal heating of the Earth's surface, and are intimately connected with the formation of *cyclones, *anticyclones, and *mid-latitude depressions. As air in the middle latitudes of the Northern Hemisphere travels west–east into a trough, it slows down, and piles up, causing *convergence just ahead of the ridge which follows. Convergence in the upper air causes a downflow to the ground, creating high pressure systems at ground level, below and just ahead of troughs in the Rossby waves. As air leaves the trough, and its passage straightens out, air speeds pick up, and the air moves very fast as it swings round the outer arc of the ridge. Air then diverges just ahead of the next trough. *Divergence in the upper air causes low pressure systems at ground level below and just ahead of ridges.

At times, the waves are few, and shallow; a pattern known as a high zonal index. In other cases, the flow becomes markedly *meridional; a pattern known as a low *zonal index. Upon occasion, the waves break down into a series of cells (**Rossby wave-breaking**; *see* BLOCKING). Woollings and Hoskins (2008) *J. Atmos. Scis* 65, 2 suggest that the low-frequency variability of the *North Atlantic Oscillation results from variations in upper-level Rossby wave-breaking events.

rotational slip The semicircular motion of a mass of rock and/or soil as it moves downslope along a concave face. Evidence for rotational slip in *cirques comes from the dirt bands observed in

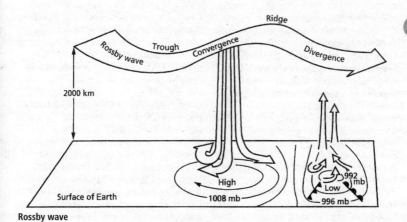

Rossby wave

cirque ice, which become progressively steeper from the back wall, but then flatten towards the cirque mouth.

r-selection, k-selection Two major strategies may be adopted for the survival of a species. r-selected plants (where r stands for maximum increase) respond swiftly to favourable conditions with most of their energies devoted to rapid maturity and reproduction. *See* OPPORTUNIST SPECIES. k-selected plants survive by putting their energies into persistence. *See* EQUILIBRIUM SPECIES.

rubified Reddened. See Hirsch et al. (2019) *Catena* 175 on rubified soils in Germany.

rugosity Roughness, wrinkliness. In ecology, more rugosity will provide more available habitat for colonization, foraging, and shelter. See Donovan (2002) *Geol. Today* 18, 4 on rugosity in karst landscapes, and Knudby et al. (2007) *PPG* 31, 4 on rugosity in coral reefs.

run-on The infiltration of water in those areas where the soil moisture capacity has not reached its maximum.

rural In, of, or suggesting the country; hence *rurality. One 'marker' of the rural is low population density. Others include rural occupations (such as farming), cultivation, nature, and wild freedom. Halfacree in P. Cloke et al., eds (2006) notes *productivist agriculture as the essence of 'modern' rurality. Chigbu (2015) *Develop. in Practice* is wonderfully clear on the nature of **ruralization**, and its possibilities for economic regeneration.

rural depopulation The decrease in population of rural areas, by migration, or through falling birth rates as young people move away. In 2004, Ioffe et al. (*AAAG* 94, 4) reported that, as a result of

farmland abandonment, no fewer than 20 million hectares of arable land were already deserted in European Russia. However, Stockdale (2002) *Int. J. Pop. Geog.* 8 asserts that the historical trend of rural depopulation is being replaced by a repopulation of the countryside.

rural geography The study of the rural landscapes of the developed world. It includes the origin, development, and distribution of rural settlement, *rural depopulation, the causes and consequences of agricultural change, patterns of recreational use of the countryside, tourism, planning, and the growing influence on rural areas of urban dwellers. For a review of the subdiscipline, see Roche (2005) *PHG* 29, 3.

rurality The values and beliefs of 'rural folk' as constructed (by both rural and urban residents) as distinct and different, and defined by particular mixes of economic exploitation, land uses, and densities; Halfacree, in P. Cloke, M. Goodwin, and B. Mooney, eds (2006) suggests a threefold and entwined structure for analysing rurality: rural localities (how places are shaped by spatial practices and processes, often linked to production or consumption); formal representations of rurality (often dominated by hegemonic interests); and everyday lives of the rural (how *agents perform, practise, and experience rurality). Try Woods (2010) *PHG* 34, 6, 835.

rural–urban continuum The belief that between the truly rural and the truly urban are many 'shades of grey'; if we actually look along a scale from the single isolated farm all the way to the *megalopolis, we do not find any clear boundaries between hamlets, villages, towns, and cities. Sheppard and Nagar (2004) *Antipode* 36, 4 state that 'it is important to reconceptualize the urban as rural–urban continuum in attending to processes of domination, co-optation and resistance, especially in Asia and Africa where populations are predominantly rural and historical patterns of colonial and postcolonial capitalist development and labor migration connect the lives in the urban, peri-urban, and rural areas in intricate and inseparable ways'.

rural–urban fringe The transition zone between the city and its suburbs, and the countryside. Certain types of land use are characteristic of this zone: garden centres, country parks, riding stables, golf courses, sewage works, and airports are common, and these are neither truly urban nor truly rural uses. They do, however, give an urban air to the countryside, which can be cited as an argument for further development— since the zone is not really 'countryside', it need not be preserved. Thus, C. Bryant et al. (1982) define the **inner urban fringe** as land transitioning from rural to urban uses—land under construction; land for which planning permission has been granted; land which will, without doubt, become urbanized.

Fragmented patterns of development in rural–urban fringe areas could be due to negative externalities that make some parcels of land unattractive. Lagarias (2007, *Cybergeo, systèmes, modél. géostat.*, 391) presents a fractal analysis of urbanization at the periphery of Thessaloniki, Greece. See Qviström and Salzman (2006) *Landscape Research* 31, 1 on the inner urban fringe in Malmö.

RUSLE Revised Universal Soil Loss Equation. See Bircher et al. (2019) *Geomorph.* 346.

rustbelt *See* SNOWBELT.

r

sackung, pl. *sackungen* An anti-slope (upward-facing) scarp feature brought about by gravitational slope-sagging and lateral spreading. See McCalpin and Corominas (2019) *Geomorph.* 337.

Sahel The **Sahelian zone** borders the southern Sahara. Vegetation—scattered grasses, shrubs, and trees—increases in density towards the southern margin. Annual rainfall is 200–400 mm, but unreliable. Severe droughts bring losses of livestock, crop failures, and famine. See *Glob. Env. Change* (2001) 11, 1, special Sahel issue. The work by D. Hummel, M. Doevenspeck, and S. Samimi (2001) *micle (Migration, Climate & Environment) working paper #1* is a superb resource on climate change, environment, and migration in the Sahel.

salient In geomorphology, a spur projecting from, for example, a beach, or the side of a landform.

saline Salty. A **saline soil** is a non-*sodic soil containing enough salt to adversely affect crop growth. Saline soils get their salts from natural or anthropogenic *salinization. **Salinity** is the degree to which water contains dissolved salts, usually expressed in parts per thousand.

salinization The accumulation of unusually high concentrations of dissolved sulphates and chlorides of sodium and calcium, often as a result of large-scale irrigation in semi-arid areas;

as irrigation water is evaporated or transpired, the salt ions are largely left behind to accumulate in the soil (Saysel and Barlas (2001) *Ecol. Model.* 139, 2–3). Salinization can reduce plant growth rate and yield, and, in severe cases, cause total crop failure. In certain circumstances, excessive salinity in soils cannot be reduced over time by routine irrigation and crop management practices (Qadir et al. (2000) *Land Degrad. & Dev.* 11).

In coastal areas, salinization can be associated with the over-extraction of groundwater, which can lower the water table and lead to the invasion of sea water (Cardonna et al. (2004) *Env. Geol.* 45, 3). Tsunamis can cause severe salinization problems.

(()) SEE WEB LINKS

• European Land Management and Natural Hazards Unit on soil salinization.

saltation The bouncing of material from and along a river bed or a land surface. The impact of a falling sand grain may splash other grains upwards so that a chain of saltating particles may be set up. Saltation upwards is the result of *lift forces (see J. S. Bridge 2003); the downward movement occurs when lift is no longer effective and the particle is subject to drag and gravity. See Stout (2004) *ESPL* 29, 10 on the relationship between saltation activity, and relative wind strength; see also Bauer et al. (2004) *Geomorph.* 59, 1–4 on shear velocity. The **saltation–abrasion model** represents river incision by saltation;

see Sklar and Dietrich (2006) *Geomorph.* 82, 1–2.

salt dome A large mass of evaporate minerals that has pierced, and risen through, denser, overlying *sedimentary rock, forming a dome-shaped arch with salt at its core. See Autin (2002) *Geomorph.* 47, 2 on salt domes in Louisiana.

salt marsh Tidal salt marshes are found along the low-energy coastlines of mid- to high-latitude regions of the major continents (except Antarctica), and lie at the *ecotone between terrestrial, freshwater, and marine systems (Greenberg et al. (2006) *BioSci.* 56, 8). As the marsh develops, *halophytes (such as marsh samphire and sea aster in Britain) pave the way for less hardy species, and the marsh becomes part of the coastland; see van Wijnen et al. (1997) *J. Coastal Cons.* 3, 1. See Santos et al. (2013) *PLoS ONE* 8, 11 e78989. DOI:10.1371/journal. pone.0078989 on salt marsh vegetation stress and recovery after the oil spill in Barataria Bay, Gulf of Mexico.

salt water intrusion The movement of saline water into fresh water aquifers. It happens to a restricted extent in most coastal aquifers, because, due to its mineral content, salt water is denser than fresh water, and can push inland beneath the fresh water. Certain human activities have contributed to salt water intrusion, which has risen in many coastal areas, as a result of pumping ground water from coastal fresh water wells. Asfahani and Abou Zakhem (2012) *Acta Geophysica*, DOI: 10.2478/s11600-012-0071-3 review all aspects of this problem, from surveying to corrective action in the Khanasser Valley, Syria.

salt water weathering (salt weathering) A form of weathering, especially important in hot deserts,

which combines crystal growth with *hydration. See Smith and Warke in D. Thomas et al., eds (1997) on the importance of the nature of salt and moisture delivery to weathering in arid areas, and Buj et al. (2011) *ESPL* 36, 3, 383 on the role of salt in the weathering of limestone. Landforms associated with arid salt weathering include columnar weathering networks (Goudie et al. (2002) *Zeitschrift Geomorph.* 46) and *tafoni. Salt weathering is closely linked to groundwater controls in enclosed arid depressions (Aref et al. (2002) *Geomorph.* 45, 3–4). See Viles and Goudie (2007) *Geomorph.* 85, 1–2 on salt weathering in coastal salt pans. Smith et al. (2005) *Geomorph.* 67, 1–2 evaluate recent simulations of salt weathering.

sand Particles of rock with diameters ranging from 0.06 mm to 2.00 mm. Most sands are formed of the mineral quartz. Sandy soils are loose, non-plastic, and permeable, and have little capacity to hold water.

sand dune A hill or ridge of sand accumulated and sorted by wind action. Sand will settle on a dune because the friction of the sandy surface is enough to slow the wind, which then sheds some of its *load. Dunes formed in the *lee of an obstacle are **topographic dunes**: *lunettes* form in the lee of a *deflation hollow, *nebkhas* in the lee of bushes, and **wind shadow dunes** in the lee of hills and plateaux. **Crescentic dunes**—sand mounds, barchans, barchanoids, and transverse ridges—are restricted to areas with unidirectional high winds and minimal vegetation (Bishop (2001) *ESPL* 26).

    Transverse dunes form when sand supply is abundant, and wind direction constant (Walker and Nickling (2002) *PPG* 26, 1). As the sand supply is reduced, the dunes transform into **parabolic dunes**—hairpin-shaped with

Sand dunes Arrows indicate wind direction

the bend pointing downwind (Mitasova et al. (2005) *Geomorph.* 72, 1–4). Where the direction is very changeable, **star dunes** form. The linear **seif dunes** form when two prevailing winds alternate, either daily or seasonally; the leeward slope is a zone of deposition, as well as a zone of erosion (Tsoar et al. (2004)

Geomorph. 57, 3–4). When sand is limited, **barchans** form with the horns pointing downwind; see Katsuki et al. (2011) *ESPL* 36, 3, 372 on a collisional simulation of barchans. The height of the slipface of a barchan is proportional to the width of the horns (Wang et al. (2007) *Geomorph.* 89, 3–2). Barchans may be changed into dome dunes, and vice versa; completely gradational forms exist between the two (Fryberger et al. (1984) *Sedimentol.* 31). **Barchanoid** dunes are undulating, continuous cross-wind dunes which may grade into long transverse dunes; see R. Cooke et al. (1993). Dune fields have recently come to be recognized as self-organizing systems that can be seen progressing from states of disorganization or randomness to uniformity; see Wilkins and Ford (2007) *Geomorph.* 83, 1–2.

sand dune dynamics The processes of dune movement; see Wiggs (2001) *PPG* 25, 1.

sand dune stabilization Techniques designed to prevent the erosion and deposition of sand include: the establishment of shelterbelts; mass tree revegetation using a synthetic water absorbing polymer; developing live hedges; and mechanical fencing. See Lomba (2008) *J. Coastal Res.* on monitoring sand dune stabilization in Israel, and Shumack et al. (2017) *Geom.* 299.

sand flux The flowing of sand. Dong et al. (2004) *Geomorph.* 57, 1–2 find that sand flux over a sandy surface increases with height in the very near surface layer, but then decays exponentially; see also Wang et al. (2008) *Geomorph.* 96, 1–2. Yang et al. (2011) *J. Arid Land* 3, 3, 199 estimate a sand flux in the Taklimakan Desert, China, to be 258.67×10^{-4} kg/m·s.

sandur A sheet, or gently sloping fan, of *outwash sands and gravels (pl. *sandar*); see Marren (2004) *Sed. Geol.* 164, 3–4,

and Magilligan (2002) *Geomorph.* 44, 1. Sandur fans are commonly found in glacial troughs (Thomas and Chiverrell (2006) *Quat. Sci. Revs* 25, 21–2).

sand-wedge polygons *See* FROST CRACKING.

Santa Ana *See* LOCAL WINDS.

sapping The breaking down and undermining of part of a hillslope, triggering small slips. Sapping is significant in valley formation (Nash (1996) *ESPL* 21, 9). See Twidale (2007) *Revista C. & G.* 21, 1–2 on sapping at a river bluff. For **sapping erosion**, see Fernández et al. (2008) *Geomorph.* 95, 3–4.

SAPRIN The Structural Adjustment Participatory Review International Network is a global network established to strengthen the organized challenge to *structural adjustment programmes.

(SEE WEB LINKS)
• The SAPRIN website.

saprolite Weathered rock *in situ* formed by deep weathering to depths of tens or hundreds of metres (Ollier et al. (2007) *Geomorph.* 87, 3).

sapropel An unconsolidated mud in estuaries, lagoons, lakes, and shallow seas, its formation depending on heavier precipitation (Rohling and Hilger (1991) *Geol. Mijnbouw* 70), and to variations in the Earth's orbit—they are thus useful in reconstructing palaeoclimates (Larrasoaña et al. (2006) *Phys. Earth & Planet. Interiors* 156, 3–4).

saprophyte A largely surface-dwelling, primarily free-living, organism, using enzymes to break down dead organic matter. Most are fungi or bacteria—fungi play vital roles in the nutrient cycles of forest ecosystems, acting as mutualists, parasites, and saprophytes of virtually all plants (Czederpiltz et al. (1999) *McIlvainea* 14).

sastrugi Ridges of ice particles lying across an ice sheet, formed by strong winds, and oriented at right angles to them. The sastrugi alter the surface roughness, and this alters the resistance to wind. Consequently, more drift snow builds up; this self-enhancing process thus forms larger sastrugi (Herzfeld et al. (2003) *Hydrol. Procs* 17, 3).

satellite town Satellite towns are established around big cities to attract population surplus and control population growth. See Ziari (2006) *Cities* 23, 6 on satellite towns in Iran. Try V. Vasudehan (2013) on life in an Indian satellite town.

satisficing The setting of sub-optimal targets to which people aspire. Satisficing has been widely observed among residential developers: 'for developers, the most important rule is to choose locations that meet a minimum profit criterion—a rule that results in satisficing . . . The main consequence of satisficing is that developers prefer projects on greenfield sites that take a shorter time to build and sell' (Mohamed (2006) *J. Planning Ed. & Res.* 26).

saturated adiabatic lapse rate (SALR) *See* ADIABAT.

saturated mixing ratio lines Lines of constant humidity mixing ratio plotted against height and pressure on a *tephigram.

saturated zone overland flow (saturation overland flow) *See* OVERLAND FLOW.

saturation (saturated air) There is a constant exchange of water molecules between liquid water, or ice, and the air, as evaporation and *condensation take place. **Saturation vapour pressure** (*es*) depicts a balance in the air between condensation and evaporation. *es* is greater in water droplets than in sheet water, and lower in impure water than in pure. Where *es* in an air parcel is greater than *e* (the *ambient *vapour pressure) there will be net evaporation; where *e* is greater than *es*, there will be net condensation.

savanna Broad belts of tropical grassland flanking each side of the equatorial forest of Africa and South America. Furley (2006) *PPG* 30, 1 reviews the determinants of savanna: climate and water, fire, soil–vegetation relationships, grazing. See also Boone et al. (2002) *Afr. J. Ecol.* 40. The savanna belts are associated with the sinking of high-level equatorial air on its return to the *inter-tropical convergence zone. Rainfall is therefore slight, and trees are modified to minimize water loss, with small leaves, and often thorny (Chidumayo (2001) *J. Veg. Sci.* 12).

The boundaries of the savanna are far from clear; there is a gradual change from tall grasses, 1–3 m high with scattered trees, to grassy woodland, and finally to the rain forest. However, where savanna vegetation has been repeatedly burnt (Balfour (2002) *Afr. J. Range & Forest Sci.* 19, 1) there can be a sharp division between this and the equatorial forest, which is less easily fired. Laris (2011) *AAAG* 101, 5, 1067, explains that anthropogenic burning differs from the one based on ecological theory in that its spatiotemporal pattern is relatively consistent from year to year.

scalar In *GIS, **scalar data** are defined by measurements based on a common standard. A particular data set may use a measurement system that may only apply to itself. 'GIS allows you to establish a scalar description for features you can't really measure any other way' (M. N. DeMers 2009) as in a ranking of landscapes by beauty.

scale A level of representation; in cartography, the ratio between map distance and distance on the ground. Geographical difference is expressed at all scales, from the inter-personal to the institutional, and from the national to the international. However, plotting geographical information on too small a scale can conceal information which only becomes apparent at larger scales and higher resolutions. The map of US persistent poverty below purports to show poverty at county level; and shows no poverty in Oregon. The map of Oregon on p. 433, also at county level, tells a different story.

Conversely an examination of geographical information at too large a scale may conceal the bigger picture. Cash and Moser (2000) *Glob. Env. Change* 10, 2 advise on matching the scale of the assessment with the scale of

management, thus avoiding **scale discordance**.

The **politics of scale** describe the ways in which scale choices are constrained overtly by politics, and more subtly by choices of technologies, institutional designs, and measurements; see Lebel et al. (2005) *Ecol. & Soc.* 10, 2, and Silvey (2004) *PHG* 28, 4. Haarstad and Fløysand (2007) *Pol. Geog.* 26, 3 criticize the debate on politics of scale 'for leaving several central questions relatively unexplored'.

For modelling scale in GIS, see E. Wentz (2003) and Florinsky and Kuryakova (2000) *Int. J. GIS* 12.

'Scale is also understood to be dynamic, created or at least shaped by political contests over control of space. It also helps relate political geography to economic, social, transnational, and natural phenomena' Gallaher (2009) in C. Gallaher et al.

scale effect (scale-dependence, spatial-scale dependence) At different

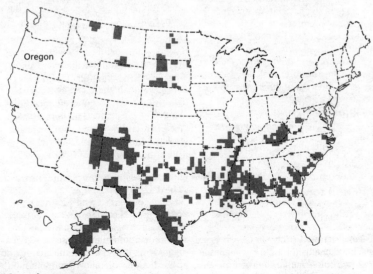

U.S.: persistent poverty

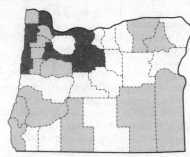

Oregon: persistent poverty

scales, certain different processes prevail. The simplest example of this is found in sediments: small particles, such as clays, are 'stickier' as cohesive forces are relatively more important when the particles get smaller; *see* HJULSTROM DIAGRAM. In biogeography, a major scale-dependence notion is that different ecological patterns and processes are dominant at different spatial scales. This notion in turn implies that changes in scale, be it the size of unit plot or the size of study area, involve a change in the correlation structure among biotic and abiotic components; see Kim et al. (2012) *AAAG* 102, 2, 276. Wei et al. (2012) *Chin. Geogra. Sci.* 22, 2, 127 show that the driving forces and mechanisms of water erosion variations are quite different at different scales.

scaling law If something is **scalable**, a change in its size will not change its nature. Thus, a triangle is always a triangle, however large or small it is. In contrast, an elephant shrew could not be scaled up into an elephant; its skeleton would not be strong enough, its metabolism unworkable; do see http://www.av8n.com/physics/scaling.htm. If there is a relationship between two scalable entities which can be expressed as a law, such as the ratio between the length of one side of a square to its area, then that relationship will hold good whatever the scale they are both at, and that law is a scaling law. With a scaling law, it is possible to compare information collected at one scale with information at a different scale, and to predict outcomes at different scales.

scarp retreat A synonym for *parallel retreat.

Schmidt hammer (Swiss hammer, rebound hammer) This device can measure the strength of a rock. See Goudie (2006) *PPG* 30, 6, 703 on the Schmidt hammer in geomorphological research.

science park An area of industrial development set up in collaboration with, and in close proximity to, a centre of higher education, in order for industry to capitalize on academic research, for jobs to be created locally, and for the educational institute to generate income. Thus, AstraZenica and Oxford generated a vaccine. Knowledge from foreign companies is also important; see Lefner et al. (2006) *Env. & Plan. A* 38, 1 on the Zhongguancun Science Park, Beijing.

Scirocco *See* LOCAL WINDS.

scoria A volcanic rock made of sharp rock fragments and full of air pockets once occupied by gases. **Scoria cones** are constructed by the accumulation of ballistically ejected clasts from discrete, relatively coarse-grained Strombolian bursts, and subsequent avalanching (Valentine et al. (2005) *Geology* 33, 8).

scour Erosion, largely by running water. The depth of a scour hole in a river bed is strongly correlated with *discharge, with scouring on the rising limb of the hydrograph, and filling as the rate of transport through the pool slows. Try Euler et al. (2012) *Catena* 91. **Scour and fill** is the erosion and subsequent filling of a channel.

scree Shattered rock fragments which accumulate below and from free rock faces and summits. Scree is formed by *freeze–thaw; see Statham (1973) *TIBG* 59. Its formation requires jointed rocks that allow water to penetrate.

sea-floor spreading The creation of new crust as *magma rises up at a *constructive *plate margin. Sea-floor spreading can be detected through the presence of magnetic anomalies; see Roest and Srivastava (1989) *Geology* 17. A **spreading axis** is a zone of active eruptive fissuring at a mid-oceanic ridge; see Wise et al. (1992) *Procs Ocean Drilling Program* 120 on the south-east Indian ridge.

sea level Modern analyses of sea-level changes due to glacial *isostatic adjustment are based on the classic sea-level equation by Farrell and Clark (1976) *Geophys. JR Astr. Soc.* 46; Mitrovica and Milne (2003) *Geophys. J. Int.* 154, 2 present a further equation. See Kendal et al. (2005) *Geophys. J. Int.* 161, 3 on theoretical approaches to computing gravitationally self-consistent sea-level changes due to ice growth and *ablation.

For something like 227 000 of the last 250 000 years, sea levels were around 10 m below their current stand (Voris (2000) *J. Biogeog.* 27). Nunn (2000) *NZ Geogr.* 56, 1 demonstrates the 'possible and/or likely' effects of a fall in sea level around AD 1220–1510 on Pacific islands: falling water-tables; the emergence of reef surfaces and the consequent reduction of nearshore water circulation; the emergence of reef islets; and the conversion of tidal inlets to brackish lakes. Heany et al. (2005) *J. Biogeog.* 32, 2 argue that late Pleistocene low sea-level shorelines, rather than current shorelines, define patterns of genetic variation among mammals on oceanic Philippine islands.

N. Bindoff et al. (2007) conclude that sea levels rose at an average rate of 1.7 ± 0.5 mm yr^{-1} during the 20th century. They estimate future sea-level rise of between 0.18 and 0.59 m for the 21st century. 'These simple figures conceal substantial spatial and temporal variability' (Church et al. (2006) *Glob. & Planet. Change* 53); see Edwards (2007) *PPG* 31, 6. For many atoll island states, sea-level rise is more than just a threat to their tourism; it also determines their survival (Wong (2003) *Sing. J. Trop. Geog.* 24, 1). These countries also have low adaptive capacity to sea-level rise, heavily dependent as they are on fragile environmental goods and services, such as coral reefs and fisheries (Nurse and Moore (2007) *Nat. Resources Forum* 31, virtual issue). Sea-level rise in Tuvalu is likely to contribute to saltwater intrusion, long-term loss of land, and damage to ecosystems, agriculture, and livelihoods (J. McCarthy et al., eds 2001). See Dawson in C. Gopalakrishnan and N. Okada, eds (2008) on sea-level rise and the Thames estuary. For a thorough review of this complex subject, see J. A. Church et al., eds (2010).

seamount A relatively small-scale (up to 250 km across) rise, sub-circular in plan view. The date a seamount is formed, and the tectonic setting they erupt into (on-ridge or off-ridge), reflect the way the lithosphere interacts with the fluid mantle beneath. For the radiometric dating of seamounts, see Hillier (2007) *Geophys. J. Int.* 168. McClain (2007) *J. Biogeog.* 34, 11 challenges the view that seamounts are oases/biodiversity hot spots.

search behaviour The way an individual, or entity, selects a location. The choice depends on the information available, and on the decision-maker's ability to evaluate the information. Try McColl-Kennedy and Fetter (2001) *J. Services Marketing* 15, 2.

S

seasonal forecasting A forecast of weather conditions for a period of three to six months, based on, in particular, sea surface temperatures. See A. Troccoli et al., eds (2005).

seawater incursion (seawater intrusion) In coastal aquifers, lighter fresh water lies above a wedge of more dense salt water, and is separated from it by a mobile transition zone. If aquifers are pumped at rates exceeding their natural capacity to transmit water, seawater is drawn into the system to maintain the regional groundwater balance. See Qahman and Larabi (2006) *Hydrogeol. J.* 14, 5 on Gaza.

secession The transfer of part of the territory and population of one state to another, whether pre-existing, or newly created. Writing on Biafra's abortive secession in 1967, Diamond (2007) *Dialec. Anthrop.* 31, 1–3 observes that 'Nigeria is not a nation but a political idea imposed by force of foreign arms. The variety of peoples within its borders will continue to seek accommodation.' See Le Breton and Weber (2001) IMF WP/01/176, on 'how to prevent secession'.

secondary air mass An *air mass which has been modified by the passage of time or by its movement to a different area. A *k* air mass is colder than the surface it moves over, and is usually unstable (*see* INSTABILITY); a *w* air mass is warmer, and usually stable.

secondary industry (secondary sector) The creation of finished products from raw materials; that is, manufacturing.

second home A property occasionally used by a household whose normal residence is elsewhere. Second homes are usually found in rural areas; the 'compensation hypothesis' describes the association between high urban density and the acquisition of a second home (Módenes and López-Colás (2007) *Tijdschrift* 98, 3). Second homes can bring increased custom to rural areas, but may drive up house prices beyond the pockets of local residents. Stressing that he is referring to Sweden only, Marjavaara (2007) *Tourism Geogs.* 9, 3 observes that second homes become a convenient scapegoat for rural problems: second home owners are often targeted as holders of alien/ unfriendly values; locals are seen as fragile victims.

sector theory The view that housing areas in a city develop in sectors along the lines of communication, from the *CBD outwards. High-quality areas locate along roads; industrial sectors develop along canals and railways, away from high-quality housing; working-class housing locates near industry. 'Basic urban pattern such as concentric zone, sectors and multi-nuclei . . . can be interpreted as "ideal" conditions or basic structure at local scale, and the real pattern could be viewed as a composition of multiple structures in the urbanization process' (Xu et al. (2007) *Landsc. Ecol.* 22, 6).

secular In physical geography, long-term; occurring once a century, or over centuries.

sedentary Fixed, not moving, as in **sedentary agriculture** where the farmer and the fields are permanently settled.

sediment 1. Material which has separated and settled out from the medium—wind, water, or ice—which originally carried it. For fluvial sediments the ability of a river to carry sediment depends on particle size as well as the river discharge. *See also* LOAD. **Sediment connectivity** is the shift of sediment

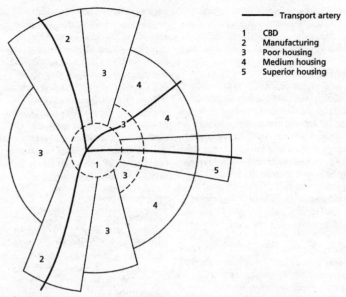

Sector theory

from a *source to a *sink, and the transfer of sediment between different zones within a *catchment.

Sediment residence time is the time an alluvial load, such as *bedload, remains in some part of a fluvial environment. A **sediment slug** is a body of rock fragments (clasts) in a fluvial system, triggered through a variety of mechanisms, from landslides to land-use change. See Gran and Czuba (2017) *Geom.* 277.

2. A **sediment cascade** describes the transport of sediment through the landscape. A **sediment budget** is the tally of inputs and outputs for a specific *open system, over a given time period; the classic paper is Trimble (1983) *Am. J. Sci.* 283. A **balanced sediment budget** is an equation of *mass conservation in which the sediment fluxes related to the sources, sinks, and storages are balanced. The concept is important in environmental management: see

Nordstrom et al. (2002) *Geomorph.* 47, 2–4.

sedimentary rock A rock composed of sediments, usually with a layered appearance (*bedding plane). **Sedimentation** is the deposition of sediment from a state of suspension in water or air, also called *siltation*.

sediment cell A length of coastline, and its associated nearshore area, within which the movement of sand and shingle is largely self-contained. The littoral sediment cell concept 'has become synonymous with pro-active coastal management practices in England and Wales' (Cooper and Pontee (2006) *Ocean & Coastal Manage.* 49, 7–8).

Sediment yield is the total mass of sediment that reaches the exit of a drainage basin. Tamene et al. (2006) *Geomorph.* 76, 1–2 find that pronounced terrain steepness, easily detachable slope material, poor surface cover, and gullies promote high sediment yields; de Vente and Poesen (2005) *Earth Sci. Revs* 71, 1–2 propose a general relation between basin area, active erosion processes, sediment sinks, and total sediment yield; and Verstraeten et al. (2003) *Geomorph.* 50, 4 develop an index model to explain variability in area-specific sediment. A **specific sediment yield** is the sediment yield per unit area of the drainage basin, usually per square kilometre, and is calculated from:

$$\text{specific sediment yield}(t \text{ km}^{-2}y^{-1})$$
$$= \text{sediment yield}(t \text{ y}^{-1})/$$
$$\text{drainage basin area}(\text{km}^2)$$

Holeman (2010) *Water Res.* 4, 4, 737 provides a table of the sediment yield of major rivers of the world. **secular** In physical geography, long-term, occurring over a century or more.

Sedimentation is a term often used to describe the blocking of an aquatic system by the deposition of sediment; see Bennett et al. (2005) *Water Resources Res.* 41, W01005, and Wang et al., *Water Resources Res.* 41, W09417. See Flemming (1965) *J. Sed. Res.* 35, 2 on **sedimentary particles**, their transport history, and their depositional environments.

segmented labour market A *dual labour market may be composed of skilled and unskilled segments. 'Welfare, tax and labour-market institutions continue to facilitate and maintain a segmented labour market with a wide range of employment systems and practices, making it virtually impossible for the growing proportion of the labour force in low-wage, service-dominated jobs to achieve independent living on an individual basis' (McDowell (2004) *PHG* 28, 2). See also Hedberg, SULCIS *W. Paper* 2008:1.

segregation The separation of a large population into subgroups—based on grounds of income, race, religion, or language—located in distinct residential areas. Smith (2012) *TIBG* 37, 3, 461 outlines ways in which distinct sub-populations of British society are becoming more transient and segregated. His work is on British coastal towns, but his conclusions apply throughout the UK. See Musterd et al. (2017) *Urban Geog.* 38, 7, on social segregation in European cities. *See also* INDEX OF SEGREGATION.

seif *See* SAND DUNE.

seismic Of an *earthquake. The **seismic focus** is the point of origin of the earthquake within the crust. The velocity of **seismic waves** varies with the density of the material through which they travel. This property is used in **seismic tomography**—the investigation of the interior of the Earth; see W. K. Lee (2003).

selective logging The felling, at intervals, of the mature trees in a forest of mixed age. Selective logging, combined with large tree retention, is recommended for minimizing ecological impacts.

selective migration This may be spontaneous, as when a particular age–sex group constitutes most of the migrants, or where healthy people move to less deprived areas and less healthy people move to or stay in more deprived areas; see Gartner et al. (2018) for a really interesting paper.

A leading cause of selective migration is the increasing skill-focus of immigration policy in a number of industrialized countries. This is likely to increase as rich countries age, and knowledge-intensive sectors build. See Hjort (2009), *Gerum Kulturgeografi* (2009) 1, notably p. 18.

self-organization A process whereby some sort of order develops from a system which was initially disorderly; a process arising from interactions within the system. For example, distinctive zones of land use may arise within a city though they have not been formally designated as such. In geomorphology, dynamic systems often experience a sequence of positive feedback driving the system towards a new state, followed by negative feedback which stabilizes the system, resulting in equilibrium. This is self-organization, and examples include the formation of evenly spaced, nearly uniform ripples on sand dunes or stream beds, or of *patterned ground in periglacial landscapes. Not every geomorphologist is convinced. Even for those who support the theory, questions still arise on how the self-organization takes place and how the changes are communicated from one part of the system to another to create that order. See Hooke (2007) *Geomorph.* 91, 3–4.

Self-organized criticality describes a system in *dynamic equilibrium near a threshold condition (tipping point). The concept arises from Bak et al. (1987) *Phys. Rev. Letts* 59, 4, who develop a simple *cellular automaton model in which sand is added, grain by grain, to a surface to form a pile. When local slopes become too steep a collapse occurs, moving sediment to neighbouring cells, which too can collapse if the adjusted slopes are too steep. See Coulthard and Van De Wiel (2007) *Geomorph.* 91, 3–4.

semiotics The study of signs; the ways in which signs and meanings are created, decoded, and transformed. Cultural landscapes are full of a collection of symbols which we can decipher, and this deciphering is an exercise in semiotics. So semiotics aid and abet our understanding of the 'dialogue' between people and their environment. Try Boogart (2001) *Middle States Geogr.* 34, 38.

sense of place Either the intrinsic character of a place, or the meaning people give to it, but, more often, a mixture of both. Societies create places, and sense includes attachment to place, national identity, and regional awareness. Thus, a dispersed settlement pattern in Norway, for example, has acquired a symbolic meaning (Cruickshank (2006) *Norsk Geograf. Tidssk.* 60, 3). Sometimes it is the intangible characteristics of places which are central to our sense of place, and the planners consciously create or preserve memorable and singular structures to make a space distinctively different— such as the UK's Angel of the North.

sequestration The removal of some substance from circulation. **Carbon sequestration** is the long-term capture of atmospheric carbon dioxide, which is then stored in geological structures, oceans, plants, and soils. The USGS carbon sequestration page is clear, and useful.

serac An ice pinnacle, or series of pinnacles on the surface of a glacier resulting from tensional failure in the more rigid upper crust when the glacier: moves over a slope; spreads out over a plain; or passes round a bend in its valley.

sere One of a sequence of developmental stages in plant succession. A primary stage is a **prisere**; a **hydrosere** develops in water; a **psammosere** on a sand dune; and a **xerosere** in an arid location. If a plant succession is interrupted, the resulting secondary stage is a **sub-sere**.

service industry Those economic activities (also called *tertiary industry) concerned with the distribution and consumption of goods and services. Proximity to clients is the major locational factor. Other factors promoting the *clustering of service firms and workplaces are: agglomeration economies, supply of infrastructure, supply of labour, and new firm formations.

Some services—particularly less specialized household services—are fairly evenly located in relation to settlement patterns. Those less evenly located include specialized household services, central government functions, and professional or business services, which concentrate in major city regions, or within regional clusters. With information and communication technology, service activities can decentralize from urban cores to the suburbs, and to more peripheral sites, both nationally and internationally.

settlement Any form of human dwelling—from a single house to the largest city.

settlement hierarchy A division of settlements into ranks, usually by population size.

settlement pattern The spatial distribution of settlements. Some settlement patterns may be seen as a reflection of cultural traditions: a dispersed settlement pattern in Norway, for example, has acquired a symbolic meaning (Cruickshank (2006) *Norsk Geograf. Tidssk.* 60, 3). *Nearest neighbour analysis may be used to test for regularities in settlement patterns.

sexuality, geographies of The spatial expressions of sexualities, interleaving discussions of sexualities with issues such as development, race, gender, and other forms of social difference, and the geographies of sexual health. For example, C. Hemmings (2002) examines bisexual spaces as places that are defined by both geographical

boundaries and cultural significance, showing how and why safe places have developed for gay, lesbian, and bisexual communities. In 1999, Binnie and Valentine (*PHG* 23, 2) argued for a move away from a simple mapping of lesbian and gay spaces towards a more critical treatment of the differences between sexual dissidents. Consult Hubbard et al. (2008) *PHG* 32, 3 and Oswin (2008) *PHG* 32, 1 to see if their call has been answered. *See also* QUEER GEOGRAPHIES.

shadow price The shadow price of any good is the increase (or decrease) in social welfare brought about by providing one more (or one less) unit of that good. Thus, the shadow price of carbon is defined as the social cost of emitting a marginal tonne of carbon (or the social benefit of abating a tonne). See Dietz (2007) DEFRA. This concept is used in *cost–benefit analysis to value intangible items like *amenity.

shanty town *See* SQUATTING.

share-cropping A farming type where the tenant pays rent in produce rather than in cash. In the post-bellum USA, lack of capital left the cropper dependent upon others for subsistence, until market time. Storekeepers furnished the cropper with food and clothing during the year, the accumulated debt to be repaid, with interest, at harvest time. The store owner was often also the buyer of the crops. Share-tenants compromise between risk and incentive; given a choice, better-off workers would prefer to be cash-tenants, while poorer workers, or recent arrivals, would enter share-cropping arrangements. See Brannstrom (2001) *Polit. Geog.* 20, 7.

shear (shearing) The deformation of a material so that its layers move laterally over each other. Shearing bends, twists, and draws out rocks along a fault or *thrust plane. A **shear plane** is the face along which shearing occurs. **Shear strength** is the ability of a rock or soil to withstand shearing; soil shear resistance is proportional to the tangent of the friction angle and therefore soils with larger friction angles exhibit a higher shear strength. Two major parameters of soil shear strength are cohesion, which is boosted by plant roots and peaks when soil moisture content is 12% to 14%, and internal friction, which is the resistance of a soil mass to sliding, itself inversely related to the amount of moisture in the soil. Slope failures are the consequence of loading conditions that exceed soil shear strength. **Shear stress** happens when the applied force acts tangentially to the surface of the body (the down-slope component). It is opposed to a normal stress (the slope-normal component) which is applied perpendicularly.

Within a glacier, the level of shear stress (τ) experienced at any point is dependent on the ice thickness and the surface slope. It can be expressed as:

$$\tau = \rho g(s - z) \sin á$$

where ρ is the density of ice, g is the acceleration due to gravity, s is the surface elevation, z is the elevation of the point within the glacier, and $á$ is the surface slope of the glacier.

sheet erosion A very slow-acting form of erosion whereby a thin film of water—**sheet wash**—transports soil particles by rolling them along the ground. Sheet erosion is a major process in the denudation of land surfaces, but perhaps it is only really effective on steep and relatively unvegetated slopes. See Chaplot and Bisonnais (1998) *Soil Till. Res.* 46 and A. Goudie (2004).

sheet flow At the base of a glacier, a thin film or sheet of water at very high water pressure, a thousandth of a millimetre to a millimetre in thickness, being thinnest over bumps in the bed because of the higher pressure upstream. Sheet flow is thought to be dominant only beneath ice masses where the majority of basal water is derived from subglacial melting due to geothermal heating, sliding, or bed deformation.

sheeting The separation and peeling away of the outer layers of a rock mass;

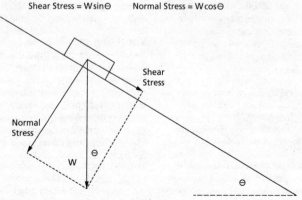

Shear Stress = W sinΘ   Normal Stress = W cosΘ

Shear Stress

Normal Stress

W

Θ

Θ

Shear stress
Source: http://uregina.ca/~sauchyn/geog323/stress.gif

see Wakasa et al. (2006) *Hydrol. Procs* 31, 10. *See also* EXFOLIATION.

shield The very old, rigid core of relatively stable rocks within a continent.

shield volcano A volcano formed of successive eruptions of free-flowing *lava which create a gently sloping, broad dome tens of kilometres across and more than one kilometre high. Mount Mazama, Oregon, is composed of several overlapping composite and shield volcanoes. See Bacon and Lanphere (2006) *GSA Bull*. 118, 11.

shifting cultivation A farming system based on the clearance of land, cultivation until soil fertility declines, followed by fallowing for up to twelve years, in order to restore the vegetation to its initial condition. 'Shifting cultivation is a truly *sustainable system, perhaps the only one for humans. It can maintain agricultural production indefinitely with little reliance on external inputs when population levels are low—roughly no more than one person per 2 hectares of agricultural land' (C. A. S. Hall (2000)).

shift-share analysis A method of estimating the relative importance of different elements in any growth or decline of regional industrial employment. See Fotopoulos (2007) *Growth & Change* 38, 1 for the methodology.

shingle Pebbles on a beach, rounded by *abrasion, and reduced in size by *attrition. The abrasion rate of shingle is linked to, in rank order: mean wave height, pebble type, and pebble weight.

shit, geographies of The studies of spatial inequalities in access to sanitation and the consequences of this for human, environmental, and economic health, and of spatial and temporal differences in cultural attitudes towards excrement. Sanitation continues to be largely absent in many developing countries, and government-backed sanitation initiatives, where they exist, often prioritize flush and discharge systems over low-cost or community-based alternatives. In 2012, the Census Commissioner of India reported that about 600 million Indians practise open-air defecation.

((()) SEE WEB LINKS

• Online version of 'Economic Impacts of Inadequate Sanitation in India'.

shopping, geographies of *See* RETAILING, GEOGRAPHY OF.

shopping centre Regional shopping centres create a significant number of jobs and wealth in the local areas they lie in, but much of the focus has been on their apparent negative impacts, despite little supporting evidence. In some cases, the regional shopping centres have given a clearer identity to an area, and helped change perceptions of it. They have often, but not always, stimulated new development and investment in unattractive areas. M. Ibrahim and P. McGoldrick (2003) examine the policies that attempt to reduce the negative impacts of decentralizing retail facilities.

shopping goods In the process of selecting and buying things, those goods which the customer compares on such bases as suitability, quality, price, and style—such as furniture and clothing. Makers of shopping goods can limit the number of wholesalers and retailers in a geographic area in order to maintain channel control and enhance the image or exclusivity of their products (Gable et al. (1995) *J. Retail. & Consum. Services* 2, 4).

shore The land adjoining a large body of water/the sea. The **backshore** is normally above the high water mark, but still influenced by the sea; the **foreshore** is the area between high and low tide marks and is exposed at low tide. The **nearshore** is seaward of the foreshore, ending at the breaking point of the waves; the **offshore** is the zone seaward of the breakers, but in which material is moved by the waves.

shoreline trend analysis The investigation of the response of a shoreline to anthropogenic and natural impacts. For a digital shoreline analysis system, See Himmelstoss, Zichini, and Ergal (2009) *US Geol. Surv.* 2008-1278.

shore platform A very gently sloping platform extending seaward from the base of a cliff, and subject to *salt weathering, alternate wetting and drying, *water-level weathering, and *quarrying, *hydraulic action, pneumatic action, and *abrasion. Their form is a function of the relative balance between wave energy and weathering. Platform gradient increases with tidal range. Waves on platforms are less capable of erosion because of their high loss of energy in shallower waters.

sial The material of the continental crust, mostly of silica- and aluminium-rich minerals.

silcrete A *duricrust cemented with silica. See Thiry (2006) *J. Sedim. Res.* 61, 1, 111.

sill An *intrusion of igneous rock which spreads along bedding planes in a nearly horizontal sheet. Sills are mostly concave, and can be fed from dykes, or the steep climbing portions of deeper sills. Both sills and dykes can provide magma to overlying volcanic fissures, and sills can feed shallow laccoliths (Thomson (2006) *Bull. Volc.* 66, 4).

silt Fine-grained minerals, 0.002–0.06 mm in diameter, ranging in size between *clay and *sand in size. Silts are often deposited by rivers when the flood water is quiet.

Silurian A period of *Palaeozoic time stretching approximately from 430 to 395 million years BP.

(((⊕))) SEE WEB LINKS
• An overview of the Silurian period.

sima The material of the lower part of the continental crust and the oceanic crust, dominated by silica and magnesium.

sink A point, line, area, or reservoir where energy or mass is removed from a system. For example, soil and trees tend to act as natural sinks for carbon.

sink hole *See* DOLINE. See also Zumpano et al. (2019) *Geom.* 332.

sinter Mostly silica and sulphate minerals, precipitated in layers from volcanic activity. Sinters provide evidence of past environments (Lynne et al. (2006) *Geology* 34, 9). **Snow sintering** is the bonding of snow particles, either by the diffusion of water molecules, or by pressure-induced creep. Lots of information can be found in Ramseier and Keeler (2017) *J. Glaciol.* 6, 45.

sinuosity The extent to which a river bends. The **sinuosity ratio** is the distance between two points on the stream measured along the channel, divided by the straight line distance between those points; it is used to determine whether a channel is straight or meandering (Aswathy et al. (2008) *Env. Monit. Assess.* 138, 1–3).

site The position of a structure or object in physical, local terms, such as a river terrace.

S

Site of Special Scientific Interest (SSSI) A site in the UK which is of particular importance because of its geology, topography, or ecology. SSSIs are graded in terms of importance from 1 to 4.

situated knowledge All knowledge is produced by people who have their own viewpoints, which reflect their time, place, and biases; their knowledge is *situated*. Knowledges are always situated, always produced by actors working in or between all kinds of locations, working up/on/through all kinds of research relationships. All of these affect what exactly gets done by whom; how and where it's done; how it's turned into a finished product, and for whom.

situation The location of a phenomenon, such as a town, in relation to other phenomena, such as other towns. *Compare with* SITE.

skid row An inner-city district characterized by a resident population that is in some way disadvantaged: through poverty, homelessness, alcoholism, or mental illness. Functions often include a local drug market, and a concentration of bars, low-rent hotel rooms, and social service facilities. Do read Collins and Loukaitou-Sideris (2016) *TPR* 87, 4 on 'skid row, gallery row and the space in between': it's easy to read, informative, and very interesting.

sky view factor A measure commonly used by urban climatologists to quantify the openness of a site within an urban setting. It has major implications for incoming and outgoing solar and terrestrial radiation (Grimmond (2007) *Geog.* 30, 3).

slaking In geomorphology, the disintegration of fine-grained rocks. Slaking is a major mechanism of rock-weathering; see Stefano and Ferro (2002) *Biosys. Engineer.* 81. **Slake durability** is the rate and amount of strength reduced in a rock after a soaking, and is an important parameter in slope stability; see Yilmaz and Kalacan (2005) *Env. Geol.* 47, 7.

slash and burn An agricultural system whereby the trees are felled, the land cleared of most of the trunks, and the remaining vegetation fired. See Tschakert et al. (2007) *Ecol. Econ.* 60 on carbon offsets in small-scale slash-and-burn agriculture.

slickenside A set of linear marks on a fault or bedding plane caused by the frictional movement of one rock body against another.

slide A form of mass movement in which material slides in a relatively straight plane, generally breaking into many blocks as it moves. Slides usually have a length far greater than the depth of the moving material, and are triggered by high water pressure, but the competence of the rock is also significant; see Warburton et al. (2003) *ESPL* 28, 5.

slip-off slope The relatively gentle slope at the inner edge of a meander.

slope *See* HILLSLOPE.

slope wash The downslope transport of sediment by an almost continuous film of water. It can be seen as three erosional processes: sheet flow, rill erosion, and gullying.

slope winds *Anabatic and *katabatic winds.

slum An area of poor housing, characterized by multi-occupance and overcrowding. Although places of marginalization, slums can also be places of great creativity and community survival. For example, slums often play an important role in a city and region's economy, with many slum residents taking part in vital informal production, services, and recycling-based services that support more affluent sections of the city; see Pyati and Kamal (2012) *Area* 44, 3, 336. **Slum clearance** is the demolition of substandard housing, usually accompanied by *rehabilitation and *redevelopment. Look for Liza Weinstein's *The Durable Slum* (2014) on Dharavi, Mumbai.

slump A form of *mass movement (also called a single rotational slide) where rock and soil rotate backwards with movement down a concave face: 'a more or less rotational movement, about an axis that is parallel to the slope contours, involving shear displacement along a concavely upward facing failure surface' (D. Varnes, 1978). Slumps are most common in thick *regoliths and large mudstone rock units.

slushflow A linear, channelled downslope movement of water-saturated snow on glaciers and in alpine stream channels during the beginning stages of the spring break-up. See Roberson et al. (2011) *Geograf. Series A*, 93, 2, 72.

small and medium enterprises (SMEs) Definitions vary across countries, but commonly enterprises with fewer than 500 employees are considered small.

smart city The objectives of a smart city are to: enhance performance; optimize resources; reduce waste, consumption and costs; and improve the quality of life of its citizens. These objectives will be aided by the use of advanced information technology. See Robinson and Franklin (2021) *TIBG* 46, 2, and Jameson et al. (2019) *Urban Geog.* 10, on Amsterdam Smart City—but neither source is wholly helpful.

smog A combination of smoke and fog. The fog occurs naturally; the 'smoke' is introduced into the atmosphere by human agency. See R. Kosobud (2006) on the cost-effective control of urban smog. *See also* PHOTOCHEMICAL SMOG.

SMSA A Standard Metropolitan Statistical Area in an urban area of the USA. This can be a town of 50 000 people, or two towns, each with more than 15 000 people, and together totalling more than 50 000; or a county with more than 75% of its population working in industry. Areas which seem, by employment, commuting, or population density, to be urban rather than rural are also included.

snow Frozen atmospheric water. A snow crystal is ice, up to 5 mm across, and variously shaped as a prism, plate, star, or needle. Snow crystals fall from stratiform clouds when the low-level air is degrees below freezing point, and the air above it is colder. When the low-level air is near 0 °C, snow crystals aggregate to form snowflakes.

Snow cover plays an important role in the hydrological cycle; see Holko et al. (2011) *Geog. Compass* DOI: 10.1111/j.1749-8198.2011.00412.x. **Snowpack** describes a sequence of layers of snow; each snow fall adds another layer which can give information on specific weather events. **Snow gliding** is a downhill motion of ground snow forceful enough to uproot vegetation and erode the soil beneath. See Höller (2014) *Natural Hazards* 71.

Snowball Earth This theory posits that, several times in Earth's history, oceans froze, ice sheets covered the tropics, and global temperatures plummeted to −50°C, so that the entire Earth may have actually frozen over, resembling a 'snowball', and potentially causing some of the most severe crises in the history of life on the planet. Such episodes may have been triggered by a reduction in greenhouse gases in the atmosphere, principally carbon dioxide and methane, which would have made the global climate colder, creating larger areas of ice and snow. This ice and snow reflects more solar radiation than does bare ground or liquid water, which creates a 'positive feedback'. If the Earth ever became approximately half-covered by ice or snow, the feedback would become self-sustaining and glacial ice would rapidly spread to the equator. That is the theory.

SEE WEB LINKS

• Snowball Earth website.

snowpack An area of naturally formed, packed snow that usually melts during the warmer months; see Caine (1995) *Geog. J.* 161, 1, 55 and *Field Guides* (2009), 15, 471.

SOC Soil organic carbon: the measurable component of organic matter in a soil.

social area analysis The analysis of a city to define **social areas** (urban areas which contain people of similar living standards, ethnic background, and lifestyle).

social capital The relations that exist between individuals within both families and communities, together with the educational, cultural, and social resources and expertise of a society. 'At the macro level, formal institutions, such as state constitutions and legal norms, make up the social capital of a society . . . at the micro level, social capital can be conceptualized as personal trust which exists in a particular

community or social network' (Batheld and Glückner (2005) *Env. & Plan. A* 37). For an overview of geographical accounts of social capital see Holt (2008) *PHG* 32, 2.

social cleavage The spatial division of society into distinct groups, thought to be linked to geographical voting patterns.

social cohesion The interdependence between the members of a society, shared loyalties, and solidarity. 'A cohesive community is a community that has naturally many cross-links, where people from different race, age, background, feel free and happy to mix together in housing, in education, [and] in leisure facilities' (UK Housing, Planning, Local Government and the Regions Sixth Report). Nagel and Staeheli (2008) *Soc. & Cult. Geog.* 9, 4 see the social cohesion agenda as 'marked by the assertion that the host society should define the terms of integration, and that the primary responsibility for integration lies with immigrants and minorities'. Yuval-Davis et al. (2005) *Ethnic & Racial Studs* 28 see the concern with social cohesion as an attempt to limit the exclusion of groups.

SEE WEB LINKS

• Website for the European Committee for Social Cohesion (CCS).

social Darwinism The application of the concept of evolution to the development of human societies over time, emphasizing the struggle for existence of each society, and the survival of the 'fittest' of them. Such pernicious ideas have been used to justify power politics, imperialism, and war.

social distance The perceived distance between social strata (different socio-economic, racial, or ethnic groups), usually measured by the amount of social contact between groups. Autant-Bernard et al. (2007) *Papers Reg. Sci.* 83, 3 define this term as the number of links between any two firms: 'firms with a higher number of partners and a smaller social distance are more likely to be engaged together in a research project . . . in this context, social distance matters more than geographical distance.'

social economy A term with various meanings, but one definition is that the social economy includes those organizations which use reciprocity to promote mutual economic or social goals, often through social control of capital. This would include cooperatives, credit unions, non-profit and volunteer organizations, charities and foundations, service associations, community enterprises, and social enterprises that use market mechanisms to pursue their ends (see BALTA, B.C.–Alberta Social Economic Alliance).

social exclusion The exclusion of part of society from any of the social, economic, political, and cultural systems which integrate someone into society. It's the denial of access, to an individual or group, to the chance of participating in the social and political life of the community, which results not only in a lower quality of life, but also in fewer life chances, fewer choices, and reduced citizenship.

Social exclusion is a more complex concept than poverty: the Transport Research Group notes that 'it is possible to be excluded without experiencing poverty; and it is possible to experience poverty, yet feel included. Exclusion may come about through a combination of problems such as income poverty; unemployment; geographical isolation or reduced accessibility to social networks, facilities, goods, and services;

poor housing; high crime environments; family breakdown; or the denial of citizenship rights and freedoms.' See the DFID Policy Paper *Reducing Poverty by Tackling Social Exclusion.* Try the *Social Exclusion Literature Review 2008.*

social formation The prevailing pattern of class structure which goes hand in hand with a particular *mode of production; try Bruneau (1981) *Antipode* 13, 3.

social geography Social geography is subdivided into three categories: The first is the spatial expression of capitalism. The second stresses the 'alternative' view of human geography; for example, studying the economically disadvantaged rather than the successful. A third category emphasizes *welfare geography.

socialism A social system based on equality and *social justice, once linked with common ownership of the *means of production and distribution.

social justice 'Putting it simply, geographers interested in justice focus on who gets what, who misses out, and where all this occurs' (Bauder (2006) *PHG* 30). These are questions that can be asked on global, national, regional, and local scales. 'Social justice requires the removal of the obstacles that inhibit individuals and groups from developing their potential . . . contrary to the orthodox view, increasing social equality improves economic efficiency' (Kitson et al. (2000) *Camb. J. Econ.* 24, 4). Also of interest to geographers is the way in which moral systems vary spatially. *See also* JUST CITY.

social physics The use of concepts from physics to illuminate aggregate human behaviour; for example, the *gravity model. In general, social physics doesn't work well.

social polarization *Segregation within a society; the gaps between individuals, households or groups of people in terms of their economic and social circumstances and opportunities (see Dorling and Woodward (1996) *Prog. in Plan.* 45, 2).

social reproduction The way classes in an unequal society reproduce their status from one generation to the next; the way the structures of dominance of one group over another are maintained. Social reproduction varies, both geographically and historically, and reflects uneven geographical development. Katharyne Mitchell, ed. (2004) has all you need on geographies of social reproduction—real, or virtual, spaces where groups of people meet, gather, and interact with each other—as opposed to *personal space. It is produced by societies according to their own spatial practices. Social space provides a framework for the behaviour of a group; it is flexible, networked, and works on many levels; a multi-layered social space defined by political, social and institutional possibilities. Thus, Gareth Southgate organizes social spaces that reject racist, sexist, classist, homophobic, and other forms of exclusionary politics. Take a quick look at Unwin (2000) *TIBG* 25, 1 for a critique of social space.

social well-being A situation where income levels are high enough to cover basic needs, where there is no poverty, and where there is easy access to social, medical, and educational services. See Roy (2018) *J. Emerging Techs. & Innovative Res.* 5, 7.

socionature (socio-nature) An *ecosystem that has been socially created; 'nature' that has been produced. This concept is used to claim that nature and society are indivisible

and shouldn't be investigated separately from each other; that social relations are inherently ecological, and that ecological relations are inherently social. See J. Anderson (2010) p. 144.

sodic soils Soils that are traditionally defined as having an exchangeable sodium percentage greater than 6%— 'but there are many soils with lower sodium levels which exhibit similar behaviour' (S. Raine and R. Loch 1995).

soil The naturally occurring, unconsolidated, upper layer of the ground, made of humus and weathered rock. Major factors affecting soil formation are: climate, relief, parent material, vegetation, and time. Soil is the medium on which we produce 99% of our food; a major store of carbon; and regulator of climate, holding two to three times more carbon than exists in the atmosphere. Soils are also a regulator of water resources, attenuating hydrological responses and removing contaminants from percolating water.

soil aggregate A clump group of soil particles that stick together more strongly than to surrounding soil particles.

soil association A cluster of soils dominated by a single soil *taxon.

soil classification A naming system to describe and identify soils.

(⊕) SEE WEB LINKS
• National soil classifications.

soil creep *See* CREEP.

soil detachment The tearing loose of soil particles; one of the major processes of *soil erosion.

soil erosion The removal of the soil by wind and water and by the *mass movement of soil downslope. The wind erosion is by *deflation; water erosion takes place in gullies, *rills, or by *sheet wash; downslope mass movement ranges from soil *creep to *landslides. A large proportion of total soil erosion over a long time period is generally due to relatively few, large storm events, according to Bagarello et al. (2017) *Catena* 155.

Accelerated soil erosion is erosion increased by human activity. Causes include: wind, tillage, war (as a result of agricultural intensification and the use of marginal land, or of scorched-earth policies), fire, urbanization, mining, and continuous cropping. However, there are doubts about the claim that overgrazing leads to soil erosion, and 'there is still some way to go before the cause and effect links are sufficiently understood to form a basis of tools for the management of grazing in a way that would minimise erosion and maximise the benefits' (see Thornes (2008) *Geogr. Res.* 46).

(((⊕))) SEE WEB LINKS
• Erosion data and erosion models.

soil flux The flow of soil over a slope, generally downslope.

soil heave The cracking and lifting of soil material on a slope which occurs when shrunken clay, in rewetting, returns to its original volume, or through the growth and melting of ice crystals. See R. Charlton (2008).

soil horizon *See* HORIZON.

soil moisture Moisture is held in the *capillary soil pores. Soil moisture controls many important processes in seedling emergence, evapotranspiration, mineralization of the soil organic fraction, surface run-off, leaching, and crop yield.

The **soil moisture budget** is the balance of water in the soil, resulting

from precipitation and *potential evapotranspiration in combination. *See also* MOISTURE INDEX.

soil pipe A horizontal tube-like subsurface cavity within the soil; *macropores greater than 1 mm in diameter and up to a metre or more in diameter. They are continuous in length such that they can transmit water, sediment, and solutes through the soil. The classic paper is Wilson and Smart (1984) *Catena* 11, 145.

soil pore Any open space within the soil. **Soil porosity** is the percentage of pore space. Water will not drain freely through the fine **capillary pores** (average diameter < 0.03 mm), which retain water through surface tension, but drains freely through the larger non-capillary pores. With 2–5% soil porosity, most water flows directly to the subsurface.

soil profile A vertical series of soil *horizons from the ground surface to the parent rock. A soil is classified according to the arrangement of its horizons. The Soil Profile Analytical Database of Europe of Measured Profiles was created to provide a common structure for storing harmonized information on typical soil profile properties of European soils (Hiederer et al. (2006) *Geografisk Tidsskrift* 106, 1).

soil respiration The movement of carbon dioxide out of a soil when it 'breathes'. As part of the *carbon cycle, soil respiration is the major means by which carbon dioxide, fixed by plants, returns to the atmosphere. For an accessible summary, see Luo and Zhou (2006) *Soil Respiration and the Environment*.

soil science The fundamental study of soil processes and properties, and the use of this information to investigate soil

S

issues, such as soil ersion, soil pollution, and crop production. Go to the website of the British Society of Soil Science—it's outstanding.

soil series A group of soils formed from the same parent rock and having similar *horizons and *soil profiles, but with varying characteristics according to their location.

((⊕)) SEE WEB LINKS

• Soil Survey Staff, Natural Resources Conservation Service, United States Department of Agriculture: Official Soil Series Descriptions.

soil structure The arrangement of soil particles into larger particles, or clumps. Platey structures are formed of thin, horizontal layers; prism-like structures are called columnar where the tops are rounded, and prismatic where the tops are level; blocky peds are bounded by flat or slightly rounded surfaces; and spheroidal structures are called crumbs if highly water absorbent, and granular if only moderately so. Two soils with the same texture can behave very differently depending on their structure. A clay soil, for example, can be easy for air, water, and roots to move through with good structure, or be almost impenetrable by roots, air, and water when its structure has been destroyed by compaction; go to Nathalie Shanstrom's outstanding *DEEPROOT* blog.

soil texture The proportions of sand, silt, and clay in any soil. Twelve different textural classes are recognized, and the structure of the soil can be determined when the percentage of these three soil constituents is plotted on a ternary diagram. TAL is a group of programs to determine soil texture classes based on major classification schemes.

soil vulnerability index A system developed by the USDA to identify the vulnerability of crop land to run-off and *leaching. The relevant data can be

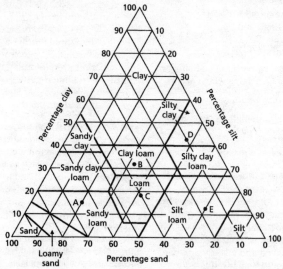

Soil texture

found on the SSURGO website, and can be used with the ArcGIS system.

solar constant The rate per unit area at which *solar radiation reaches the outer margin of the Earth's atmosphere. This fixes the energy supply for the *atmospheric heat engine. For variations in the solar constant, see Gribben (1991) *New Scientist* 132, 1796.

solar energy 1. Of all the sunlight that passes through the atmosphere annually, only 51% is available at the Earth's surface to do work. This energy is used to heat the Earth's surface and lower atmosphere, melt and evaporate water, and run photosynthesis in plants. Of the other 49%, 4% is reflected back to space by the Earth's surface, 26% is scattered or reflected to space by clouds and atmospheric particles, and 19% is absorbed by atmospheric gases and clouds. The geography of a place determines how much solar energy is absorbed by the earth, how much is stored and how readily it is released to the atmosphere; see R. Thomas and R. Huggett (1980) and Martin et al. (1998) *Geophys. Res. Lett.* 25, 23 on the distribution of solar energy at the Earth's surface.

2. Any energy source based directly on the sun's radiation. See Zweibel et al. (2008) *Scientific American* for a wide-ranging review.

solar radiation The electromagnetic waves emitted by the sun, varying in wavelength from long-wave radio waves, through infrared waves and visible light, to ultraviolet waves, X-rays, and gamma radiation. Earth gets only 0.0005% of the sun's radiation. Most solar radiation passes straight through the *atmosphere without warming it, but it is received and absorbed by the Earth. Changes in solar radiation are known as solar forcing.

sol brun lessivée A type of *brown earth from which some clay has been *leached, forming an E *horizon.

solfatara A vent through which steam and volcanic gases are emitted.

solifluction The slow, downslope movement of water-saturated debris in *periglacial regions and other areas with cold climates. Rates of movement may be 0.9 cm yr^{-1} on gentle slopes, 12–25 cm yr^{-1} on steeper slopes. Temperature is the most important factor controlling morphometry of solifluction landforms, explaining about 22% of the variability (Ridefelt (no date) ISSN 1650–6553, 70). Solifluction may develop in areas without *permafrost; winter freezing of the sub-surface layer may be sufficient. See Matsuoka (2001) *Earth-Sci. Revs* 55, 1–2. **Solifluction terraces** are step-like, linear, patterned-ground formations, made of fine-grained material. Heights and widths are about 1 m, and, with very fluid material, terraces show rough bedding (*tumultuous bedding*). **Solifluction lobes** are bulges in the slope profile that do not break the surface. The major environmental controls on solifluction lobes are gradient and soil moisture; see Ridefelt and Boelhouwers (2006) *Perm. & Periglac. Procs* 17.

solonchak An *intrazonal saline soil found in hot, arid climates. Evaporation of soil moisture brings saline groundwater to the surface where it, too, evaporates. The sodium and calcium chlorides and sulphates which have been translocated remain as a grey surface crust.

solonetz An *intrazonal, formerly saline soil. Periodic rainfall has leached the salts from the surface layer and these accumulate in the B *horizon.

solum In a soil, the layer above the parent material. The solum is strongly influenced by vegetation.

solution In geomorphology, the picking up and dissolving of particles by a fluid, usually water or carbonic acid. *See also* CARBONATION; AGGRESSIVITY.

sorptivity A measure of the ability of a porous medium, such as a soil, to absorb or 'desorb' liquid by capillary action.

sorting In geomorphology, the deposition of sediments in size order; smaller grains advance faster, while heavier grains trail behind. Sorting occurs in both fluvial and aerial systems. River beds become finer grained downstream through a combination of grain abrasion, size-selective transport, and deposition. *See also* VARVE.

soundscape The landscape of sounds in which we live. Sound can give meaning to spaces and places, and can help orientate human lives. But to be sensed, sound has to be embodied.

Southerly Burster *See* LOCAL WINDS.

southern annular mode (SAM)
Also known as the Antarctic Oscillation, this term describes the north–south movement of the westerly wind belt that circles Antarctica, dominating the middle to higher latitudes of the Southern Hemisphere. It is a zonally symmetric or annular structure, with a large low pressure anomaly centred on the South Pole and a ring of high pressure anomalies at mid-latitudes.

In a positive SAM event, the belt of strong westerly winds contracts towards Antarctica. This results in anomalously dry conditions over southern South America, New Zealand, and Tasmania, and wet conditions over much of Australia and South Africa. The stronger westerlies above the Southern Ocean also increase the insulation of Antarctica. As a result, there is less heat exchange between the tropics and the poles, leading to a cooling of Antarctica and the surrounding seas.

sovereignty Conventionally, political authority exercised by a state over a given territory. **Domestic sovereignty** includes both authority and control but only within a state, while **international legal sovereignty** concerns the authority (government) of the state as recognized by other sovereign states. This legal personality can survive territorial and internal changes. **Westphalian sovereignty** implies the territorial organization of the state as an inviolate realm, free from intervention by other states, while **interdependent sovereignty** relates wholly to the state's control of its borders and its exposure to external influences. Sovereignty in the modern era of multi-tiered governance (devolution) is often murky and shifting' (see Rodden (2004) *PSOnline*). **Graduated sovereignty** is a spatially selective form of territorial governance, in which some slices of territory are more fully integrated into the world economy than others; or a state's differential treatment of the population when the state treats different citizens differently, based on ethno/class differences. See Ong (2000) *Theory, Cult. & Soc.* 17, 4; see also Park (2005) *Pol. Geog.* 24, 7.

sovereign wealth fund (SWF) An investment fund that is owned and controlled by a national government. SWFs are mainly located in Asia and in the Middle East. In Asia, they are generally 'non-commodity' funds whose resources originate from current and financial balance of payments surpluses and/or from fiscal (tax) surpluses. In the Middle East and Africa, they are primarily 'commodity' funds managing the incomes of oil and mineral exporting countries. However, the world's largest fund is in fact European—the Norwegian government pension fund—and it is considered to be one of the most honest, transparent sovereign funds in the world.

space The extent of an area, usually expressed in terms of the Earth's surface. The term **spatial** derives from this meaning; and spatial relationships are at the heart of geography.

Y.-F. Tuan (1997) contends that a space requires a movement from a place to another place. Similarly, a place requires a space to be a place. 'Place' refers to the process by which spaces are organized.

'**Absolute space** is fixed and we record or plan events within its frame . . . the **relational view of space** holds there is no such thing as space or time outside of the processes that define them' (Harvey 2006). Relational spaces only exist through the presence of other entities—ordinary things, or the basic things that make up the universe (Leibnizian monads). Processes define their spatial frame (the spaces they occupy). Absolute space is geometrically indexed, and **relative space** is object oriented.

For most, if not all, human cultures, there are different kinds of spaces, applied to different situations or phenomena that are thought of in different ways. And every society uses its space differently, both artistically and technologically: 'Europeans have a notion of time and space that is generally assumed by them to be universal. This is not necessarily so.

space-for-time substitution

Certain processes in geomorphology are too slow to be wholly investigated in one individual's lifetime, so the development of such a process may be observed by finding other sites where that process is exhibited at a different developmental stage. In ecology this means that past or future trajectories of ecological systems may be inferred from current spatial patterns. However, this approach assumes that the sites only differ in time and not in disturbance, or other underlying factors unrelated to time.

space of flows The communicational space created by new techno-economic networks; a new type of space that allows distant, but simultaneous, real-time interactions.

space–time 'Space and time are treated as sticky concepts that are difficult to separate from each other' (Dodgshon (2008) *Geografiska B* 20, 1). We need *time* to tell how an individual place developed, and *space* to understand the way different places develop. Yet Galton (2004) *Spat. Cogn. & Comput.* 4, 1 explains that in the real world there is rarely a clean separation between (spatial) objects and (temporal) events. 'For example, a meteorologist tracking a tropical cyclone will naturally view it as an object which comes into existence at a certain place and time and moves along a well-defined path, eventually fading away and ceasing to exist. From the point of view of an inhabitant of a town in its path, the cyclone is much more like an event: the sudden onset of strong winds and rain, bringing destruction to the town and then dying away leaving the inhabitants to pick up the pieces.'

Massey (1999) *TIBG* 24, 3 argues that space 'is not static, not a cross-section through time; it is disrupted, active and generative. It is not a closed system; it is constantly, as space-time, being made.'

For *geomorphologists, Raper and Livingstone (1995) *Int. J. GIS* 9 propose a **relational space–time**, since the identity of a geomorphological phenomenon is generated by the spatio-temporal relations between causal processes and the local nature of the environment. 'The relational notion of space-time implies internal relations; external influences get internalized in specific processes or things through time' (D. Harvey 2005). For a visualization of space and time in three dimensions using a hybrid (CAD/GIS) system, see Neutens et al. (2007) *Int. J. GIS* 21, 10.

spatial To do with geographic space; with distribution or location across a landscape or surface. See B. Warf and S. Arias (2008) on the **spatial turn** in the humanities and social sciences, an emphasis on the role of space and place. Fairly obvious to geographers, you might say, but there has also been a spatial turn for some historians; go to Torre (2008) *Annales. Histoire, Sciences Sociales* 63, 5.

spatial analysis (spatial data analysis) A geographical analysis which seeks to explain patterns of human behaviour and their spatial expression in terms of mathematics and geometry *See also* GEOCOMPUTATION.

spatial autocorrelation A clustering pattern in the spatial distribution of some variable which seems to be due to the very fact that the occurrences are physically close together, that is, that they are in geographical proximity. Spatial autocorrelation is widespread: rich people move to areas where other rich people live; people only go to parties because other people go, and so on. In any exercise using sampling, the problem with spatial autocorrelation is that if observed values are spatially clustered, samples will not be independent; see Ehlschlaeger (2000) *Computers Env. & Urb. Sys.* 24, 5 on measuring spatial autocorrelation.

spatial differentiation Geography can be defined as the study of spatial differentiation. Within geography, spatial differentiation is seen as:

- a process (see Krevs (2004) *Dela* 21 and Li and Wu (2006) *Housing Studs* 21, 5),
- a causal mechanism (see Walker (1978) *Rev. Radical Polit. Econs* 10, 3),
- a response (see Spark and Burt (2002) *Soil Sci. Soc. Amer. J.* 66).

The identification and measuring/mapping of spatial differentiation is fraught with difficulties. The European Union has proposed seven criteria for the differentiation of the European territory. Go to the Study Programme on European Spatial Planning—Final Report, 2001, where you will find the list of methods, together with the problems.

spatial diffusion The spreading of a phenomenon, such as a language, through geographic space. Models may give adequate statistical explanations for the volume, distance, and direction of geographical diffusion flows, but they do not reveal the causal factors.

spatial-interaction theory The view that the movement of people between places can be expressed in terms of the characteristics of each place, like population or employment rates. Look for Oshan (2020) *Prog. Hum. Geog.*, who explains four spatial interaction models; and see Greenwood and Hunt (2003) *Int. Reg. Sci. Rev.* 26, 1—it's older than Oshan's work, but is still an excellent review.

spatiality The effect of space on actions, interactions, entities, and theories. Spatiality is a social construct, not absolute but a product of the political economic system; it's a product of different forms of political power. The two major dimensions of spatiality are reach (extent) and intensity. The term is also used as a synonym for distribution, or spatial expression.

spatial monopoly The monopoly of a good enjoyed by a supplier over a marked area where no competitor exists. Spatial monopolies happen when firms agree to split the market spatially, or through the privatization of public utilities, as in the case of the British Water Boards.

spatial ontology The form of knowledge used in representing spatial concepts; the many ways people make geography. Martin (2013) *Prog. Hum. Geog.* 38, 3 writes of knowledge systems: 'each contains its own spatial ontology, its own rhythms, its own theories of matter, systems of circulation, and rules of exchange'.

spatial preference The choice of one spatial alternative, such as a housing area, or a shopping centre, rather than another. Rushton (1969) *Annals Assoc. Amer. Geogs* 2 is old, but very informative.

spatial randomness A distribution pattern of some variable where the observed spatial pattern is no more, nor less, likely than any other.

spatial science The measurement, management, analysis, and display of spatial information describing the Earth; both its 'natural' and socially constructed features. It emphasizes the study of aggregate spatial patterns, including spatial behaviour, within clear theoretical frameworks, using *quantitative methods to evaluate models and hypotheses. As such, it uses *cartography, *geodesy, *geographic information science, *hydrography, photogrammetry, remote sensing, or surveying.

spatio-temporal fix Capitalism is an imperfect system, which creates problems that need fixing. For example, over-*accumulation can result in surpluses, *dumping, idle capacity, lack of investment opportunities, and unemployment. The capitalist system fixes these problems by developing new (*temporal*) sites (*spatial*) that can absorb surplus labour and/or capital. In other words, a spatio-temporal fix is simply a term referring to a particular type of solution to capitalist crises via geographical relocation.

special economic zone (SEZ) An area within the territory of a *state where normal regulatory practices, such as foreign exchange and remittance restrictions, and export levels, do not apply. By 2002, there were more than 5000 SEZs in China aided by foreign investment. *See also* FREE TRADE ZONE.

speciation The concept of ecological speciation attributes Darwinian natural selection to the generation of biodiversity and, specifically, reproductive isolation. Speciation and specialization, long studied as distinct disciplines, one evolutionary, the other ecological, could well be one and the same thing.

species–area relationship The relationship between area and number of species: larger islands, for example, contain more species than smaller islands. As area increases, the number of species inhabiting those ecosystems increases; rapidly at first, but then more slowly. But Báldi (2008) *J. Biogeog.* 35, 4 argues that a heterogeneous habitat overrides the species–area relationship.

specific humidity The actual mass of water vapour present in a kilogram of moist air; in general terms, the mixing ratio.

spectacle A striking visual presentation. For example, Cinquegrani

(2010) looks at the ways in which early films, as much as photography and colonial exhibitions, transported the spectators to a version of India created by unchecked British imagination; the 'Imperial Spectacle'. Thomas (2007) *Cult. Geogs* 14, 3, 369 on the clothing (the 'lived garment') of Lady Curzon, Vicereine of India 1898–1905 is strangely fascinating.

speleothem A collective noun for depositional features such as *stalactites and *stalagmites. Most are made of calcareous rock. See Mattey et al. (2008) *Geophys. Res. Abstr.* 10, EGU2008–A–00461 on seasonal controls on speleothem growth.

sphere of influence A field/hinterland, or area of control.

The term is also used for the *urban field of a city.

spheroidal weathering The weathering of jointed rocks by water along the joints, so that shells of decayed rock surround isolated, unaltered corestones. See Viera et al. (2018) *Catena* 163 on the development of spheroidal weathering.

spillover In economic geography, an overflow, or spreading, of information from elsewhere; an indirect, unpaid-for effect of economic activity (also called an *externality), such as spillovers between university *research and development, and firms. **Knowledge spillovers** mean that firms can acquire knowledge and information created by others through expenditure on industrial R&D without payment. Movements of labour between firms, informal contacts, business meetings, close customer–producer relations, and face-to-face contacts are all mechanisms fostering spillovers. Academic spillovers are more localized than industrial spillovers. *See* PROXIMITY.

spit A ridge of sand running away from the coast, usually with a curved seaward end. Breaking waves produce a turbulent flow of water that dislodges sediment from the beachface, and create the *longshore currents that transport the sand along the beachface. Since the water beyond the end of the spit is usually deep, waves do not break, but are refracted and diffracted around the spit, curving it.

'The strategy of coastal management that seems more appropriate to the spit–tidal flat sedimentary system is a "do-nothing option". The sea has to be left free to overwash the spit, supplying sand to the tidal flat. The spit will start again to migrate uniformly westwards, even if this means that for some time it will be transformed into an island' (Ciavola (1997) *Catena* 30).

spodosol A soil of the *US soil classification. *See also* PODZOL.

spontaneous settlement An unsatisfactory term for a *squatter settlement, since it implies that squatter settlements are unplanned. They often are planned, although rarely legally.

spread (lateral spread) The *slumping of clayey sediment at the edge of *ice sheets.

spread effect The filtering of wealth from central, prosperous areas, to *peripheral, needier areas. The spread effect is the spatial equivalent of trickle-down economics. Spread-backwash effects were outlined by G. Myrdal (1957) and A. Hirschman (1959) in the late 1950s. 'Generally, spread effects are the positive effects of urban proximity for communities, and backwash effects are the negative consequences of proximity. According to Hirschman, the most important spread effect is the purchase and investments of the affluent region in the outlying region' (Ganning 2010).

spring tide See TIDE.

squall A storm characterized by sudden and violent gusts of wind. A **squall line** is a cluster of storm-bearing *convection cells along a non-frontal line, or belt. Squall lines form along an *occlusion, along an *inversion created by a flow of maritime air over a cool surface, or below an upper-air trough. Mohr et al. (2003) *J. Hydromet.* 4, 1 define squall line as a convective line with a trailing stratiform cloud anvil.

squatting The unlawful requisition of housing or land. **Squatter settlements** are illegally occupied, which necessarily involves insecurity of tenure, irrespective of the quality of the buildings and the prevailing physical conditions. Squatter settlements are not necessarily *slums because not all houses in a squatter settlement have to be substandard for it to be a classical squatter settlement. *Compare with* SLUM.

Squatter settlements grow because demand for cheap housing outstrips supply. Houses are often flimsy, and sanitation is grossly inadequate (in Indian urban areas, scarcely 20% of the population has access to water/flush toilets, power may be unavailable, roads are not metalled, and education and medical facilities are severely limited. Brazil's squatter settlements are characterized by a lack of access to credit and funding for urban infrastructure, and alarming levels of violence, particularly in poor areas.

It's heartening that Klaufus (2000) *J. Housing & Built Env.* 15, 4 focuses on the inhabitants of squatter settlements as individuals; similarly, Datta (2007) *Gender, Place & N. Cult.* 14, 2 describes the way different women in a Delhi squatter settlement experience different forms of spatial control, using multiple modes of resistance to different spatial controls.

stability The state of a parcel of air which, if displaced vertically, will return to its original position. Thus, if a parcel of air cools more on rising than the air which surrounds it, it becomes denser than its surroundings and therefore sinks. The atmosphere is **absolutely stable** when the environmental *lapse rate is less than both the dry and saturated *adiabatic lapse rates. Atmospheric stability is reinforced by *inversions. See Peppier (1988) *SWS Miscellaneous Publication* 104.

stable ecosystem An ecosystem that will maintain or return to its original condition after any disturbance. A stable ecosystem is one that is capable of constraining its own fluctuations within certain bounds. However, Bowker in A. Ong and S. Collier, eds (2005) argues that 'the stable ecosystem in equilibrium is a myth'.

stable population A population where fertility and mortality are constant. A stable population will show an unvarying age distribution and will grow at a constant rate. Stable population theory is based upon measuring the reported age–sex distribution against that of an appropriately chosen stable population, in order to validate census information.

stack An isolated islet or pillar of rock standing up from the sea bed, close to the shore. Initially, caves develop; some ultimately meet to form an *arch. With the collapse of the roof of the arch, a stack is left.

stadial A time when glaciers advanced and periglacial conditions extended, but not as significantly as in a *glacial. Stadial–interstadial transitions in the North Atlantic were most probably caused by changes in the thermohaline circulation in the North Atlantic Ocean.

S

stalactite A limestone column hanging from a cave roof. It grows as an underground stream deposits its dissolved load of calcium carbonate, and may extend far enough to meet a *stalagmite, thus forming a continuous column. Stalactites retain a record of environmental conditions; see Mayer et al. (2003) *Geophys. Res. Abstr.* 5, 12952.

stalagmite A limestone column, formed on a cave floor when underground water evaporates, depositing the calcium carbonate dissolved within it. R. Boch (2008) provides useful detail on stalagmite growth dynamics.

standard industrial classification A grouping of industries classified by a government.

(()) SEE WEB LINKS

• British government standard industrial classification.

standardized mortality ratio The ratio of observed to expected deaths. Julious et al. (2001) *J. Public Health Med.* 23, 1 criticize this statistic.

standard of living, index of An assessment of living standards using indicators such as access to health care, standard of education, house ownership, car ownership, take-home pay, and employment rates. As the standard-of-living index is a constructed measure, it does not have an absolute interpretation, so it's better to use it hierarchically.

standing crop The *biomass present at a given time in a given area.

standpoint The way in which something is thought about or considered. *See* SITUATED KNOWLEDGE; see also McDowell (1992) *TIBG* 17, 4, 399 on feminist standpoint theory.

star dune *See* SAND DUNE.

state A territorial unit with clearly defined and internationally accepted boundaries, an independent existence, and responsibility for its own legal system. The state thrives through simplification, making social relations legible so they can be controlled. 'Processes of globalization increasingly breach state borders and have led to questioning of the degree to which "fixed" territorial structures of the modern state are adapted to organizing, describing and analysing the contemporary world . . . The spatial foundations of the state are also undermined by formal international integration processes, most notably those of the supranational European Union' (Wise (2006) *Area* 38, 2).

steady state landscape A landscape whose input, output, and properties remain constant over time. Over long time scales a hilltop can indeed be in a geomorphic steady state. See Willet and Brandon (2002) *Geology* 30, 2 on steady states in mountain belts. *See also* DYNAMIC EQUILIBRIUM.

steam fog A shallow, wispy, smoke-like fog, formed when cold air passes over warmer water and is rapidly heated. Convection currents carry moisture upwards, which quickly recondenses to form fog. *See also* ARCTIC SEA SMOKE; FOG.

stem flow *See* INTERCEPTION.

step An abrupt, but short, break of slope, especially in a *glacial trough. Steps tend to be smoothed and striated upstream, as the ice grinds the bedrock, and craggy downstream, because of *quarrying. Step–pool sequences are common in stream channels; see Chin and Wohl (2005) *Prog. Phys. Geog.* 29, 3.

steppe The wild grasslands of Central Eurasia stretch from the Danube, through modern Ukraine and southern Russia, to the Caspian. The natural vegetation has by now been removed or much altered by cultivation and grazing.

stepped leader *See* LIGHTNING.

stepwise migration A type of migration which occurs in a series of movements, for example, moving to a town larger than the home town, but not directly to one of the city regions.

sticky Describing cities and regions which can attract and anchor income-generating activities. For the conditions under which some places manage to remain 'sticky', see Markusen (1996) *Econ. Geog.* 72, 3.

sticky spot An area of high drag at the glacier bed.

stillstand A time of tectonic inactivity.

stochastic Governed by the laws of probability.

stock An irregular igneous *intrusion, cutting across the strata of the *country rock. A stock is less than 10360 hectares in area.

stocking rate The number of livestock per unit area, a rate used to measure the ecological pressure of livestock.

stoichiometry *See* ECOLOGICAL STOICHIOMETRY.

stone pavement In a *periglacial landscape, pavements of large boulders on saturated land in valley floors.

stone stream A linear arrangement of rocks, seemingly developed through *freeze–thaw.

stone stripes *See* PATTERNED GROUND.

stoping The assimilation at depth of *country rock by an igneous *intrusion.

The heat of the intrusion melts the country rock which then mingles with the *magma. Žák et al. (2006) *Int. J. Earth Sci.* 95, 5 hold that roof pendants, stoped blocks, and discordant intrusive contacts are evidence of magmatic stoping.

storm beach A beach ridge situated well above the normal limit of high tides, formed as unusually large waves fling ashore shingle, cobbles, and boulders.

storm surge (sea surge) Sea surges occur when a *depression passes over the sea, subjecting the water to lower *atmospheric pressure than its surroundings. This causes the water level to rise and the surrounding water to sink; a fall in pressure of 1 mb will produce an increase in height of almost 1 cm. Such rises in the sea surface can be compounded by high winds.

Storm surges associated with *hurricanes are common in the relatively confined locations of the Gulf of Mexico and the Bay of Bengal. Climatic variability is a fundamental element in explaining the changing frequency and intensity of sea surges in coastal British Columbia.

stoss and lee Glacial landforms with a pronounced asymmetric profile—an abraded slope on the up-ice side (**stoss**), and a steeper rougher quarried slope on the down-ice side (**lee**) that has both abraded and quarried surfaces. Individual examples are commonly known as roches moutonnées.

The term may be applied to sand ripples.

strategic environmental assessment The consultation and monitoring of the likely impacts of any strategies on the environment. It is a key component of sustainable development, establishing important methods for protecting the environment, and extending opportunities for public participation in decision-making. 'SEA has been promoted as a promising instrument expected to be able to provide better informed, more credible and more broadly beneficial strategic initiatives, as well as more timely and clearer guidance for subsequent undertakings' (Kirchoff (2011) *UWSpace*). *See* ENVIRONMENTAL IMPACT.

strath terrace A wide, flat-floored valley, shaped mainly by degradation. See Ortega-Becerril et al. (2018) *Geom.* 319.

stratified Showing distinct layers. Glacial till is often stratified through sorting and redeposition by *meltwater.

stratigraphy The study of rock layers and their formation. It is thus mostly concerned with *sedimentary rocks, but not invariably.

strato-, stratus Layered cloud. *See also* CLOUD.

stratosphere A layer of the Earth's atmosphere, above the *troposphere, 50 km thick.

stratovolcano A cone-shaped volcano with a layered internal structure.

streamer Aeolian sand transport occurs primarily in the form of streamers—'wriggles' of sand—that are driven by distinct eddies travelling down toward the surface through the boundary layer and scraping across the bed while triggering saltation along their trail.

streamflow Discharge that occurs in a natural stream channel. Streamflow velocity varies with distance from bed; with distance from banks, downstream, and over time. Channel geometry affects turbulence, and thus velocity, and hydraulic jumps and drops affect

velocity and depth. Variations in streamflow affect the amount and type of geomorphic work a river can do. The driving force is gravity, which varies with channel gradient; the resisting forces are the viscosity of the water; the *Reynolds number; *Manning's roughness coefficient; and the *Froude number.

See Rantz et al. (1982) *US Geol. Surv. Water-Supply Paper* 2175, 631 on the measurement and computation of streamflow.

stream gauge (stream gage) A location equipped to measure the water surface elevation and/or the *discharge of running water. Equipment used includes: weirs, flumes, and cableways from which to suspend a hydrographer and/or current meter over a river.

stream order A method of classifying the different segments of a drainage system in order to investigate any possible relationship between these segments. Horton (1945), *Geol. Soc. Amer. Bulln.* 56, 3, 275 promoted stream ordering, but it is Strahler's modification of Horton's ordering system (Strahler (1957), *Trans. Amer. Geophys. Union* 38, 6, 913) that has been most widely adopted. Despite the variable sizes and nature of stream patterns, each stream network has an ordered internal composition. *See also* BIFURCATION RATIO.

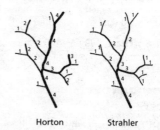

Horton   Strahler

Stream order

stream power This describes the energy available, measured in watts, in a particular area of channel. It is calculated at any given cross-section in a river by the equation:

$$\Omega = \rho g Q S$$

where Ω is stream discharge, ρ is the density of water (1000 kg/m^3), g is acceleration due to gravity (9.8 m/s^2), Q is discharge (m^3/s), and S is the channel slope.

Total stream power can be related to rates of bank or bed erosion.

Unit stream power (P) describes the energy available per unit width of channel, and provides a more detailed insight into morphological processes:

$$P = \frac{\rho g Q S}{w}$$

where w is water width in metres and the other variables are in the earlier stream power equation.

Unit stream power can be used to predict when particles are entrained and the rate at which they move downstream. It may also be used to assess the likely morphological impact of altering one or more independent fluvial variables.

⊕ SEE WEB LINKS

• EPA public access website with RIVERMorph program, which may be used to determine changes in unit stream power and thence to predict any resulting responses by the channel.

streamsink See DOLINE.

strength In geomorphology, the resistance of a rock mass to rupture under stress. Strength varies with, in order of importance: the spacing of joints, the cohesion and *frictional force of the rock, the *dip of any fissures, the state of weathering of the rock, the width of fissures, the movement of water in or out of the rock mass, the continuity of the fissures, and the amount of infilling of soil within the fissures. See Jackson (2002) GSA Today 12, 9 on the 'jelly sandwich' view of the continental lithosphere.

stress The force applied to a unit area of a substance measured in newtons per square metre. **Compressive stress** crushes the rock; see Bunds (2001) GSA Bull. 113, 7. **Tensile stress** is a force which tends to pull a rock or soil apart and which may cause fractures and pores to open; see Wiprut and Zoback (2000) Geology 28, 7.

striations Long scratches on a rock, a few millimetres across, formed by glacial abrasion, usually in clusters, often well developed on the stoss side of roches moutonnées. Striations can show the direction of glacier movement (Fjellanger (2006) Geomorph. 82, 3–4).

strip mining The removal of the overburden to extract minerals. See E. Reece (2006) on strip mining and the devastation of Appalachia.

Strombolian eruption Volcanic activity, relatively frequent and mild, in which gases escape at intervals, producing small explosions.

structural adjustment Changes to a nation's economy, usually imposed upon debtor nations by their institutional creditors; for example the *International Monetary Fund. Changes include: slimming down of government employment, cuts in taxation to stimulate investment, a reduction in public services, privatization of the economy, the promotion of exports; liberalization, and a reduction in government subsidies in order to bring domestic prices more in line with world prices and combat inflation (with great success in Bolivia, for example: Kohl

(2006) *Antipode* 38, 2). However, in countries that have undergone structural adjustment programmes, the majority of citizens have seen their standard of living fall.

Structural adjustment policies may be adopted unilaterally; see Wong (2007) *China & World Econ.* 15, 2.

structuralism This is based on the thesis that behind local cultural variations there are deep-laid foundations (structures): 'all manner of practices, objects, events, and meanings, are taken to exist not as discrete entities, but as parts relationally embedded within, and constituted by, underlying wholes' (Dixon and Jones in J. S. Duncan, N. C. Johnson, and R. H. Schein (2004)). For example, one society might kill to stop schoolgirls being educated while another bans female popes. In both cases, the deep structure is *patriarchy. See* POSTSTRUCTURALISM.

structuration theory A theory of the creation and reproduction of social systems that is based in the analysis of both structure and agents, with neither being more important than the other. This is the **duality of structure**, where both the agent and structure interact to bring about social change. Structure is not the main object of social research, but acts of structuration are. Lippuner and Werlen in N. Thrift and R. Kitchin, eds (2009) explain at length, but the abstract should be enough.

subaerial Occurring on land, at the Earth's surface.

subduction The transformation into *magma of a denser *plate as it dives under another, less dense, plate. A **subduction zone** occurs where rocks of an oceanic plate are forced to plunge below much thicker continental crust. As the plate descends it melts and is released into the *magma below the Earth's crust. Tectonic erosion of the overriding plate by the downgoing slab is believed to occur at half the Earth's subduction zones; see Vannucchi et al. (2008) *Nature* 451 who provide excellent diagrams.

subglacial Formed or deposited beneath a glacier. When an ice sheet flows over unfrozen sediment, it may cause **subglacial bed deformation**, as the water in the sediment pores increases enough to reduce the resistance between the individual grains. **Subglacial water flows** are maintained by high water pressure, rather than by melting of the subglacial channel walls. More than 70 **subglacial lakes** exist in Antarctica; the largest, L. Vostok, lies beneath 3 km of ice and is 230 km long.

subgraph In *network analysis, the graphs, or networks, forming an unconnected part of a whole graph or network.

sublittoral zone The zone from the lowest mark of ordinary tides to the end of the *continental shelf.

submarine slide A form of underwater *mass movement. 'Earthquakes are among the most frequently suggested triggering mechanisms for submarine slides . . . sliding in various forms is a major agent in the evolution of passive continental margin morphology' (Bugge et al. (1987) *Geo-Marine Letts* 7). See Yamamoto et al. (2007) *Island Arc* 16, 4; and Dawson and Stewart (2007) *PPG* 31, 6 on tsunamis generated by submarine slides.

sub-optimal location A satisfactory, but not the best, location.

subrosion Subsurface mechanical erosion.

subsequent streams Rivers running down the strike of usually weak *strata, or along the line of a fault, at right angles to *consequent streams.

subsistence farming A farming system where almost all of the output feeds and supports the household. See Pandi et al. (2007) *Ag. Sys.* 92, 1–3 for a detailed, practical analysis.

subsurface flow *See* THROUGHFLOW.

subtropical anticyclones Areas of high pressure formed when air, which has risen in the tropics, subsides in subtropical areas. The air is warmed adiabatically as it descends; therefore rainfall is unlikely. The subtropical anticyclone over the North Atlantic, often called the 'Azores' or the 'Bermuda high', influences weather and climate over the eastern United States, western Europe, and north-western Africa. Miyasaka and Nakamura (2005) *J. Climate* 18, 23 suggest a local land–sea–atmosphere feedback loop associated with a subtropical high.

suburb A one-class community, located at the edge of the city, with low rates of housing per hectare. Suburbs were originally made up of housing types not much different from those in the city, but the 'garden suburb' tradition came to dominate Western cities during the twentieth century. Suburbs became more regulated, and more middle class. Recently, many North American suburbs have adopted land-use controls that promote upper-class sprawl by reserving large areas for the construction of small numbers of expensive homes on spacious lots. This change in planning controls may be the result of pressure from the suburbs' residents, or because local government officials are keen to do as other communities are doing.

succession (plant succession) A series of groups of plants, at a particular site, over time. A complete plant succession begins with bare earth, but Sokołowska et al. 2020, *Catena* 187, detail an interesting succession from meadow to forest. The term **successional dynamics** applies to the shifts between successional stages of an *ecosystem.

Suess cycle *See* DE VRIES CYCLE.

sunbelt In the USA, the southern and western states, such as New Mexico, Arizona, and Florida.

sunspot A dark patch on the surface of the sun, usually occurring in clusters, and lasting about two weeks. Sunspot activity fluctuates in an eleven-year cycle. It has been suggested that the sun is 1% cooler when it has no spots, and that this variation in solar radiation affects the climates of the Earth; see Schrier and Vesteeger (2001) *Geophys. Res. Letts* 28, 5.

superadiabatic lapse rate A *lapse rate over the 9.8 °C/1 000 m dry *adiabatic lapse rate, rare in the free *atmosphere, but common just above land surfaces emitting strong terrestrial *radiation. When the environmental lapse rate is superadiabatic, the density of the atmospheric parcel relative to its surroundings results in a buoyancy force in the same direction as original vertical movement.

superimposed drainage If a drainage pattern develops on a non-resistant mass covering a buried, resistant bedrock, and if base level falls enough, the drainage will cut into the bedrock in the established pattern; see Douglass and Schmeeckle (2007) *Geomorph.* 84, 1–2.

S

superimposed ice Water that refreezes at the base of the snowpack. See King et al. (2002) *Annals Glaciol.* 34, 1, 335 on the detection of superimposed ice in northern Norway.

Super Output Areas (SOAs) A new geographic hierarchy designed to improve the reporting of small area statistics in England and Wales. The layers of SOA are:

Lower layer	Minimum population 1000; mean 1500. Built from groups of OAs (typically 4 to 6) and constrained by the boundaries of the Standard Table (ST) wards used for 2001 Census outputs.
Middle layer	Minimum population 5000; mean 7200. Built from groups of lower layer SOAs and constrained by the 2003 local authority boundaries used for 2001 Census outputs.
Upper layer	To be determined; minimum size c.25000.

(⊕) SEE WEB LINKS

- Names and codes for Super Output Area geography.
- Set of slightly revised Super Output Areas (SOAs) on NISRA Geography website.

superterrane *See* TERRANE.

supraglacial On the surface of a glacier. See Benn et al. (2000) *Int. Ass. Hydrol. Scis* 264 on the early stages of **supraglacial lake** growth.

surf The waves of the sea as they break on a shore or reef. On beaches, surf is formed between the breaker and *swash zones. See S. H. Sweeny (2008) on the geography of surfing.

The **surf similarity parameter** is based on a consideration of the wave characteristics and the gradient of the beach and may be used to predicted breaker type via the formula:

$$\xi = \frac{\tan \beta}{\sqrt{H_b/L_o}}$$

where $\tan \beta$ is the gradient of the beach and the subscripts $_b$ and $_o$ indicate breaker and deep-water conditions respectively. Small values for ξ are attained when the beach has a gentle gradient and the incident wave field is characterized by a large wave height and a short wave length (or a short wave period). Large values of ξ are found when the beach is steep and the incident wave field is characterized by a small wave height and a long wave length (or a long wave period). Spilling breakers occur for $\xi < 0.4$; plunging breakers for $\xi = 0.4$ to 1; and surging breakers for $\xi > 1$ (Masselink in J. Holden 2012).

surface In a *GIS, a surface has three dimensions—width, length, and a third dimension determined by the character of the surface.

surface boundary layer *See* BOUNDARY LAYER.

surface process model *See* LAND SURFACE PROCESS MODELS.

surge phenomenon *See* EDDY.

surging glacier A faster-flowing glacier. Normally, glaciers flow at 3–300 myr^{-1}, whereas a surging glacier may reach 4–12 kmyr^{-1}. Some glaciers always flow at surging speed; others surge periodically, suddenly accelerating to 10–100 times their normal velocity. Surging glaciers are closely crevassed, with compressional flow at the leading edge and *extending flow at the rear. The net result is a deepening of the long profile of the glacier. The surge will continue as long as high water pressure sustains rapid sliding.

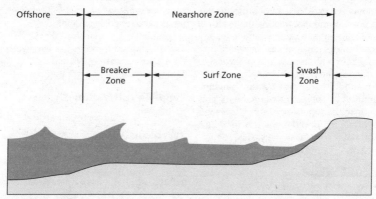

Surf zone terminology

susceptibility The degree to which an environment, landform, or organism is sensitive to a particular factor. A landslide zonation map divides the land surface into zones of varying landscape susceptibility, based on the estimated significance of causal factors in inducing instability. Such maps identify and delineate potentially unstable areas, so that suitable mitigation strategies can be used, and help planners choose favourable locations for siting development schemes. See Sandercock et al. (2007) *PPG* 31, 2 on a **vegetation-susceptibility index**, defining the force required to bend vegetation to 45° from the vertical. See Märker et al. (2008) *Geomorph.* 93, 1–2 on **land degradation susceptibility**.

suspension The state in which small, insoluble particles are evenly distributed within a fluid. Particles are carried upwards when *turbulence outstrips the force of gravity.

sustainable development
Development that avoids breaking the environmental cycles of renewal; conservation and generational concern. This would be development that Article 9 of the UN Declaration of Human Rights (1948–66) states that people should promote all over the world to assure dignity, freedom, security, and justice for all people. Sustainable development meets the needs of present generations without compromising the ability of future generations to meet their own needs. Thus, it requires a balance between economic prosperity, social equity, and the environment, to ensure quality of life now without damaging the planet for the future.

Advocates of '**strong sustainability**' argue that natural capital—the range of functions the natural environment provides for humans and for itself—should be afforded special protection, while those who espouse '**weak sustainable development**' hold that natural capital can be substituted by other forms of capital, especially produced capital. Weak sustainable development has been criticized for being an anthropocentric world-view, typically seeing sustainable development in terms of an environmental question, rather than also a developmental one.

'A **sustainable community** is one in which improvement in the quality of human life is achieved in harmony with improving and maintaining the health of ecological systems; and where a healthy economy's industrial base supports the quality of both human and ecological systems' (Indigo Development). A **sustainable drainage system** aims to mimic as closely as possible the natural drainage of a site to minimize the impact of urban development on the flooding and pollution of waterways. See Jones and MacDonald (2007) *Geoforum* 38, 3.

( ) SEE WEB LINKS

- UK government sustainable indicators.
- Indigo Development.

Sustainable Urban Drainage Systems (SUDS) Programmes designed to reduce the risk of small and medium-sized floods, and the pollution of urban rivers. Strategies include the creation of ponds, wetlands, and porous paving that can increase the storage of runoff in the catchment, and the improvement of water quality by physical, chemical, and biological processes.

sustained-yield resource A resource that can be extracted in a mutually advantageous long-term partnership with nature. Forestry may be managed as a slow-growing but renewable resource of fuel *and* timber and a sustained-yield recreational resource.

swale A long, narrow, usually shallow trough between ridges on a beach, running parallel to the coastline.

swallow hole *See* DOLINE.

swash The water moving up a beach from a breaking wave. The **swash zone** involves a moving land–water boundary travelling across the beach; at least some part of the beachface is usually unsaturated. Water depths in the swash zone are very shallow, and sediment is often transported as complicated, single-phase, granular-fluid flows. **Swash-aligned beaches** are parallel to the incoming wave crests and net *longshore drift is at a minimum.

swidden cultivation *Shifting cultivation.

symbiosis An association of two participants whereby both partners benefit, as when flowering plants rely on insects for pollination and the insects feed on their nectar. **Industrial symbiosis** describes organic relationships between dissimilar industries; see Lyons (2005) *Local Env.* 10, 1 on the use of waste products from one industry as input to another. See also Wu and Wall (2005) *Int. J. Tourism Res.* 7, on economic and environmental symbiosis in tourism.

symbolic capital The resources owned by an individual that serve as a marker of her or his status; the material display of social status. For the role of symbolic capital in Moscow in the transformation of Russian national identity, see Forset and Johnson (2002) *AAAG* 92, 3.

symbolic places A symbol is a concrete, tangible thing, like a building, or a statue, which delivers something intangible, like a feeling, or an idea. So symbolic places are places invested with meaning. Missingham (2002) *Urb. Studs* 39, 9 describes Government House, Thailand, in a time of popular protest by the poor: 'the streets outside Government House are both particular and symbolic. The symbolic and metaphorical meanings of mobilising on the streets and pavements served to highlight the poor's exclusion from the places of authority and power within the walls of Government House.'

sympatric Living in the same region; a term used in ecology to specify separate species whose territories overlap.
Sympatric speciation happens when two groups of the same species live in the same place, but evolve differently until, eventually, they are different species that cannot interbreed. And this without the aid of geographical barriers like rivers, deserts, or mountain ranges.

synchronic analysis The study of the internal linkages of a system at a given time, as in cross-sections in *historical geography (Conedera et al. (2007) *J. Hist. Geog.* 33, 4).

syncline and synclinorium *See* FOLD.

synekism The geographical relationships that create and give importance to cities; the power of urban agglomeration (E. Soja 2000). See Blake (2002) *J. Soc. Anthrop.* 2, 2 for Soja's explanations.

synoptic Overall, or relating to a large-scale or broad view. Synoptic meteorology is primarily concerned with large-scale weather systems, such as extratropical cyclones, but not on a global scale.

system Any set of interrelated parts; an abstraction that is assumed to exist in reality, unlike a *model which does not and is not intended to mirror reality.

Systems can be classified as open, closed, or isolated. **Open systems**—such as the ocean—allow energy and mass to pass across the system boundary.

A **closed system** allows energy but not mass across its system boundary. The Earth system *as a whole* is a closed

system (the boundary of the Earth system is the outer edge of the atmosphere). The Japanese *keiretsu* model represents a relatively closed system of vertically networked firms; see Gross et al. (2005) *J. Japanese Int. Economies* 19.

In a **cascading system**, a series of small sub-systems are linked from one system to another. Thus, the drainage basin system is made up of the weathering, slope, channel, and network subsystem, where the output from one is the input of other, and one subsystem affects another subsystem.

Terrestrial hydrology involves a dynamic, cascading system consisting of complex feeds forward and feedbacks, with interactions between any disrupted inputs or stores. In geomorphology, a **morphological system** is a subsystem of morphological components.

systematic geography The study of a particular element in geography, such as agriculture or settlement, seeking to understand the processes which influence it and the spatial patterns which it causes. Ye and Cai (2012) *Geog. Res.* 31, 5, 771 claim that 'systematic geographers stress that geography is a science pursuing the general rules, but regional geographers regard the region as the centre of geographical research'. This claim is worthy of debate.

systems analysis The study of the way that the component parts of a system work within it, and the way they interact. R. Inkpen (2005) points out that 'systems are still, however, a simplification of reality, not reality as it really is'. See R. Hugget (2005) for an excellent summary.

S

tacit knowledge The hidden, unspoken, unwritten, knowledge that we all possess, which we cannot pass on through writing, like riding a bike, or recognizing a face. This is knowledge that can only be transferred by face-to-face contact. Howells (2002) *Urb. Studs* 39, 5–6, 872 is very helpful on this.

tafone A hollow on a sheer face, produced by localized *weathering, mainly through *granular disintegration (pl. *tafoni*). Canopy-shaped, overhanging cave roofs are typical. In semi-humid to semi-arid regions, tafoni weathering is among the most effective weathering processes operating on surfaces of plutonic rocks such as granite. Data gathered by Roqué et al. (2013) *Geomorph.* 196, 94 support a subterranean origin of some tafoni. Hejl (2005) *Geomorph.* 64, 1–2 describes a prehistoric wall painting which displays stages of tafoni development.

taiga The predominantly coniferous forest, south of the *tundra in northern continents, also known as boreal forest. See Kuris and Ruskule (2006) *Baltic Env. Forum* on management of boreal forest habitats.

talik Within a *permafrost zone, the layer of unfrozen ground that lies between the permafrost and the seasonally thawed *active layer. Talik most often occurs below rivers and lakes, or where groundwater upwelling occurs along fractures. See Grosse et al. (2007) *Geomorph.* 86, 1–2. For the role of talik in the formation of pingoes, see Gurney (1998) *PPG* 22, 3.

talus A *scree slope formed of *frost shattered rock *debris which has fallen from the peaks above. A talus slope is usually straight, and at an angle of 34–35°. Slope angle in a talus cone reflects the balance between the supply of sediment from upslope (usually by physical weathering of bedrock) and the processes moving the sediment downslope while spreading it out. Lithology is also an important control See Obanawa and Matsukura (2006) *Computers & Geosci.* 32, 9 for a mathematical model for the topographic change of a cross-section of a talus landform. See also Shumack et al. (2017) *Geom* 299 on the quantification of controls on talus deposition. A **talus flatiron** is a slope profile, commonly in *Badlands, where soft sediments are covered by a caprock of resistant material. This upper section has a vertical free face, while the sediments below form a debris slope. See Desir, Marin, and Guerrero, *6th International Conference on Geomorphology*.

The slow, downslope movement of talus is **talus creep**, initiated either by the shock of new fragments falling on the scree or by the movement of individual particles resulting from heating and cooling.

taphonomy The study of the processes impacting an organism, especially fossilization, between the time of its death and its later discovery.

tariff A list of duties or customs to be paid on imports. Tariffs may be imposed to increase the cost of imported goods in relation to domestic production, thereby reducing the volume of imports and keeping the *balance of payments in credit, or to protect domestic industry from foreign competition. **Preferential tariffs** reduce import duties on products of a certain type or origin. In a highly technical paper, Mai et al. (2006) *CIRJE-F*-435 show that, when the transport cost is sufficiently small, tariff competition, coupled with the re-location of companies, leads to a core–periphery economy.

tax havens and offshore financial centres (THOFC) The term 'tax haven' applies to a small, low-tax jurisdiction specializing in providing corporate and commercial services to non-resident offshore companies, and for the investment of offshore funds, together with connotations of financial secrecy and tax avoidance. Despite the use of the acronym THOFC, a tax haven is *not* the same as an offshore financial centre, since many of the latter have no statutory banking secrecy, and most of which have adopted tax information exchange protocols to allow foreign countries to investigate suspected tax evasion.

taxocene *See* TROPHIC LEVEL.

taxon *See* TROPHIC LEVEL.

taxonomy A classification system, especially of organisms; the method by which scientists, conservationists, and naturalists classify and organize 'the vast diversity of living things on this planet in an effort to understand the evolutionary relationships between them' (Fenneman, E-Flora BC).

(()) SEE WEB LINKS
• Electronic Atlas of the Flora of British Columbia.

Taylorism A system of production characterized by the division of factory work into the smallest and simplest jobs, while closely coordinating the sequence of tasks in order to achieve maximum efficiency, as, for example, on a production line. Skilled managers and technicians oversee semi-skilled or unskilled workers who are engaged in simple, repetitive chores. Most people would say that the biggest change in work towards the end of the last century was the abandonment of Taylorism.

tear fault A *fault characterized by lateral movement, transverse to the strike of the rocks. See Huang et al. (2004) *Marine Geophys. Res.* 25, 1–2.

technics Technology; see Davies (2019) *PHG* 43, 3.

technological revolution The development in the last 25 years in information and telecommunications technology. It has made possible the worldwide integration of economic processes while allowing organizational structures to rapidly respond to changes in markets and technology; information technology has pushed forward *globalization. However, unilateral advantages do not last long, and latecomer institutions and capital are not locked into older technologies, and so can be quicker to take up, develop, and diffuse new technologies. S. Amin (2003) even refers to the 'obsolescence of capitalism' due to the technological revolution, as it will a make a lot of labour redundant. **Technological determinism**, simply put, is the idea that technology has important effects on our lives. Prominent in popular imagination, for example, is the idea that the internet is revolutionizing economy and society. Search for 'lecture notes on technological or media determinism' written by Daniel Chandler (which also feature daleks).

technopole A global, rather than local, *growth pole. Technopoles are of three major types: the industrial complex of high-tech firms; those emerging spontaneously, and technopoles linking research and development with manufacturing. See Fontan et al. (2004) *Canadian J. Urb. Res.* 13, 2 on the Angus Technopole in Montreal: 'a new type of community initiative which combines endogenous and community development with business oriented projects.'

technoscience The marriage of technology and science, with each of these using the other for support. A. Nordmann (2013) refers to a 'growing number of fields such as nanotechnology, synthetic biology, and climate studies where it is not possible to separate the scientific from the technological'. **Geographies of technoscience** consider the way that space matters in the production of science and technology; the situating of science and technology, and technosciences, in their social, cultural, economic, and political context.

technosol A soil largely formed by human activity, such as mineral processing; an artificial soil containing 25% or more of artefacts, or capped with technic rock within 5 cm of the soil surface. See Santini and Fey (2016) *Catena* 136.

tectonic Of, or concerned with, the processes acting to shape the Earth's crust.

telecommunications The electronic transmission of information over long distances. Geographers are interested in telecommunications because of the way they 'fold' (shorten) space, and because of their impacts on various economic,

political, and social phenomena. See Rimmer in F. C. Lo and Y.-M. Yeung, eds (1998) on transport and telecommunications among world cities, and try Guldman (2020) *Urban Science*, on the urban/rural telecommunications divide.

teleconnections Causal links between patterns of weather in two locations, or between two atmospheric occurrences, which are far apart, such as the *El Niño–Southern Oscillation, where unusually high sea surface temperatures off the west coast of South America can be connected with droughts in Indonesia, and the blocking-like pattern over the British Isles in summer which co-occurs with a low over south-eastern Europe. Teleconnections are usually expressed statistically by the amount of shared variance between the two locations or systems. See Simpson and Erkal (2004) *Papers Applied Geog. Confers* 27 on using teleconnections to predict temperatures.

teleology The theory that actions, objects, states of affairs, and so on should only be engaged in, and can only be justified, in terms of their end results. In this view, 'rightness' is not intrinsic in an action or process, but is dependent only on its consequences. H. Obenzinger (1999) claims that the field of 'Holy Land' studies has been controlled by colonialist teleology in much American and Israeli scholarship. S. Elden and E. Mendieta (2011) is worth a try, particularly chapter 10.

temperate Describing those locations and climatic types falling between subtropical and subarctic.

temperate glacier A glacier with basal temperatures of around 0 °C. Warmth is provided by friction with the bedrock, by shearing within the ice mass, and from *geothermal heat.

Glacier flow is rapid because warm glaciers are lubricated at their base, and velocities of 20–30 m per day have been recorded. Warm glaciers are the most effective agents of *glacial erosion.

temporality In contrast to the measurable, and calculated, idea of time, temporality is concerned with the way in which a sequence of events, of history, is physically experienced by those who live through them or experience them. In this way the passing of time is not treated as a neutral dimension, but is constituted by social practices. See Ho (April 2021) *Prog. Hum. Geog.*

tephigram An *aerological diagram, of two axes aligned at 45° to each other. The nearly horizontal lines show *atmospheric pressure, together with the heights at which they are found; the lines at 45° to the isobars show temperatures, running from bottom left to top right. Superimposed upon these two are three sets of guidelines: those running from

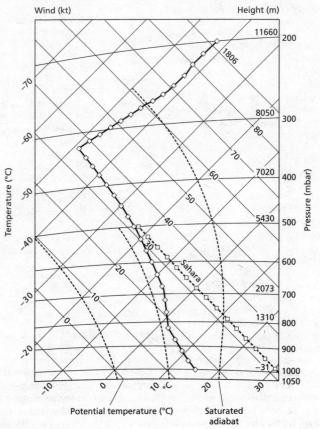

Tephigram

bottom right to top left show the dry *adiabatic lapse rate (DALR); those running from bottom left to top right, at about 60° to the base, show constant humidity (mixing ratio lines); and convex, curved lines indicate the saturated adiabatic lapse rate (SALR). With the help of these lines, the behaviour of a rising parcel of air may be predicted. First, the readings from a *radiosonde are plotted to show the environmental lapse rate. Next, from the same source, dew point temperatures are plotted against height. From the temperature and height of the air parcel under investigation, a plot is made of its temperature fall at the DALR, parallel to the DALR guidelines. Similarly, the dew point change of the air parcel with height is plotted, parallel to the mixing ratio lines. Where these two plots intersect, the parcel will begin to cool at the SALR, and a plot is made of the parcel parallel to the SALR guidelines. At the point where this plot intersects the environmental lapse rate line (above), the height of the top of any cloud formed by the parcel may be established.

tephra A deposit made of fragments of rock—from so-called 'bombs' to dust and *ash—shattered by an explosive volcanic eruption. The coarser, heavier particles fall out close to the volcano vent, while the finer dust may be carried hundreds of kilometres.

Tephrostratigraphy is the study of sequences of tephra layers and related deposits and their relative ages. It may be used as a tool for correlation in the study of stratigraphic sequences. See Hall and Mauquoy (2005) *Holocene* 15 on tephra dating, and Xia et al. (2007) *Env. Geol.* 51, 8 on correlating tephra layers.

terminal moraine *See* MORAINE.

terms of trade The ratio of the index of export prices to the index of import

prices. If export prices rise at a faster rate than those of imports, then there has been an improvement in the terms of trade.

A deterioration in the terms of trade resulting from higher or lower export prices can lead to either a reduction of investment, an increase in government indebtedness, or higher unemployment and lower wages. Go to the UNCTAD (Trade and Development) site.

Broda (2002) *Staff Report, Federal Reserve Bank of New York*, 148, on terms of trade and developing countries, is useful.

terrace *See* RIVER TERRACE.

terracette A small staircase-like landform generally less than 1 m in tread width and riser height, and as long as 300 m, running transversely along slopes. Explanations of terracette formation include animal disturbance, soil creep, solifluction (gelifluction), slumping and rotational slippage, regolith and vegetation control, subsidence, or through the impacts of people or animals (sheep like walking along terracettes!). See Brandon and Scroder (2011) *Zeitschr.* 55, 1, 45.

terracing The construction of a series of horizontal levels built on a hillside, especially in upland or piedmont zones, in order to retain water and reduce soil erosion. Terracing technology is intimately related to irrigation and the availability of water. Slopes of less than 8° are more suitable than steep slopes for changing into terraces.

terrain analysis (digital terrain analysis) The use of *remote sensing satellite data for mapping various aspects of terrain, such as land cover, land use, and soils. Software may then be utilized to derive terrain parameters, such as aspect, catchment area, and wetness index, which are then used to describe the morphology of the landscape and the influence of topography on environmental processes. See I. Florinsky (2011).

terrains of resistance The material and/or symbolic ground upon which action by social groups takes place. It is simultaneously both metaphoric and literal. R. Neumann (1998), for example, shows how Arusha National Park, Tanzania 'embodies all the political-ecological dilemmas facing protected areas in Africa'.

terrain skeleton A depiction of a valley created by drawing the valley lines that connect the deeper points in the valley, and the ridge lines that connect the highest points of the ridge. See Lui et al. (2018) *Geom.* 314 for some most helpful maps and diagrams, and Gűlgen and Gökgöz, Yidriz Tech. Univ. (2004) for a run-down of the methods.

terrain surface The landscape which may be simulated in a terrain surface display system. Smith and Mark (2003) *Env. & Plan. B* 30 identify six factors as topographic existences: elevation, terrain surface shape, topographic position, topographic context, spatial scale, and landform.

terrane In geology, a fragment or block of crust with an individual geological history that differs from the surrounding areas, and is usually bounded by faults. **Accreted terranes** are those that become attached to a continent by tectonic processes. If a relatively small crustal fragment is accreted to a larger continent, it is an **exotic terrane** (originating elsewhere). **Superterranes** are composite terranes, composed of individual terranes and other assemblages, which share a distinctive tectonic history; see Mezger et al. (2001) *J. Metam. Geol.* 19, 2.

SEE WEB LINKS
• Cordilleran accretionary terranes.

terra rossa A red intrazonal soil developed in Mediterranean regions on weathered limestone.

terrestrial magnetism *See* GEOMAGNETISM.

terrestrial radiation The heat radiated from the Earth. Short-wave *solar radiation reaching the Earth does not heat the atmosphere it passes through, but does heat the Earth's surface. In turn, and particularly on clear nights, much of this heat is radiated out from the Earth. The Earth also absorbs terrestrial radiation reflected from the overlying opaque atmospheric layer. It is by long-wave terrestrial radiation that the *atmosphere is heated. Almost one-third of the solar radiation intercepted by Earth is radiated back into space. In urban areas, a reduction in the *sky view factor decreases the loss of terrestrial radiation, thereby contributing to the creation of *urban heat islands (Grimmond (2007) *Geography* 30, 3).

terrier A written survey of an estate, recording the value and extent of the land. The earliest English example dates from AD 900, but they were extensively produced from the 16th century onwards. See Fletcher (1998) *TIBG* 23, 2.

territoriality The need by an individual or group to establish and hold an area of land is an urge, supported by force if necessary, to define a territory for mating and food supply. The individual needs security and identity, and this is shown most clearly in relation to the home. Additionally, the community requires a suburb or small town with which to associate. The importance of territory extends to larger units; the reorganization of the counties of Britain always causes distress. Territoriality is most often associated with nation-states, which have the formal power to demarcate, defend, and control their borders. Between 1816 and 1992, over half of all disputes began between neighbours, over a quarter involved territorial issues, and both contiguity and territory were involved in over half of all full-scale wars. See Bufon (2015) 'Territoriality and political geography'.

territorial justice The application of ideas of *social justice to an area of territory; that is, the identification by a government of areas of need, followed by a deliberate policy of redressing an imbalance ('levelling up'). Ideas of social justice vary according to the *mode of production and the prevailing ideology, and so territorial justice varies globally. See Morgan (2006) *Publius* 36 on territorial justice and the North–South divide.

territorial seas (territorial waters) The coastal waters together with the sea bed beneath them and the air space above them, over which a state claims *sovereignty. Most countries claim twelve nautical miles. In 1983, the Law of the Sea Convention proposed a 200-nautical-mile exclusive economic zone, but most nations are less than 400 nautical miles apart, so a median line is drawn between the baselines of the states concerned.

territory 1. In ecology, the living space of an animal which it will defend from the forays of other territorial animals. Animals need space in which to reproduce and their territory can be some or all of the following: a source of food, a source of mates, and a breeding area.
  2. The area claimed or controlled by a state or other political actor. A 'national territory'—the area of land seen to be inhabited by a nation—is based on claims to a particular space, usually backed up by references to history (actual or invented) and central to the nation's being. 'Territory comprises ancestors, knowledge, the use of plants, their evolution, the perception of the cosmos, customs and community and living history' (Ceceña (2004) *Antipode* 36, 3).

terroir A rather small area, whose soil and microclimate impart distinctive qualities to food products, and which is tied in with the marketing and cultural branding of food through its association with place. Gade (2004) *AAAG* 94 uses the French term *patrimonialisation* to describe a mesh of authenticity, heritage, and food as manifested in regional cuisine, and the protection of rural landscapes.

terrorism The use of unauthorized and spectacular violence and intimidation for political purposes.

(🌐) SEE WEB LINKS
• Website of the Handa Centre for the Study of Terrorism and Political Violence.

tertiary The first period of the *Cenozoic era, from about 65 to 1.8 million years ago, subdivided into the Palaeocene, Eocene, Oligocene, Miocene, and Pliocene *epochs.

tertiary industry Economic activity concerned with the sale and use of economic goods and services; service industry. Economically advanced countries have a high proportion of employment in services.

Tethys An ocean which developed during Palaeozoic and Mesozoic times, running from the coast of southern Spain to South-East Asia. Great thicknesses of sediment were formed since the sea floor kept subsiding at the same rate as deposits were laid down. These sediments were compacted, subjected to volcanism, and then uplifted and deformed by the earth movements. See Schreiber et al. (2011) *Geodin. Acta* 24, 1, 21 on the subduction of the Alpine Tethys.

text A group of practices for signalling meaning(s). This commonly means written texts, but has recently included economic, political, and social institutions, paintings, landscapes, and maps. Anthropologists view culture as a text. The main foci of texts in geography have been the interpretation of maps as cultural texts (see Kokkonen (1998) *Cartographia* 35, 3–4 on cartography in the Baltic Sea) and the nature and usefulness of texts as a metaphor.

textural analysis In *geomorphology, the study of the roughness of land surface. Nothing to do with literature! See Bugnicort et al. (2018) *Geom.* 317.

thalweg The line of the fastest flow along the course of a river. This usually crosses and recrosses the stream channel. Channel morphology can undergo rapid transitions along the thalweg, depending on changes in topography, bedrock structure, water, and sediment. For techniques to quantify thalweg profiles, see Bartley and Rutherfurd in F. Dyer et al. (2002).

theography The study of goddesses and gods. A geographer might study the effects of a theology, on a subject's spatial imagination, and their praxis. See Sutherland (2017) *PHG* 41, 3; not an easy read.

thermal erosion *See* THERMO-EROSION.

thermal expansion (insolation weathering) The rupturing of rocks and minerals mainly as the result of large daily temperature changes. The exterior of the rock expands more than the interior. Whether thermal expansion is effective in an environment with no water is open to question; see Koch and Siegesmund (2002) *Spec. Publ. Geol. Soc. Lond.* 205. See Wigley and Raper (1987) *Nature* 357 on the thermal expansion of sea water associated with global warming.

thermal inertia The slowness with which the temperature of a body approaches that of its surroundings. 'The Earth's climate system has considerable thermal inertia. The effect of the inertia is to delay the atmospheric and oceanic response to various climate forcings. The existence of thermal inertia implies that still greater climate changes will be in store for present increased atmospheric CO_2 levels, which may be difficult or impossible to avoid' (Lockwood in J. Holdem 2012).

thermal low An intense, low-pressure system caused by local heating of the Earth's surface, and leading to the rising of air by convection. Heavy rainfall will result if the air rises and cools enough for condensation to occur. The Iberian thermal low is thought to be the most prominent example of its type; see Portela and Castro (2006) *Qt. J. Royal Met. Soc.* 122, 529.

thermal pollution The contamination of cold water by warm water. Sources of heat include water used for cooling in electricity stations, the *urban heat island, and the construction of reservoirs. That warmer temperatures lower dissolved oxygen in the water, increase respiration rates of organisms, and increase fish and wildlife susceptibility to disease, parasites, and toxic chemicals is generally agreed.

Interestingly, Beser's 2007 paper (*Digital Repository U. Maryland*), in a study of the effects of the thermal effluent in Chesapeake Bay, finds that thermal effluent does 'not cause diversity differences between submersed aquatic macrophyte communities in different areas of the thermal regime'.

thermal stress The strains set up as a result of temperature changes. The term is also used to describe the impact of temperature changes on humans; Yan (1997) *Sing. J. Trop. Geog.* 18, 2 analyses climatic thermal stress on the people of Hong Kong.

thermal wind Not a real wind, but an expression of *wind shear for a given layer of *atmosphere; the *vector expressing the difference between the *geostrophic winds at the bottom and top of the layer. It is proportional to the thickness of the layer, and is directed along the isotherms, with cold air to the left in the Northern Hemisphere, and to the right in the Southern. However, the term is used to denote a wind developing as follows: the *pressure gradients which produce surface winds may be due to the presence of cold and warm air masses. The fall in pressure with height is rapid in cold air, and much less rapid in warm air. Thus, at height, air pressure in the cold air will be less than that in the warm air. This creates a high-level pressure gradient and, therefore, a wind, often described as the 'thermal wind'. The

strength of this wind is a function of its height and the temperature difference between air masses; the greater the difference, the stronger the wind.

Since there is a marked *meridional temperature gradient in the *troposphere, influenced at height by a powerful westerly factor, thermal winds are very strong at the point where the temperature gradient is greatest; at the polar *front. The result is the polar front jet. The force of a thermal wind may be strengthened by any pressure gradient at ground level. See Chu-shih (1963) *Acta Met.* 33 on thermal wind in a *baroclinic atmosphere.

thermocirque A large hollow on a hillside formed from the coalescence of *nivation hollows. See Kizyakov (2005) *Atlas Conferences Inc, Document #* capx-22 for the formation and retreat of thermocirques in the Arctic.

thermocline Also known as the metaliminion, the thermocline is the depth at which the temperature gradient of a large body of fluid water changes rapidly with depth. In an ocean, it is the transition layer between the mixed layer at the surface, which is of almost uniform temperature, and the deep water layer below. Within the lower

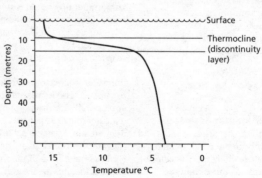

Oceanic thermocline

atmosphere, a thermocline occurs at the boundary between the *troposphere and the *stratosphere.

thermo-erosion The combined thermal and mechanical activity of running water in *periglacial conditions. See Etzelmüller (2000) *Zeitschrift Geomorph.* 4, on thermo-erosion in proglacial areas.

thermohaline circulation Also known as the **global thermohaline conveyor belt**, the Atlantic thermohaline circulation begins when the Gulf Stream (and its extension, the *North Atlantic Drift) bring warm, salty water to the north-east Atlantic, warming western Europe. The water cools, mixes with cold water coming from the Arctic Ocean, and becomes so dense that it sinks, both to the south and east of Greenland. This current is part of a larger system, connecting the Atlantic with the Indian and Pacific Oceans, and the Southern Ocean. Further sinking of dense water occurs near to Antarctica. Water from the two main sinking regions

spreads out in the subsurface ocean affecting almost all the world's oceans at depths from 1000 m and below. The cold, dense water gradually warms and returns to the surface, throughout the world's oceans. These surface and subsurface currents, the sinking regions, and the return of water to the surface form a closed loop. This phenomenon is probably more accurately termed the Meridional Overturning Circulation (MOC), for distinguishing that part of the circulation which is driven by salinity and temperature from that driven by salinity alone (as opposed to other factors such as the winds and tides) is difficult. See Wunsch (2002) *Science* 289.

thermokarst A landscape characterized by irregular hummocky topography and interconnected linear ridges, caused by the irregular heaving and melting of ground ice under periglacial conditions. The exact form of the depressions depends on the original distribution of ice segregations, the subsurface movement of water during warmer periods, and the presence or

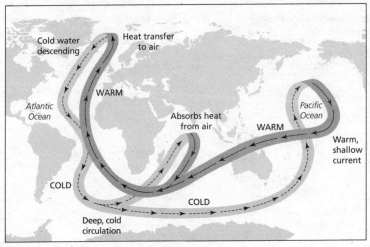

Thermohaline circulation

absence of water in the hollows, since water-filled hollows tend to perpetuate themselves. (The term 'karst' is used to indicate the numerous features formed by subsidence, and does not imply the presence or development of a limestone landscape.) Jones et al. (2013) US Geological Survey Open-File Report is a really useful resource.

thermosphere That part of the atmosphere, starting at about 85 km above the Earth (the top of the mesosphere), extending to the uttermost fringe of the atmosphere. Here, temperatures increase with height. Search for thermosphere on the excellent Climate Policy Watcher site.

thick description Describing a human action not merely to explain, but, additionally, to comment on and interpret it in the light of its context and structures that are complexly layered one on top and into each other so that each fact might be subjected to criss-cross interpretations. At least half of Gibson-Graham (2014) *Current Anthropology* 55, 59 should be of use.

Thiessen polygon A subdivision of a drainage basin, containing a rain gauge.

Polygons are constructed by first siting the rain gauges, plotting them on a base map, and connecting the sites by straight lines. The lines are bisected with perpendiculars, which meet to form the polygons. The areas of the polygons are calculated and expressed as fractions of the total area. Each fraction is multiplied by the precipitation recorded by its rain gauge. The sum of these calculations represents total precipitation over the catchment area. For an account, see Yoo et al. (2007) *J. Hydrol.* 335, 3–4.

thin region As compared to a region with *institutional thickness, a less favoured region. A region may be 'thin' if it is located far from relevant knowledge organizations; for example, the Finnish sub-region of Seinäjoki. Functionally, it may have small industrial milieux, a lack of relevant organizations, or its firms may lack collective learning. This functional thinness can be a consequence of a region's decision-making powers, financial resources, or policy orientation. See Zimmerbauer (2013) *Eur. Urb. & Reg. Studs* DOI: 10.1177/0969776413512842.

third food regime Also termed the *post-productivist transition, the third

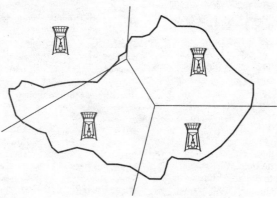

Thiessen polygon

food regime is organized by global corporations, profiting from internationally coordinated agri-food chains. It is characterized by the delivery of healthy and/or fresh food, and by supermarkets selling ready-meals and other own-brand goods. The third food regime is highlighted by the increasing globalization of production and consumption markets reflected, in part, by corporations which are geared to global or regional, rather than national, markets, and characterized by flexible production and adaptation to niche markets. The development of organic food production (and its creation of social and economic space) is an expression of the third food regime. See Birch and Lawrwence (2009) *Agric. Hum. Values* 26.

third space theory If first and second spaces are two contrasting, and possibly conflicting, spatial locations where people interact physically and socially, such as home (everyday knowledge) and school (academic knowledge), third spaces are the in-between, or hybrid, spaces. There, the first and second spaces work together to generate a new third space. Linehan and Gruffudd (2004) *TIBG* 29, 1 apply the term to the tented camps set up in inter-war Wales for retraining miners ('isolated from their wives and families, and separated from the rural society around them'); and, in a cheeky usage, Starbucks propose their premises as 'third space experience: the place between work and home'. See Moje et al. (2004) *Reading Res. Qt.* 39, 1 on third space in an educational context.

third world Originally, a synonym for those nations that aligned themselves with neither the West nor with the Soviet bloc during the Cold War. Today, the term is used to denote nations with the smallest UN Human Development Indices (HDI). There is no objective definition of Third World or 'Third World country', and these countries are also referred to as 'the South', 'developing countries', and 'least developed countries'. See Sidaway (2007) *PHG* 31, 3 on the decline of third worldism.

 SEE WEB LINKS
• The Third World Network.

thread A river channel. *Anastomosing (braided) river channels are multi-threaded. Richardson and Thorne (2001) *Geomorph.* 38, 3–4 argue that the division of the water in a single-channel river into multiple velocity threads is a prerequisite for development of *braiding.

threshold In a system, the 'tipping point' which brings about lasting, non-reversible changes, after which the system must develop a new equilibrium condition, adjusted to the characteristics of the altered controlling factors. If, thereafter, the system reverts to the original, balanced condition, it has not crossed a threshold. **External thresholds** are crossed as a result of a change in one of the external variables. For geomorphic thresholds on landslides see Mandal and Ramkrishna (2014) SpringerNat; for thresholds and gully incision, see Vandaele et al. (1996) *Geom.* 16, 2. Schumm (1979) *TIBG* 4, 4 argues that if the threshold conditions can be recognized, incipiently unstable landforms can be identified and their future development predicted, aiding the work of land managers and engineers.

threshold hillslope A hillslope steepened to its maximum stable, or threshold, angle. *See* CLIMATE THRESHOLD.

threshold population The minimum population needed to justify the provision of a certain good or service. This may be expressed crudely, in population numbers, but purchasing power may be a better yardstick for commercial goods or services. Note that threshold populations may vary regionally, and certainly vary nationally; the threshold population for a baker's shop is far lower in France than in the UK, for example. The actual evaluation of the threshold population for most goods and services is difficult. However, Coon and Leistritz (2002) *North Dakota State U., Agribus. & Appl. Econ. Misc. Rept* 191 have, with many caveats, attempted to do so.

throughfall *See* INTERCEPTION.

throughflow Also known as subsurface flow, this is the movement diagonally downslope of water through the soil, as opposed to the vertical movement known as *percolation. It may follow natural *percolines in the soil. Throughflow involves micro-*pores (less than 0.1 mm in diameter) and macro-pores, natural *soil pipes, displacement flow, and groundwater discharge. Throughflow is a major factor in the hydrology of a drainage basin where the rocks underlying the soil are impermeable. See Borga et al. (2002) *Hydrol. Procs* 16.

throw Of a *fault, the vertical displacement of strata along a fault line.

thrust A movement causing the formation of a reverse *fault of a very low angle. The **thrust plane** is the low-angle fault face over which movement occurs. Rubin et al. (2001) *Eos* 82, 47 write on Taiwanese **thrust faults**. A **thrust belt** is a geological formation caused by compressional tectonics, which ultimately results in the formation of large mountain ranges. The Patagonian fold and thrust belt is an example (Suarez et al. (2000) *Geol. Mag.* 137, 4).

Thrust belts present enormous potential for tapping into oil and gas deposits; see Johnson et al. (2007) *The Leading Edge* 26, 2.

thunder When a stroke of *lightning passes through the atmosphere, the air becomes intensely hot, perhaps to 30000 °C. The violent expansion thus caused makes a shock-wave heard as thunder.

thunderstorm A storm including strokes of *lightning, which cause the thunder, and draw off electrons earthward as part of the atmospheric electrical cycle.

tidal energy Energy based on the motions of the tide. See Howell and Drake (*FERN technical report# 2012.1*) on the socio-economic effects of tidal energy.

tidal prism The volume of water exchanged between a lagoon or estuary and the open sea in the course of a complete tidal cycle. **Tidal asymmetry** describes differences in magnitude between ebb- and flood-tidal currents. The tidal prism is also significant in the size and configuration of ebb-tidal deltas (Fitzgerald et al. (2002) *Geomorph.* 48, 1–3).

tidal range In coastal geomorphology, the difference in height between high and low tides. **Micro-tidal range** is less than 2 m, **meso-tidal range** is 2–4 m, and **macro-tidal range** is more than 4 m.

tidal wave *See* TSUNAMI.

tide Tides result from the gravitational pull of the moon and sun on the Earth (which also affects land masses, but the reaction of the water is greater and more apparent). The greatest effect of this pull is on each side of the Earth as it faces the moon. The moon 'pulls out' two bulges of water, which are fixed—the Earth moves through them, causing high water twice daily. The sun also attracts water. When the effects of both sun and moon coincide, twice monthly in the second and fourth quarters of the moon, high **spring tides** occur. When the sun and moon seem to be at right angles to each other from the Earth, the forces of moon and sun are opposed to each other, and lower, **neap tides** result. The **dynamic tide theory** suggests that the global tidal water motion is broken up into a large number of tidal systems, constrained by the coastal topography, and known as amphidromes. The tides travel around the centre of the amphidrome as a wave.

A **tidal notch** is an incision running along the base of a cliff, cut by wave or bio-erosion. Tidal notches may be used as indicators of tectonic movement and sea level change. See Moses et al. (2017) *EGU General Assembly*.

till The sediment deposited directly below a glacier, which exhibits a wide range of particle sizes, from fine clay to rock fragments and boulders. The rate of **till flow** beneath a glacier is thought to depend on basal shear stress and on the effective pressure, defined as the difference between overburden pressure and the interstitial pore water pressure within the sediments. See Evans et al. (2006) *Earth Sci. Revs* 78, 1–2. Where glaciers overlie soft sediments, deformation may rearrange and reorientate particles to form **deformation till**, and the examination of these tills can reveal structures that allow the strain history to be reconstructed. See Murray (2012) in J. Holden, ed., p. 490.

tilt-block *See* BLOCK.

time–space compression The impact of new systems of transport and communications as experienced by the individual. The major periods of time–space compression are from about 1850 to 1914, and from the late 20th century onwards, when there occurred a radical restructuring in the nature and experience of both time and space. Janelle (2003) UCSB colloquium provides the following data for Los Angeles to Santa Barbara:

- 500 minutes apart in 1901
- 100 minutes apart in 2001.

time–space convergence Places are separated by absolute distance and by time. With improvements in communication systems and methods of transport, this time-distance diminishes. Some geographers express this as the 'folding' of time/space. In essence, time–space convergence means that the *friction of distance—a concept fundamental to conventional *central place theory, *diffusion theory, and *location theory—is lessening. Speed permeates the history of transport and modernity, but it does so in multiple ways.

time–space geography Time–space geography provides a method of mapping spatial movements through time, and may be based on four propositions:

- space and time are scarce resources which individuals draw on to achieve their aims
- achieving an aim is subject to *capability constraints, *coupling constraints, and authority constraints
- that these constraints interact to demarcate probability boundaries
- that choices are made within these boundaries.

Ahmed and Miller (2007, *J. Transp. Geog.* 15) propose 'analytical formulations for basic time geography entities and relations, specifically, the space–time path, prism, composite path-prisms, stations, bundling, and intersections'. Raubal et al. (2004) *Geografiska B* 86 also propose a spatio-temporal theory of location-based services, which they claim to be closer to the individual user and more plausible with respect to their daily life. A **time–space prism** is a representation of the constraints limiting the time within which the individual can act.

TNC *See* TRANSNATIONAL CORPORATION.

Tobler's first law of geography This 'law' holds that everything is related to everything else, but near things are more related than distant things. See Klippel et al. (2011) *AAAG* 101, 5, 1011 for an enquiry into this 'law'.

tolerance The ability of an organism to survive environmental conditions. The prefixes *eury-* and *steno-* refer to wide and narrow ranges of tolerance respectively. An organism can be widely tolerant of one factor, such as temperature, but narrowly tolerant of another, such as salinity. The **climatic tolerance thesis** is that boundaries to the distribution of life forms and species often coincide with isometric lines of climatological variables. See Ruggiero (2001) *J. Biogeog.* 28, 10.

tombolo A spit, resulting from *longshore drift, which joins an offshore island to the mainland.

topography The arrangement of the natural and artificial physical features of an area: a **topographic(al) map** indicates, to scale, the natural features of the Earth's surface, as well as human features, with features at the correct relationship to each other.

topological map A map designed to show only a selected feature, such as the stations on the London Underground. Locations are shown as dots, with straight lines connecting them. Distance,

scale, and relative orientation are not important. The London Tube map is just one example of a topological map, for the connections are important, and not the precise location of nodes and routes. Try Remolina and Kuipers (2004) *Artific. Intell.* 152 on a general theory of topological maps.

topology This explicitly defines spatial relationships between connecting or adjacent features in geographic data; the arrangement that defines how point, line, and polygon features share coincident geometry. In geography, topology generally refers to the location of nodes and their connectivity to other nodes to form *networks. In a geographic information system, **topology** is a set of rules that define the relationship between points, lines, and polygons. ESRI enables topology generation within its geodatabase feature classes.

toponymy The study of place names, and what they can reveal about the history or nature of a place. Street names and signage help in encountering and experiencing a city and its socio-political and cultural contexts. See Bigon and Zuvalinyenga (2021) *Urban Geog.* 42, 2.

topophilia The feeling of affection which individuals have for particular places. Environmental designers have long exploited the basic ideas of topophilia to create attractive surroundings, based on 'comforting' materials, sensory stimuli, and images of place and environmental settings that are soothing and/or associated with healing. See White (2007) Fac. Env. Earth, & Resources, U. Manitoba. Look, too, for González (2005) *Space & Culture* 8, 193 on topophilia and its reverse, topophobia.

topple The forward rotation, out of the slope, of a mass of soil or rock about a point or axis below the centre of gravity of the displaced mass. See Spreafico et al. (2017) *Geomorph.* 288.

topsoil The cultivated soil; the surface soil as opposed to the subsoil. **Topsoil erosion** occurs when the topsoil layer is blown or washed away; it takes approximately 500 years for one inch of topsoil to be deposited, but there are 25 billion tons of topsoil lost each year. See Evans (2006) *CABReviews* 1, 30 for a review of sustainable practices to limit soil erosion.

tor An upstanding mass of rocks or boulders which rises above the gentler slopes which surround it. Tors were probably formed as a result of deep weathering and stripping in preglacial times, with additional sediment removal by mass movement under periglacial conditions in interglacial and interstadial times (Stroeven et al. (2002) *Geomorph.* 44, 1–2; see also Hätterstrand and Stroeven in the same issue).

tornado A destructive, rotating storm under a funnel-shaped cloud which advances over the land along a narrow path. The tangential speed of the whirling air may exceed 100 m s^{-1}, the core may perhaps be 200 m across, the duration of the storm about 20 minutes. Within a tornado, the central pressure is around 100 mb below that of the exterior; this may cause buildings to explode. This type of storm is very common in 'tornado alley', extending from northern Texas through Oklahoma, Kansas, and Missouri, with as many as 300 tornadoes a year. The exact mechanism of its formation is not fully understood, but tornadoes are associated with intense local heating coupled with the meeting of warm, moist air from the Gulf of Mexico and cold air from the *basin and range area of the

western United States. See H. B. Bluestein (2006).

SEE WEB LINKS
• The enhanced Fujita scale, which uses 28 tornado damage indicators.

total fertility rate *See* FERTILITY.

tourism Making a holiday involving an overnight stay away from the normal place of residence. Tourism has grown to become the world's second largest industry, directly accounting for 3.8% of global GDP. The mushrooming of **international tourism** may be explained by high levels of disposable income and longer holidays in *more economically developed countries; the development of package holidays, which reduce risk; cheap, mass air transport; and place myths, which persuade the tourist that the local culture they see represents the 'real thing'. Of critical importance has been the internationalization of finance: credit and debit cards, travellers' cheques, and hotel vouchers. **Creative tourism** means allowing visitors to participate in events specific to the city or area visited; see G. Richards and J. Wilson (2007).

tower karst Limestone towers, 30–200 m high, with nearly vertical walls and gently domed or serrated summits. The towers stand above large, flat *flood plains and swamps and show undercutting from rivers and swamps. Tower karsts are thought to represent the last remnant of a limestone outcrop. See Zhu and Chen (2006) *Cave & Karst Sci.* 32, 2–3 on the tower karst (*fenglin*) in southern China.

trace As defined by J. Anderson (2010), traces are marks, residues, or remnants left in place. They may be on material such as buildings, signs, statues, or graffiti; or non-material such as activities, events, performances, or emotions.

traction load *See* BED LOAD.

trade The movements of goods from producers to consumers. The classic explanation for trade is expressed in terms of *comparative advantage. Comparative advantage models suggest that an abundant human capital endowment, as well as a large domestic market, improves the quality of manufacturing exports, while lowering trade costs leads initially to increased concentration and then to dispersion of production. When a pattern of comparative advantage exists, integration may lead to international specialization of production. 'This may be good news for peripheral countries, which may be able to retain industry despite the attraction of the core' (Forslid and Wooton (2003) *Rev. Int. Econ.*).

Baier and Bergstrand (2005, *Atlanta W. Paper* 2005-3) answer the question 'do free trade agreements increase members' international trade?' with a *yes!* See Hummels and Hillberry (2005) *NBER W. Paper* 11339 on the impact of distance on trade.

trade bloc A large area of *free trade, formed through tariff, tax, or trade agreements, such as the *EU. 'Mercosur' is an economic and political agreement among Argentina, Brazil, Paraguay, Uruguay, and Venezuela (suspended December 2016) to promote the free movement of goods, services, and people among member states. Mercosur's primary interest has been eliminating obstacles to regional trade, such as high tariffs and income inequalities (Klonsky et al. (2012) *Council on Foreign Relations*). *See also* BLOC.

trade wind inversion A temperature *inversion, found in the tropics at heights of between 3000 and 2000 m, caused by descending air of the *Hadley circulation. This inversion acts as a ceiling for pollution; *see* PHOTOCHEMICAL SMOG.

trade winds The tropical *easterlies, blowing towards the equator from the *subtropical anticyclones at a fairly constant speed. They are at their strongest around the equatorial flank of these *highs. Trades are not true *zonal winds; their zonal character is weaker in summer because of the south-west *monsoon. They are most regular around 15°, and are associated with fine weather resulting from the *anticyclonic subsidence of the Hadley cell. Over the equator and the western oceans, **trade wind weather** is rainier; see Snodgrass et al. (2006) *12th Conf. Cloud Phys.* Liu (2002) *Science Daily* suggests that shifts in Pacific trade winds may support *El Niño formation.

traffic The movement of people and vehicles along a route. **Traffic capacity** is the maximum number of vehicles which can pass over a route in a given time, while **traffic density** is the existing number of vehicles. See Barthelemy et al. (2007) *Advances Complex Sys.* 10, 1 on the role of geography and traffic in complex networks.

traffic segregation The subdivision of towns and cities into certain units where road traffic is restricted and pedestrians predominate. See Bickerstaff and Walker (2005) *Urb. Studs* 42, 5 on local transport planning.

tragedy of the commons G. Hardin (1968) described an increase in the use of common land by a number of graziers, with each grazier continually adding to his stock of animals for as long as the marginal return from each animal is positive, even though the average return for each animal is falling, and even though the quality of the grazing deteriorates. Hardin used this metaphor to describe any situation where the interests of the individual do not coincide with the interests of the community, and where no organization has the power to regulate individual behaviour. Bloom (2002) *Agric. Hist.* 76, 3, 497 writes movingly of an *American Tragedy of the Commons: the Cherokee Nation*, 1870–1900. See also Hawkshaw et al. (2012) *Sustainability* 4.

transferability The capacity of a good to be transported. Transferability is largely determined by transport costs: as economic distance increases, transferability decreases. Since *economic distance and *intervening opportunities vary, transferability may change over time. See Chang (2004) *J. Geographical Sys.* 6, 1.

transfer costs Total transport costs involved in moving a cargo including extra costs such as tariffs and insurance. Transfer costs are highest for people, because of the very steep cost of insurance. If there is just one market and just one material source, transfer costs can be minimized by locating the processing unit at one of those two points and not at any intermediate point; see McCann (2005) *J. Econ. Geog.* 5, 3.

transform fault *Faults which are parallel to the arc of sea-floor spreading. See Greg et al. (2006) *Geology* 34, 4.

transfrontier zone A space which straddles a land or sea frontier—whether political, environmental, or socio-economic. See Bunnell et al. (2006) *Int. J. Urb. Reg. Res.* 30, 1 on transfrontier zones and resistances in Bintan, Indonesia, and Duffy (2006) *Polit. Geog.* 25, 1 on the politics of **transfrontier conservation** areas.

transgender people (trans people) People whose sense of personal identity and gender does not correspond to their sex at birth, or does not otherwise conform to conventional notions of sex and gender. Dodd (2021) *Geog. Compass* 15, 4 explores the way trans people's lives have been conceptualized and researched in human geography.

transgressive In geomorphology, describing a rise in sea level relative to the land.

translocal There seems to be little or no consensus on this term: 'across space, but with no prioritising of either the local or the global'; 'enduring, non-linear, close'; 'specifically local, yet spatially global'; 'not confined to the nation-state'; 'a diverse spatial concept'. Frankly, the last one is your best bet.

transmission loss The loss of flow from a river channel, due to downward *percolation and high rates of *evaporation.

transnational capitalist class (TCC) The owners of the leading worldwide means of production as embodied in the transnational corporations; an ungrounded and deterritorialized class.

Transnational corporate interlocking is less about intercorporate control than it is about the construction of an international business community, and these elites have long been associated with the global city: embedded at work, and disembedded at home. However, Burris and Staples (2012) *Int. J. Comp. Soc.* 53, 4, 323 conclude that a transnational capitalist class is very far from being realized on a global scale.

transnational corporation (TNC) A firm which owns or controls production facilities in more than one country, through direct foreign investment.

Transnationals are made possible by improved international communications which provide rapid containerized trans-shipment and foreign travel; easy communication of information; and international mobility of capital. TNCs have the potential to 'take advantage of geographical differences in the availability and cost of resources, and in state policies, and to switch and re-switch operations between locations' (Dicken (2004) *TIBG* 29, 1), and when one market is saturated, the transnational can rapidly develop others, since foreign investment cuts transport costs, and makes possible a rapid response to local markets. TNCs are probably the major force affecting worldwide shifts in economic activity, since the largest have a turnover greater than the *GNP of many *less economically developed nations.

However, the ability of transnational corporations to move capital across frontiers, and to manipulate the transfer prices at which their component firms exchange goods and services, has been held against them. Transnationals have also been criticized for undermining national cultures through intensive use of advertising to substitute their synthetic and standardized goods for natural and distinctively local alternatives.

Some claim that TNCs undermine states, or, indeed, the state system, but Dicken (2004, *TIBG* 29, 1) argues that 'states still have power over what they permit'.

transnationality There seems to be no general agreement on the definition of transnationality; its aspects include international cooperation on a level other than exclusively between nation states; cooperation not only with neighbouring countries, but rather with regions that suffer from similar difficulties and problems; and relating to a transnational region, such as the North Sea Region (see the Northern Periphery Programme).

A **transnationality index** is composed of the number of private transnational relations, the number of times an individual had been abroad in the twelve months before the time of the interview, and the total time the individual had lived abroad for a longer period. For a company, the UNCTAD transnationality index (an indicator of the share of a company's activities taking place in a country other than its own) is the unweighted average of the following ratios: foreign assets/total assets, foreign sales/total sales, and foreign employment/total employment.

(((●))) SEE WEB LINKS
• The Northern Periphery Programme.

transnational municipal networks (TMN) Networks of municipalities that operate nationally and transnationally, directly representing and involving cities in policy issues at the international and European levels, and across national borders: the transnational activities of municipal urban governments. TMNs became a pivotal feature of mainstream policy in the European Union from the mid-1980s. **Transnational urbanism** represents the ways in which cities are ever more defined by all sorts of connections to faraway places (A. Latham et al. 2009).

transport costs As long as transport costs are low, firms can benefit from concentrating in a single location and delivering products to farmers in another location. Alonso-Villar (2005) *J. Econ. Geog.* 5, 5 finds that the effects of cost reductions in transporting final goods are different from those in intermediate goods. For **feeder costs** (the costs of transferring freight from large ships/vehicles/aircraft to smaller ones) and **handling costs** (the costs of physically shifting the freight), and marginal cost/revenue analysis, see Konings (2007) *J. Transp. Geog.* 15, 6.

Knowles (2006) *J. Transp. Geog.* 14, 6 adds the concept of **cost/space convergence** to *time–space convergence: time/space and cost/space have both plummeted in the last 50 years, due to innovations in air, rail, road, and sea transport technology, communications, and handling technology. Small but incremental changes can have significant consequences for time/cost–space convergence or divergence, for tardy responses and inefficient decision-making retard change. The differential spatial effects of cost/space convergence reflect a more unequal world, and are largely due to unequal investment in modal capacity, and routes and terminals at local, national, and international scales. This tends to enhance the importance of the largest demand centres in developed countries, and of nodal centrality. See Link (2005) *J. Transp. Geog.* 13, 1 on the **social costs of transport**: pollution, noise, and accidents among them.

transport geography A branch of human geography concerned with: transport policy practice and analysis, especially the impacts of deregulation, privatization, and subsidy control; infrastructure impact on trip-making, the spatial economy, and regional development; technological innovation in transport and telecommunications and global and regional economic integration; the growing mobility gap between rich and poor and differential accessibility to jobs and services; transport, environment, and energy; travel, recreation, and tourism; and spatial and behavioural aspects of modelling transport demand.

travel-to-work area (TTWA) In theory, a self-contained labour market area is one in which all commuting occurs within the boundary of the area. In practice, there are entirely separate labour market areas. For those involved in labour market analysis and planning, it is useful to be able to define zones in which the bulk of the resident population also work, so, by applying a multi-stage allocation process, the UK Office for National Statistics has defined 'Travel to Work Areas' as approximations of self-contained labour markets. The basic criteria used for defining a TTWA are that, of the resident economically active population,

- at least 75% actually work in the area,
- of everyone working in the area, at least 75% actually live in the area.

The definitive minimum working population in a TTWA is 3500, but many are much larger—indeed, the whole of London and surrounding area forms one TTWA.

((⊕)) SEE WEB LINKS
- National Statistics, Travel to Work Areas.

trellised drainage *See* DRAINAGE PATTERNS.

triangular irregular network (TIN) A representation of a surface in a *GIS created through digital data. It is made up of irregularly distributed nodes and lines with three-dimensional coordinates (x, y, and z). See Lee (1991) *Int. J. Info. Systems* 5, 3, 267 for methods of building TIN models from grid *digital elevation models.

triangulation In the social sciences, a form of cross-checking, using multiple people's perspectives, or multiple research gathering methods, to check that your interpretation is robust and not wholly a subjective construct. Forms include data, investigator, method, and

Triangular irregular network
Source: http://www.ncgia.ucsb.edu/cctp/units/
unit06/tin.gif

theoretical triangulation; see K. Hoggart
et al. (2002) on researching human
geography, and Farmer et al. (2006)
Qual. Health Res. 16 on developing and
implementing a triangulation protocol.

Triassic The oldest period of *Mesozoic
time stretching approximately from 225
to 190 million years BP.

SEE WEB LINKS
• An overview of the Triassic period.

trigger In geomorphology, an event
which puts an end to the balance
between driving and resisting forces and
thus sets the mass moving. P. Fookes
et al. (2005) consider the main
predisposing and triggering factors of
geomorphological surface processes,
including climate and weathering,
sedimentology, tectonics, and
stratigraphy. Jones and Omoto (2000)
Sedimentol. 47, 6 outline the criteria that
can be used to identify a **seismic
triggering** agent. See also Moro et al.
(2007) *Geomorph.* 89, 3–4. For an
example of atmospheric triggering, see
Miniscloux et al. (2001) *J. Appl. Climat.*
40, 11. *See also* THRESHOLD.

trope A figurative or metaphorical turn
of phrase, used in social constructions of
gender, race, imperialism, and so on.
Almost a cliché.

trophic level An individual layer on
the *pyramid of numbers which
represents types of organisms living at
parallel levels on *food chains. All
herbivores live at one level, all primary
carnivores on the next level, all
secondary carnivores on the next level,
and so on. The animals on each level are
remarkably distinct in size from those on
other levels; there is a clear jump in size
between an insect and a bird, for
example.

The trophic biology of a taxon
(species/family/class) fundamentally
constrains its ability to convert
productivity into individuals. Predator
taxa require at least an order of
magnitude more *net primary
productivity (NPP) to support an
individual than the trophic level upon
which it feeds; the more trophic links
between NPP and a consumer taxocene
(taxonomically related set of species
within a community), the more energy is
necessary to support viable populations
of its individuals. If taxocene abundance
is energy limited, consumer taxocenes
that occupy broad NPP gradients should
show a predominance of lower trophic
levels at low NPP. Higher trophic levels
should accrue as NPP increases, and
taxocenes comprised largely of
predators should be under-represented
at lower NPP.

tropical cyclone Also known as a
hurricane, or tropical storm, this is a
disturbance about 650 km across,
spinning about a central area of very low
pressure, with winds over 140 km per
hour. The violent winds are
accompanied by towering clouds, some
4000 m high, and by torrential rain;
150 mm (6 inches) frequently fall within

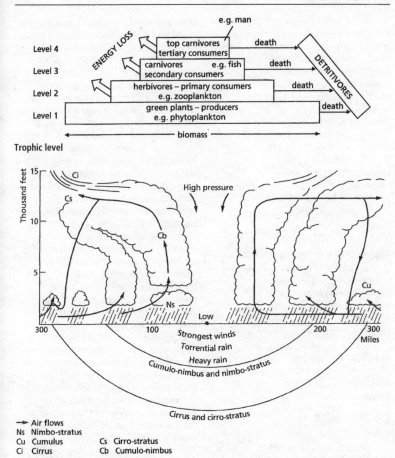

Trophic level

The structure of a tropical hurricane

→ Air flows
Ns Nimbo-stratus
Cu Cumulus   Cs Cirro-stratus
Ci Cirrus   Cb Cumulo-nimbus

the space of a few hours. There is, as yet, no complete understanding of how these storms develop; they can begin when air spreads out at high level above a newly formed disturbance at low levels. The upper-level outflow acts rather like a suction pump, drawing away the rising air at height and causing low-level air to be pulled in. The winds spiral into the centre because they are affected by the Earth's rotation. The intense energy of

these storms comes from the warmth of the tropical seas over which they develop. Thus, an extensive ocean area with surface temperatures of over 27 °C is necessary for hurricane formation.

The source regions must be far enough away from the equator—5° at least—for the *Coriolis force to have an effect. The removal of air at height may be along the eastern limb of an upper air trough. Moisture-laden air spirals into the centre and rises, condensing to form a ring-like tower of *cumulo-nimbus clouds. With this condensation, *latent heat is released which causes the air to rise further and faster. The condensation also causes torrential rain. In the upper *troposphere water droplets freeze and form cirrus clouds, which are thrown outwards by the spin of the storm.

At ground level, the temperature at the centre, or eye, of the storm is only slightly warmer than that at the margins, but, at heights of around 5000 m, the centre can be 18 °C warmer than the margins. This warm core maintains the low pressure which drags in the winds. *See* TROPICAL STORM.

tropical dry forest A tropical forest type that goes through an annual dry season. Drier soils sometimes produce pockets of tropical dry forest within a tropical rain forest. R. Robichaux and D. Yetman, eds (2002) provide the ur-text.

tropical forest Forested areas which often extend beyond the tropics, and consist of *tropical rain forest and *mangrove forest.

(⊕) SEE WEB LINKS
• The European Tropical Forest Network.

tropical montane forest Forest found in mountainous regions of the tropics, where rainfall can be heavy and persistent. The trees are typically short and crooked, and sites for mosses, climbing ferns, lichens, and *epiphytes. See Bruijnzeel and Veneklaas (1998) *Ecology* 79, 1 on climatic conditions and tropical montane forest productivity. Martin et al. (2007) *J. Biogeog.* 34, 10 examine relationships between climate-disturbance gradients, vegetation patterns, and *ecotones on subtropical mountains.

tropical rain forest A tropical forest of trees characterized by buttress roots, long, straight, lower trunks, and leathery leaves (see I. Turner 2001). The vegetation shows distinct layering: the canopy, or upper layer, at around 30 m; the intermediate layer at 20–25 m; and the lower layer at around 10–15 m. Undergrowth is poorly developed but *epiphytes and lianas are common. Davidar et al. (2005) *J. Biogeog.* 32, 3 indicate that seasonality, and not annual rainfall, is the variable that drives tropical rain forest tree diversity. The range of plant and animal species is immense. Hill and Hill (2001) *PPG* 25, 3 evaluate the following theories explaining the species richness of tropical rain forests: cumulative evolution, productivity, genetic drift and isolation, pleistocene refugia, intermediate disturbance, *niche packing, *habitat heterogeneity, seedling and foliar herbivore, recruitment limitation, and temporal resource partitioning.

The felling of the rain forest causes soil erosion and the destruction of potentially useful species, and the reduction of oxygen from photosynthesis, while smoke from burning logs increases the quantity of *aerosols and carbon dioxide in the atmosphere.

tropical storm An intense low-pressure cyclonic system that does not contain *fronts between warm and cold air masses, developing when the upper 50 m of tropical ocean warms to over 26.5 °C in regions over 500 km from the equator. See DiMego and Boshart (1982) *Monthly Weather Rev.* 110, 5 on the transformation of a tropical storm into a *tropical cyclone.

tropics The *Tropic of Cancer* lies approximately along latitude 23° 30′ N. Around 21–2 June, the sun's rays are perpendicular to the ground along this line and the sun exerts its maximum strength in the Northern Hemisphere. Conversely, the sun is overhead at the approximate latitude of 23° 30′ S, the *Tropic of Capricorn*, on 22–3 December when the sun's heat is at its maximum in the Southern Hemisphere. Between these two lines of latitude lie the tropics. In America this zone is known as the **neotropics** and in Africa and South-East Asia as the **palaeotropics**.

tropopause The upper limit of the *troposphere, above which very few clouds, except for the nacreous and noctilucent, form.

troposphere The lower layer of the *atmosphere, extending to 16 km above ground level at the equator, 11 km at 50 °N and S, and 9 km at the poles. Most clouds and *precipitation, and, indeed, weather events, occur within this layer. Increasingly, it is understood that air movements in the upper troposphere greatly influence weather systems in the lower troposphere. *See also* JET STREAM; ROSSBY WAVES.

Within the troposphere, temperatures of rising air fall, at varying *lapse rates, because the air expands and, therefore, cools. Sinking parcels of air experience a corresponding heating. Most of the water vapour in the troposphere is concentrated in the lower, warmer zone; there is little above the temperature falls below about −40 °C because moist air, rising through *convection or *turbulence, condenses out as ice crystals form.

truncated spur A steep bluff on the side of a *glacial trough, protruding between tributary, possibly *hanging, valleys. This landform is the result of mainly vertical *glacial erosion.

tsunami A huge sea wave, or series of waves. Most are formed from

earthquakes of 5.5 or more on the *Richter scale. Other causes include the eruption of submarine volcanoes, very large landslides off coastal cliffs, or the calving of very large icebergs from glaciers in fiords. Fault-generated tsunamis may spread thousands of kilometres across ocean basins with velocities up to 800 km h^{-1}. In mid-ocean the height of a tsunami may be no more than 30–50 cm, but this can increase up to 500 m when the tsunami hits the shore. The most active source region of tsunamis between 1900 and 1983 was along the Japan–Taiwan island arc, where over a quarter of all tsunamis were generated. Wikipedia is very good on the Japanese 2011 Tohoku earthquake, which triggered tsunami waves up to 40.5 m high, and travelled up to 10 km inland. The earthquake which caused the tsunami on 26 December 2004 was caused when a portion of the Indian tectonic plate slid under the Burma plate. The US Geological Survey estimates that the ensuing rupture was more than 1000 km long, displacing the sea-floor above by perhaps 10 m horizontally. The great volume of the ocean displaced thereby created the tsunami, which hit the coastline of eleven Indian Ocean countries, with wave heights up to 15 m. Estimates of the death toll range between 188000 and 320000. See Krom in J. Holden (2012), pp. 42–3 for an excellent summary of this tsunami.

tsunami deposits Many former tsunamis have left sedimentary fingerprints; see Dawson and Stewart (2007) *PPG* 31, 6.

tundra The barren plains of northern Canada, Alaska, and Eurasia. Temperatures and rainfall are low, so tundra vegetation is restricted by the intense winter cold, insufficient summer heat, and waterlogged soil, and characteristically comprises herbaceous perennials with scattered trees, mosses, lichens, and sedges. Arctic tundra ecosystems are warmth-limited and can respond to temperature dynamics on a relatively short time scale. They therefore are highly sensitive to both environmental fluctuation and directional climate change, and play a significant role in biospheric feedbacks to global climate. Alpine tundra does not contain trees because it is at high altitude, and differs from arctic tundra, as its soils are generally better drained than arctic soils. See Mallinson (2011) *Phys. Geog.* 32, 2.

tundra soil A dark soil with a thick, peat layer of poorly decomposed vegetation, which is usually underlain by a frozen layer of soil. Translocation is limited and there is, therefore, little development of *horizons. Tundra soils range from *brown earths in the more humid areas to polar desert soils in arid areas. Climate change may cause tundra soil to dry—accelerating oxygen uptake and hence heterotrophic respiration, which would result in increased CO_2 and methane emissions (Grant et al. (2003) *Glob. Change Biol.* 9, 1).

tunnel valley A large, elongate, *overdeepened depression cut into bedrock or glacigenic sediment, up to 4 km wide, over 100 m deep, and 30–100 km long. In long profile they contain enclosed hollows or isolated and elongated basins, sometimes displaying convex-up or undulating longitudinal channel profiles and anabranching channel networks. Ó'Cofaigh (1996, *PPG* 20) attributes their formation to the creep of deformable subglacial sediment into a subglacial conduit, followed by the removal of this material by meltwater flow.

turbosphere That part of the *troposphere where the mixing of the atmosphere is achieved more by *convection than by *diffusion. Its upper limit is the **turbopause**, an ill-defined layer about 100 km above the Earth's surface.

turbulence (**turbulent flow**) A gustiness in the three-dimensional flow of a fluid, irregular in both space and time, and characterized by local, short-lived rotation currents known as vortices. Turbulence is hierarchical; large *eddies produce smaller ones, and so on, down a series of smaller and smaller scales. In meteorology, it develops because of disturbances in air flows, the most important of which is *wind shear; air parcels caught in a wind shear tend to roll over and over. The *Reynolds number for turbulent flow is 2000 to 2500. In geomorphology, turbulent flow is classified according to the *Froude number of a stream.

turbulent boundary layer *See* BOUNDARY LAYER.

UAV An unmanned aerial vehicle used in geomorphology to access a terrain. A **UAS** refers to an unmanned aircraft system.

ubac That side of a valley which receives less *insolation; the shaded side, termed **back ridge** in Jamaica (Beckford and Barker (2007) *Geog. J.* 173, 2).

ultisol A soil of the *US soil classification. *See* FERRUGINOUS.

umland A *sphere of influence, *catchment area, tributary area, or urban field.

unauthorized colony (UC) In India, a residential area that has evolved mostly on private land. Roughly a quarter to one-third of Delhi's population live on UCs. These are technically illegal locations for residential development, and consequently are excluded from Delhi's network of basic urban services (water, roads, and electricity), and face potential demolition. See Lemanski et al. (2013) *TIBG* 38, 1, 91.

uncertainty A term which has subtle differences in meaning according to its academic context; F. Knight (1921) distinguishes between *risk* (randomness with knowable probabilities) and *uncertainty* (randomness with unknowable probabilities). The measurement of uncertainty is imperative for the production of knowledge; as Goodchild in S. Shekhar and H. Xiong (2008) notes: 'all geographic information leaves some degree of uncertainty about conditions in the real world'. The two broad questions of uncertainty in geography are the estimation of experimental error, and the question of 'how we come to know'.

The first theme is less problematic: for useful guidelines for assessing the value of information, see D. W. Hubbard (2007), and for a model used to assess uncertainty in analyses of US census tracts, using aggregated data from incompatible zonal systems, see Schroeder (2007) *Geogr. Analysis* 39, 3.

The second theme relates to the production and status of 'knowledge'. 'In assessing uncertainty, there is a need to question both "what is known" in geography, where knowledge and reflections about the "status" of knowledge differ between individuals and groups of scientists, and "how we came to know", where knowledge and uncertainty are closely related to research tradition and practice' (Brown (2004) *TIBG* 29, 3). Reed and Peters (2004, *ACME* 31, 1) draw on contemporary systems ecology to design research practices which 'embrace the uncertainty and partiality of knowledge creation'—one of their wise practices is 'preparing for surprise'. See also Ekinsmyth et al. (1995, *Area* 27, 4) on the uncertainty of *economic geography, G. Foody and P. Atkinson's e-book on *GIS and uncertainty, and Hunsaker et al., eds (2001) on spatial uncertainty in *ecology.

underclass The poor, the unemployed, the lowest social class, for whom the primary predicament is joblessness, reinforced by growing social isolation, inadequate access to jobs and job networks, and the lack of involvement in quality schools.

underdevelopment The original meaning of the term indicated that existing resources had not been exploited. The word is now close in meaning to 'poverty', although some oil-rich underdeveloped countries have high incomes that are enjoyed by the few. Indicators of underdevelopment include: high birth rates, high infant mortality, undernourishment, a large agricultural sector and small industrial sector, low per capita *GDP, high levels of illiteracy, and low *life expectancy.

Underdevelopment has been brought about by factors such as unfair trade and agricultural dumping of subsidized products. See Power and Sidaway (2004) *AAAG* 94, 3 for a review of explanations for underdevelopment; it's easy to read with useful references and footnotes. Desmet and Ortán (2007) *Scand. J. Econ.* 109, 1 describe what they call **rational underdevelopment** whereby the advanced region protects itself against low-wage competition, thus preventing the backward region from taking off, and ensuring its dominant position. See Bhattacharyya (2007) Padjadjaran University WoPEDS 200704 l for the root causes of Africa's current state of underdevelopment.

underfit stream (misfit stream) A stream that flows with narrower meander belts and shorter meander wavelengths than are appropriate to the valley. If the valley was initially formed by a river of much greater *discharge, an underfit stream is evidence of climate change. However, while Dury (in K. J. Gregory, ed. 1977) developed mathematical relationships between the valley meander wavelength and the stream meander wavelength on the Ozark Plateaus to define the underfit condition, Shepherd et al. (2011) *Phys. Geog.* 32, 2 were unable to replicate his findings in the same locations.

underpopulation A situation where there are too few people to develop fully the economic potential of an area or nation; a larger population could be supported on the same resource base. Underpopulation is a largely hypothetical concept.

understorey In a forest, the layer between the *canopy and the forest floor. Understorey plants represent the largest component of biodiversity in most forest ecosystems and play a key role in forest functioning. See Mestre et al. (2017) *Forest Ecosys.* 4.

unemployment Being involuntarily out of work. **Structural unemployment** occurs when the labour market no longer requires a particular skill; in the case of **fractional unemployment**, jobs are available but not taken up because of immobility or the lack of information. A *core–periphery in real wages is associated with and magnified by regional unemployment disparities. Measures to reduce unemployment may cut out-migration, but only during an era of low unemployment. Okun's law states that unemployment falls with increases in the ratio of real output (GDP) growth to *potential* real output growth. The **unemployment rate** is the number unemployed as a percentage of the total population of working age.

uneven development On a multitude of scales, the condition of an economy that has not benefited equally from development either in a spatial sense and/or within classes in society. The most striking fact about the

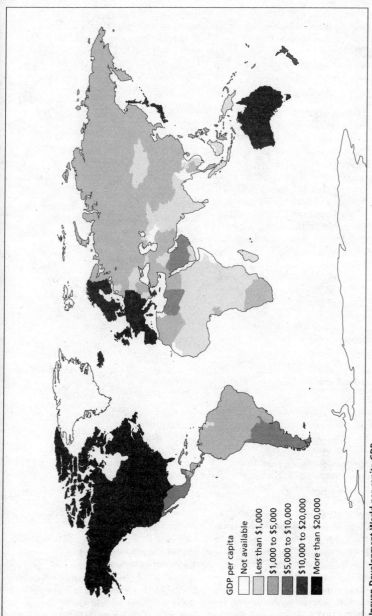

Uneven Development World per capita GDP
Source: World Bank

GDP per capita

- Not available
- Less than $1,000
- $1,000 to $5,000
- $5,000 to $10,000
- $10,000 to $20,000
- More than $20,000

u

Uneven Development in China

economic geography of the world is the uneven spatial distribution of economic activity, including the coexistence of economic development and underdevelopment. High-income regions are almost entirely concentrated in a few temperate zones, half of the world's GDP is produced by 15% of the world's population, and 54% of the world's GDP is produced by countries occupying just 10% of the world's land area. The poorest half of the world's population produces only 14% of the world's GDP, and 17 of the poorest 20 nations are in tropical Africa.

Global uneven development may be seen to be a result of capitalism, based as it is on competition and *accumulation, but inequality is not unique to capitalism. For a national case study, see Hsu and Cheng (2002) *Reg. Studs* 36, 8 on uneven development in post-war Taiwan. Location and climate have large effects on income levels and income growth, through their effects on transport costs, disease burdens, and agricultural productivity.

Within the UK, business development is very uneven because of local transaction costs, variations in *institutional thickness, *agglomeration, *external and urbanization economies, and the continued importance of face-to-face contacts. Transport costs are still important. Peripheral location appears to affect economic development strongly in China. Wei has shown that the per capita gross value of industrial output of the coastal provinces, such as Guangdong, was double that of provinces deep inland, such as Xizang (Tibet). J. Y. Lin (2005, East Asian Bureau of Economic Research *Development Economics Working Papers* 656) proposes that flawed development strategy is responsible for the increasing disparities in economic development among Chinese provinces.

uniformitarianism The view that the interpretations of Earth history can be based on the present-day evidence of natural processes. From this comes the maxim 'the present is the key to the past'.

universal soil loss equation An empirical model of the relationships between the various components of soil loss and rates of erosion. It has the form:

$$A = R \times K \times L \times S \times C \times P$$

where A is the mean annual soil loss, R is a rainfall erosivity factor, L a slope length factor, S a slope steepness factor, C a crop management factor, P an erosion control practice factor, and K the soil erodibility index.

Universal Transverse Mercator (UTM) A two-dimensional coordinate system that divides the Earth into sixty zones, each a six-degree band of longitude. The system is then used to pinpoint a location. It is not a single map projection, and *not* the same as the traditional method of latitude and longitude.

unloading The removal, by erosion, of rock or, by ablation, of ice. See Bonow et al. (2006) *Geomorph.* 80, 3–4; and Cossart et al. (2008) *Geomorph.* 95, 1–2

on paraglacial stress release in the immediate aftermath of glacial unloading.

Upton Warren An *interstadial, between 43 000 and 40 000 years BP. Faunal and floral evidence suggests that the landscape was devoid of trees. *See also* PLEISTOCENE.

upwelling The rise of sea water from depths to the surface, bringing nutrients for plankton. Many of the world's best fishing grounds are located at such points. See Ramp and Bahr (2008) *J. Phys. Oceanog.* 38, 1 on upwelling-favourable winds. *See also* THERMOHALINE CIRCULATION.

urban In medieval Europe, for example, there was a clear-cut boundary between urban and rural, usually in the form of a wall. Subsequently, towns have progressively overflowed their municipal boundaries. In the 1960s, UNO introduced the concept of the **urban agglomeration** (UK—urban area; France—*unité*; Germany—*Verdichtungsraum*) in order to harmonize the statistical definitions of urban settlements from one country to the next. The extensive suburban sprawling based on mass car ownership led to the concepts of *daily urban systems, or *SMSA (Standard Metropolitan Statistical Areas). Later, the French statistical system defined the *aire urbaine*: a contiguous urban core and its commuter belt. About 360 *aires urbaines* were defined (Le Jeannic (1996) *Cybergeo écon. & stat.* 294/295).

urban assets The urban talent in the *knowledge-based economy, believed to 'energize a new capitalist epoch as well as urban economic renewal' (Amin and Thrift (2007) *PHG* 31, 2).

urban climate Urban centres are subject to climatic conditions that represent a significant modification of the pre-urban climatic state, arising as a result of the modification of radiation, energy, and momentum exchanges resulting from the built form of the city, together with the emission of heat, moisture, and pollutants from human activities. 'For the increasingly urbanized population of the world these effects are already of similar to, or of greater magnitude than, the climatic changes predicted to occur at larger scales as a result of the enhanced *greenhouse effect' (McKenry (2002) *PPG* 37, 4).

Increasingly, the focus of urban climate research is on understanding the fundamental processes that generate urban climates, not just the resultant effects. The rationale for much urban climate research is human health and well-being or energy and water consumption. See Assimakopoulos et al. (2003) *Atmos. Env.* 37, 29 on pollution in 'street canyons'. Mills (2006) *Theoretical and Applied Climatology* 84, 69–76 provides a useful summary of ways to modify urban climates, including outdoor landscaping, street orientation, zoning, and transport policy.

urban commons The principle that urban communities should have access to, and produce, the range of goods and services to sustain them, particularly for the most vulnerable populations. This would include: sharing, collaboration, civic engagement, inclusion, equity, and social justice. Urban commons would be created and managed by: participants from local communities, government, business, academia, and local nonprofit organizations. In this way, the city is a platform utilized and optimized by citizens from all backgrounds and social statuses. See Chatterton (2016) *TIBG* 41, 4.

urban containment Limiting *urban sprawl. New Zealand's Regional Growth Forum has developed a strategy for regional urban containment, matched by urban intensification. Urban containment has been promoted by US state governments since the 1980s but only statewide growth management programmes (with the stipulation that local plans must coincide with a state plan or with geographically contiguous local plans), have been effective at reducing the square mile size of US urban areas.

urban density Cities are characterized by high population density. See Chen et al. (2020) *Urban Geog.* 41, 10 on the meanings and implications of density in cities today. There are other relevant articles in the same issue.

urban diseconomies The financial and social burdens arising from an urban location. These include constricted sites, high land prices/local taxation/commuting costs, traffic congestion, and pollution. Firms concentrate geographically in order to exploit benefits of *agglomeration which are traded off against urban diseconomies in determining city sizes.

urban ecology The application of the principles of *ecology to a study of urban environments. *See also* HUMAN ECOLOGY. Urban ecologists look at individual areas of the city in the context of the whole city, and focus on the way a population organizes itself, and the way it adapts to change. Just as ecologists study the way in which an ecosystem seeks to re-establish equilibrium after a sudden alteration, so urban ecologists assume that people will try to re-establish equilibrium after sudden change. The journal *Urban Ecology* examines relationships between socioeconomic and ecological structures in urban environments, landscape architecture, meteorology and climate, policy, and urban planning. Urban ecology has been criticized by many; among them E. Fowler (1996), but Longley (2005) *PHG* 29, 1 detects some rehabilitation in urban ecology and social area analysis.

urban geography The spatial study of the city, including urban policy; race, poverty, and ethnicity in the city; international differences in urban form and function; historical preservation; the urban housing market; and provision of services and urban economic activity. Pacione (2003) *Urb. Geog.* 24, 4 writes that 'it is now difficult, if not impossible, to place analytical boundaries around the city'.

Many of the concerns formerly encapsulated within urban geography are now studied under different banners, but Lees (2002) *PHG* 26, 1 argues that urban geography has taken on board the interpretative turn, engaging with language, discourse, culture, the immaterial, and so on.

urban governance The exercise of political, economic, and administrative authority in the management of a city. Proudfoot and McCann (2008) *Urb. Geog.* 29, 4 discuss how street-level bureaucrats negotiate the constraints and pressures inherent to their practice while also exercising a degree of discretion, arguing that these micro-level concerns are important to understanding how cities are produced, but must also be linked with analyses of wider processes that shape contemporary urban development.

urban heat island (UHI) The temperature differentials between urban and rural areas, occurring on the ground, at the surface, and at various heights in the air. The UHI may actually be described in multiple ways with various methodological approaches to investigate each type, and is divided into: **subsurface heat islands** (shallow aquifers existing below many northern European cities have groundwater temperatures 3–4 °C higher than in surrounding rural areas) and urban heat islands in the air. Each city's unique land cover patterns will influence the spatial character of its UHI. Spatial and temporal patterns of CO_2 concentrations are useful in understanding urban climate; see Wentz et al. (2002) *AAAG* 92, 1.

urban hydrology The study of the consistent transformation of hydrological regimes that results from urban development. These changes have been attributed to the construction of impervious surfaces and the compaction of pervious surfaces, which promote rapid surface run-off from rainfall, and the efficient drainage of surface run-off through storm sewers to the river network. The percentage urban land cover/percentage impervious cover/percentage connected impervious cover within a catchment provide simple but very effective predictors of changes in streamflow characteristics resulting from urban development. See Gurnell et al. (2007) *Geog. Compass* 1, 5 for an excellent review of the topic.

urbanism A way of life associated with urban dwelling (suggesting that urban dwellers follow a distinctly different way of life from rural dwellers), a hallmark of modernism, progress, development, and the metropole—the opposite of provincialism. **Neoliberal urbanism** is seen as the production of self-governing, consumption-oriented, autonomous urban citizens; see Wilson (2008) *Urb. Geog.* 29, 3.

The concept of **planetary urbanism** argues that urbanism can happen anywhere on the globe, and **political urbanism** considers the political

u

bases—of local or national governments—on urban social organization.

urbanization The increase in the proportion of the population residing in towns, brought about by migration of rural populations into towns and cities, and/or the higher urban levels of natural increase resulting from the greater proportion of people of childbearing age in cities. R. Bilsborrow (1997) discusses the problems in the definition of urbanization, and the difficulties of making international comparisons. Start with Solecki and Leichenko (2006) *Environment* 48, who give an excellent overview, while also discussing urbanization and sustainability. See also T. Hall (2006).

Urbanization indicates a change of employment structure from agriculture and *cottage industries to mass production and *service industries. This backs up the view that urbanization results from, rather than causes, social change. This is most notable in the development of *capitalism and its attendant *industrialization. Others argue that urbanization is the inevitable result of economic growth, with the rise of specialized craftsmen, merchants, and administrators. A further view stresses the importance of *agglomeration economies; cities offer markets, labour, and capital with a well-developed infrastructure, all of which increase their *comparative advantage. It is argued that urbanization economies have been the primary driving force for the geohistorical development of human societies over the past 12000 years.

Globalizing cities in the 'south' exhibit a colonial legacy, an imposed neoliberalism, and mass urban migration, creating extreme pressure on infrastructure and service provision.

T. Champion and G. Hugo (2003) argue that there is no longer any difference between the urban and the rural; instead, there are new forms of urbanization. Gandy (2005) *Int. J. Urb. & Reg. Res.* 29, 1 argues that an emphasis on **cyborg urbanization** (a 'dialectically conceived version of urban metabolism relating technical developments to a broader political and cultural terrain') 'extends our analysis of flows, structures and relations beyond so-called "global cities" to a diversity of ordinary or neglected urban spaces'.

urbanization curve A model of the progress of urbanization, based on empirical evidence from Europe. In a traditional society, urbanization is below 20%, and the rate of urbanization is slow, so the curve starts gently. With industrialization, and a rise in the importance of manufacturing and services, the pace of urbanization quickens, but the curve slackens after about 75%. While most developed countries have reached this third stage, the countries of the developing world are still on a rising curve of urbanization, often with a steeper gradient than is characteristic of advanced economies.

Some geographers believe that *counter-urbanization constitutes a fourth stage, when urban populations percentages fall; others that counter-urbanization is temporary.

urbanization economies The benefits of locating in an urban environment; the positive effects on productivity and utility created endogenously (*in situ*) by larger cities. These economies include proximity to a market, labour supply, good communications, and financial and commercial services such as auditing, stockbroking, advertising, investment, industrial cleaning, and maintenance. Localization *economies*, which derive from being located close to other firms in the same industry, differ from

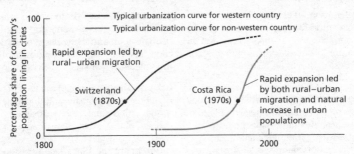

Urbanization curve

urbanization economies, *which are* associated with closeness to overall economic activity. Larger cities have a greater *comparative advantage than small ones, though the relationship between city size and urbanization economies is non-linear. Demand-side explanations for this argue that the bigger the city, the more likely a match between specific demands and suppliers (an explanation with echoes of *threshold population), while supply-side explanations cite the provision of traded, and non-traded, inputs which increase innovation and variety, and lower costs.

urban morphology The form, function, and layout of the city. Geographic attributes—including shape, density, and pattern of land use categories—all reflect the outcome of the *urbanization process. For an account and explanation of urban morphology through time, see C. Boone and A. Modarres (2006).

urban planning An attempt to manage the city, often in order to avoid, or alleviate, common urban problems such as inner-city decay, overcrowding, traffic and other forms of congestion. In order to understand the role of visioning in contemporary urban politics and in policy-making outcomes, we must recognize the sociospatial context in which it is deployed.

urban political ecology If one component of *political ecology is the study of political struggles for control over natural resources, *urban* political ecology is easy to define, and Swyngedouw and Heynen are its begetters: 'the political programme of urban political ecology is to enhance the democratic content of socio-environmental construction' ((2003) *Antipode* 35, 5).

urban regeneration (urban renewal, urban revival) The attempt to reinvigorate a run-down urban area, through strategies including the redevelopment of *brownfield sites, rehabilitation of the existing building stock, and enhancement of public spaces and infrastructure. In the *slums of the developed world, substandard housing is broken up by urban regeneration projects, and low-income households, 'unable to gain access to the shrinking stock of public rental housing find themselves increasingly restricted to pockets of low quality housing in the private rental sector' (Randolph and Holloway (2005) *Urb. Policy & Res.* 23). Public art has become wedded to urban regeneration in developed world cities, and Bell (2007) *PHG* 31, 1 finds that hospitality has become central to

u

regeneration. In Houston's urban revival, physical upgrading has been accompanied by the displacement of the community's traditional population and the destruction of an historic neighbourhood.

urban social geography The study of social and spatial dimensions of city life; the study of the social patterns and processes of cities; the social interpretation of the spatial structure and functioning of the urban environments. See P. Knox and S. Pinch (2000).

urban sprawl The extension of the city into the countryside, particularly associated with improvements in mass transport. Population growth is one of the most important engines of urban sprawl; another possible cause is negative *externalities between developments. The implication is that the undeveloped land that is adjacent to sprawl is less valuable in residential use and less likely, *ceteris paribus*, to be developed residentially in the future. Bartl (2011) *Urbaniz. & Water* has some interesting thoughts on urban sprawl in Indian cities. *See* URBAN CONTAINMENT.

urban system Any network of towns and cities, and their hinterlands, which can be seen as a system, since it depends on the movements of labour, goods and services, ideas, and *capital through the network. Abdel-Rahman (2002) *J. Reg. Sci.* 42 links the structure of an urban system with income disparities: economies that are dominated by a large metropolis, with a number of low-technology cities, will have large income disparities.

urban theory Theorizations of *urbanization which stress the logic of capital in *uneven urban development, appraise the tensions between modernity and postmodernity as conceptual frameworks for understanding 'the urban', explore the *globalization and the concept of world cities, and analyse the complexities of multiple and hybrid urban identities.

urban village A residential area of the city containing a cluster of individuals of similar culture and/or interests. The concept of the urban village in the UK was first promoted by the Urban Villages Group in the late 1980s as a means to achieve more human-scale, mixed-use, and well-designed places.

USGS DEM A *geospatial file format developed by the United States Geological Survey for storing a raster-based digital elevation model. It is an open standard and is used throughout the world (K.-T. Chang 2014).

U-shaped valley Most U-shaped valleys—valleys with a parabolic cross-section—are *glacial troughs. Brook et al. (2008) *Geomorph.* 97, 1–2 find that valley cross-profiles become U-shaped with *c.* 400,000–600,000 years of glacial occupancy. When no longer supported by ice, the steep sides of U-shaped valleys are prone to rock fall. Valleys with this form are also encountered in non-glaciated chalk topography.

US soil classification The US Department of Agriculture recognized ten major soil groups in the Seventh Approximation. Alfisols are relatively young and acid soils with a clay B horizon. Aridisols are semi-desert and *desert soils. Entisols are immature, mainly *azonal soils. Histosols are primarily organic in content, developing in marshes or peat bogs. Inceptisols are young soils with weakly developed horizons. Mollisols are characteristic of grassland, high in bases, and with a thick, organically rich A horizon. Oxisols occur in tropical and subtropical areas. They are well weathered and often have a layer of *plinthite near the surface. Spodosols have been *podzolized. Ultisols develop where summers are wet and winters are dry. They are quite deeply weathered and are often reddish-yellow in colour. Vertisols are clay soils characterized by deep, wide cracks in the dry season.

utility The satisfaction given to an individual by the goods and services used.

vadose Describing the zone immediately below the ground surface and above the *water-table in which the water content varies greatly in amount and position.

SEE WEB LINKS
• The *Vadose Zone Journal*, online.

validation In academic life, the methods used to check the validity of a finished piece of research. 'Validation in quantitative methods involves the calculation of statistical errors, whereas in qualitative methods validation methods include cross-checking or triangulation with other methods; theoretical saturation, or the point at which no new information is gained; and adherences to a code of conduct believed to ensure accurate results' (Nightingale, in R. Kitchin and N. Thrift 2009).

valley train See SANDUR.

valley wind An *anabatic air flow generated as a valley floor is heated by the sun, and the warm air moves upslope. Valley winds are at their strongest in valleys of a north–south orientation. See Weigel et al. (2007) *Boundary-Layer Met.* 123, 1.

value The importance, worth, or usefulness of something; its material or monetary worth. The **governmentalization of value** is the proliferation of tactics (programmes, laws, codes, standards, and institutionalized knowledges) that regions.

vapour pressure In meteorology, the pressure exerted by the water vapour in the air. This is a partial pressure since pressure is also exerted by the air itself. Moisture content is commonly measured by *relative humidity. The absolute value of the water content is not usually the most important issue but its value compared with saturation vapour pressure; thus relative humidity is more relevant than specific humidity.

variable source areas See *DYNAMIC CONTRIBUTING AREAS.

variegated capitalism (varieties of capitalism) *Capitalism varies according to the way it is embedded in different sorts of states. The major varieties of capitalism are: *welfare states, *neoliberal states, *developmental states, and *socialist states; see P. A. Hall and D. Kosice (2001) for the seminal work.

varve A pair of coarse and fine deposits that reflect seasonal deposition in *proglacial lakes. In summer, *meltwater streams deposit silt- to sand-sized coarse sediments; in winter the lake surface is frozen, the water is calm, and the fine deposits settle out in a thinner layer than the summer sediments. A **varve couplet** represents the total *fluvio-glacial deposition on a lake floor for one year.

vector In *geographic information systems (GIS), **vector data** represent features as discrete points, lines, and polygons. In computer graphics, **vectorization** refers to the process of converting raster graphics into vector graphics. Vectorization is not well defined, meaning there is not a single correct method. Many different algorithms exist, and each gives different results, as vector representations are more abstract than pixels (K.-T. Chang 2014).

veering Of winds in the Northern Hemisphere, changing direction in a clockwise motion, e.g. from westerly to northerly. The *Coriolis effect causes freely moving objects to veer towards the right in the Northern Hemisphere and to the left in the Southern Hemisphere.

velocity reversal hypothesis An explanation of the occurrence of coarse sediments in riffles, and fine sediments in pools, by variations in *discharge. At high discharge, high velocities would move coarse sediment rapidly through the pools, while relatively low velocities in riffles would cause its deposition there. The fine particles are winnowed from riffles and deposited in the pools. Booker et al. (2001) *ESPL* 26 think that velocity reversal is an inadequate explanation of pool-riffle sequences.

vent In geomorphology, an opening in the *crust through which volcanic material flows. **Vent conglomerate** consists of rounded blocks of juvenile material, derived from the collapse of the cone walls during a volcanic eruption, filling a volcanic vent. See Oppenheimer (2003) *PPG* 27, 2.

venture capital Private equity (finance) offered to outside investors in new/growing businesses. Venture capital investments are usually high risk, but offer the possibility of above-average returns. In the UK, venture capital investments were concentrated in

Greater London and the South-East in the 1980s, but this regional concentration has been greatly reduced since the 1980s, and, although it has recently experienced a marked increase in activity, the venture capital industry is much less developed in Europe than it is in the USA.

vertex In *network analysis, this is the place joined by two or more *links (also called *node*).

vertisols Soils of the *US soil classification found in regions of high temperature where bacteria destroy organic residues; hence the humus content is low. Alternate wet and dry seasons lead to the alternate swelling and shrinking of these soils. By these processes, horizons become mixed or inverted.

vicariance (vicariance biogeography) The geographical separation and isolation of a subpopulation, resulting in the original population's differentiation as a new variety or species. When a continuous population (or species) is divided by a **vicariance event**, such as the formation of a new river or a new mountain range, two populations or species are created. See Wiley (1988) *Ann. Rev. Ecol. Syst.* 19, 513, but Wallis and Trewick (1998), now published in *Systematic Biol.* 50, 4, 602, find fault with vicariance.

violence, geographies of Violence can be seen as social practice where human needs and human securities are contested and fought over. This would see the geographies of violence as places where human freedoms and rights are struggled for, negotiated, lost, and won.

viscosity The resistance to flow exhibited by a material; the parameter describing the flow characteristics of a fluid. Easy flows result from low viscosity.

vital rates Rates of those components, such as birth, marriage, fertility, and death, which indicate the nature of and possible changes in a population.

volcanic ash Finely pulverized fragments of rock and lava which have been thrown out during a volcanic eruption. Volcanic ash forms a thick coating on the landscape adjacent to a volcano, and can travel great distances. A **volcanic bomb** is a block of lava ejected into the air from a volcano. As it is thrown out, it cools and spins; see Francis (1973) *GSA Bull.* 84, 8 on cannonball bombs.

volcanism (vulcanicity, vulcanism) All the processes associated with the transfer of magma and volatiles from the interior of the Earth to its surface. *Magma beneath the crust is under very great pressure. Deep in the crust, faults and joints develop downward, reach the magma, and allow it to rise up and intrude the crust. The magma then rises in conduits, depressurizes, forms bubbles, and can give rise to explosive volcanism; see Gaonac'h et al. (2003) *Geophys. Res. Letters* 30, 11, but also see Gonnermann and Manga (2003) *Nature* 426. **Intrusive volcanicity** describes magma injected into the crust which fails to reach the surface, and cools slowly underground forming *batholiths, *laccoliths, *dykes, *pipes, and *sills. **Extrusive volcanicity** describes magma which forces its way to the Earth's surface and erupts as *lava from volcanoes or fissures, forming cones or plateaux.

Current volcanicity is confined to regions of the Earth where lithospheric *plates converge, diverge, or pass over possible mantle *hot spots. In a *subduction zone, water is trapped in the descending plate. As the plate is heated, water is released, rising into the wedge of mantle above the plume, causing it to melt. This process forms chains of volcanoes; the cordillera of North America is an area of volcanism related to the subducting eastern margin of the Pacific Ocean. Where one oceanic plate is subducted beneath another, a chain of volcanic islands forms; Japan is an example (Bunbury in P. L. Hancock and B. J. Skinner, eds 2000). When a mantle plume rises through the lithosphere, and eventually erupts at the surface, it is often called *hot-spot volcanism*. About 10% of volcanoes arise through hot-spot activity. Mass movements on volcanic islands may generate large-magnitude tsunamis.

In recent years the focus of volcanology has shifted from the physical aspects of why, when, and how volcanoes erupt to incorporate studies on the short- and long-term effects volcanic eruptions have on society and the atmospheric, aquatic, and terrestrial environments; see J. Marti and G. Ernst, eds (2005).

volcano An opening in the crust out of which *magma, ash, and gases erupt. The shape of the volcano depends very much on the type of lava. *Cone volcanoes are associated with thick lava and much ash. *Shield volcanoes are formed when less thick lava wells up and spreads over a large area, creating a wide, gently sloping landform. Most volcanoes are located at *destructive or *constructive plate margins.

volume On a *GIS triangulated irregular network, three-dimensional space measured by length, width, and depth, or height.

volumetric In geography referring to the particular arrangement of dimensions and capacities above and below the surface; for example, in relation to the surface of a city. Verticality, the surface, and the subterranean reflect the power and political economy of cities. See McNeil (2020) *PHG* 44, 5.

von Thünen models Johann von Thünen (1783–1850) had two basic models: the first postulates that the intensity of production of a particular crop declines with distance from the market; the second is concerned with land use patterns as they relate to transport costs. See Barnes (2000) *PHG* 27, 1 for a summary.

vortex (pl. *vortices*) A whirling mass of fluid or air.

The strong **polar vortex** in the Antarctic atmosphere isolates the polar air mass for much of the year. Over the Arctic region the polar vortex is weaker—resulting, in part, from the uneven distribution of land and water. **stratospheric polar vortex**.

vorticity In meteorology, a measure of the local spin of a part of the *atmosphere. The local spin of the atmosphere relative to the Earth has opposite signs in *cyclones and *anticyclones; conventionally the cyclonic direction is taken as positive. A major principle governing the vorticity change of flowing air is the *conservation of angular momentum: as air spreads out horizontally, the rate of spin falls; as it contracts horizontally, its rate of spin rises. **Absolute vorticity** is the sum of the **planetary vorticity**, which is associated with the rotation of Earth, and **relative vorticity**, which is associated with the fluid motion relative to the surface of Earth.

vulcanian eruption Also known as a **Vesuvian eruption**, this is marked by periodic lulls during which gas pressure builds up behind the lavas that clog the vent. This blockage is removed by an explosion which throws off masses of *pyroclasts.

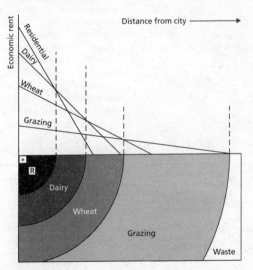

R Residential
* City centre

von Thünen model

wadi In a hot desert, a steep-sided, flat-floored valley, very occasionally occupied by an intermittent stream. Wadis were probably cut during the rainier *pluvials; see Morin et al. (2007) *Adv. Geosci.* 12 on wadis. See also F. El-Baz, ed. (2003).

Walker circulation A major atmospheric system, which is strongly tied to the equatorial Pacific Ocean. It stretches across the whole of the tropical Pacific, and consists of: the trade winds, which are blowing from east to west; rising air over the western Pacific, south-east Asia and northern Australia through convection; high-level winds blowing counter to the trades; and descending air over the eastern Pacific. See Lockwood in J. Holden (2012), p. 104. Changes in the Walker Circulation are linked to the

*El Niño–Southern Oscillation; see Power et al. (1999) *Climate Dyn.* 15.

wandering river A fluvial channel with alternate stable single-channel reaches, and unstable multi-channel sedimentation zones. This term has also been applied to channels which are midway between meandering and braiding.

waning slope *See* PEDIMENT.

war on terrorism (war on terror) An operation initiated by the United States government under George W. Bush, using legal, military, personal, and political actions to limit the spread of terrorism after 9/11. US cities have been reworked in order to construct them as 'homeland' spaces, and re-engineered in the interests of national security, and the world has been mapped into 'civilized

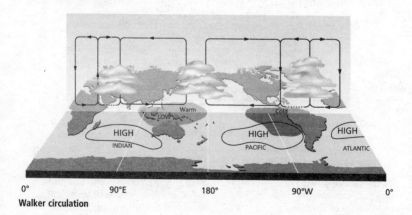

Walker circulation

core' and 'dangerous periphery'. See Amore (2009) *Antipode* 41, 1, 49.

wash board moraine Synonymous with De Geer *moraine.

wash load An alternative term for *suspension load.

washover The supply of sediment across a coastal barrier to the barrier floodplain, driven by storms at sea. A 'washover opening' would appear to be a breach in the barrier. See Wesselman et al. (2019) *Geomorph.* 327.

water balance The balance at any location between the input (*precipitation, P), and the outputs: *evapotranspiration (E) and *run-off (R):

$$P = E + R$$

If water balance is computed for a number of years, *groundwater storage (S) is held to be constant, but for a single year, groundwater fluctuations are taken into account:

$$P = E + R \pm S$$

Storage and run-off tend to be higher in winter, and evapotranspiration is higher in summer. See Arnell (1999) *J. Hydrol.* 217 on a simple water balance.

waterfall A site on the *long profile of a river where water falls vertically. Waterfalls may be found at a band of more resistant rock, or where deposition has occurred. The principal factor in determining the rate of waterfall recession is the ratio of the erosive force of stream to the bedrock resistance. For an equation connecting the force/resistance ratios and the rates of waterfall recession, see Hyakawa and Matsukora (2003).

water harvesting The capture and storage of rainwater through the construction of channels, check-dams,

micro-dams, and percolation pools. Try Grum et al. (2017) *Catena* 159.

water-level weathering The development and enlargement of tidal pools by *weathering, and by the action of rock-grinding animals. Water-level weathering is seasonal, with higher rates in the summer months. See Inkpen (2007) *Area* 39, 1.

water quality This can be registered as a measure of the ecological health of the water, determined by the level of indicator species in relation to a predetermined critical threshold, derived from toxological data; the biochemical oxygen demand can also be used. Land cover change and the expansion of modern agricultural practices has significantly increased leaching of chemicals to the surface waters, leading to the degradation of aquatic ecosystems worldwide. See Beasley and Kneale (2002) *PPG* 26, 2 on pollution from urban surface run-off.

watershed In the UK, the boundary between two river systems. Horton (1945) *GSA Bull.* 56, 3 uses the term *watershed line*, and refers to drainage basins as separate entities, but watershed is now commonly used in North America to mean drainage basin.

water-table The level below which the ground is saturated; the upper surface of the *groundwater.

water transfer The movement of water from a *drainage basin or catchment where there is more than enough water for human needs, to another catchment where there is a deficit.

wave A ridge of water between two depressions. **Wave height** is the difference in elevation between the wave crest and the wave trough. The height of

w

a wave is generally proportional to the square of wind velocity.

The **wavelength** is the distance between successive crests (or troughs), and the **wave period** is the time it takes for the wave to travel a distance equal to its wavelength. **Wave steepness** is given by the ratio of wave height to wavelength H/L. **Significant wave height** H_s is the average of the highest one-third of the waves, and can be obtained by recording a time series of water depth and multiplying the associated standard deviation by four. *See* LINEAR WAVE THEORY.

Wave shoaling is the process whereby waves change in height as they travel into shallower water and make contact with the seabed. See Grilli et al. (1994) *J. Waterw. Port, Coast. Ocean Eng.* 120, 6 on the shoaling of solitary waves on plane beaches.

wave convergence The focusing of wave rays so that they come together, increasing in energy and height. This occurs when waves travel over a localized area of relatively shallow water such as a shoal on the sea floor, causing an increase in *wave energy and wave height.

wave-cut platform *See* SHORE PLATFORM.

wave energy Energy generated by the force of ocean waves. Wave energy is expressed as the amount of energy per unit area ($N\ m^{-2}$) and is more appropriately referred to as **wave energy density**. Wave energy depends on the square of the wave height. A doubling of the wave height therefore results in a four-fold increase in wave energy. Total wave energy E is given by:

$$E = \frac{1}{8}\ \rho g H^2$$

where ρ is the density of water, g is gravity, and H is the wave height.

See Masselink in J. Holden, ed. (2012). **Wave energy flux** is the rate at which wave energy is carried along by moving waves and is the product of the amount of energy associated with the waves and the speed at which this energy travels.

wave refraction The change in the approach angle of a wave as it moves towards the shore. As water becomes shallow, waves slow down. This change in speed causes the *orthogonals of a wave to 'bend' so that the line of the wave mirrors the submarine contours. Refraction causes waves to converge on headlands and diverge in bays. This means that the energy of the waves is concentrated on the headlands rather than on the beaches.

wave set-up When waves break, water piles up against the shore. This results in a slope of water, with higher water pushed nearer the shore and this 'set-up' is sufficient to oppose the shoreward wave stresses (Masselink in J. Holden, ed. 2012).

waxing slope The *convex element at the foot of a hillslope.

waypoint A reference point in physical space used for purposes of navigation, including GPS and *GIS.

weathering The breakdown, but not the removal, of rocks. **Chemical weathering** involves chemical change, and includes *carbonation, *hydration, *hydrolysis, *oxidation, and some *organic weathering. See A. Turkington et al. (2005); see also the special issue of *Geomorphology* 67, 1–2. **Mechanical weathering** is the physical disintegration of the rock, as in *pressure release, crystal growth, *salt weathering, *thermal expansion, and some organic weathering such as *chelation, and bacterial reduction as in *gley soils. Particularly in summer, but often in

Wave

Orthogonal

Wave refraction

winter, rock temperatures support mechanical *and* chemical weathering, if water is present. **Biological weathering** is the disintegration of rocks through the chemical and/or physical agency of an organism.

Weathering rates vary with the physical structure of the rock (particularly on crystalline rocks—see Borrelli et al. (2006) *Geomorph.* 87, 3); bond strength (Velbel (1999) *Am. J. Sci.* 299), and chemical composition (Yokoyama and Matsukura (2006) *Geology* 34).

The **weathering front** is the zone of contact of the *regolith with the underlying rock. In humid environments, increasing soil thickness increases weathering rates, primarily due to moisture storage and biological effects as soil develops. Eventually the rock weathering rate slows down with increasing distance from surface moisture. See Gracheva et al. (2001)

Quat. Int. 78. **Weathering pits** are depressions on a flat surface, usually on very soluble rocks, and varying in shape and ranging in size from a few centimetres to several metres in width. See Domínguez-Villar (2006) *Geomorph.* 76, 1-2. **Honeycomb weathering** is a grouping of many, closely spaced pits; see Pye and Mottershead (1995) *Int. J. Rock Mechan. Mining Scis & Geomech.* 33, 8. *Tafoni are weathering pits which are cut into near vertical rock faces. **Weathering rind** is the chemically altered 'skin' of rock around an unaltered core; see Etienne (2002) *Geomorph.* 47, 1.

Web Erosivity Model, WERM A web-based system used to estimate the erosivity of a soil. See Risal et al. (2016) *Catena* 147.

welding The bonding of grains; see, for example, Anthony et al. (2006) *Geomorph.* 81, 3-4 on dune accretion.

welfare While welfare might be equated with well-being, within human geography it refers mostly to factors within the control of societies: environmental quality, security, and access to commodities and services. It therefore incorporates income, standard of living, housing, employment, and access to educational, health, and social services; see Sen's **welfare index** (1976) *Rev. Econ. Studs* 43. Within the EU, regions with high welfare tend to have advanced services, high productivity, and good human capital. Low-welfare regions tend to higher unemployment, and a larger agricultural sector.

Welfare capitalism is the provision of welfare by corporations aiming to foster loyalty among the workforce. S. Jacoby (1998) provides a comprehensive history of 20th-century welfare capitalism.

Welfare geography studies *inequalities in *social well-being and *social justice. It focuses on those factors which affect the quality of human life: crime/lack of crime, poverty/wealth, housing/ homelessness, and the provision/lack of educational, health, leisure, and social services, looking for spatial variations in welfare, and seeking to explain them. Current explanations link inequality with *uneven development as an inevitable consequence of *capitalism; or with an uneven spatial pattern of service provision. Measures to bring about a fairer distribution of resources and opportunities may be suggested.

westerlies (westerly winds) Winds blowing *from* the west, most often in mid-*latitudes.

Westphalia The Peace of Westphalia treaty, 1648, is considered a watershed event in traditional *geopolitics. It led to the creation of the modern nation-state system by allowing *hegemons (political powers) to connect their power to discrete (separate), definable territorial units for the first time in history. As the system was concretized and formalized, it also paved the way for inter-state struggles for power. See Gallaher, in C. Gallaher et al. (2009).

wetland Any land which is intermittently or periodically waterlogged, including *salt marshes, tidal *estuaries, marshes, and bogs. Wetlands are rapidly disappearing habitats.

In the United States a **constructed wetland** is intentionally created from non-wetland sites for wastewater or storm water treatment; the Constructed Wetland Association (CWA) website has a reading list. The concept of a **designer wetland** emphasizes the life history strategy of species as the major factor in restoring a wetland type, with no fixed endpoint; see M. Reuss (1998). For **wetland restoration**, see Tooth and McCarthy (2007) *PPG* 31, 1 and Holden et al. (2004) *PPG* 28, 1; see also Van Lonkhuyzen et al. (2004) *Env. Management* 33, 3. A **wetland mitigation bank** site is a property purchased and developed by a public agency or utility to earn credits to compensate for adverse impacts to wetlands due to development activities of other agencies, utilities, or in specific instances, private-sector developers.

(⊕) SEE WEB LINKS
• The Constructed Wetland Association reading list.

wetness index (*wi*) Also called the *compound topographic index*, the wetness index describes the effect of topography (slope and aspect) on the location and size of areas of water accumulation in soils. In its simplest form, the wetness index is calculated as: $wi = \ln(\alpha / \tan \beta)$, where α is upslope contributing area and β the local surface

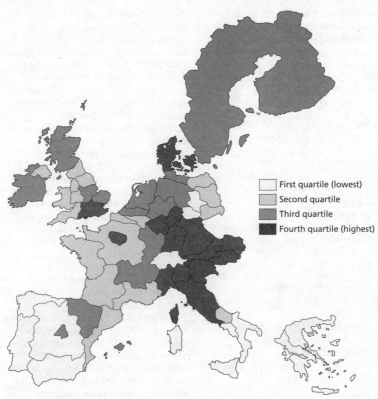

Welfare variations in the EU

gradient. See Sulebak et al. (2000)
Geografiska A 82, 1.

wetted perimeter In a cross-section
of a river channel, the line of contact
between water and bed. *See also*
HYDRAULIC RADIUS.

whaleback A streamlined rock knob
with symmetrical longitudinal profiles
caused by abrasion of both stoss and lee
sides. Small whalebacks can form under
only a few hundred metres of ice; larger
ones under deep ice streams. In theory,
whalebacks tend to form in areas of
deep ice with rapid flow velocities.

See Glasser and Harrison (2005)
Geograf. Annal. A 87, 3.

whiteness 'White' is used as a
synonym for people of European origin
who have a light-coloured skin, and
'whiteness' may be defined as
normalization of white people's social
norms and cultural practices as the
'natural' and 'proper' way of society.
Simply put, a higher value has been
attached to people who are classed as
white.

For anti-racists, whiteness is a
problem because it works to
discriminate against non-whites, either

implicitly or explicitly, and suppresses other cultural forms.

Geographies of whiteness consider that spaces, places, landscapes, natures, mobilities, and bodies are assumed to be white or are in some way structured, though often implicitly, by some notion of whiteness.

wilderness An area which has generally been affected more by natural forces than by human agency; a region little affected by people. Some 77 000 km^2 of the USA have been designated as **wilderness areas**, under the Wilderness Act of 1964. These are, ideally, areas which have never been subject to human manipulation of the ecology, whether deliberate or unconscious, and which are set aside as nature reserves to which human access is very severely restricted. Roads, motor vehicles, aircraft (except in an emergency), and any economic use are all forbidden.

The *idea* of wilderness is socially constructed; enmeshed in late capitalist economics and politics, and structured to reproduce class distinction. See Baker (2002) *Space & Cult.* 5. Hintz (2007) *Eth. Place. Env.* 10, 2 writes that 'wilderness, we are told, can no longer be seen as a scenic playground for weary humans—it is, rather, an ecological necessity for the conservation of biodiversity'.

Some would say that a true wilderness has no people in it. This would entail rewilding the landscape, aiming not just to save existing wilderness, but to restore much of what has been lost. This view has led to the expulsion of indigenous people from designated wilderness areas (W. Cronon 1995). But the belief that nature, to be natural, must be human-free, is rooted in a binary separation of nature and culture which is now widely regarded as false.

Wilson cycle A description of the aggregation and break-up of supercontinents. Initially, a *hot spot rises up under a *craton, heating it, causing it to swell upward, stretch, and thin, not only splitting a continent in two, but also creating a new divergent plate boundary. A new ocean basin is generated between the two new continents, and as this widens (possibly by thousands of miles), wedges of divergent continental margins' sediments accumulate on both new continental edges. All of the above is the opening phase of the Wilson cycle. The closing phase of this cycle begins when a new *subduction zone forms at the margin of one of the continents.

Once this zone is active, the ocean basin begins to be subducted, causing the two continents to draw nearer. Magma is generated deep in the subduction zone, rising to the surface to form a cordillera of volcanoes, accompanied by *metamorphism, folding, and faulting. When the two continents finally collide, the closing phase of the Wilson cycle is technically over. Because the subduction zone acts as a ramp, the continent with the subduction zone slides up over the edge of the other one. After this collision, the cordillera will be eroded to a *peneplain. Most of the upper continent will be eroded away, and the lower continent will eventually return to the Earth's surface. See Russo and Silver (1996) *Geology* 24 for further explanation on cordillera formation and the Wilson cycle.

wind energy Power generated by harnessing the wind. Drawbacks include the inconstant nature of the wind, construction difficulties, and finding a suitable site. 'Wind energy is economically competitive at good wind sites, can be rapidly deployed compared to conventional sources, and contributes to the goal of meeting increasing electricity demand while reducing emissions of greenhouse gases' (Barthelmie (2007) *Geog. Compass* 2). See Bassi et al. (2012) on the case for and against onshore wind energy in the UK.

(⊕) SEE WEB LINKS
• The London School of Economics/Grantham Institute pages on Climate Change and the Environment where Bassi et al. can be accessed.

Windermere interstadial An interstadial of about 13 000 to 11 000 years BP, including the *Bølling, Older *Dryas, and *Allerød chronozones of Scandinavia. See Holloway et al. (2002) *Scot. J. Geol.* 38, 4.

wind shear The local variation of the wind vector, or any of its components, in a given direction. A change in wind speed and/or direction with height is the **vertical shear**, given by the *thermal wind equation, if the wind is *geostrophic.

window In *meteorology, the ability of radiation of wavelength around 10 μm to escape absorption by the Earth's *atmosphere. This wavelength nearly coincides with Earth's peak radiation, and allows some of the outflow of terrestrial radiation to be lost to space, thus, in part, upholding the thermal equilibrium of the atmosphere, which is also achieved by *convection. See, for example, Gube et al. (1996) *J. Climatol.* 3, 4.

work, geographies of The geography of work, whether at global, national, regional, or local scale, varies over time and space with the nature of the economy, the level of technology, and spatial variations in *class structure. 'To stake an agenda for "geographies of work" rather than for "labor geographies" means, in the most general

sense, mounting a critique of the putative unitariness of value on three fronts: its basis—ostensibly, the production process; the homogeneity of its extraction—ostensibly, the wage/labor relation; and its frequently posited autonomy from use-values—ostensibly, because the qualities of things, including labor, are effaced when values truck as commodities or exchange-values' (Gidwani and Chari (2004) *Env. & Plan. D* 22, 4).

World Bank The International Fund for Reconstruction and Development, popularly known as the World Bank, and now generally concerned with aid to the developing world. The Bank acquires its funds partly through government subscription, but mainly through borrowing, and voting power is proportional to capital subscription, so the bank is effectively controlled by the rich countries. It has made loans of over $US 200 billion.

(⊕) SEE WEB LINKS
• The *World Bank Economic Review* and the *PovertyNet Newsletters*.

world city A city that has a substantial, straightforward effect on world affairs: economically, socially, and politically. Key descriptors include: centrality—a concentration of international activities; externality—a relatively high global role, compared to its national role; and connectivity—as measured by the numbers and intensity of transactions. 'World cities are not merely outcomes of a global economic machine; rather, they are key structures of the world economy itself' (Taylor et al. (2002) *Urb. Studs* 39. See Taylor (2007) *GaWC Res. Bull.* 238 for an admirably clear summary of GaWC's work on world cities.

(⊕) SEE WEB LINKS
• GAWC Inventory of world cities.

w

world city network An interlocking network with cities at a nodal level, the network at inter-nodal level, and business service firms as a sub-nodal level. These last 'interlock' cities through their myriad networks of offices. In this new city system, it is the functions of a city that largely determine its role and importance nationally, regionally, and globally. See Beaverstock et al. (2000) *Applied Geog.* 20 on the conceptualization and measurement of the network connectivity of cities.

world-systems theory A theory arguing that macro-level patterns govern social and economic change; that history is not about singular events but about materially structured ways of life. See I. Wallerstein (1974, 1983) and S. Babones and C. Chase-Dunn (2012).

The *core* consists of the most economically developed countries, the *periphery* the less economically developed countries, and the nations of the *semi-periphery*, such as the newly industrialized countries, have aspects of both core and periphery— actively part of the capitalist world system but with limited authority therein.

World Trade Organization (WTO) An international organization established in 1995. Its stated aims are: expanding free-trade concessions equally to all members; establishing freer global trade with fewer barriers; making trade more predictable through established rules; and making trade more competitive by removing subsidies. The *Congressional Research Service* will give you a frequently updated overview of the WTO, and some indication of its future direction.

xeroll A soil of the *US soil classification. *See also* CHESTNUT SOIL.

xerophyte A plant which is able to grow in very arid conditions because it has adapted to restrict any water loss. Such adaptations include dense hairs or waxy leaves and shedding leaves at the start of the arid season. See Lauterbach (2019) *Ecol. Evol.* 9, 6. **Succulent xerophytes** incorporate water into their structure.

xerosere A plant succession developed under dry conditions such as bare rock, or sand.

yardang In a *desert landscape, a long ridge which has been isolated by the removal of rocks on either side. Yardangs can be 100 m or more in height and can stretch for many kilometres. See Hu et al. (2017) *Geom.* 299.

yazoo A tributary stream which does not join the main stream directly but runs parallel to it for some distance. When streams raise their courses above the level of their flood plains, their tributaries cannot join the main stream. Usually, the tributaries will then flow along the side of the flood plain until they reach some point farther downstream where the main stream swings against the valley wall. Many of the tributaries of the Mississippi behave in this manner.

Younger Dryas A phase of reduced temperatures and glacial advance in north-west Europe between 11000 and 10000 years BP. There is evidence that other regions, such as the Mediterranean, Russia, and North America, also underwent climatic deterioration. During this phase, these non-glaciated areas experienced periglacial conditions. This climatic cooling has been attributed to increased *precipitation, and therefore cloudiness, resulting from changes in the locations of the polar and oceanic Atlantic fronts.

young fold mountain *See* FOLD MOUNTAIN.

zebra stripes In arid environments, bands of angular gravel, grain-sorted and of a specific wavelength, running parallel to the contours. See May et al. (2019) *Geomorph.* 344.

Zelinsky *See* MOBILITY.

zero population growth The ending of population growth, when birth and death rates become equal. This would require an average number of 2.3 children per family. In 1993, 58 members of the World's Scientific Academies stated that 'there is no doubt that the threat to the [global] ecosystem is linked to population size and resource use . . . Family planning could bring more benefits to more people at less cost than any other single technology . . . Success in dealing with global social, economic and environmental problems cannot be achieved without a stable world population . . . We must achieve zero population growth within the lifetime of our children' (1993, Royal Society, London). This is a contentious subject.

zero-sum game A *formal game whereby, on choosing a particular strategy, one competitor's gain is his opponent's loss. Belief in a zero-sum game has been described as an antagonistic belief about social relations in the struggle for limited resources. According to the Columbia Business School site, trade is *not* a zero-sum game; have a look at it.

zeuge (pl. *zeugen*) An upstanding rock in a *desert landscape capped with a harder stratum and undercut by wind at the base, indicative of *differential erosion.

Zipper-rift model A model of rifting across the supercontinent Rodinia, proposed by Eyles and Januszczak (2004) *Earth Science Revs* 65, 1, who hypothesize that Rodinia became progressively 'unzipped' after *c.*750 Ma by two large 'megarift' systems that developed along the western and eastern margins of Laurentia, and that the duration of rifting varied from one part of the rift system to another.

zonal Referring to phenomena occurring in bands roughly parallel with lines of *latitude.

zonal index A measure of strength of the mid-latitude westerly winds, usually given as the horizontal *pressure difference between latitudes 35° and 55° N, or as the corresponding *geostrophic wind.

zonal model *See* CONCENTRIC ZONE MODEL.

zonal soil A soil where differences in local rock formation and *lithology are largely masked by the overriding effects of climate. The major zonal soils are *tundra soils, *podzols, Mediterranean soils, *chernozems, *chestnut soils, and ferrallitic soils (*see* FERRALLITIZATION).

zonal winds Winds, such as the *trade winds or the *westerlies, which are associated with particular *latitudinal zones.

zone of assimilation The area which increasingly develops the functions of the *CBD; the CBD of the future, characterized by whole scale redevelopment of shops, offices, and hotels. The zone of discard is that area that was once a part of the CBD but is now in decline and characterized by low-status shops and warehouses, and vacant property.

zone of overlap 1. An area served by more than one urban centre, i.e. within two or more different urban fields.

2. A transition zone between two floral/faunal communities. See, for example, Wolf and Flamenco-S (2003) *J. Biogeog.* 30, 11.

zoogeomorphology The study of animals as geomorphic agents—as in the creation and collapse of gopher tunnels. The classic text is by D. R. Butler (1995). And how could you resist Maré et al. (2019) *Geomorph.* 327 on baboons 'rocking the landscape'? See also Butler (2018) *Geomorph.* 303 on zoogeomorphology in the *anthropocene.